Sedimentation in the African Rifts

GEOLOGICAL SOCIETY SPECIAL PUBLICATION NO 25

Sedimentation in the African Rifts

EDITED BY

L.E. FROSTICK
Department of Geology
Royal Holloway and
Bedford New College
University of London

R.W. RENAUT
Department of
Geological Sciences
University of Saskatchewan
Saskatoon

I.REID
Department of Geography
Birkbeck College
University of London

J.J. TIERCELIN
GIS Océanologie et Géodynamique
Université de Bretagne Occidentale
Brest

1986

Published for
The Geological Society by
Blackwell Scientific Publications
OXFORD LONDON EDINBURGH
BOSTON PALO ALTO MELBOURNE

Published by
Blackwell Scientific Publications
Osney Mead, Oxford OX2 0EL
8 John Street, London WC1N 2ES
23 Ainslie Place, Edinburgh EH3 6AJ
52 Beacon Street, Boston, Massachusetts 02108, USA
667 Lytton Avenue, Palo Alto, California 94301, USA
107 Barry Street, Carlton, Victoria 3053, Australia

First published 1986

Printed in Great Britain by
the Alden Press Ltd, Oxford

DISTRIBUTORS

USA and Canada
Blackwell Scientific Publications Inc.
PO Box 50009, Palo Alto
California 94303

Australia
Blackwell Scientific Publications
(Australia) Pty Ltd
107 Barry Street
Carlton, Victoria 3053

British Library Cataloguing in Publication Data

Sedimentation in the African rifts.—
(Geological Society special publication,
ISSN 0305-8719; no. 25)

1. Geology—Africa, Eastern—Rift Valley
I. Frostick, L.E. II. Geological Society of London
III. Series
556.76 QE326

ISBN 0-632-01534-9

Contents

Continental Rift Basins

African Rift Basin Development

Siliciclastic, Chemical, Pedogenic and Organic Sediments in Contemporary Rift Environments

Sedimentary History of African Rift Basins

Sedimentation and the Preservation of Fossil Faunas

List of contributors

P.I. Abell *Dept of Chemistry, University of Rhode Island, Kingston, Rhode Island 02881, USA*

D.A. Adamson *School of Biological Sciences & Quaternary Research Unit, Macquarie University, North Ryde, New South Wales 2113, Australia*

Getaneh Assefa *Faculty of Sciences, Addis Ababa University, PO Box 1176, Addis Ababa, Ethiopia*

B.H. Baker *Center for Volcanology, University of Oregon, Eugene, Oregon 97403, USA*

R. Bonnefille *Laboratoire de Géologie du Quaternaire, CNRS—Case 907 Luminy, 13288 Marseille, Cedex 9, France*

C.F. Burgess *Dept of Geology, Duke University, Durham, North Carolina 27706, USA*

J. Casanova *Laboratoire de Géologie du Quaternaire, CNRS—Case 907 Luminy, 13288 Marseille, Cedex 9, France*

G.R. Chapman *British Geological Survey, 154 Clerkenwell Road, London EC1R 5DU, UK*

J. Chorowicz *Dept de Géotectonique, Université de P. et M. Curie, 4 place Jussieu, 75252 Paris, Cedex 05, France*

A.S. Cohen *Dept of Geology, The Colorado College, Colorado Springs, Colorado 80903, USA*

R. Crossley *Dept of Geography & Earth Science, Chancellor College, University of Malawi, PO Box 280, Zomba, Malawi*

G. Curtis *Dept of Geology & Geophysics, University of California, Berkeley, California 94720, USA*

G. Delibrias *Centre Faibles Radioactivités—CNRS, 91/90 Gif-sur-Yvette, France*

C. Denys *Laboratoire de Paléontologie des Vertébrés et de Paléontologie Humaine, Université de P. et M. Curie, 4 place Jussieu, 75252 Paris, Cedex 05, France*

S.J. Derksen *Marathon Oil Co., PO Box 269, Littleton, Colorado 80160, USA*

R. Drake *Dept of Geology & Geophysics, University of California, Berkeley, California 94720, USA*

C. Elenga, *Ecole Normale Supérieure, 5 rue Boucicaut, 92260 Fontenay-aux-Roses, France*

H.P. Eugster *Dept of Earth & Planetary Sciences, Johns Hopkins University, Baltimore, Maryland 21218, USA*

J.D. Fairhead *Dept of Earth Sciences, The University, Leeds LS2 9JT, UK*

D.S. Ferguson *Dept of Geology, The Colorado College, Colorado Springs, Colorado 80903, USA*

L.E. Frostick *Dept of Geology, Royal Holloway & Bedford New College, Egham Hill, Egham, Surrey TW20 0EX, UK*

P.M. Gram *Dept of Geology, The Colorado College, Colorado Springs, Colorado 80903, USA*

A.T. Grove *Dept of Geography, The University, Cambridge CB2 3EN, UK*

A. Hamilton *Dept of Environmental Sciences, University of Ulster, Coleraine, Co. Londonderry, Northern Ireland*

R.L. Hay *Dept of Geology, 245 Natural History Building, 1301 W. Green Street, University of Illinois, Urbana, Illinois 61801, USA*

J.P. Herbin *Institut Français du Pétrole, BP18, 92502 Rueil Malmaison, Cedex, France*

A. Hill *Dept of Anthropology, Yale University, New Haven, Connecticut 06520, USA*

S.L. Hubler *Dept of Geology, The Colorado College, Colorado Springs, Colorado 80903, USA*

M.A. Khan *Dept of Geology, University of Leicester, Leicester LE1 7RH, UK*

J.J. Lambiase *Marathon Oil Co., PO Box 269, Littleton, Colorado 80160, USA*

A.M. Lezine *Laboratoire de Géologie du Quaternaire, CNRS—Case 907 Luminy, 13288 Marseille, Cedex 9, France*

P.M. Lorber *Dept of Geology, Duke University, Durham, North Carolina 27706, USA*

J.P. McClory *Dept of Chemistry, University of Rhode Island, Kingston, Rhode Island 02881, USA*

J. McGill *Dept of Geology, Duke University, Durham, North Carolina 27706, USA*

C.K. Nyamweru *Dept of Geography, Kenyatta University, PO Box 43844, Nairobi, Kenya*

R.B. Owen *Dept of Geography & Earth Science, Chancellor College, University of Malawi, PO Box 280, Zomba, Malawi*

G. Perinet *Laboratoire de Géologie du Quaternaire, CNRS—Case 907 Luminy, 13288 Marseille, Cedex 9, France*

M. Pickford *Institut de Paléontologie, 8 rue Buffon, 75005 Paris, France*

H.G. Reading *Dept of Earth Sciences, The University, Parks Road, Oxford OX1 3PR, UK*

I. Reid *Dept of Geography, Birkbeck College, London WC1E 7HX, UK*

R.W. Renaut *Dept of Geological Sciences, University of Saskatchewan, Saskatoon, Saskatchewan S7N OWO, Canada*

D.J. Reynolds *Dept of Geology, Duke University, Durham, North Carolina 27706, USA*

C. Robert *Laboratoire de Géologie du Quaternaire, CNRS—Case 907 Luminy, 13288 Marseille, Cedex 9, France*

B.R. Rosendahl *Dept of Geology, Duke University, Durham, North Carolina 27706, USA*

R.J.G. Savage *Dept of Geology, University of Bristol, Bristol BS8 1SS, UK*

D. Scott *Dept of Geology, Duke University, Durham, North Carolina 27706, USA*

K.W. Sims *Dept of Geology, The Colorado College, Colorado Springs, Colorado 80903, USA*

N.J. Skinner *Dept of Physics, University of Botswana, Private Bag 0022, Gaborone, Botswana*

C.J. Swain *Dept of Physics, University of Zimbabwe, PO Box MP 167, Harare, Zimbabwe*

M. Taieb *Laboratoire de Géologie du Quaternaire, CNRS—Case 907 Luminy, 13288 Marseille, Cedex 9, France*

D. Taylor *Dept of Environmental Sciences, University of Ulster, Coleraine, Co. Londonderry, Northern Ireland*

N. Thouveny *Laboratoire de Géologie du Quaternaire, CNRS—Case 907 Luminy, 13288 Marseille, Cedex 9, France*

J.J. Tiercelin *GIS Océanologie et Géodynamique, Université de Bretagne Occidentale, 29287 Brest, Cedex, France*

A. Vincens *Laboratoire de Géologie du Quaternaire, CNRS—Case 907 Luminy, 13288 Marseille, Cedex 9, France*

R.T. Watkins *Dept of Geology, University of Cape Town, Rondebosch, Cape Town, South Africa*

L.A.J. Williams *Dept of Environmental Sciences & Museum of Comparative Zoology, University of Lancaster, Lancaster LA1 4YQ, UK*

M.A.J. Williams *Dept of Geography, Monash University, Clayton, Victoria 3168, Australia*

P.G. Williamson *Dept of Geological Sciences & Museum of Comparative Zoology, Harvard University, Cambridge, Massachusetts, USA*

R.F. Yuretich *Dept of Geology & Geography, University of Massachusetts, Massachusetts 01003, USA*

Preface

This book arises out of an international meeting held at the apartments of the Geological Society of London in September 1984. The meeting attracted considerable interest from academics working on both African Rift geology in particular, and fault-controlled sedimentation in general. In addition, a large group of oil company representatives attended, reflecting the general increase in exploration interest in non-marine basins and the recent recognition of the importance of non-marine sediments as sources of oil and gas.

The main aim of the meeting and of this book is to provide an opportunity for the presentation of recent results and ideas that have been produced by many individuals working on various aspects of African Rift sedimentation. The research has been carried out in universities and other establishments in more than ten countries spread throughout four different continents. Fieldwork associated with this research has been conducted in eight different African countries—Chad, Ethiopia, Kenya, Malawi, Nigeria, Sudan, Tanzania and Uganda. Geographical representation is very wide, but it is not comprehensive. This is partly explained by the difficulties of access that have arisen because of political unrest. Some sectors of the rift have been effectively closed to research in recent years. There has been a concentration of effort in the Kenyan, Ethiopian and Tanzanian sectors of the African Rift in the last two decades. This has arisen as geologists have adopted the role of camp followers to the exciting fossil hominid discoveries of Louis, Mary and Richard Leakey, and of Don Johanson and others.

However, in the last few years there have been both southward and westward trends in interest and more research has been conducted in the Lake Tanganyika and Lake Malawi sectors of East Africa and in the Benue trough and its extensions (see Fig. 1).

Another aim of both the meeting and this volume is to provide a clearer picture of the interrelationship of structure and patterns of sedimentation during the early and middle phases of continental rift development. The models developed here for the African Rift will undoubtedly find application in the interpretation of ancient rift sediments worldwide. This is not only of academic importance. There are significant economic implications attached to the current interpretation of non-marine petroleum source and reservoir rocks.

In rifts, as in all sedimentary basins, the nature and quality of clastic reservoirs are largely controlled by the rate and nature of tectonic adjustment. Fault movements govern the size, location and subsidence of sedimentary basins, as well as the pathways of sediment transport and the thickness, orientation and interdigitation of the facies belts. Saline lakes are a common feature of rift basins and are often sites of high organic productivity. Much of the kerogen that evolves is sapropelic. This, combined with rapid burial and high thermal gradients, favours rapid maturation of hydrocarbons which can then migrate into flanking coarse clastic alluvial fan and deltaic deposits. In closed lake basins the rapid fluctuation of water level results in an interdigitation of coarse and fine lithofacies near the shoreline, and this provides the excellent stratigraphic traps for oil and gas.

The volume opens with an introduction by Reading which places the African Rifts in a global context. Grove follows with a broad view of the present landforms and the relationship between the rift basins. These set the scene for later more local analyses of these important sediment sinks.

As tectonic adjustment is paramount in controlling rift basin sedimentation, the next section of the book focuses on the structural evolution of both the western (Fairhead) and the eastern (Rosendahl *et al.*; Baker; Williams & Chapman; Swain *et al.*; Watkins) branches of the African Rift. Baker, Williams & Chapman, and Watkins concentrate on surface expressions of faulting and volcanism, relating these to basin location and size, as well as to controls on sediment provenance. In contrast, Fairhead and Rosendahl *et al.* use geophysical data to elucidate subsurface rift basin structure and the thickness of sedimentary fills.

The third section is concerned with contemporary patterns of sedimentation. The first three papers (Reid & Frostick; Frostick & Reid; Cohen *et al.*) consider various aspects of clastic sediment sources, transport and deposition in the Turkana basin, following the debris train from source right through to present sink. Then follow papers dealing with chemical processes and diagenesis in modern rift lakes (Yuretich; Renaut *et al.*; Eugster), and Abell & McClory discuss the application of isotopic analyses of sedimentary carbonates to stratigraphic and palaeoenvironmental interpretation. The section closes with a consideration of organic controls of sediment deposition. Crossley examines the effect of a pedogenic subaerial process, Casanova investigates the lacustrine biosediments, and Hamilton & Taylor review the accumulation of plant material in marginal wetlands.

The modern patterns discussed in the third section are analogues that assist interpretation of older preserved equivalents and are a natural precursor for the fourth section. This consists of a series of case studies of sediments from various parts of the rift, ranging in age from late Mesozoic to Holocene. These case studies focus attention on the Ethiopian (Tiercelin; Williams *et al.*; Bonnefille *et al.*), the Kenyan (Williamson & Savage; Hill *et al.*; Nyamweru; Owen & Renaut; Vincens *et al.*) and the Tanzanian (Thouveny & Taieb) Rifts. Each paper provides an important insight into the ways in which long-term lake level

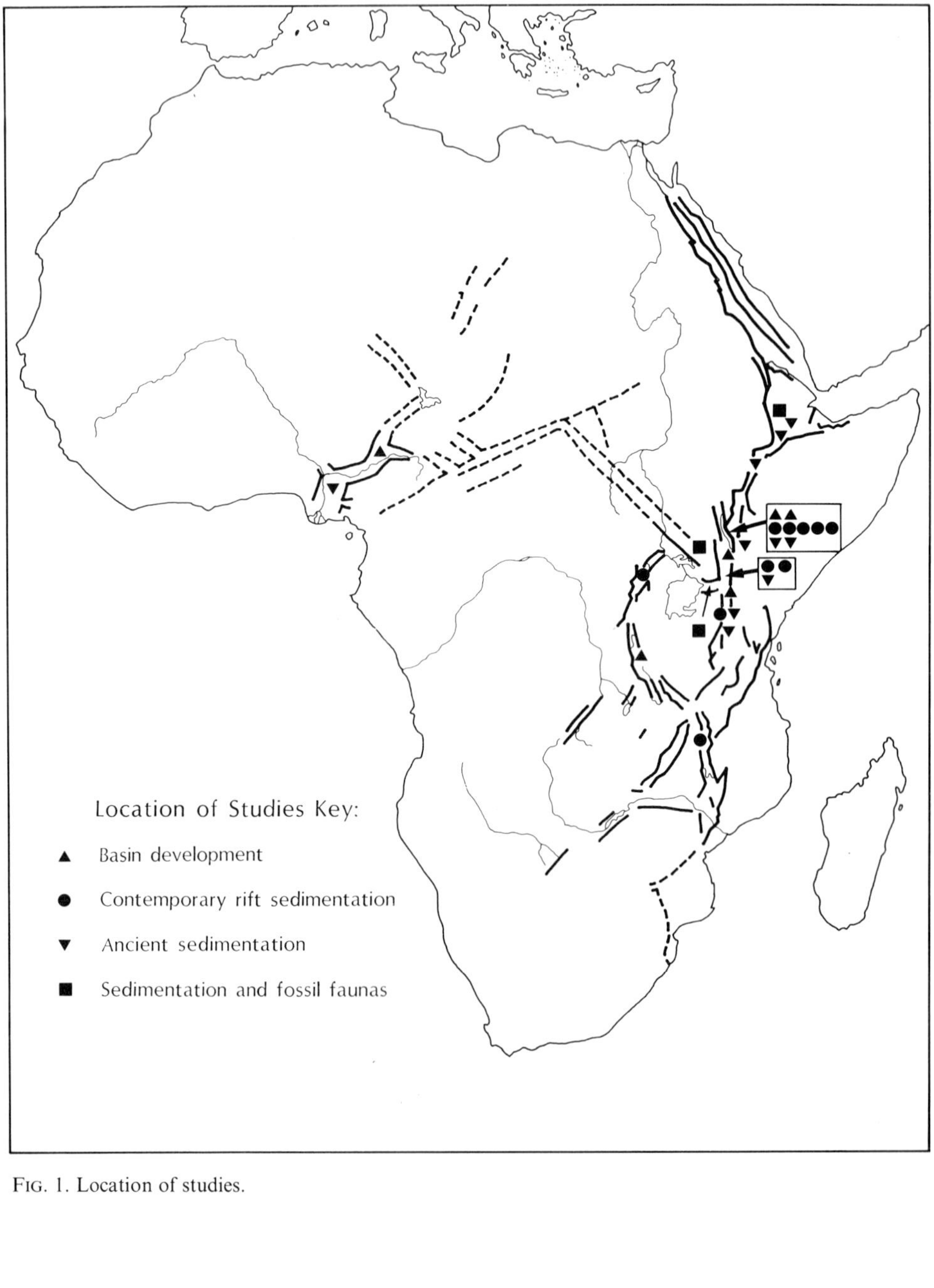

FIG. 1. Location of studies.

fluctuations, responding to tectonic and climatic controls, influence patterns of sedimentation.

One important aspect of the sediments of the African Rifts which has drawn them particularly to world attention, is the prolific and well-preserved vertebrate faunas, especially the early hominid remains. The volume would be incomplete if they were not mentioned. As a result, the final section deals with some of the volcanic and sedimentary factors that lead to good fossil preservation (Hay; Pickford; Denys *et al.*). Rift basins provide sites in which rapid burial and early mineralization preserves a fine record of the evolution of species, including our own. Bill Bishop, late friend and colleague of many of the contributors to this volume, often lectured under a favourite title: 'Rift basins are a good place to die!' This is a good summary of the value of the deposits to palaeontologists in general, but to palaeoanthropologists in particular.

This book provides an overview of sedimentation in a single, extensive continental rift system. However, the wider implications of the work that has been, and continues to be, carried out in Africa are considerable. There are strong similarities in both structure and sediment character between the African Rifts and others which are proven sources of oil and gas (e.g. the North Sea basins, the Ta Chung basin of China, and the Sirte basin of Libya). Models developed for the African Rifts may find wider applications in the exploration of new, as yet untapped, sources of hydrocarbons in little-known rift basins. Besides this, the research in Africa clearly enhances our understanding of already productive basins, and thereby helps industry optimize exploration during further development.

Saskatoon
August 1985

L.E. Frostick
R.W. Renaut
I. Reid
J.J. Tiercelin

Obituary

In the period that has elapsed between the meeting at the Geological Society and the final stages in the preparation of this Special Publication three eminent professors of the African Rift System have passed away. A glance through the acknowledgements of the papers that follow will reveal how much we are indebted to the enthusiasm, the insight and the support that each of these men gave to succeeding generations. Basil King, Emeritus Professor of the University of London died at his retirement home in his beloved Isle of Arran, Scotland. Glynn Isaacs, Professor of Archaeology, University of California, Berkeley, and more recently of Harvard University, was a victim of sudden illness. Brian Baker, Professor of Geology at the University of Oregon, died in Nairobi close by the rift he loved so much. We shall miss all three.

Continental Rift Basins

African Rift tectonics and sedimentation, an introduction

H.G. Reading

Rifts are one of the most spectacular features of global morphology. Within oceans they separate plates as new oceanic crust is created. Within continents they may form deep valleys such as the Rhine graben, within which runs one of the world's busiest waterways. They are often floored by deep lakes, such as Lake Baikal, the deepest lake in the world today whose floor lies 1700 m below the surface of the lake. Ancient rifts are the sites of petroleum accumulations beneath the margins of the Atlantic, in the North Sea and in China, of coals and oil shales and of minerals, including phosphates, barite, Cu-Pb-Zn sulphides and uranium (Robbins 1983). Of all the world's rifts none can match in scale and diversity the Great Rift System, which runs for over 7000 km and includes the East African Rift System and its extension through the Red Sea into the Dead Sea, to form a unique feature of global geology. Its study is important not only for what it can tell about the nature and origin of present-day rifts, the thermal, magmatic and tectonic processes which gave rise to them, climatic changes, and sedimentary, particularly lacustrine and volcaniclastic, processes. It is also essential for the understanding of the processes which formed passive continental margins, all of which are underlain by rifted sedimentary basins, and failed rifts whether they formed within continents or at the junction of continents and oceans.

Early workers on rifts were impressed by the wide valleys, 40–50 km across, filled by young alluvial and lacustrine sediments and bordered by steep escarpments rising in some cases a few hundred metres, in others up to 3–4000 m above the valley floor. Many rifts are the sites of volcanic activity, sometimes of rather unusual composition. These unusual volcanic source rocks may in turn be the cause of peculiar lake water chemistries, many rift lakes containing unique suites of evaporitic minerals.

Although at one time some people considered rifts to have been formed by compression (ramp-valley hypothesis), discussion today revolves principally around the extent to which rift valleys are extensional features or are the result of oblique-slip movements between two laterally moving blocks.

There is no doubt now that some rift valleys are the result of strike-slip movement. The Dead Sea/Gulf of Elat rift lies along a lineament that separates two lithospheric plates and, as Arabia moves northwards relative to the Palestinian (Mediterranean) block and faults curve or side-step each other, extensional pull-apart basins are formed, such as the Dead Sea whose surface lies further below sea level than any other place in the world. A feature of these strike-slip rift valleys is that the extensional valleys pass, along the rift, into mountainous zones of compression and uplift. Rapid erosion from these mountains leads to substantial sediment supply for the basins from within the fault system itself. Ancient strike-slip rift basins are therefore characterized by very thick, rapidly accumulated basin fills with areas of contemporaneous compressional tectonics and unconformities not far away.

Rift valley basins that are extensional features on a regional scale tend to sink rather more slowly, have less sediment entering them, and show no evidence of synchronous compressional tectonics in their neighbourhood. Cross-sections through them show normal faulting, probably listric in form.

The main controversy with regard to extensional rift valleys is whether magmatic activity in rifts reflects an underlying heat source such as a convection plume in the mantle or hot spot, the rifting then being the *consequence* of expansion and uplift brought about by thermal activity, or whether the magmatic activity is merely the *result* of regional or local extensional stresses which have passively allowed magma to come to the surface.

Some rifts that penetrate continents are clearly failed rifts or aulacogens that began as one arm of a triple junction. As ocean floor spreading developed along two arms, one arm was left as a 'failed rift' in the sense that new oceanic crust was not created. Perhaps the best known is the Benue trough within which the Niger delta developed. However, the northern (Viking) graben of the North Sea has been claimed to be one (Whiteman *et al.* 1975), and so has that part of the East African Rift known as the Afar triangle where the Red Sea and the Gulf of Aden Rifts successfully drifted, whilst the Afar triangle 'failed'.

Molnar & Tapponnier (1975) have argued that other rifts are the result of continental collision, in particular that the impact of India with Asia caused not only major lateral displacement along

From Frostick, L.E. *et al.* (eds) 1986, *Sedimentation in the African Rifts*, Geological Society Special Publication No. 25, pp. 3–7.

faults such as the Red River and Altyn Tagh strike-slip faults, but also E–W extension to produce the Baikal graben. The Rhine graben has also been interpreted as the result of stresses generated in the Alps.

However the rifts in Africa are not the result of continental collision, and there is little evidence for any substantial degree of strike-slip movement. However, strike-slip movement may occur locally and some African rifts lie along major shear zones or lineaments extending for thousands of kilometres, which may have had a considerable amount of lateral movement at some time in the past.

Although the East African Rift System is the best known of African rifts, other extensive rifts occur especially in West and Central Africa (Fairhead, this volume). Some parts have been well studied, for example the Benue trough of Nigeria because of its petroleum resources, and the Volcanic Cameroon Line is also well known. Other parts, though, have little topographic expression and therefore have only recently become known, due to hydrocarbon exploration. In particular the nature of the geology across the Central African Shear Zone has not been understood.

This complex pattern of shear zones, associated rifts and rift valleys, known as the West and Central African Rift System (Fairhead, this volume), is apparently older than the East African Rift System in that at the present day it is largely inactive except for some gentle subsidence. The main rift phase was the Cretaceous when substantial and rapid stretching of the lithosphere caused thinning of the crust and passive upwelling of hot asthenosphere. It developed as a result of large extensional stresses, continental disruption and high sea-floor spreading rates. Volcanism was minor and although there was probably a large anomalously low density body within the upper mantle during the Cretaceous it did not produce domal uplift, except perhaps at the oceanward end of the Benue Rift. Instead deep rifts formed, open to the sea. This was due to the major crustal extension which occurred at this time, perhaps some four times that of the East African Rift System. Towards the end of the Cretaceous some lateral displacements took place along the Central African Shear Zone, and during the Tertiary wider and more extensive sedimentary basins than the original rift basins formed as the original thermal anomaly decayed and the lithosphere contracted. Unfortunately, apart from the Benue trough, little is known about sedimentation in these stress-related rift basins.

The East African Rift System is significantly different from the West and Central African Rift System, and its southern and older portion is very different again from the northern Kenyan to Ethiopian parts. The dominant feature of the East African Rift System is the wide zone of low density upper mantle which is expressed topographically by the localized Ethiopian and Kenyan domes where volcanic activity is most active. In the northern portion the rifting appears to be relatively young (late Tertiary) though there is evidence that rifting commenced during the latest Mesozoic (Williamson *et al.*, this volume). The amount of crustal extension is relatively small, about a quarter of that in the West and Central African Rift System (10–35 km, Fairhead, this volume), and there seems little doubt that the rift is the result of thermal activity in the upper mantle and consequent thermal uplift. Support for thermally-induced rifting comes from the migration of some rifts eastward with volcanism preceding doming, that in turn precedes rifting. The reason it has not opened yet may simply be that regional stresses are not yet conducive to opening. Were they to be so it would probably open.

Sedimentation in the northern portion is commonly away from the domes and main rift. For example the Turkana basin is to the W of the present active rift, though it probably started within a late Miocene–Pliocene rift before the rift migrated eastward (Cerling & Powers 1977). Sedimentation may also take place between rifts in sag basins such as Lake Victoria. Sedimentary thicknesses are not too great; volcanic accumulations are considerable, and volcanic source areas give rise to unusual and varied sedimentary compositions.

The southern portion of the East African Rift differs from the northern portion in that volcanic activity is limited. Although there is still a wide, low density zone on either side of the rifts, there is no thermal doming. The initiation of rifting appears to be earlier, and rifts follow Precambrian lineaments. Sediments in the rift basins are thicker and up to 3–4 km of sediments may have accumulated.

From the early days of the plate tectonics hypothesis, the East African Rift has been considered to be a present-day example of the earliest stage of continental breakup, that is the *rift* stage which precedes the *drift* stage represented by the Gulf of Aden and the Red Sea (e.g. Drake 1972). It was then supposed that in a relatively short geological time the East African Rift would itself open and a new ocean would be created along its length.

This view, however, was opposed by those who pointed to the relatively long-lived nature of some parts of the East African Rift, stressing in particular structural evidence in the exposed basement rocks, particularly in the central and southern portions of the rift (McConnell 1980). It was argued that the whole 5000 km length of the African Rift System, from the Afar triangle to the Orange river in South Africa, was a Precambrian taphrogenic lineament which has since been frequently reactivated. On this view the East African Rift is a more or less permanent feature within a continent which is unlikely to be split apart.

To some extent both views are correct since there is evidence in some part of Africa that: (1) rifting frequently takes place along lines of older lineaments which have long geological histories; (2) the type of movement and rifting reflects global stress patterns; (3) mantle processes can cause both linear and radial zones of high heat flow, low density upper mantle, volcanic activity and topographic uplift, as in the Red Sea, Gulf of Aden and the northern East African Rift; and (4) global stresses on their own, without heat sources from the upper mantle, may themselves form rifts, as for most of the West and Central African Rift System.

As with all linear or curvilinear features on the globe, rifts are not the continuous simple features they appear to be when looked at from a satellite. They are invariably broken into segments by transverse features. The larger the scale at which they are seen, the more complex they become. Individual segments vary considerably in their structural style. This segmentation is probably a prime factor in controlling sedimentation since it controls the rate of basin subsidence and the relative subsidence within the basins, and the amount of sediment supply and its composition.

Although most rift models were initially developed as essentially symmetrical models, deepening towards the centre, it has gradually become apparent in the past few years from detailed studies and geophysical profiles that continental rifting is an asymmetrical process (see Bally 1983). This is the consequence of rift geometry. However, not only are rifts asymmetric in cross-section, but the direction of asymmetry appears to vary systematically along a rift with the direction of asymmetry alternating between segments. Between each segment are transverse structures called transfer faults, transform zones or accommodation zones. In some cases these zones are ridges and they may inherit pre-existing structures.

Many areas of the East African Rift System are known to display segmentation from detailed mapping (e.g. Williams & Chapman, this volume). In Lake Tanganyika seismic profiles (Rosendahl *et al.*, this volume) show alternating asymmetrical basins *c.* 50–70 km long and 20–40 km wide. This alternation of two overlapping and interfacing sub-basins can result in apparently symmetrical basins with a ridge in the middle being seen in some cross-sections. In plan view Lake Tanganyika has a sinusoidal pattern resulting from this overlapping of arcuate or crescent-shaped sub-basins. The sedimentological effect of the division of the lake into sub-basins separated by ridges is that for the most part the lake is divided into a number of individual depocentres more or less independent of each other, probably with reversed facies transitions except where spillover can occur between sub-basins.

Ancient volcaniclastic sediments are notoriously difficult to interpret. The normal models of sedimentation used to interpret sediments formed in non-volcanic terraines no longer seem to apply. The reasons are: (1) the diverse, compositionally immature, volcanic source rocks; (2) the rapid volcano-tectonic movements that cause complex patterns of uplift and subsidence and rapidly changing water levels; and (3) the build-up of volcanic edifices, both subaerial and subaqueous, on the scale of large mountains thousands of metres high to small, but sedimentologically important, lava flows. These lava flows can, in one instant of time, transform the topography and form both physiographic barriers which isolate depositional systems and temporary local sources of sediment.

As shown by Cohen *et al.* (this volume), a spectrum of rift models can be postulated, based on the degree of volcanism and rift subsidence. Where the rate of volcanism exceeds that of rift propagation rifts are totally filled by volcanics and their derived sediments. Where there are moderate to high extrusion rates relative to rate of rift propagation, shallow lakes will form. Where volcanism is absent or minor, deep rift lakes develop. Rifts can thus be divided into: (1) those that are volcanically filled, in which lakes do not exist, e.g. the Gregory Rift, N of latitude 1°N (Williams & Chapman, this volume); (2) volcanically-dominated rift basins with lakes, e.g. the Turkana depression; and (3) non-volcanic rift basins such as Lakes Malawi and Tanganyika. Non-volcanic lakes are almost devoid of lake margin sediments, most sediment being transported direct to the lake centre. However, in volcanically-dominated lakes volcanic barriers form both across the main rift dividing it into sub-basins, and parallel to the lake margin to form sediment traps where lake margin sediments accumulate, particularly coarser-grained material as around Lake Turkana. In non-volcanic basins Cohen *et al.* (this volume) predict there should be a higher proportion of coarse clastic deep water turbidites.

Thus the study of volcanism and its effects in rifts helps towards the understanding of ancient volcaniclastic sediments, the rapidly changing sedimentary environments and the influence of varied source areas. In addition, another effect of volcanics, as pointed out by Eugster (this volume), is that a volcanic floor to a basin is often extensively fractured and very porous. Consequently streams are short and much water is supplied to the lakes by perennial springs, rather than from surface flow.

A final consequence of volcanism is ash fall. Firstly, ash falls are the most reliable and short-lived chronostratigraphic markers that exist in geology. Therefore, a feature of volcanically-dominated rifts is the abundance of good stratigraphic markers which make complex sedimentary facies patterns intelligible. Secondly, subaerial ash falls can, as with lava flows, form build-ups of considerable thickness. Thirdly, and probably the most importantly, they preserve fossils, including footprints and soft bodies, in places where fossils do not normally form. In continental environments fossils are normally only preserved at lake margins, in anoxic lakes where plants and animals can settle undisturbed and be preserved from oxidation; as detrital particles (bones and teeth) in fluvial sediments; and as burrows or traces in palaeosols or peat build-ups.

Volcanic ash, however, preserves faunas in all types of environments (Bishop 1968) including savannas. Of particular importance is the chemistry of the ash (Hay, this volume; Pickford, this volume). The optimum conditions are those of low silica and high lime. East African volcanoes are famous for their carbonatites, and most of the eruptive carbonatite magma was probably alkali rich or natrocarbonatite. As pointed out by Hay (this volume), when natrocarbonatite ash is altered by contact with water it can provide: (1) an instant cement of sodium carbonate (=trona); (2) an alkaline environment in which bone apatite and calcium carbonate fossils are relatively insoluble; (3) a fine-grained matrix for preserving footprints and moulds of soft-bodied organisms, including the muscles of birds and even the tongue of a lizard; and (4) a supply of calcium carbonate for calcification and coating of bones, replacement of plant material and filling of fossil bones. It is no coincidence that East Africa is the domain of so many beautifully preserved continental animals (particularly primates and other mammals, and reptiles), especially soft-bodied ones.

Although sedimentation in rifts is primarily controlled by tectonics and the degree and type of volcanism, climate is almost as important. It not only governs the amount, rate and periodicity of runoff, and the composition of the sediment supplied, but also influences the environment of deposition in the lakes and interfluvial areas that make up so large a part of the depositional areas of rifts. In these environments, the productivity and nature of the plants and animals living in water or forming soils, and the chemistry of inflowing surface and spring waters and of lake waters, are crucial to sedimentation. For this reason also, ancient rift sediments are one of the best indicators of palaeoclimate. Work on East African rift sediments by Pleistocene stratigrahers has given us a measure of Pleistocene climatic changes matched only by research on northern hemisphere glacial deposits and, more recently, on oceanic sediments. Fortunately the East African Rift System crosses all the tropical environments, from the high rainfall equatorial belt, through the fluctuating climates of the subtropical zones, to the undoubted deserts to the N and the S. A range of sediments can thus be studied, from those forming in freshwater, biologically very productive lakes, to those forming under conditions of diverse salinity and composition in both perennial and ephemeral lakes.

For the sedimentologist, the virtues of studying lakes are manifold, especially where the sources of the sediments are so varied. Lakes are microcosms of larger marine basins. Within a relatively small area one can study rivers, deltas, shorelines, and shallow and deep water sedimentation. The proximity of active normal faulting allows one to compare sediments forming at the present day with those uplifted a few thousand or more years ago, within a tectonic and sedimentary context not so different from that which is currently operating.

The Lake Turkana basin, for example, is now one of the best studied lake basins in the world. The lake is the largest (7500 km^2) closed-basin lake and the fourth largest natural lake in Africa. In the past it has had a much larger extent. Several papers in this volume describe the processes involved in the interaction of sediment supply from short-lived ephemeral streams, a major delta, and shoreline and cliff erosion, with the slightly saline (2500 ppm) and alkaline lake waters (pH 9.2) no deeper than 125 m.

Thanks largely to the work of Eugster and his colleagues (Eugster, this volume) and its use as a comparison with the Green River Formation in the United States, Lake Magadi is, to geologists, the best known African lake. It is a moderately saline, alkaline ephemeral closed-basin lake which has served as a model for the formation of sodium carbonates and Magadi-type cherts in many ancient sedimentary rocks. Of increasing importance, however, is Lake Bogoria (Renault *et al.*, this volume; Vincens *et al.*, this volume). Though similar in many ways to Lake Magadi, Lake Bogoria is perennial and is fed not only by hot springs but by seasonal streams. It is saline ($>$40,000 ppm) and highly alkaline (pH 10.3), and within its basin a range of carbonates form.

Non-marine carbonates, though at one time sadly neglected, are now established as a major lithology that occurs extensively in ancient sedimentary rocks. Although lacustrine carbonates are not widely developed in alkaline soda-rich lakes, stromatolites are known from many basins in the African Rift System, even if some types are not forming today (Casanova, this volume). They form as lacustrine stromatolites, as oncolites and as travertine chimneys and pool-rim dams, each with its own characteristic morphology. Oncolites grow on floodplains during inundation, and travertines of several distinct forms develop at springs, at waterfall dams and in rivers.

There is thus no region of the world to match African lakes for the study of the biochemical processes in lakes, especially alkaline lakes within a range of volcanic and tectonic backgrounds.

References

Bally, A.B. 1983. Structural styles and the evolution of sedimentary basins. *Am. Assoc. Petrol. Geol., Educ. Short Course* 198 pp.

Bishop, W.W. 1968. The evolution of fossil environments in East Africa. *Trans. Leicester Lit. Philos. Soc.* **62,** 22–44.

Cerling, T.E. & Powers, D.W. 1977. Paleorifting between the Gregory and Ethiopian Rifts. *Geology* **5,** 441–4.

Drake, C.L. 1972. Future considerations concerning geodynamics. *Bull. Am. Assoc. Petrol. Geol.* **56,** 260–8.

McConnell, R.B. 1980. A resurgent taphrogenic lineament of Precambrian origin in eastern Africa. *J. Geol. Soc.* **137,** 483–9.

Molnar, P. & Tapponnier, P. 1975. Cenozoic tectonics of Asia: effects of a continental collision. *Science* **189,** 419–26.

Robbins, E.I. 1983. Accumulation of fossil fuels and metallic minerals in active and ancient rift systems. *Tectonophysics* **94,** 633–58.

Whiteman, A., Naylor, D., Pegrum, R. & Rees, G. 1975. North Sea troughs and plate tectonics. *Tectonophysics* **26,** 39–54.

H.G. Reading, Dept of Earth Sciences, The University, Parks Road, Oxford OX1 3PR, UK.

Geomorphology of the African Rift System

A.T. Grove

Sedimentation in African rift valleys depends firstly on their structure, especially the recency of tectonic activity and volcanism, secondly on climatic conditions, particularly rainfall, and thirdly on the relief and drainage, the geomorphology.

'Low Africa', the N and W of the continent, mostly below 500 m, is characterized by extensive structural basins with intervening updomings locally capped by volcanics. 'High Africa', in the S and E, mostly over 1000 m, consists largely of upland plains, dissected by major rivers in the S, interrupted by fault-troughs towards the N, and with extensive volcanic lavas in Ethiopia and Kenya.

The Great Rift Valley System is best developed topographically in the region extending northwards from near the mouth of the Zambezi and Lake Malawi (or Nyasa) via Lake Tanganyika to Lake Mobutu (or Albert) and also across the very large updomings thickly covered by volcanics in southern Kenya and southern Ethiopia. To the NE the Ethiopian Rift opens out into the Afar depression, the Gulf of Aden and the Red Sea (Fig. 1).

Whereas the northern part of the system is essentially the outcome of late Cenozoic rift faulting, the relief of the southernmost part is associated with much older dislocations (King 1978a). South-central Africa experienced rift faulting in early Mesozoic times. Karroo strata were let down into troughs where they were preserved when later erosion stripped rocks of this age from the areas left upstanding. When crustal rupturing later in the Jurassic resulted in the eastward departure from Africa of Madagascar/India and of Antarctica/Australia, the proto-Zambezi and proto-Limpopo, running down to the newly formed Indian Ocean, maintained their courses across the rising Lebombo and Sabi monoclines, exploiting the less resistant Karroo rocks within the rifts. Tributaries of the proto-Zambezi excavated the Malawi, Rukwa and Tanganyika valleys, and in these troughs Cretaceous and Cenozoic sediments accumulated and still remain. Rift faulting in much more recent times, corresponding to some degree with the ancient dislocations, has rejuvenated the trough forms, notably of the Rukwa basin.

Geophysical work and geological mapping over the last few decades have made it clear that rift structures underlie a much larger part of Africa than the Great Rift Valley region of the E. From the Bight of Benin, rift structures extend up the Benue valley and via the Chad basin towards Tibesti. Fairhead, in this volume, shows such structures running across the Cameroon Highlands and Adamawa Plateau and further to the ENE. South of the Jebel Marra they meet other rift structures extending from western Kenya NW across Uganda and the southern Sudan. Rift structures have also been traced running SW from Lake Tanganyika across Zaire, and from Malawi towards the Makadikgadi depression in the central Kalahari. The Congo or Zaire basin thus lies within a semicircle of rift structures, with the Great Rift Valley System tangential to it.

The two phenomena, the rift structures and the rift valleys, are very different geomorphologically and within each there is great morphological variety (Fig. 2). This attempt to characterize the main components of the whole complex adopts a regional treatment.

The Bight of Benin Rifts

The rift structures extending ENE from the Bight of Benin are, it is generally agreed, related to the opening of the South Atlantic when South America broke away from Africa in the Lower Cretaceous (King 1950; Wright 1978). A triple junction comparable to the one at the southern end of the Red Sea, was centred on the Niger delta. As the Atlantic gulfs opened to S and W, the arm pointing ENE failed to develop fully and little or no new crust was generated within it. A linear depression extended along the line of the Benue valley with subsidiary arms running E to Mamfe and later WNW along the line of the Niger from Lokoja to Jebba. The linear Benue depression forked into a northerly trough along the line of the lower Gongola and then NE into the Chad basin, and an easterly trough, followed by the upper Benue, extended E to Yola and Garoua.

As much as 5 km of sediments underlie the centre of the middle and lower Benue trough. They include not only thick fluvio-deltaic sandstones but also marine beds of Albian and Turonian age, laid down in seaways that are believed to have linked the Gulf of Guinea to the relatively short-lived Saharan seas extending S

From Frostick, L.E. *et al.* (eds) 1986, *Sedimentation in the African Rifts*, Geological Society Special Publication No. 25, pp. 9–16.

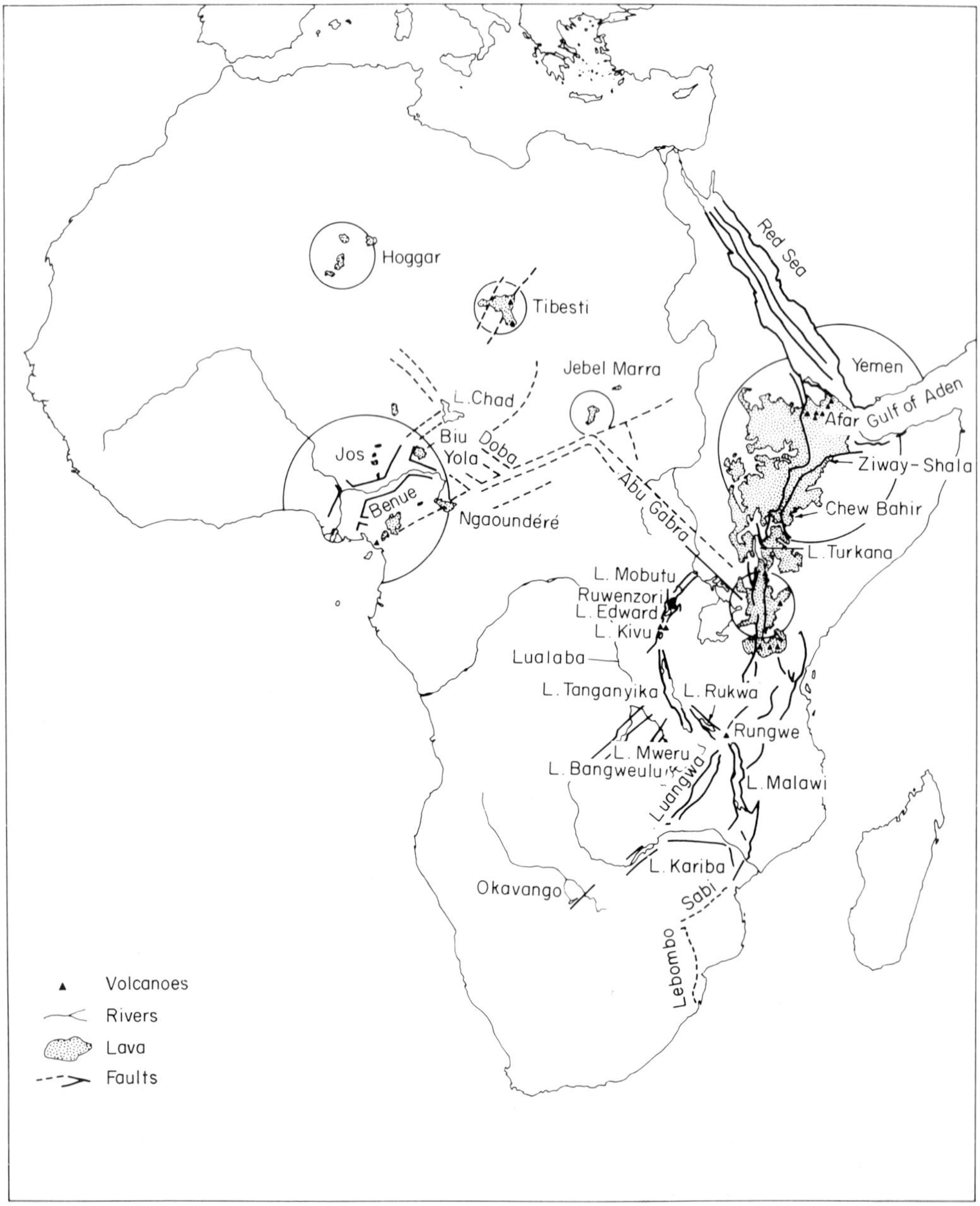

FIG. 1. The Mesozoic rift system of western 'Low Africa' and the Cenozoic rift valley system of eastern 'High Africa'. Sabi and Lebombo monoclines shown by dashed lines. Circles = updomings.

from Tethys. There seem to have been two periods of folding, the earlier, possibly Santonian, affecting mainly the shaley beds in the extreme S; a later one produced broader folds in rocks of Maastrichtian age and older. Whether such folding was the result of a few kilometres of compression or was the result of draping of the beds over vertical dislocations in the underlying basement is not known (Wright 1981; Benkhelil 1982).

A number of features distinguish the Benue from other rift valleys. The main scarps, rising over 1000 m high on either side, are wide apart, with 170–260 km separating them. On the SSE side of the depression, trough valleys extend far into the Cameroon Highlands, and basement

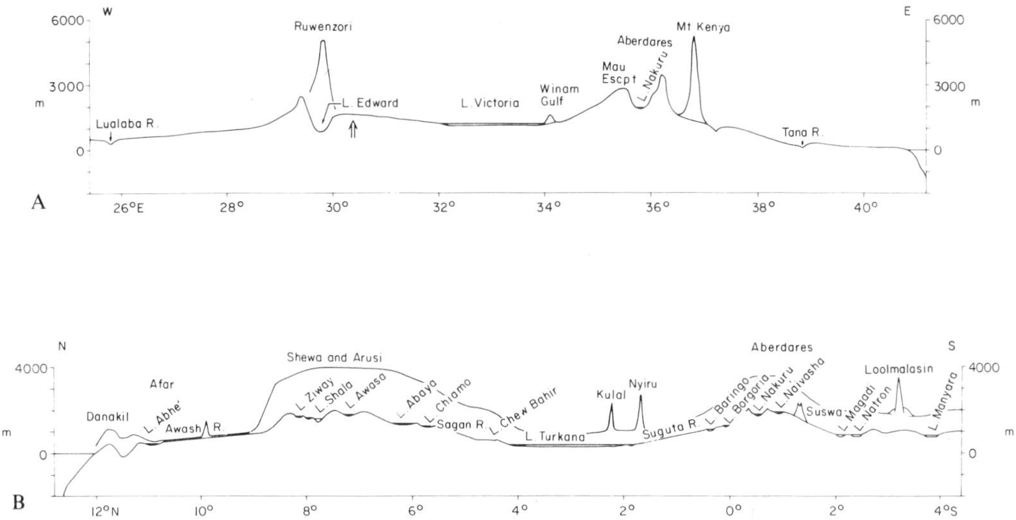

FIG. 2. A. Cross-section of East Africa and the rift valleys; vertical exaggeration *c.* ×40. B. Longitudinal section of the eastern rift and highlands within about 100 km of it; vertical exaggeration *c.* ×40.

crystalline rocks forming the scarps extend far out onto the plains. Neither topographic breaks nor recognizable fault planes distinguish the outcrop of sedimentary rocks in the centre of the trough floor. On the N side of the Benue, E of the Jos Plateau, thick beds of sandstone amongst the folded Cretaceous rocks give anticlinal arches, sharp straight ridges, and inward facing escarpments. Narrow valleys cut across these structures; the drainage seems to have been superimposed on the landscape, possibly from lava flows of which remnants are preserved on crests forming the Longuda Plateau.

East of the lower Niger and N of the eastern sector of its delta, Cretaceous and Tertiary rocks, gently folded into an anticlinorium (the Abakaliki anticlinorium) with its axis dipping to the SSW, have been dissected and the eastern limb eroded away to give an eastward-facing escarpment, deeply gullied between Nsukka and Udi, with the Cross River plains extending to the E. From Okigwi a ridge runs E and then S through Bende, with the Imo basin to the W. North of the Imo basin, another scarp in weak Tertiary beds, which include thick beds of clay, has been strongly gullied in the headwaters of the Mamu River (Grove 1951). The question of the age of the gullies and the progress downstream of the sediments derived from them, deserve more attention than they have so far received.

The Cameroon Highlands and the Mambila Plateau on the one side, and the Jos Plateau on the other, rising to about 1500 m above the floor of the Benue trough, can perhaps be regarded as the remnants of Cretaceous updoming across which the Benue Rift developed. The scarp of the dome remnant on the SE side of the trough has been indented by rivers rising on the Cameroon Highlands. The scarp of the Jos Plateau is less broken. The surface of the plateau, preserved by granite bastions, with granite cupolas rising above the general level and Tertiary lava flows of various ages, declines gently towards the N. The Kaduna River rising on the plateau, first flows NW before swinging round to run SW, cutting through gorges, to enter the Niger. The Gongola, the only river of any size entering the Benue from the N, is directed towards the Chad basin before it loops S to cut across fold structures in the vicinity of Bima Hill. In the vicinity of its northward loop it traverses Palaeocene Kerri-Kerri beds, unfolded, non-indurated sandstones derived from the erosion of the folded Cretaceous rocks.

Borings in connection with bridge construction have shown that the Niger at Onitsha and the Benue at both Makurdi and Yola are underlain by deep buried channels, presumed to have been excavated when sea levels were lower in glacial periods of the Pleistocene. Fluvial sedimentation in the late Quaternary, the building up of extensive floodplains on either side of the Benue mainstream, and the formation of the modern Niger delta are probably the outcome of climatic events at the end of the Pleistocene, when desert dune sands were swept downstream by the overflows from Megachad and from the region of the inland Niger delta.

At the present day, the Benue at low water

meanders between large sandbanks which are submerged between July and November beneath floodwaters occupying a very gently curving, broad flood channel. The smaller tributaries spread out over backswamps behind levées alongside the mainstream. Larger tributaries from the S have constructed small deltas at their confluences with the Benue. Most of the discharge of the mainstream comes from rivers draining basement rocks and the Tertiary lavas of the Mambila Plateau and Cameroon Highlands. Much of its bedload appears to be derived from the Faro river which rises in the Alantikas. Above the Faro confluence the Benue is a meandering stream running at a gentle gradient between clayey banks; immediately downstream of the confluence it is converted into the sandbank encumbered river that joins the Niger at Lokoja (Grove 1985).

The extension of the Benue-Gongola rifting to the NE is obscured by an 800 m thickness of Cenozoic Chad Group clays with intervening sand beds. They appear to have suffered little dislocation. The decline in altitude towards the N is probably as much a matter of climate zoning and the southerly source of sediments as any tilting. The rift extends beneath the Bodélé-Djourab depression, which was occupied in the early Holocene by Megachad, a lake some 150 m deep (Servant & Servant-Vildary 1980). The more broken erosional country towards Ounianga Kebir was, in that last pluvial period, occupied by swamps and lagoons fed by streams rising in Tibesti.

A chain of Tertiary to Recent volcanoes, which has been called the Volcanic Cameroon Line by Fitton (1980), stretches from the island of Annobon and Cameroon Mountain ENE across the highlands of the western Cameroon Republic. It splits into two branches giving a Y-shaped feature, one arm running N towards the Biu Plateau and the other ENE through Ngaoundere. Fitton sees the similarity between the Y forms of the Benue trough and the Cameroon Line as being the result of the displacement of a Y-shaped hot zone in the asthenosphere from beneath the former to beneath the latter. Fairhead (in this volume) shows the extension of the Cameroon Line to the ENE as the Central African Shear Zone; with no topographic expression it skirts the southern margins of the swampy Sara-Salamat plains of the Chari system, near the border between Chad and the Central African Republic.

The Afar to Tanzania Rifts

On the E side of Africa at about the same latitude as the Benue valley, the triple junction centred on Djibouti is an appropriate starting point for an outline of the regional geomorphology of the Great Rift Valley System of East Africa. The rift scarps overlooking Afar from the W and S, the Ethiopian and Southeastern Escarpments as they are called by Gasse *et al.* (1980), are unusually high and steep, rising 2000–3000 m above the rift floor, with peaks a few kilometres behind reaching above 4000 m.

In plan, the SW corner of the highlands of southern Yemen would fit snugly against the scarps. The threefold rifting splits a great dome about 1000 km in diameter that included most of Ethiopia and southern Yemen. It was thus about twice the size of the Benue dome and several times the size of the Kenyan, Tibesti and Jebel Marra updomings. The margins of the two most active arms, the Red Sea and Gulf of Aden Rifts, are now about 300 km apart, with oceanic crust between. The main trough floor of the Red Sea is little more than 200 m deep, but a central trough about 50 km wide goes down more than 2000 m. Within 100 km of the Red Sea coast, highlands rise on either side to 2000 m above sea-level in the S and 1000 m in the N. Almost the whole of the Gulf of Aden is more than 2000 m deep, and the altitudinal range from the sea-floor to the bordering highlands exceeds 4 km over distances of less than 100 km.

Both the Red Sea and the Gulf of Aden are starved of terrigenous sediments. No large rivers enter them at the present day and, with the possible exception of the Wal Masitah in wet periods of the late Quaternary, none may have done so since their formation. The climate of the coastal areas is arid and the long slopes of the bordering highlands are away from the rift troughs. The sill at the southern end of the Red Sea is little more than 100 m deep. Movement of water from the Gulf of Aden must have been restricted in glacial times but it seems unlikely that the Red Sea was converted into a closed basin, at least in the late Pleistocene, for coral reefs are conspicuous features of the coastal waters even in the Gulf of Akaba.

Oceanic material on the floor of the Afar depression is generally obscured by marine beds, basaltic or other lava flows and by fluvio-deltaic and lacustrine sediments of late Cenozoic age. The rift walls at the margin of the depression, sharpened by renewed faulting in Quaternary time, are bordered by enormous debris fans fed from great gullies notching the margins of plateaus sloping away from the rift. In the N, the floor of the Afar triangle descends over 100 m below sea-level with the Danakil Alps separating the Assele basin from the southernmost Red Sea.

The only river of any size entering the Afar

depression is the Awash, which rises W of Addis Ababa and was fed towards the end of the Pleistocene and in the early Holocene by overflow from the expanded lakes of the Ziway-Shala basin. Lake Abhé (Fig. 2), the terminal lake, has varied in size very markedly in the course of the late Quaternary, leaving behind an unusually full record of environmental fluctuations in the Awash catchment (Gasse *et al.* 1980).

Almost the entire western half of the Ethiopian dome is drained by rivers directed towards the Nile. They carry sediments down to the Sudan plains which are derived from volcanic rocks capping the plateau, from Jurassic rocks exposed in the Abbai and Takazzé gorges, and from basement rocks near the Sudan border. The drier eastern side of the dome drains towards the Indian Ocean, though little more than a trickle of water survives evaporation losses to enter the sea at Kismayu.

The rift valleys crossing southern Ethiopia and Kenya have much in common so far as geomorphology and sedimentation are concerned. Rift faulting in the Miocene was accompanied by the outpouring of great volumes of lavas over both areas. Low angle volcanic piles were built up, both at the margins of the rift and within it, in the late Miocene and Pliocene. Faulting, volcanism and caldera formation were concentrated in the Pleistocene along the central axis of the rift.

The floor of the rift falls from about 1690 m at Lake Awasa on the water-parting within the Ethiopian Rift, to 375 m at the surface of Lake Turkana (see Fig. 2). It rises to 1890 m at Lake Naivasha and falls again to 600 m at Lake Natron. Lying between the two domes and easily the largest lake in the Eastern Rift, Lake Turkana was fed in the late Pleistocene and early Holocene not only by the Omo, Turkwell and Kerio Rivers but also by an overflow from Chew Bahir, bringing in the waters of the Sagan River which drains Chiamo and Abaya within the Ethiopian Rift (Grove *et al.* 1975; Frostick & Reid 1980), and by the Suguta River running N in the Kenya Rift (Truckle 1976). The Suguta overflowed westwards to bypass the late Middle Pleistocene Teleki volcanic barrier (which cuts off the southern end of the Turkana depression) and entered the lake by way of the lower Kerio River. For most of the period from 8500 to 3000 BP, Lake Turkana stood about 80 m higher than it does now and drained W and then N to the Pibor tributary of the White Nile (Butzer 1980; Harvey & Grove 1982).

The rift margins commonly consist of almost vertical fault complexes, with slices of rock a kilometre or so thick and several kilometres long with surfaces dipping to the N or S (giving railway platform faulting) and tilted in some cases towards, and in other cases away, from the adjacent main rift wall. The impression given by such a landscape is of the tops of variously sized thick paperbacks that have slid a little way along a shelf (see King 1978b). On the floor of the rift such structures are usually concealed beneath some kilometres of sediment, only rarely do near vertical blocks poke through the surface, as seen in Chew Bahir. According to geophysical interpretations of gravity and other data, the sedimentary fill is usually deeper on one side of the rift than the other. The rift troughs are almost invariably asymmetric, with one wall higher than the other, and the deepest sediments usually on the side of the highest wall.

The floor of the Ethiopian and Kenyan Rifts is compartmentalized into basins, each of which is about 50 km wide and 100 km long. The basins are usually separated from one another by volcanic features, with a slight change in trough direction from one basin to the next. In the Ethiopian Rift, the Alutu volcano rises between the Ziway and Abiyata basins; Chubbi separates Shala and Awasa. In Kenya, the Menengai caldera separates the Baringo-Bogoria basin from the Nakuru basin; when Lake Nakuru rose almost 200 m in the early Holocene it overflowed into Menengai's crater. Naivasha at its highest levels overflowed a col between the rift wall and the Longonot volcano which, with Suswa, separates the Naivasha and Magadi-Natron basins.

Most of the basin lakes in the Eastern Rift are relatively small because of the low rainfall (only a few hundred millimetres annually) on the rift floor at the present day and the limited size of the catchments of the influent streams. Many of the lakes lack outlets and are saline in consequence. Rivers bring in heavy loads of sediment as can be seen, for instance, from the appearance of Lake Abaya on satellite imagery. In Turkana, fed by the Omo which drains the highlands S of the Blue Nile, the rate of sediment accumulation has been given by Yuretich (1982) as between 1000 and 3000 mm per 1000 years (*c.* 10 million t a^{-1}), as compared with the Nile's sediment transport of about 10 times that amount. In the clearer lakes diatomites have accumulated. Stromatolites belonging to more than one generation have formed along shorelines where wave action has been marked, filamentous algae have grown on the rocks and springs are close by. Evaporites are also, in part, derived from springs. The bulk of the infill consists of material of volcanic origin, much of it airfall pumice and ash derived directly from eruptions, the rest transported by rivers. Such materials, which have accumulated to great depths, are highly pervious so that streamflow is limited and lakes are commonly expressions of the local water table.

South into Tanzania the rift diverges into asymmetric structures holding Lakes Manyara and Eyasi. Other arms extend SE under Meru and Kilimanjaro and SW towards the Rungwe volcano which separates the Rukwa and Malawi troughs. Other structures radiate from the central areas of the Kenya updoming W of Mt Kenya. The Kavirondo or Winam Gulf Rift extends WSW under Lake Victoria. Browne & Fairhead (1983) show on their schematic map faults running towards Mt Elgon and then extending NW, as the Asua (Achwa) wrench fault with the Abu Garaba, *en echelon*, reaching towards Jebel Marra.

The Tanganyika to Malawi Rift

Whereas the Bight of Benin and the Afar–Tanzania Rifts can be regarded as the failed arms of triple junctions associated with the breakup of Gondwanaland, the western line of the Tanganyika–Malawi rift cannot be so readily set in a context of sea-floor spreading. Dixey (1956) emphasized the long history of development of the East African Rift System, with Tertiary fractures following the lines of Jurassic and earlier (Ubendian) orogenic belts. He saw erosion as having been guided by Karroo rocks let down between more resistant basement crystallines by Jurassic rifting. He also pointed to flights of erosion surfaces as evidence of successive periods of downcutting, and Cretaceous dinosaur beds as evidence of intervening periods of sedimentation within rift troughs. The fact remains that there are many fault features, such as the Livingstone escarpment overlooking Lake Nyasa (or Malawi) and its prolongation along the NE side of Lake Rukwa, which are evidently quite recent and still active. Girdler's map (1983) of seismicity in Africa for the period 1964–75 shows many more earthquake epicentres around the western Tanganyika–Malawi Rift than in the vicinity of the eastern Afar–Tanzania Rift (except for the region near Mt Kilimanjaro).

The absence of volcanic lavas in and around the rift valleys of the W contrasts strongly with their great extent and thickness in Kenya and Ethiopia. There are exceptions, for example, Rungwe has been mentioned. The Virunga volcanics blocked the upper basin of the Rutshiru draining to Lake Edward and the Nile system in the Pleistocene, thereby forming Lake Kivu which overflowed S by the Ruzizi to supply Lake Tanganyika, which drained into the Lualaba and the Zaire (or Congo).

Yuretich (1982) has argued that the sparsity of volcanics in the vicinity of the Western Rift has resulted in the lake basins there being starved of sediments. Certain rivers of moderate size, such as the Malagarasi entering Lake Tanganyika from the E, rise on upland plains of crystalline rocks and transport little sediment. Yuretich gives the rate of sediment accumulation in Lake Tanganyika as only 300–500 mm ka^{-1}. This may help to account for the great depth of the Western Rift lakes: Tanganyika is 1400 m deep and Malawi 700 m deep, both reaching far below sea-level. On the other hand, geophysical studies indicate a thickness of some 6 km of sediment in Rukwa and 2.5 km or more in the Mobutu (Albert) trough. It would not be very surprising to find comparable thicknesses of sediment underlying Lakes Tanganyika and Malawi, suggesting that sedimentation has been in progress there for some millions of years.

Fault dislocations have continued into the present, deepening the rifts and subdividing the floors of Lakes Tanganyika and Malawi into distinctive basins, each about 120 km long and 40 km wide. Rosendahl *et al.* (this volume) interpret these basin forms as half-graben produced by extensional stresses. Basin downfaulting alternately to the E (giving a high western wall, concave in plan form and cross-section) and to the W (with a high eastern wall) gives a sinusoidal trough. Large rock shelves or platforms within the basins are described as 'showing basement' and as being 'adjacent to, and associated with, drainage systems into the lake' (p. 28) This, it is suggested, shows that drainage systems are a response to the existence of the platforms, with most fluvial clastics entering the basins via these platforms, a view which should prompt research along these lines elsewhere.

The disposition of Lake Victoria, occupying the gentle basin-like depression between the two main arms of the East African Rift System, appears to be somewhat incidental. It has been argued that in Cenozoic times most of the East African region was drained towards the W across the site of Lake Victoria by rivers rising on the crest of the Kenya dome, before it was breached by faults that produced the current rift *c.* 2 Ma. The implication is that the eastern shoulders of the rift in the region of Lakes Edward and Mobutu (Albert) were much lower at that time, and that rivers could escape N to the Nile or W to the Zaire (Congo) system. Certainly the through valleys of the Kagera, Katonga and Kafu provide support for the comparative recency of the uplift of the eastern shoulder, which is considerably lower than the western one. Exactly when the drainage was reversed and Lake Victoria was ponded back is not known. Ollier (1981) comments that as the oldest sediments in the rift give the age of the faulting as Miocene, the old drainage pattern must be older than this. However, the interruption of the N-flowing rivers by upwarping along a line some 30 km E of the rift fault could have been much later, conceivably in the Pleistocene, as Bishop & Posnansky proposed (1960).

Rift faults running SW about 250 km and 500 km W of the southern end of Lake Tanganyika define the parallel shores of Lake Mweru and the Lualaba lake basin between Manone and Mukama. Signs of NE–SW faulting also appear in Lake Bangweulu. Southwest of Rungwe, the Muchinga escarpment overlooks the upper part of the rectilinear Luangwa valley, a major tributary of the Zambezi. Rift structures can also be traced SW, with some uncertainty, by way of Kariba into the Kalahari where blown sand and Kalahari beds obscure any topographic features. However, geophysical studies and satellite imagery make it clear that NE–SW faulting defines the southeastern, distal limit of the Okavango delta, and that the delta itself is underlain by a rift. The region continues to be seismically active (Reeves 1972).

Conclusions

In some areas rifting is concealed at the surface, in others it has produced trough forms about 35–100 km wide with a vertical relief of as much as 3 km, or even more exceptionally as in the vicinity of Ruwenzori where the vertical relief exceeds 4.5 km. Usually the rift walls are no more than 2 km high as compared with vertical displacements of the crust of up to 7 km. The thickness of the lithosphere in East Africa, according to Girdler (1983), is about 35–55 km, about a half of the thickness over much of the rest of the continent.

The spectacular relief of the Great African Rift Valley would seem to be largely the outcome of dislocations in Quaternary times, which involved vertical displacements of about 1 km. The Central African Rifts remain filled with sedimentary rocks largely concealed beneath later Cenozoic sediments. The Benue valley is the expression of long continued denudation of marine and continental rocks within a Cretaceous rift, and evacuation of the resultant sediments to the Niger delta and the Atlantic. The Karroo Rifts of south-central Africa have been partially re-exhumed in the course of a long history of rejuvenation, sedimentation and erosion and have also been transected by the southern extension of the Great Rift Valley System.

References

Benkhelil, J. 1982. Benue trough and Benue chain. *Geol. Mag.* **119,** 155–68.

Bishop, W.W. & Posnansky, M. 1960. Pleistocene environments and early man in Uganda. *Uganda J.* **24,** 44–61.

Browne, S.E. & Fairhead, J.D. 1983. Gravity study of the central African Rift system: a model of continental disruption. 1. The Ngaoundéré and Abu Gabra rifts. *Tectonophysics* **94,** 187–203.

Butzer, K.A. 1980. The Holocene lake plain of North Rudolf, East Africa. *Phys. Geogr.* **1,** 42–58.

Dixey, F.H. 1956. The East African Rift System. *Colonial Geol. Mineral Resources, Suppl. Series,* **1,** 1–71.

Ebinger, C.J., Crow, M.J., Rosendahl, B., Livingstone, D.A. & Le Fournier, J. 1984. Structural evolution of Lake Malawi, Africa. *Nature* **308,** 627–9.

Fitton, J.G. 1980. The Benue trough and Cameroon line—a migrating rift system in West Africa. *Earth Planet. Sci. Lett.* **51,** 132–8.

Frostick, L.E. & Reid, I. 1980. Sorting mechanisms in coarse-grained alluvial sediments: fresh evidence from a basalt plateau gravel, Kenya. *J. Geol. Soc.* **137,** 431–41.

Gasse, F., Rognon, P. & Street, F.A. 1980. Quaternary history of the Afar and Ethiopian Rift Lakes. In: Williams, M.A.J. & Faure, H. (eds) *The Sahara and the Nile.* Balkema, Rotterdam, pp. 361–400.

Girdler, R.W. 1985. Processes of planetary rifting as seen in the rifting and break up of Africa. *Tectonophysics* **94,** 241–52.

Grove, A.T. 1951. Soil erosion and population problems in south-east Nigeria. *Geogr. J.* **117,** 291–306.

—— 1983. Evolution of the physical geography of the East African Rift Valley region. In Sims, R.W., Price, J.H. & Whalley, P.E.S. (eds) *Evolution, Time and Space: The Emergence of the Biosphere.* Academic Press, London, pp. 115–55.

—— (ed) 1985. *The Niger and its Neighbours.* Balkema, Rotterdam.

——, Street, F.A. & Goudie, A.S. 1975. Former lake levels and climatic change in the Rift Valley of Southern Ethiopia. *Geogr. J.* **141,** 177–202.

Harvey, P. & Grove, A.T. 1982. A prehistoric source of the Nile. *Geogr. J.* **148,** 327–36.

King, B.C. 1978a. A comparison between the older (Karroo) rifts and the younger (Cenozoic) rifts of eastern Africa. In: Ramberg, I.B. & Neumann, E.R. (eds) *Tectonics and Geophysics of Continental Rifts.* Reidel, Dordrecht, pp. 347–50.

——, 1978b. Structural and volcanic evolution of the

Gregory Rift Valley. In: BISHOP, W.W. (ed) *Geological Background to Fossil Man*. Scottish Academic Press, Edinburgh, pp. 29–54

KING, L.C. 1950. Outline and disruption of Gondwanaland. *Geol. Mag.* **87,** 353–9.

OLLIER, C. 1981. *Tectonics and Landforms*. Longman, London.

REEVES, C.V. 1972. Rifting in the Kalahari? *Nature* **237,** 95–6.

SERVANT, M. & SERVANT-VILDARY, S. 1980. L'environnement quaternaire du bassin du Tchad. In: WILLIAMS, M.A.J. & FAURE, H. (eds) *The Sahara and the Nile*. Balkema, Rotterdam, pp. 133–62.

TRUCKLE, P.H. 1976. Geology and late Cainozoic lake sediments of the Suguta Trough, Kenya. *Nature* **263,** 380–3.

WRIGHT, J.B. 1978. Origins of the Benue Trough. In: KOGBE, C.A. (ed) *Geology of Nigeria*. University of Ife, Ile Ife, Nigeria, pp. 309–17.

—— 1981. Review of the origin and evolution of the Benue trough in Nigeria. *Earth Evol. Sci.* **2,** 98–103.

YURETICH, R.F. 1982. Possible influences upon lake development in the East African Rift valleys. *J. Geol.* **90,** 329–37.

A.T. GROVE, Dept of Geography, The University, Cambridge CB2 3EN, UK.

African Rift Basin Development

Geophysical controls on sedimentation within the African Rift Systems

J.D. Fairhead

SUMMARY: Gravity and seismological data are used, together with unpublished oil company data, to investigate the lithospheric structures associated with the East, and the West and Central African Rift Systems. Although these two rift systems are the result of extensional tectonics, they appear to be tectonically very different, with uplift and volcanism in East Africa, and subsidence and sedimentation in West and Central Africa. These differences can be explained in terms of the amounts of crustal extension (and thinning) that these rift systems have undergone and the isostatic response of these crustal structures during, and subsequent to, the extensional phase. In West and Central Africa, gravity modelling indicates crust extensions in the order of 70 km, while in East Africa extension across the Kenya Rift could be as little as 10 km. The implications of these lithospheric models for the formation of sedimentary basins are discussed, as well as how the rift structures relate to the crustal structure of continental margins.

Since the early Cretaceous, two major rift systems have developed within the African continent, the Tertiary to Recent rift system of East Africa and the Cretaceous to early Tertiary rift system of West and Central Africa. Although these systems appear to be tectonically different (uplift and volcanism in East Africa compared to subsidence and sedimentation in West Africa), they are both the result of extensional tectonics and link, via triple junctions, with successful rift structures that now form passive continental margins bounding the Red Sea, Gulf of Aden and South Atlantic. These continental margins were once part of intracontinental rifts, thus facets of their crustal structure should be reflected in the crustal structures associated with the failed rift arms within Africa.

Studies of passive continenal margins (*Oceanol. Acta* 1981) indicate that they are generally associated with (1) anomalously thin crust extending 50–200 km landward of the continental margin (e.g. Chenet & Montadert 1981); (2) pre-rift and rift phases of non-marine and marine sedimentation (e.g. Falvey & Middleton 1981; Keen *et al.* 1981); and (3) a rift phase mainly devoid of major volcanism. In these respects, the crustal structures associated with the West and Central African Rift System are more closely related to the above crustal model than those of the East African Rift System.

This contribution uses the available gravity and seismological data to investigate the lithosphere structure associated with the East, and the West and Central African Rift Systems, and then shows how the isostatic compensation of these rift systems controls both the regional and local sedimentary environment. An attempt is also made to explain why these rift systems have developed in such contrasting ways.

The East African Rift System

The East African Rift System is the least developed arm of the much larger Afro-Arabian Rift System which encompasses the embryonic oceanic basins of the Red Sea and Gulf of Aden. These rifts link together in the Afar depression of Ethiopia and join with the World Oceanic Rift System via the ridge system at the eastern end of the Gulf of Aden. Since the Afro-Arabian Rift System exhibits youthful stages in the breakup of a continental mass, the lithosphere model for the East African Rift System is considered by many to be typical of the structures that develop at incipient plate boundaries.

Geophysical studies (see Gass *et al.* 1978) indicate that the East African Rift System is underlain by a broad (*c.* 1000 km wide) zone of hot, low density upper mantle which is shallowest beneath the Kenya and Ethiopia domal uplifts. This body is interpreted by Girdler *et al.* (1969) and others to represent thinned lithosphere and replacement by lower density asthenosphere. The approximate spatial extent of this inferred zone of lithosphere thinning is shown in Fig. 1 by the −100 mGal Bouguer gravity contour. This contour outlines a zone of low gravity, extending SSW from Ethiopia, which gradually decreases in importance southwards, as does the surface expression of rifting. This zone is isostatically compensated with respect to the adjacent lithosphere, a fact reflected in its high topographic relief. The localized domal uplifts of Kenya and

From FROSTICK, L.E. *et al.* (eds) 1986, *Sedimentation in the African Rifts*, Geological Society Special Publication No. 25, pp. 19–27.

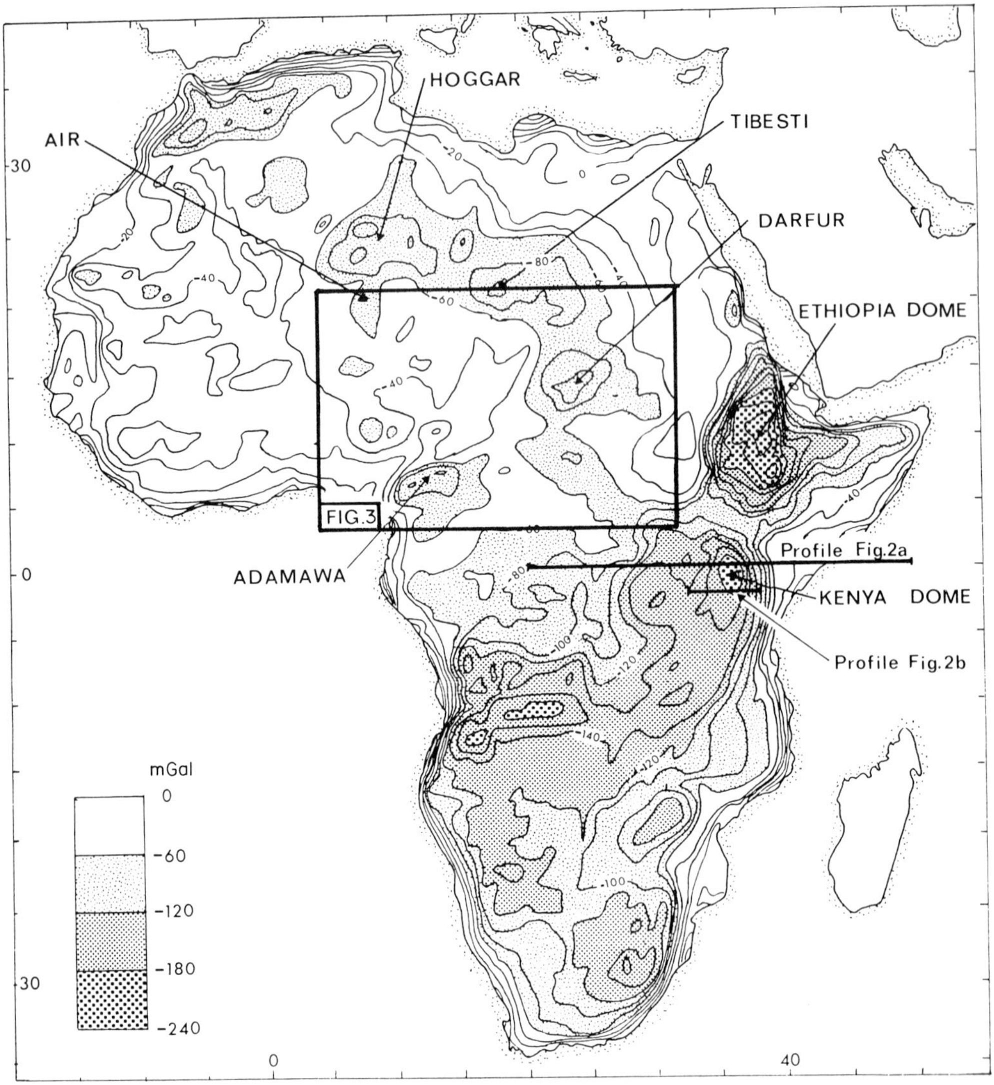

FIG. 1. Bouguer gravity map of Africa after Slettene *et al.* (1973), contoured at 20 mGal intervals and shaded at 60 mGal intervals to enhance the negative Bouguer anomaly associated with the East African Rift System. The domal uplifts of Tertiary age are indicated, as are the profiles across East Africa.

Ethiopia are associated with the most negative gravity values within this zone, and it is these areas which form the focus of volcanic activity and where the rift valleys are best developed.

The main features of the regional gravity field across East Africa are shown in profile form in Fig. 2A together with their possible interpretation. This simplistic model is consistent with the results of teleseismic P-wave travel time residuals (Fairhead & Reeves 1977), surface waves (Long *et al.* 1972) and heat flow (Williamson 1975) studies. Seismicity studies (see Fairhead & Stuart 1983) indicate that within the Kenya and Ethiopia Rifts the major part of the tensile stress is currently being released by micro-earthquakes along their axial zones, where geothermal activity and recent volcanism occur. Crust extension is estimated to range from about 10 to 35 km (see Searle 1970; Fairhead 1976) with most of it being confined to the 60–80 km wide rift valley. Seismic array studies (Maguire & Long 1976) further show that normal crustal thicknesses (*c.* 40 km) are present beneath the rift shoulders, whereas within the rift seismic refraction studies (Griffiths

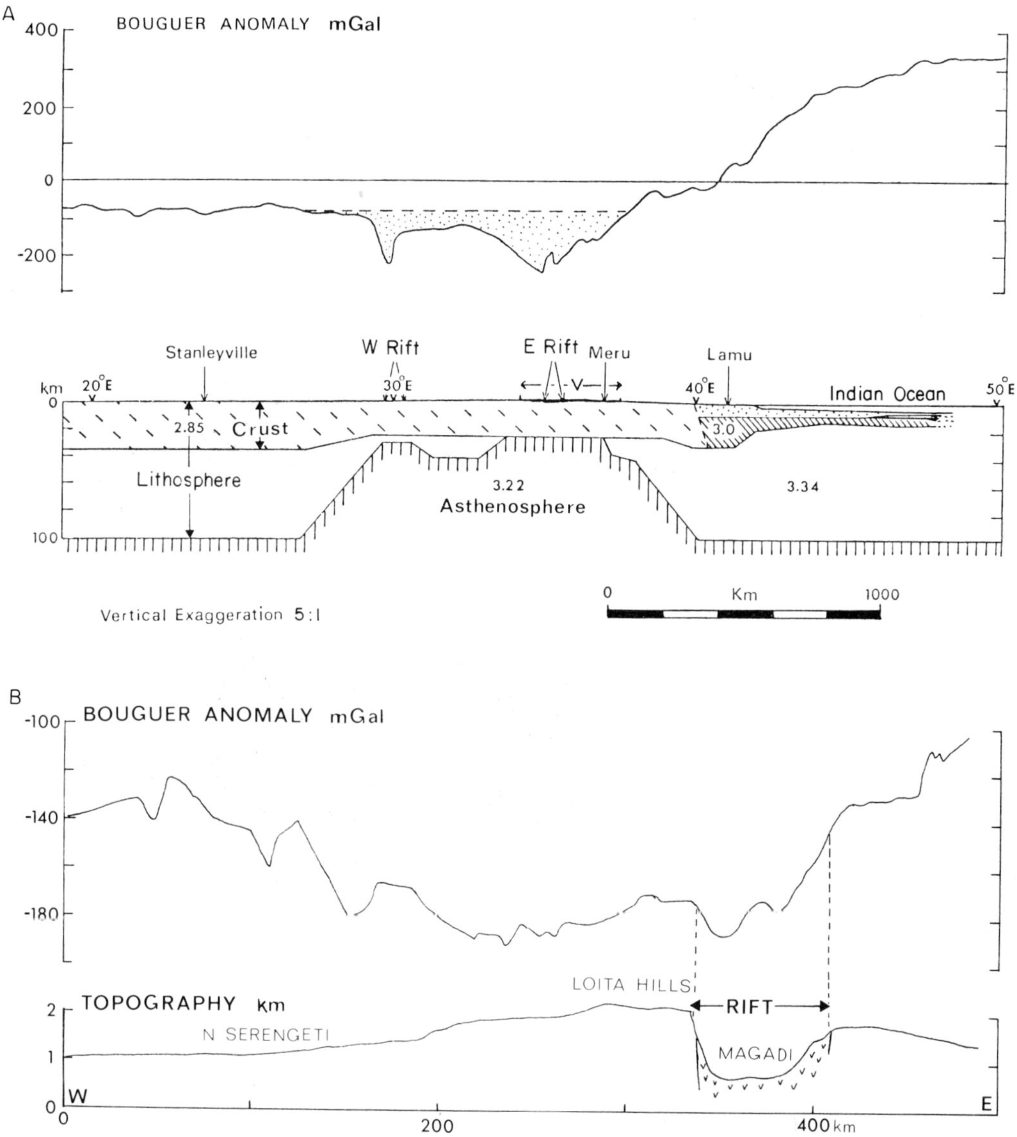

FIG. 2. A. Regional Bouguer anomaly profile across East Africa (see Fig. 1 for location) together with lithospheric model, after Girdler *et al.* (1969). B. Gravity and topography profile across the southern part of the Kenya dome at latitude of 1.8°S, after Fairhead (1976). The gravity profile shows a strong negative correlation with topography over a section of the dome where the volcanics and sediments are confined to the rift trough.

et al. 1971) indicate a crustal thickness of about 20 km.

A more detailed gravity and topography profile (Fig. 2B) crossing the Kenya Rift at the latitude of Magadi shows the relation between the gravity field and topography over the Kenya dome. This profile was chosen because the volcanics are confined to the rift valley and do not influence the regional gravity field. The topography of the Kenya dome is clearly not conducive to the development of sedimentary basins, except within parts of the 60–80 km wide rift where faulting and volcanism provide barriers to the flow of water out of the rift to the N and S. The main sedimentary areas associated with the East African Rift System are away from the Kenya and Ethiopia uplifts, and lie within the rift basins (Lakes Tanganyika and Malawi) along the block faults of Tanzania (Lakes Manyara and Eyasi), and within the shallow crustal depression between the eastern and western rifts (Lake Victoria).

The West and Central African Rift System

The rift system of West and Central Africa has not developed the spectacular topographic features found in East Africa. This is due to the contrasting nature of its formation and the major role that sedimentation has played. The evolution of the rift system is contemporaneous with the opening of the South Atlantic during the Cretaceous and structurally related to it via the development of a triple junction on the site of the Niger delta (Burke *et al.* 1972). Two arms of the triple junction have developed into the Equatorial Shear Zone and the South Atlantic (Bonatti & Crane 1984), while the third arm, the Benue trough of Nigeria, failed to develop beyond the initial extensional phase that lasted about 30 Ma (Albian to Santonian).

During this period the lithosphere is considered to have undergone rapid stretching which caused thinning of the crust and passive upwelling of the hot asthenosphere. The resulting isostatic compensation of these structures was surface subsidence, i.e. the subsidence effect of the thinned low density crust was greater than the buoyancy effect of the low density asthenosphere (see McKenzie 1978). Subsidence was contemporaneous with the stretching phase, allowing marine sediments to be deposited in the Benue trough throughout this period, before being terminated by a period of deformation in the Santonian. Following this deformation, the marine sediments were eroded and deltaic sediments spread throughout the trough. Continental facies sediments persisted until the end of the Cretaceous, apart from a short-lived extensive marine incursion in the Maastrichtian (Carter *et al.* 1963; Burke *et al.* 1972). This indicates that much of the basin topography of the West African Rift System was close to sea-level up until the Maastrichtian. When South America finally broke away from Africa in the late Cretaceous, this presumably caused a major change in the lithosphere stress regime within Africa, which resulted in the cessation of crustal/lithosphere extension in West Africa.

The late Cretaceous – early Tertiary was a time of rapid sea-floor spreading and changes in the relative motion between Africa and South America resulting in a major reorganization of the Equatorial Shear Zone (Larson & Pitman 1972; Sibuet & Mascle 1978). The effects of the reorganization have been recorded in the Benue trough and along the Central African Shear Zone by sinistral and dextral shear displacements (Benkhelil 1982; Browne & Fairhead 1983; Cornacchia & Dars 1983). Since the Cretaceous (65 Ma to present) the thermal anomaly caused by the upwelling of asthenosphere has slowly decayed through heat conduction to the surface. This has resulted in thermal contraction of the lithosphere and its return to a more normal thickness and density (Jarvis & McKenzie 1980). The isostatic response of the crust to the changing mantle density has been a further phase of subsidence causing downwarping over an area more extensive than the Cretaceous sedimentary basins. This resulted in the formation of Cenozoic sedimentary basins generally 1–2 km deep in Central Africa (Esso, pers. comm.), thus concealing the nature and extent of the Cretaceous rift system.

The post-tectonic sedimentary phase is absent for much of the rift system in Nigeria and Cameroon, which undoubtedly reflects the role of external drainage helped by the development of the Tertiary Adamawa uplift in central Cameroon. Where the Cretaceous rift basins are not exposed it has only been extensive regional gravity mapping (Louis 1970; Browne & Fairhead 1983), seismic reflection surveys, and exploration drilling by the major oil companies (Conoco, Chevron, Elf, Esso and Texaco) that have revealed their presence (Fig. 3). The extent and nature of the Cretaceous rift system makes it comparable in size to, if not larger than, the East African Rift System.

Hydrocarbon exploration in Sudan by Chevron has revealed that clastic sediments infilling the Abu Gabra and Unity basins of the southern Sudan Rift are normally compacted and have no major density inversions, so that the gravity data tend to reflect the shape of the basins (Flege 1982). This uniform increase in density with depth is observed from exploration wells throughout Central African (Conoco, pers. comm.) and Niger regions (Esso, pers. comm.). Drilling within these rift basins also indicates a general lack of volcanism during their formation.

Gravity and topographic profiles across different sections of the rift system are presented in Fig. 4. The locations of the profiles together with a simplified tectonic model based on regional geophysical and oil company data are shown in Fig. 3. The profiles, although typical of their respective rifts/shear zones, have been chosen so that they follow gravity traverses in order that the higher frequency parts of the gravity field are

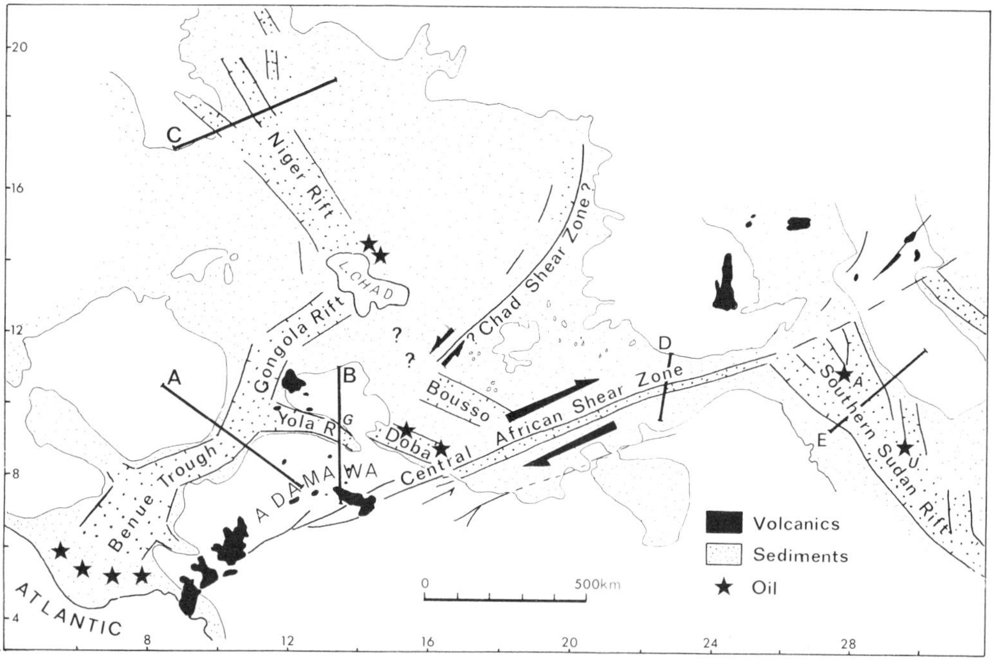

FIG. 3. Sketch map of the main fault-controlled sedimentary basins of the West and Central African Rift System. The location of profiles in Fig. 4 are shown together with location of areas where oil has been found. The question marks indicate areas of uncertain subsurface geology and structures. *A* & *U* represent the Abu Gabra and Unity basins of the southern Sudan Rift. *G* represents the Garoua basin at the eastern end of the Yola Rift.

preserved. The common features on all profiles are: (1) the broad, smooth positive anomaly, (2) the narrow high-gradient negative anomalies superimposed on the broader positive anomalies; and (3) the lack of topographic relief over the basins.

Of the five profiles shown in Fig. 4 the Benue and Yola Rift profiles cross exposed Cretaceous basins, while the remaining three profiles cross areas covered by Cenozoic sediments. Evidence that the residual negative anomalies are caused by fault-controlled sedimentary basins is good and is consistent with aeromagnetic, seismic reflection and drilling data. For example, seismic reflection data indicate sedimentary thickness at least 3–4 km thick for the Garoua basin of the Yola Rift (Elf Serepca, pers. comm.), the Niger Rift (Esso & Elf, pers. comm.) and the southern Sudan Rift (Chevron, pers. comm.), and this is confirmed by drilling (Nicod 1983). The shape of the positive anomaly has been drawn to follow the observed Bouguer gravity field over the exposed basement areas, while over the basins it is controlled by estimating the gravity effect of the sedimentary infill using borehole density logs where possible.

The only controls on the interpretation of the broad, smooth positive anomaly are the seismic refraction results from the Garoua section of the Yola Rift (Dorbath *et al.* 1984; Stuart *et al.* 1985). These seismological results indicate the crust has thinned from 34 to 23 km beneath the rift with a corresponding decrease in Pn velocities from 8.0 to 7.8 km s^{-1}. These crustal thicknesses are used in Fig. 5 to control the interpretation of the positive anomaly (after the regional anomaly has been removed) in terms of a simple crustal thinning model. An unexpected result of this modelling was the requirement that the crust–upper mantle density contrast has to be as small as $+0.17$ g cm^{-3}. Assuming the volume of crystalline crust is conserved during the stretching process then the amount of crustal extension can be estimated from the model cross-section areas of the basin and necked region. Such estimates for the Garoua section of the Yola Rift (Fig. 4) give *c.* 55 km of extension. Using a similar density contrast, extension estimates based on gravity modelling give 95 km for the middle Benue (Fig. 4), 65 km for the Gongola Rift (Fairhead & Okereke 1986), 58 km for the Niger Rift (Fig. 4),

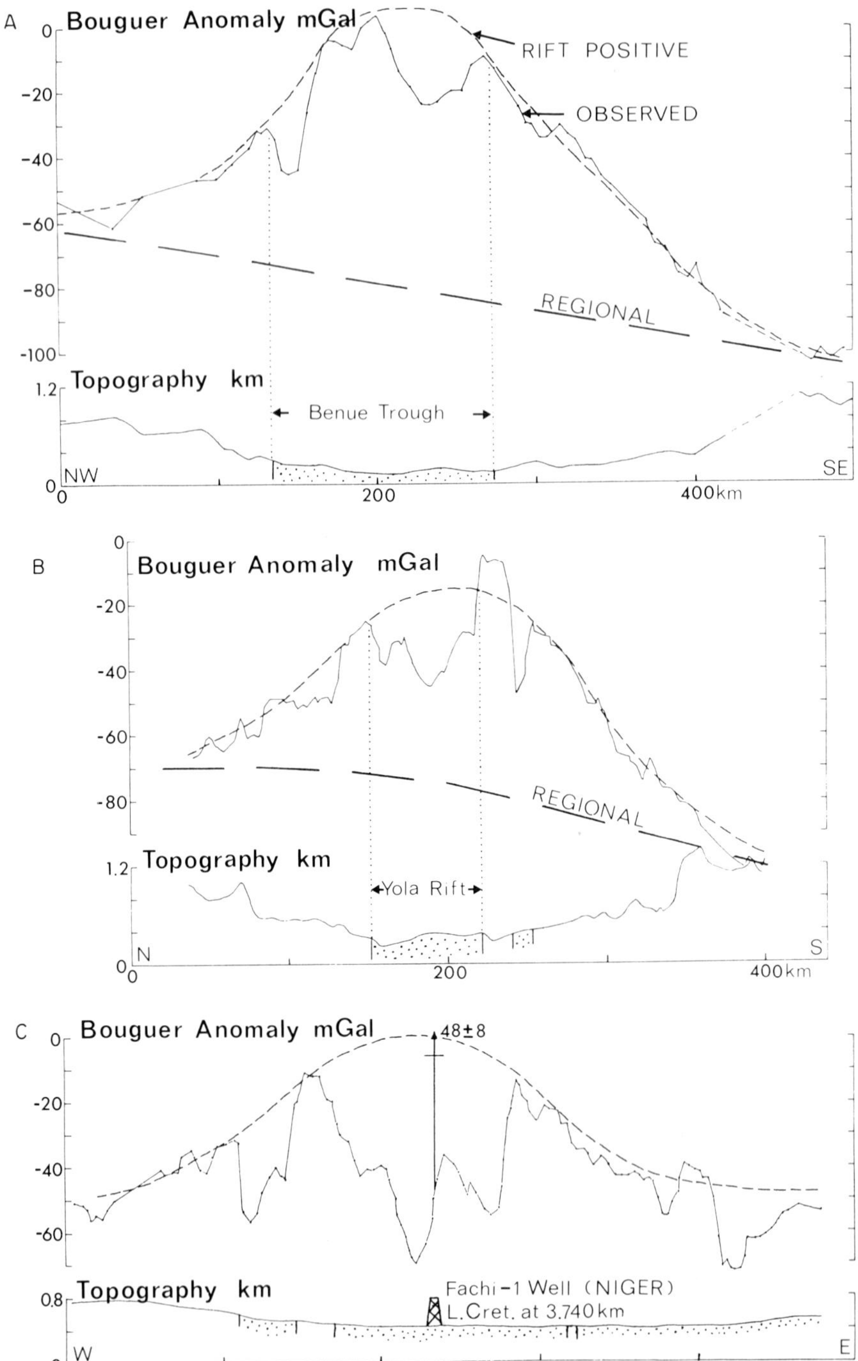

FIG. 4. Five Bouguer gravity and topography profiles across various sections of the West and Central African Rift System. The location of the profiles are shown in Fig. 3. Gravity data sources: A. Fairhead & Okereke (1986); B. Collingnon (1968); C. Rechenmann (1969); D. (opposite page) Louis (1970); E. (opposite page) Browne & Fairhead (1983). Stippled areas are sediments. The gravity value from the Fachi well (profile C) is calculated from the well density log (Esso, pers. comm.). The regional Bouguer trend in profiles A & B are due to the Adamawa uplift in central Cameroon.

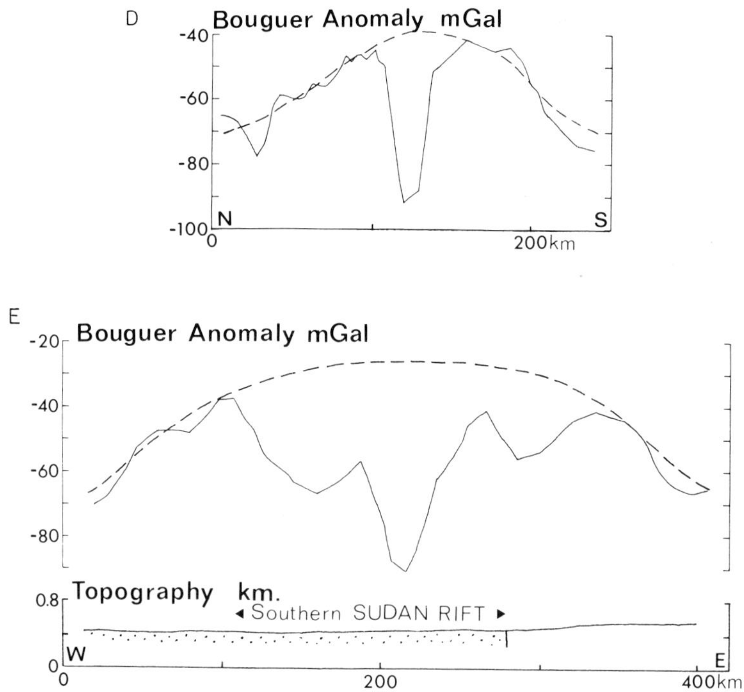

FIG. 4. Continued.

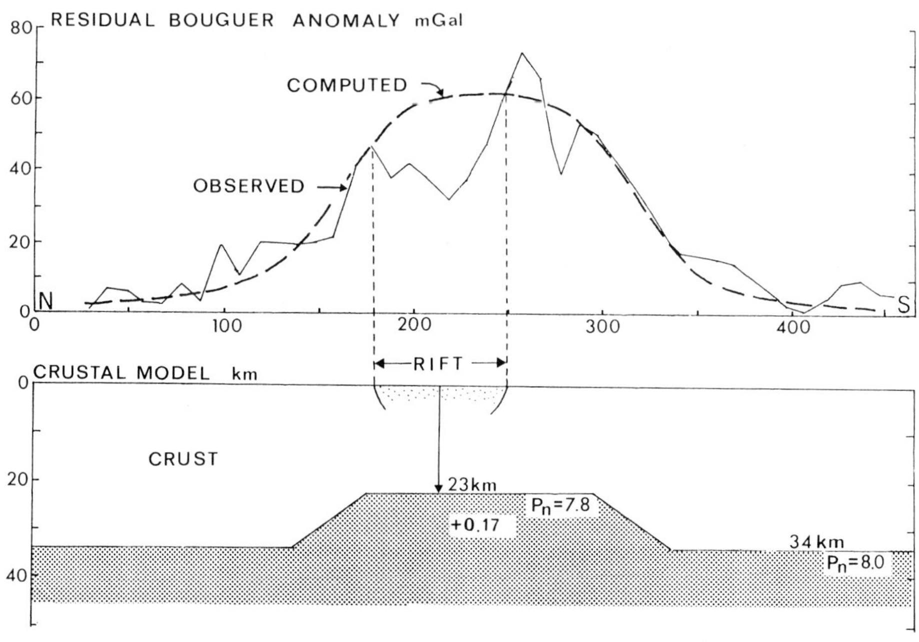

FIG. 5. Interpretation of profile B (Fig. 4), after removal of regional anomaly, across the Yola Rift at Garoua (Cameroon). The simple crustal model is controlled by seismic refraction data (Stuart *et al.* 1985) which give crustal thicknesses of 23 km beneath the rift and 34 km beneath the Adamawa Plateau further S.

in excess of 13 km for the Central African Shear Zone (Fig. 4), and 48 km for the southern Sudan Rift (Fig. 4) (Browne & Fairhead 1983).

Evolutionary differences of the rift systems

The main differences between the rift systems have been illustrated in Figs. 2 and 4. The active rifting in East Africa is associated with domal uplift and volcanism, with only minor sedimentation, whereas the West and Central African Rift System is dominated by subsidence and clastic sedimentation with only minor volcanism. The uplift in East Africa is clearly related to the presence of a large, anomalously low-density body within the upper mantle. For West and Central Africa, the gravity data do not indicate the presence of any anomalous upper mantle body beneath the much older and now inactive Cretaceous rift system. However, the post-tectonic subsidence that has taken place along much of the West and Central African Rift System suggests that a hot low-density upper mantle body did exist during the Cretaceous, but since that time (last 60 Ma) the lithosphere has cooled and reverted back to more normal densities. The presence of such an anomalous upper mantle body does not imply that during the Cretaceous period there was domal uplift. There is good geological evidence (already cited) that Cretaceous marine sediments are present in all the rift troughs except, possibly, the eastern end of the Central African Shear Zone and southern Sudan Rift. Thus the topographic differences between these rift systems were present even during their active phases.

It thus appears that the fundamental difference between these rift systems is the amount of crustal extension (and thinning) each system has undergone, and how these structures were isostatically compensated. Crustal extension estimates, although based on simple models, indicate that the extension across the West and Central African Rift System is greater than that of the East African Rift System by at least a factor of four.

Conclusions

The implications of this study are that the structural differences between the rift systems of Africa are controlled by the amount (and probably by the rate) of crustal extension. In East Africa the lithospheric tensional stresses appear to have been much smaller during the Tertiary than across West and Central Africa during the Cretaceous, such that crustal stretching in East Africa has been minimal and that isostatic compensation of the asthenospheric body has controlled the physiography of the surface. Tensional stresses generated by domal uplift in East Africa may in fact account for most of the observed tectonic structures. The lack of lithospheric stress may explain why East Africa has been unable to separate, although the asthenospheric body is of similar dimensions to that found beneath oceanic ridges. Such an argument would also explain the presence of massive surface volcanism, which at a divergent plate boundary such as the Red Sea and Gulf of Aden, would go into making new oceanic crust.

The West and Central African Rift System developed during the Cretaceous at a time of high rates of sea-floor spreading and major continental disruption. This suggests that the lithospheric plates were subjected to large tensional stresses, especially in West Africa where the shape of the continent would produce stress concentration. The post-tectonic subsidence of the rift system suggests that the lithosphere was anomalously thin during the Cretaceous, but due to changes in the tensional stress regime did not progress to the sea-floor spreading phase of dyke injection, except possibly in the lower Benue trough. Even here there is little evidence of major surface volcanism, suggesting that the intrusive zone was confined to the basement beneath the Cretaceous sedimentary cover.

The West and Central African Rift System has attracted much attention from the hydrocarbon industry since the Cretaceous sedimentary basins are deep, extensive, have formed along zones where the geothermal gradients are enhanced, and have been subsequently blanketed by thick Cenozoic sedimentary cover which has further enhanced oil maturation. Existing oil shows and discoveries (Nicod 1983) are shown as stars in Fig. 3. Sedimentation within the East African Rift System has, by comparison, been restricted by the uplift of the region to the development of sedimentary basins along the block faults in Tanzania and within the rift grabens of Lakes Tanganyika and Malawi. All these basins lie well outside the area affected by the localized domal uplifts.

In conclusion, the crustal structures associated with these rift systems appear to represent end members of a range of possible incipient plate margin structures. The West and Central African Rift System is associated with large amounts of crustal extension, and its crustal structure is typical of many continental margins. On the other hand, the upper mantle structure beneath the East African Rift System is probably typical of an active incipient plate margin. Such an upper mantle structure is likely to have existed beneath the West and Central African Rift System in the Cretaceous. Since the rift system of East Africa is still active, it is this rift system which is the more likely to develop into a plate margin, if lithospheric stresses allow.

References

BENKHELIL, J. 1982. Benue trough and Benue chain. *Geol. Mag.* **119,** 155–68.

BONATTI, E. & CRANE, K. 1984. Oceanic fracture zones. *Sci. Am.* **250,** 40–51.

BROWNE, S.E. & FAIRHEAD, J.D. 1983. Gravity study of the Central African Rift System: a model of continental disruption. 1. The Ngaoundere and Abu Gabra Rifts. *Tectonophysics* **94,** 187–203.

BURKE, K.C., DESSAUVAGIE, T.F.J. & WHITEMAN, A.J. 1972. Geological history of the Benue valley and adjacent areas. In: DESSAUVAGIE, T.F.J. & WHITEMAN, A.J. (eds) *African Geology*. Ibadan University Press, pp. 187–205.

CARTER, J.D., BARBER, W. & TAIT, E.A. 1963. The geology of parts of Adamawa, Bauchi and Bornu Provinces in northeastern Nigeria. *Bull. Geol. Surv. Nigeria* **30,** 1–108.

CHENET, P.Y. & MONTADERT, L. 1981. Rifting tectonics on Galicia Portugal and North Biscay margin. *Conference on the Processes of Planetary Rifting*, Lunar Planet, Instit., Houstin, Texas, 47–50.

COLLIGNON, F. 1968. Gravimetrie du reconnaisance Cameroun. *ORSTOM*, Paris, 35pp.

CORNACCHIA, M. & DARS, R. 1983. Un trait structural majeur du continent africain. Les lineaments centreafricains du Cameroun au golfe d'Aden. *Bull. Soc. Géol. Fr.* **25,** 101–9.

DORBATH, L., DORBATH, C., STUART, G.W. & FAIRHEAD, J.D. 1984. Tectonique-Structure de la croute sous le plateau de l'Adamaoua (Cameroun). *C. R. Acad. Sci. Paris, Ser. II* **298,** 539–42.

FAIRHEAD, J.D. 1976. The structure of the lithosphere beneath the Eastern Rift, East Africa, deduced from gravity studies. *Tectonophysics* **30,** 269–98.

—— & OKEREKE, C.S. 1986. A regional gravity study of the West African Rift System in Nigeria and Cameroon and its tectonic interpretation. *Tectonophysics* (in press).

—— & REEVES, C.V. 1977. Teleseismic delay times, Bouguer anomalies and inferred thickness of the African lithosphere. *Earth Planet. Sci. Lett.* **36,** 63–76.

—— & STUART, G.W. 1983. The seismicity of the African Rift System and comparison with other continental rifts. *Geodynamics Ser.* **8,** 41–61.

FALVEY, D.A. & MIDDLETON, M.F. 1981. Passive continental margins: evidence for a pre breakup deep crustal metamorphic subsidence mechanism. *Oceanol. Acta*, Colloque **C3,** 103–14.

FLEGE, R.F. 1982. Regional gravity and magnetics in the Sudan: applications to petroleum exploration. *Abs. 52nd Ann. Int. Meet.* Soc. Explor. Geophys., Dallas, Texas, PF2.4.

GASS, I.G., CHAPMAN, D.S., POLLACK, H.N. & THORPE, R.S. 1978. Geological and geophysical parameters of mid-plate volcanism. *Phil. Trans. R. Soc. Lond.* **288A,** 581–97.

GIRDLER, R.W., FAIRHEAD, J.D., SEARLE, R.C. & SOWERBUTTS, W.T.C. 1969. Evolution of rifting in Africa. *Nature* **224,** 1178–82.

GRIFFITHS, D.H., KING, R.F., KHAN, M.A. & BLUNDELL, D.J. 1971. Seismic refraction line in the Gregory Rift. *Nature* **229,** 66–71.

JARVIS, G.T. & MCKENZIE, D. 1980. Sedimentary basin formation with finite extension rates. *Earth Planet. Sci. Lett.* **48,** 42–52.

KEEN, C.E., BEAUMONT, C. & BOUTILIER, R. 1981. Preliminary results from a thermo-mechanical model for the evolution of Atlantic-type continental margins *Oceanol. Acta*, Colloque **C3,** 123–8.

LARSON, R.L. and PITMAN, W.C. III 1972. World-wide correlation of Mesozoic magnetic anomalies and its implications. *Bull. Geol. Soc. Am.* **83,** 3645–62.

LONG, R.E., BACKHOUSE, R.W., MAGUIRE, P.K.H. & SUNDARLINGHAM, K. 1972. The structure of East Africa using surface wave dispersion and Durham seismic array data. *Tectonophysics* **15,** 165–78.

LOUIS, P. 1970. Contribution géophysique à la connaissance géologique du bassin du Lac Tchad. *Mém. ORSTOM* **42,** 311 pp.

MAGUIRE, P.K.H. & LONG, R.E. 1976. The structure of the western flank of the Gregory Rift (Kenya) part 1: the crust. *Geophys. J. R. Astr. Soc.* **44,** 661–75.

MCKENZIE, D. 1978. Some remarks on the development of sedimentary basins. *Earth Planet. Sci. Lett.* **40,** 25–32.

NICOD, M.A. 1983. Oil and gas developments in North Africa in 1982. *Bull. Am. Assoc. Petrol. Geol.* **67,** 1795–826.

OCEANOLOGICA ACTA 1981. *Colloque C3: Geology of Continental Margins.* 26th Congres Geologique International, Paris, 1980, 294 pp.

RECHENMANN, J. 1969. Cartes gravimetriques du Niger. *Not. Explic. ORSTOM.* **36,** 11.

SEARLE, R.C. 1970. Evidence from gravity anomalies for thinning of the lithosphere beneath the rift valley in Kenya. *Geophys. J. R. Astr. Soc.* **21,** 13–31.

SIBUET, J-C. & MASCLE, J. 1978. Plate kinematic implications of Atlantic equatorial fracture zone trends. *J. Geophys. Res.* **83,** 3401–21.

SLETTENE, L., WILCOX, L.E., BLOUSE, R.S. & SANDERS, J.R. 1973. *A Bouguer Gravity Anomaly Map of Africa.* DMAAC tech. paper 73–003, Defence Mapping Agency, St Louis, USA.

STUART, G.W., FAIRHEAD, J.D., DORBATH, L. & DORBATH, C. 1985. A seismic refraction study of the crustal structure associated with the Adamawa Plateau and Garoua Rift, Cameroon, West Africa. *Geophys. J. R. Astr. Soc.* **81,** 1–12.

WILLIAMSON, K.H. 1975. *Terrestrial heat flow studies in Kenya.* Unpublished PhD thesis, University of London.

J.D. FAIRHEAD, Dept of Earth Sciences, The University, Leeds LS2 9JT, UK.

Structural expressions of rifting: lessons from Lake Tanganyika, Africa

B.R. Rosendahl, D.J. Reynolds, P.M. Lorber, C.F. Burgess, J. McGill, D. Scott, J.J. Lambiase & S.J. Derksen

SUMMARY: The fundamental building blocks of Lake Tanganyika are half-graben that tend to be arcuate or crescent-shaped in plan view. When combined with the asymmetric subsidence of the half-graben, this geometry creates horizontal components of motion which are expressed as shearing and rotation along the ends of the units. The key to understanding rift-basin morphology is recognizing the various ways in which these fundamental units can be linked together in time and space. Although many modes of linking occur, the basic theme is one of alternating half-graben polarities along the strike of the rift. Where adjacent half-graben face in opposite directions and do not overlap with each other, they are usually separated by interbasinal ridges trending oblique to the rift axis. In cases where facing half-graben overlap, the apparent morphology can be that of a full-graben with a central high or a half-graben abutting a structural platform, depending upon the orientation of the observation line relative to the overlap geometry. Mechanically, such full-graben should be viewed as opposing half-graben which hinge away from the central high. Examination of the linking models in Tanganyika has enabled us to derive an empirical model of rifting. The model is extremely useful in understanding the stratigraphy of Tanganyika, and it seems to explain many of the styles and patterns of deposition which are peculiar to rift basins. If the model has general applicability to continental rifts, it can also be useful to those who deal with the mechanics of rifting and the deep structure of passive margins.

Research workers dealing with the tectonics of rifts have done so mainly from the perspective and scale of the rift system; when confronted with the sedimentological side of the picture there is a tendency to treat rift basins as holes in the ground into which sediments are dumped indiscriminately. Research into the stratigraphy of rifts has tended to brush aside and even dismiss the tectonocist's perspective as too broad and vague to be of any specific value.

This is not a condemnation of either group, because the pursuit of a 'middle ground' has been hampered by a basic lack of information on the structure of rifts at the basinal scale, on the early evolution of rift basins, and on how both come together to influence the depositional histories of rifts. The databases that do exist have been largely acquired by the petroleum industry. Too often, this information is divided into prospect-sized packages, even within individual companies. It is usually unavailable to the academic researcher and irretrievable, at least as a whole, to the petroleum explorationist. Things are beginning to change on almost all fronts, evidenced in part by some of the discussions in this volume and by a new willingness of the petroleum industry to fund field efforts which might reveal the tectono-depositional secrets of rift environments. The work discussed in this paper is an outgrowth of just such a willingness on the part of a consortium of petroleum companies.

This report begins with a description of the basinal morphology of Lake Tanganyika, progressing from the lake bed to the earliest identifiable units of rifting. We then construct a tectonic model of rifting in Tanganyika and discuss those elements which seem to have the greatest impact on depositional processes in the lake. We conclude with some statements on the applicability of our model to other areas and problems.

Fig. 1 is the key illustration with regard to the overall morphology, structure, and acoustic character of Tanganyika. Most of the other figures are derived wholly or partly from information and inferences contained in the seismic profiles, from which the line drawings in Fig. 1 were made.

Structural expressions and relations in Lake Tanganyika

General morphology

The modern geomorphology of Lake Tanganyika is shown in Fig. 2. Four sedimentary basins are identified: the Ruzizi, Kigoma, Kalemie, and Marungu-Mpulungu basins. These basins are

From Frostick, L.E. *et al.* (eds) 1986, *Sedimentation in the African Rifts*, Geological Society Special Publication No. 25, pp. 29–43.

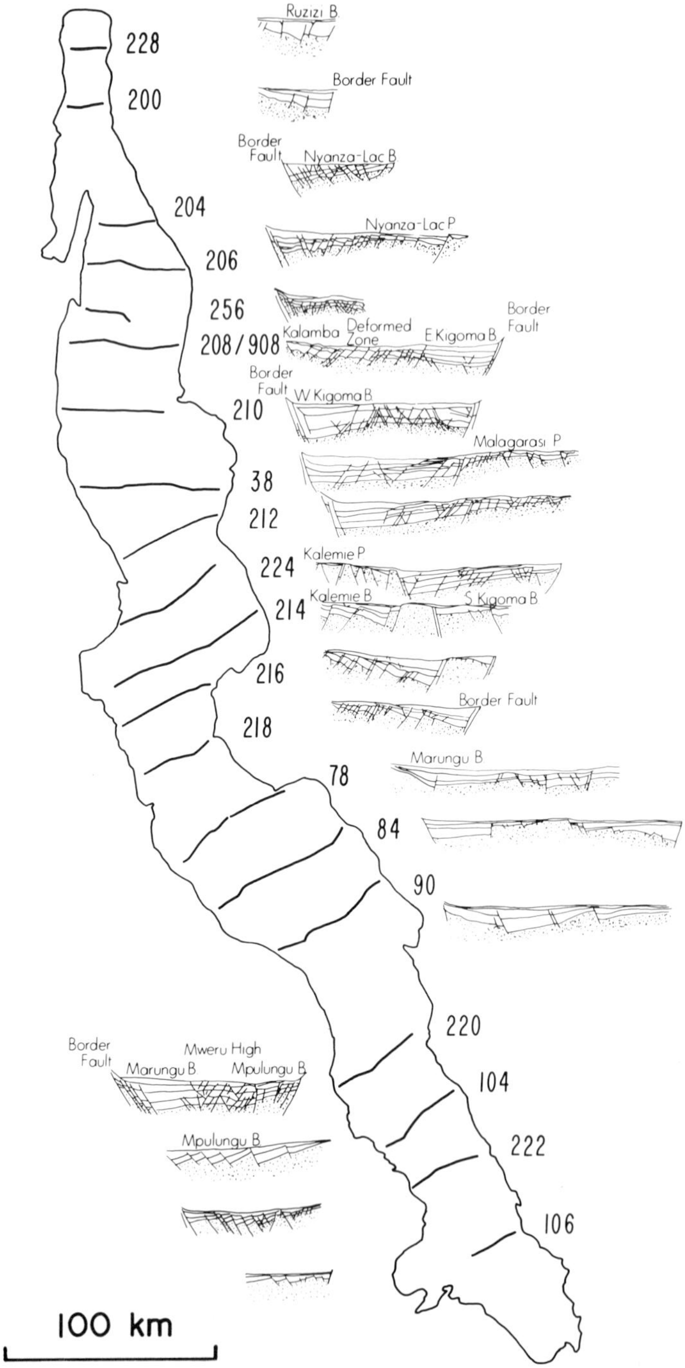

FIG. 1. Multifold seismic coverage of Lake Tanganyika and line drawings of selected profiles. Total seismic coverage including single-channel data (not shown) exceeds 5000 km. Examples of key profiles are shown in Figs. 7, 9 and 11. The top of the stippled pattern corresponds to a reflection event termed the Nyanja Event, which is the acoustic basement in most locations. B = basin; P = platform.

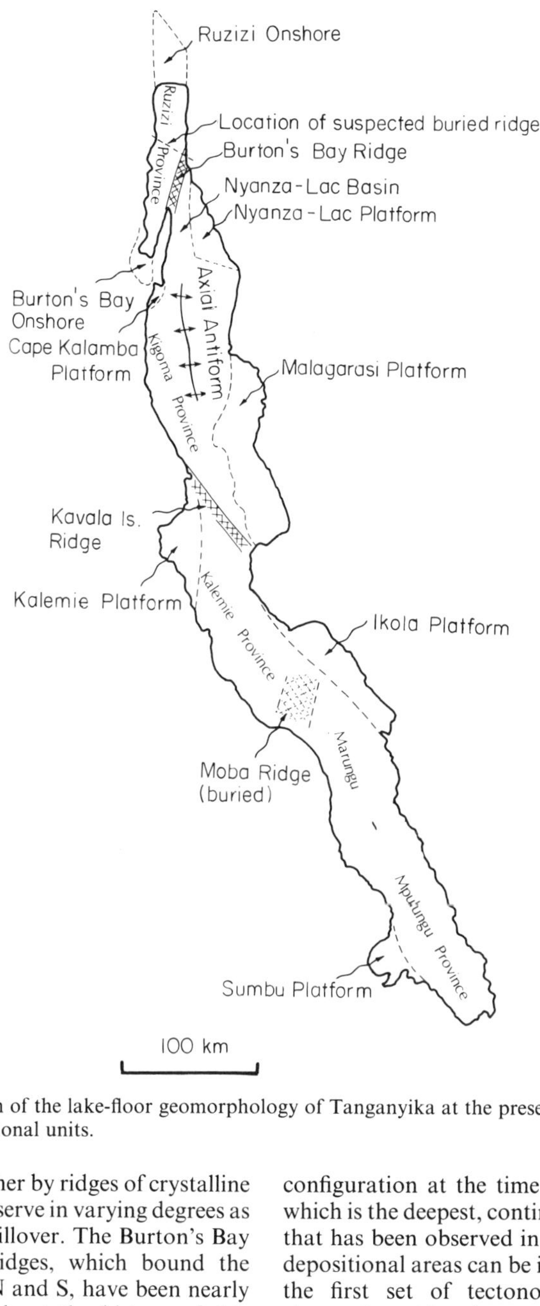

FIG. 2. Schematic diagram of the lake-floor geomorphology of Tanganyika at the present time. Provinces represent modern depositional units.

separated from each other by ridges of crystalline basement rocks, which serve in varying degrees as barriers to sediment spillover. The Burton's Bay and Kavala Island Ridges, which bound the Kigoma basin on the N and S, have been nearly perfect barriers throughout the history of this basin. This is evidenced by the lack of acousto-stratigraphic continuity between the Kigoma basin and its neighbouring basins. Other dividing ridges, such as the Moba Ridge, are now buried by sediment.

The early geomorphology of Tanganyika can be visualized from the tectonic map shown in Fig. 3. This is essentially the border fault-basin-ridge configuration at the time of the Nyanja Event, which is the deepest, continuous, seismic reflector that has been observed in the lake. At least nine depositional areas can be identified. This leads to the first set of tectono-stratigraphic conclusions—depositional areas in Tanganyika have increased in size as rifting has progressed, via burial of some of the interbasinal ridges. One stratigraphic consequence of this is the predicted occurrence of something that could be called a 'spillover' facies, which should be more broadly distributed and possibly thinner and more fine-grained than pre-spillover sedimentary packages.

Another geomorphologic element of Lake

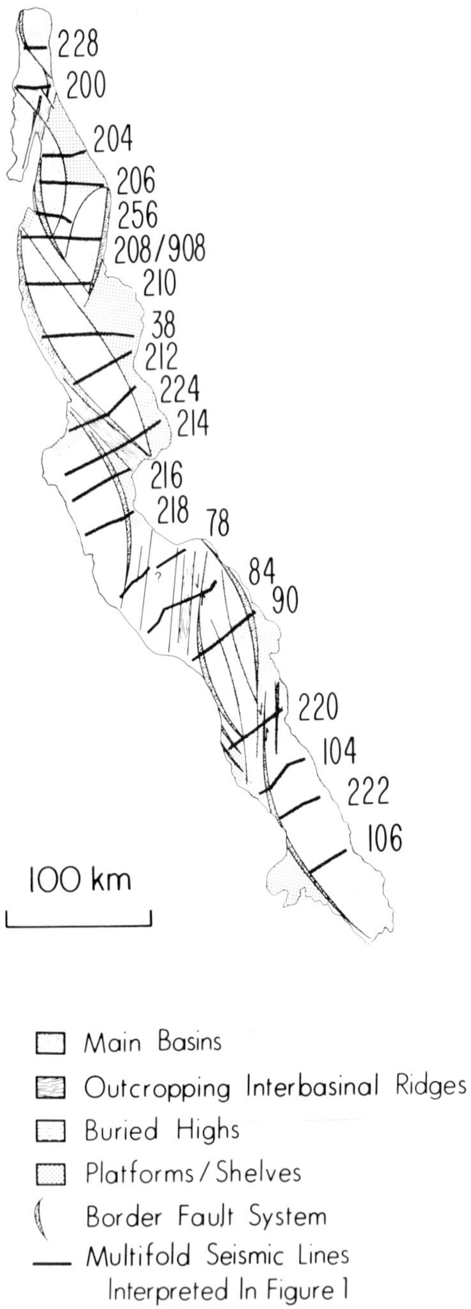

FIG. 3. Simplified tectonic map of Lake Tanganyika. Structure has been stylized in order to emphasize the main elements of the original fracture geometry of the lake. Numbered lines refer to locations of line drawings shown in Fig. 1 and seismic profiles cited in the texts.

Tanganyika is an admixture of elevated rift fringes we term platforms. They occur throughout the history of Tanganyika (Figs. 2 and 3) and they tend to occupy positions near the ends of basins. All the large platforms are adjacent to, and associated with, drainage systems into the lake. For example, the Malagarasi platform is associated with a large watershed, drained now by the Malagarasi River and in the past by two other rivers (Patterson 1983).

Key observations

Because the structure of Lake Tanganyika is exceedingly complex, we have produced a schematic diagram of the tectonics in Fig. 3. This might be considered as the product of a filter which removes the details of faulting within the basins but retains the overall patterns, especially those which repeat throughout most of the basins. In order to interpret the geological story contained in this diagram, it is useful to divide it into key observations and themes.

Perhaps the most obvious structural expression in Fig. 1 is the asymmetry in basinal forms. The vast majority of the profiles show half-graben, each of which has subsided mainly along one border fault system. True full-graben forms (approximately equal subsidence along two opposing border fault systems) are seen in only four of the profiles shown in Fig. 1 and probably characterize less than 20% of the lake. The occurrence of half-graben in extensional terrains is not a new discovery (e.g. Harding & Lowell 1979; Bally 1980), although the predominance of this form to the extent noted here is not commonly recognized outside the exploration industry (Bally 1985). The prevalence of half-graben should not be surprising in juvenile rifts such as Tanganyika or Malawi because this morphology is more energy conservative than full-graben expressions.

What is surprising, even perplexing, is the alternation of the direction of half-graben asymmetry between basins in Tanganyika. This alteration, or scissoring, seems to be systematic in the sense that whatever the polarity of a basin might be, those basins adjacent to it usually display the opposite direction of asymmetry. The expression here is reminiscent of what one observes on either side of an oceanic ridge–ridge transform fault.

The nature of the transitions between downthrow-to-the-E and downthrow-to-the-W half-graben has been the subject of much study and debate among our group. The key observations in this context seem to be that: (1) full-graben forms, such as those shown in lines 210 and 220 of Fig. 1, occur within the transition intervals of the alternating half-graben; (2) full-graben forms always seem to show some variety of axial antiform or high; and (3) the border faults which bound the deep sides of half-graben are concave inward in map view, and are usually contiguous with the ridges that bound Tanganyikan basins on their northern and southern ends (e.g. Fig. 3). Indeed, the main difference between border faults and interbasinal ridges is their location relative to the basin proper. A case in point is the western border fault of the Kigoma basin, which is continuous with the northeastern flank of the Kavala Island Ridge (Figs. 2 and 3).

The theme

No single concept explains all of the above structural relations and styles, but most can be envisaged as variations on a central theme. The main tenet of this theme is that half-graben are the fundamental rift unit, or building block, of Lake Tanganyika. There is a large variety of ways in which these units can link together in time and space to form a rift system, but they generally do so in a fashion which minimizes the work necessary to relieve the extensional stresses. For reasons discussed by Reynolds (1984), the linking is often accomplished by a sinusoidal alternation of opposing half-graben (Fig. 4).

The main characteristics of an ideal fundamental unit are shown in plan view in Fig. 5. There are many important consequences of this geometry. Paramount is the realization that virtually all properties of rift basins—amount of subsidence and extension, sediment infill, nature of faulting, etc.—are spatially dependent. Another consequence of the arcuate geometry is that border fault systems must also be arcuate in cross-section, flattening with depth. As the basin subsides, this results in a component of horizontal motion which is manifested in shearing along the extremities of the unit. The border fault system evolves from dip-slip to strike-slip expressions as it progresses from the centre to its ends, with a concomitant decrease in relative amounts of subsidence (or extension).

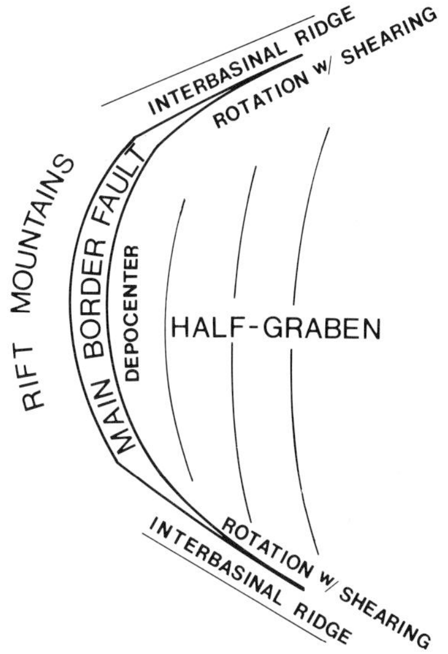

FIG. 5. Main characteristics of the fundamental unit. Note that the arcuate form in map view requires listric border fault geometries in cross-section, which in turn mandate shear and rotation along the ends of a fundamental unit.

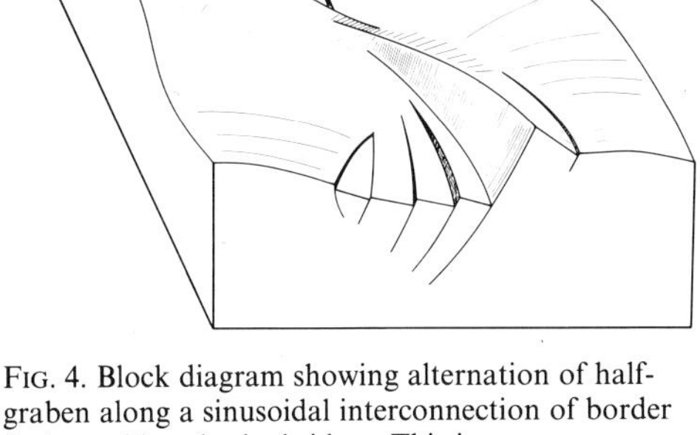

FIG. 4. Block diagram showing alternation of half-graben along a sinusoidal interconnection of border faults and interbasinal ridges. This is a common way in which half-graben are linked together, but Fig. 3 shows that other modes are possible. The interbasinal ridges often extend well beyond the rift proper.

In addition to this spatial dependence, there are a number of different morphologies which can be observed along any particular orientation of observation line. For example, the 'end-member' morphologies that have been observed along the bisections of fundamental units are given in Fig. 6. Fig. 1 suggests that cases A_1 and A_2 predominate in Lake Tanganyika, and there is some indication that there is a progression from simple to complex border fault systems as the amount of subsidence along the border faults increases. The occurrence of intrabasin antithetic faulting to the extent depicted in cases A_3 and A_4 probably occurs as a byproduct of one type of linking of half-graben (see below). If so, cases A_3 and A_4 should not be viewed as a type-morphology of an isolated half-graben unit. Fig. 7 shows seismic sections which illustrate the above cases.

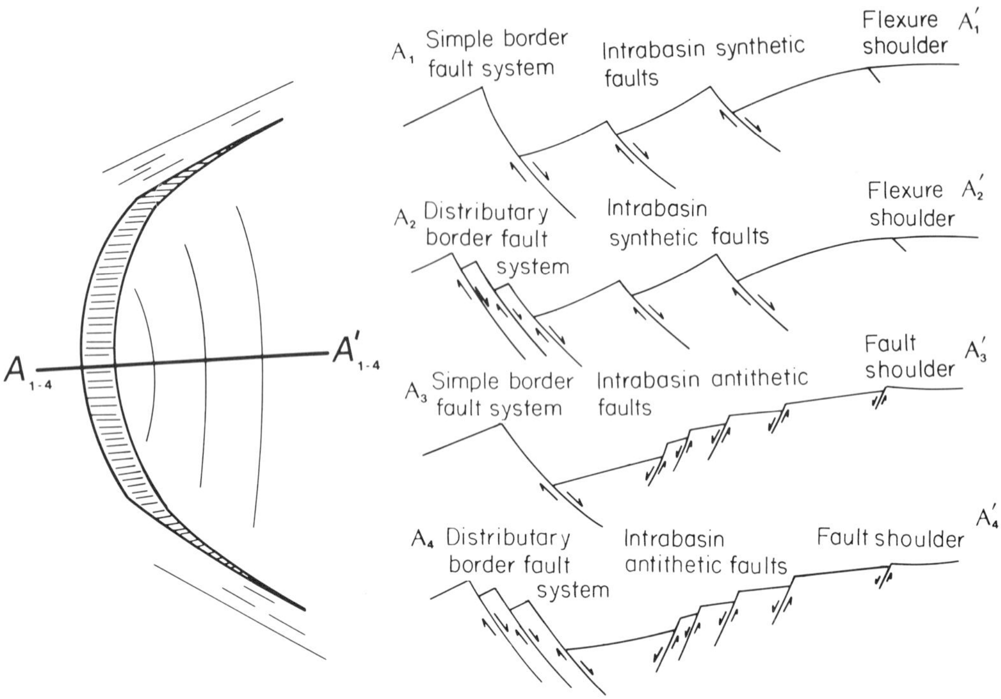

FIG. 6. Range of cross-sectional morphologies observed along A–A′. Note that the top two cases are probably the type-expressions across the middle of relatively isolated units. Cases A_3 and A_4 are probably expressions of a certain type of overlapped linking of fundamental units. The average width of intrabasin synthetic blocks is about 4 km, that of intrabasin antithetic fault blocks perhaps only 2 km. As used here, synthetic faults refer to intrabasin normal faults which parallel the master border faults, both in map and cross-sectional views. Antithetic faults have opposing dips to the master system in cross-section and may arc in the opposite direction in map view.

Linking of fundamental units

Another key to understanding the geological story contained in Fig. 3 is the recognition of the morphologies that result from the various ways in which half-graben units are linked together in time and space. One common mode of linking is shown in Fig. 4. Some of the possible cross-sectional morphologies of this mode of linking are given in Fig. 8; an actual seismic example is shown in Fig. 9 (line 214). Another mode of linking is shown in Fig. 10, along with some of the possible cross-sectional morphologies. Several examples of seismic expressions are provided in Fig. 9 and 11.

In many ways these illustrations are self explanatory, but a few points are worth emphasizing. Firstly, interbasinal ridges and hinged highs originate at different positions within the rift from somewhat different stress conditions. Secondly, hinged highs often terminate or merge into Malagarasi-type platforms where the linking geometry is as shown in Fig. 10. Finally, it is important to understand that the above modes of linking are simply examples of what must be a large family of possible combinations. Most of the above models and seismic sections show linked half-graben at similar stages of structural evolution (i.e. similar amounts of subsidence in both units). This need not always be the case. For example, the antithetic platform in line 206 of Fig. 7 is actually a hinged high of the type shown in line 210 of Fig. 9. The difference here is that subsidence of the two basins is very dissimilar in the former section, but approximately equal in line 210. A close examination of Fig. 3 might give the reader further insight into the range of variations we have actually observed in Tanganyika.

Even the few simple examples described above go a long way toward explaining the structural styles that occur in the Tanganyika Rift. Combined with an understanding of the different modes of linking which can occur, we have

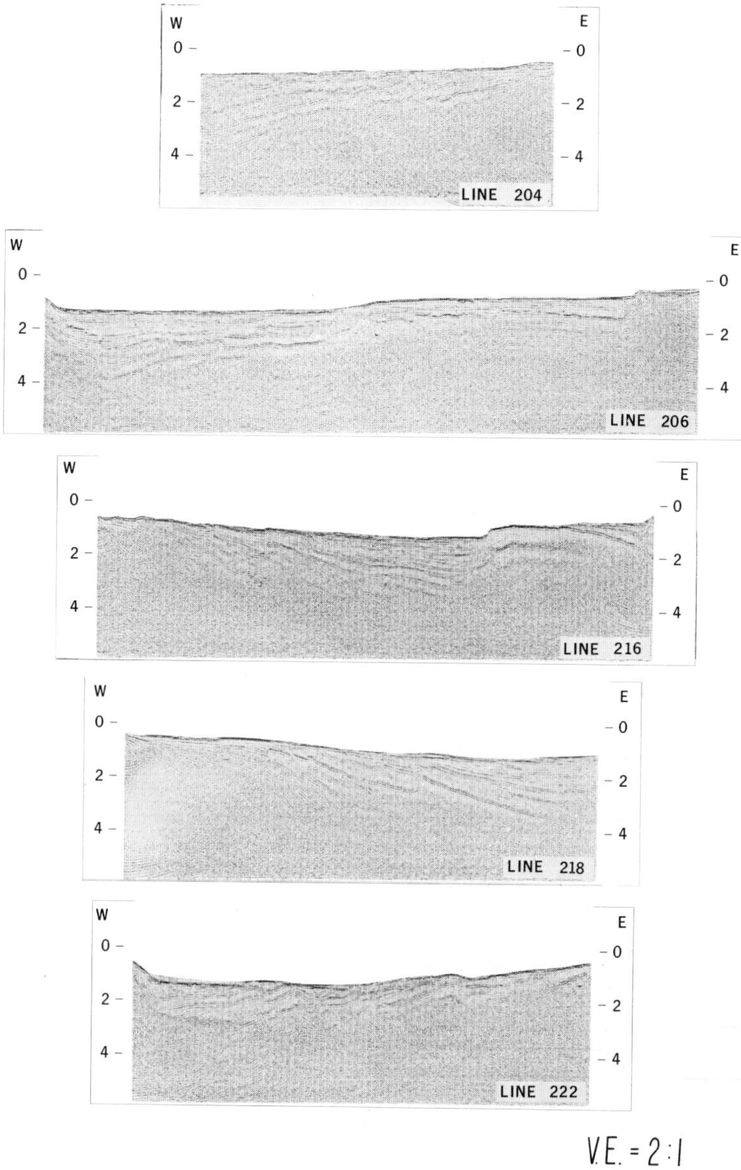

FIG. 7. Seismic sections illustrating the morphologies shown in Fig. 6. Lines 204, 216, 218 and 222 are examples of cases A_1 or A_2 of Fig. 6; line 206 is an example of case A_4. Note that subsidence along the newer border fault near the right-hand side of line 206 has created the Nyanza-Lac platform (Figs. 2 and 3), which evolved into a hinged high further S. Also note that many of the internal faults are antithetic to the original border fault and tend to be rotated toward the platform. See Figs. 1 and 3 for line locations. All profiles are 24-fold, migrated sections with W on the left. Vertical scale in km.

essentially derived an empirical model of rifting for this particular part of East Africa. There are important similarities and differences between our model and those derived by other researchers. Although this subject will be tackled in detail in a future paper, we would be remiss not to mention here the efforts of Bally (1985) and Gibbs (1984). Both of these authors recognize the importance of basinal units and both also recognize that there must be some mechanism (and presumably an

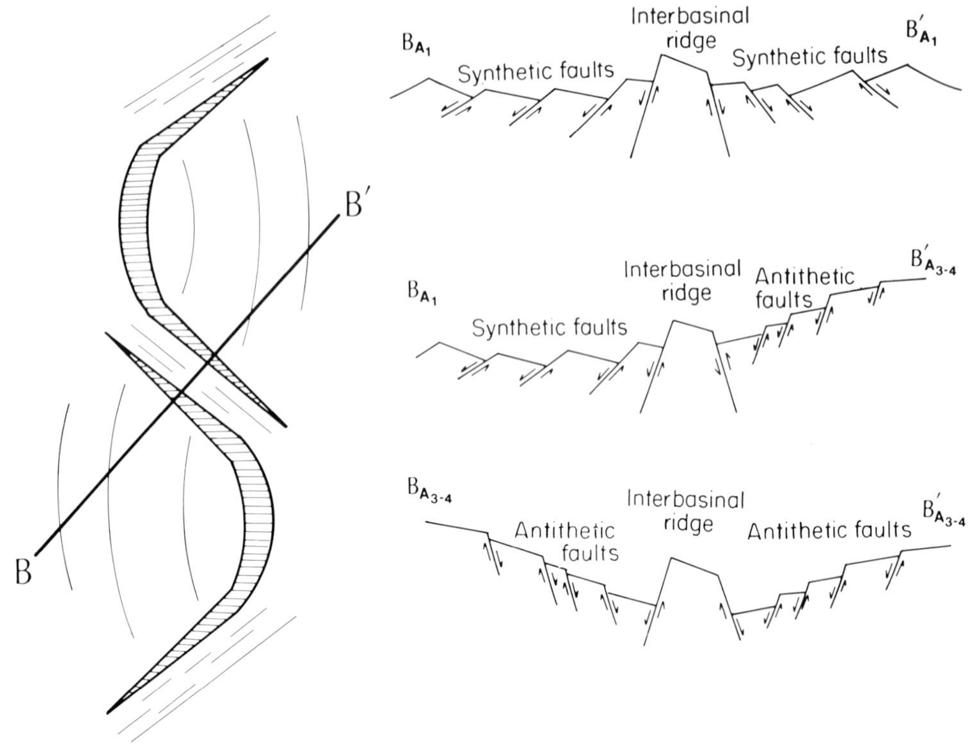

FIG. 8. Some possible cross-sectional morphologies observed along B–B′. Note that the angles of synthetic faults are markedly lower than in Fig. 6 because the observation line is more oblique to intrabasin fractures. Faulted margins of interbasinal ridges are relatively high-angle, oblique-slip or even strike-slip faults. The bottom case is not observed in Tanganyika, but is hypothetically possible. Volcanics are thought to exist along two interbasinal ridges in Tanganyika.

expression) of transferring strain from one basinal unit to another. In our model, we do this via oblique shear along interbasinal ridges and hinged highs, whereas Gibbs (1984) uses 'transfer faults' and Bally (1985) uses 'transform faults'. Several of the authors of our report also have used the term 'accommodation zone' (e.g. Reynolds & Rosendahl 1984).

There are many differences between the Tanganyika model and the concepts of Bally (1985) and Gibbs (1983, 1984), but most can be explained if the latter are taken to be special cases of the former. For example, Bally's orthogonal transform offsets of adjacent half-graben could be viewed as one type of linking mode, albeit a rare one in the context of our databases. Gibbs' arrangement of parallel transfer faults trending oblique to boundary faults is closer to what we see in Tanganyika. However, parallelism of adjacent interbasinal ridges (or transfer faults in the terminology of Gibbs) is the exception, not the rule. If fundamental differences exist between these three models, they probably relate to the ways in which we derive rift morphology from the various modes of linking of half-graben.

Effects of structural expressions on depositional processes

If our model of rifting in Lake Tanganyika is valid, it could profoundly influence the way researchers view the synrift depositional histories of rifts. We include a short discussion of this subject here in the hope that those with more expertise will carry the task forward, particularly with regard to other continental rift systems. Our type examples shall again be taken from Lake Tanganyika and to some extent Lake Malawi.

Sediment lithologies

Synrift sedimentation in these rift lakes can be viewed as a complex interplay between fluvial clastic input and blanket deposition of biogenic

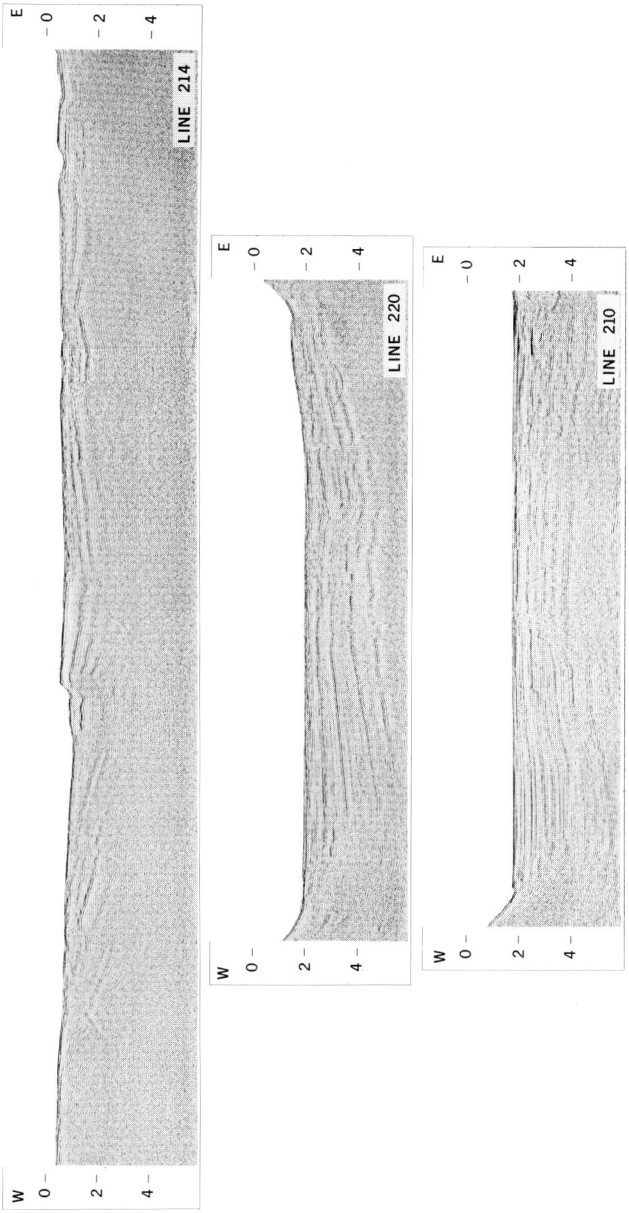

FIG. 9. Seismic sections illustrating some of the morphologies shown in Figs. 8 and 10. Line 214 is an example of the top case in Fig. 8; lines 220 and 210 are examples of the top and bottom cases in Fig. 10. Note that the structural high shown in line 214 originates from a different type of linking than the hinged highs shown in lines 220 and 210. See Figs. 1 and 3 for line locations. All profiles are 24-fold, migrated sections with **W** on the left. Vertical scale in km; vertical exaggeration ×2.

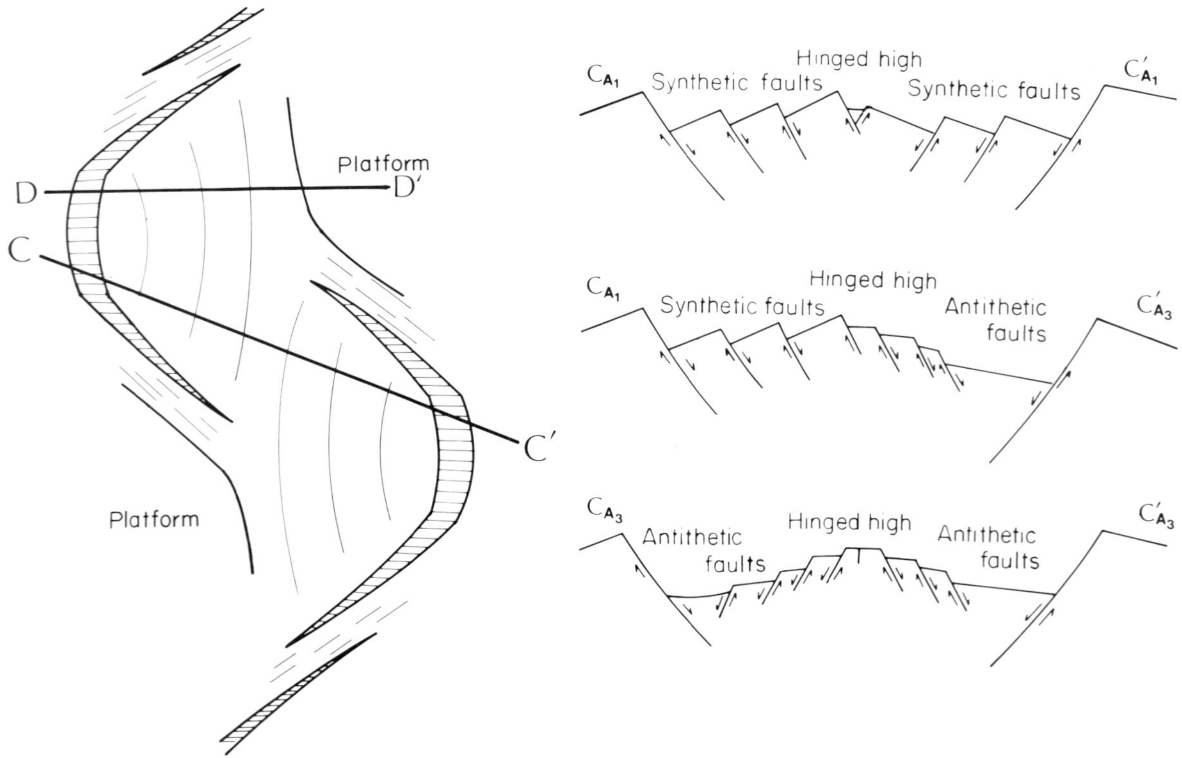

FIG. 10. Some possible cross-sectional morphologies observed along C–C′. Note that apparent full-graben forms originate from an orientation of the observation line across the overlapping portion of two facing half-graben. In Tanganyika these forms are always associated with antiformal highs. The terms hinged ridge or hinged high have been used here to denote the mechanical significance of the antiforms. In this concept of rifting full-graben forms do not carry the same level of morphological significance as half-graben. True platforms of the Malagarasi-type usually occur in the non-overlapping areas. See Fig. 9 for seismic examples of full-graben and Fig. 11 for seismic examples along D–D′ (Malagarasi-type platforms).

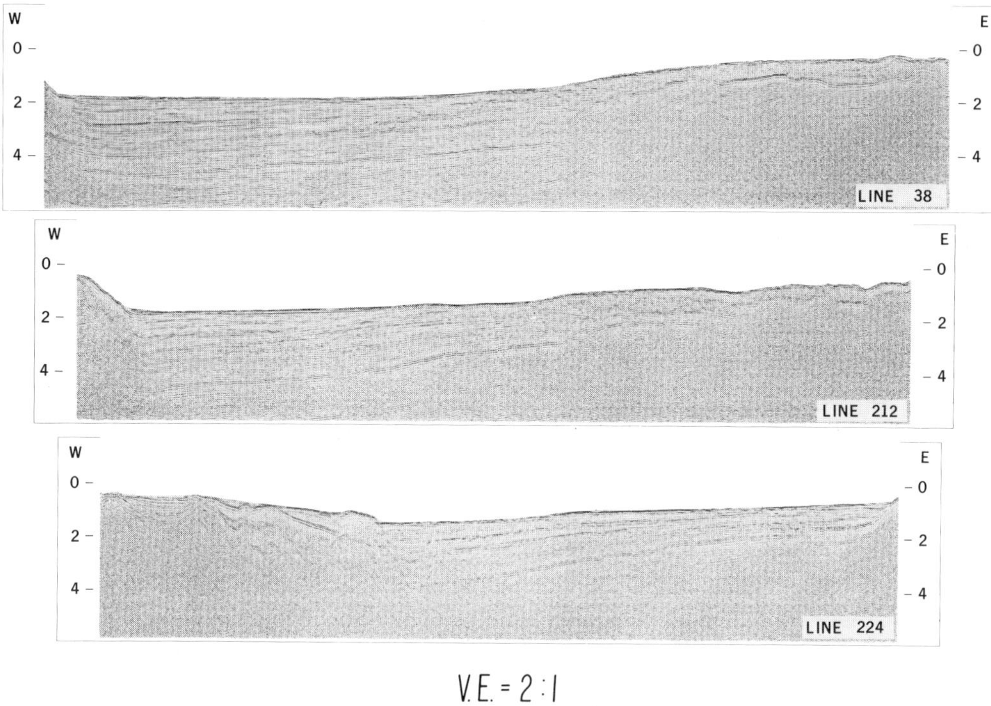

FIG. 11. Seismic sections, illustrating morphologies observed along D–D′ in Fig. 10. The platform on the right-hand side of the profiles is associated with the Malagarasi drainage system, but the platform is clearly an elevated structural block, not a true delta. Note the extensive erosion of the section along the top of the platform on lines 38 and 212, which indicates that the platform is only an ephemeral repository for fluvial clastics. The highly reflective and stratified appearance of the basinal sediments is probably due to proximity to platform-derived clastics (compare to the basinal stratigraphy of line 210, Fig. 9). See Figs. 1 and 3 for line locations. All profiles are 24-fold, migrated sections with W on the left. Vertical scale in km.

ooze, with local accumulations of border fault fanglomerates (Patterson 1983; Rosendahl & Livingstone 1983; Ebinger *et al.* 1984, Lorber 1984). For a first approximation we can take biogenic 'rain' to be a geographic constant, so the lithology at any given place is largely a function of the extent to which the biogenic component is diluted by fluvial clastics and/or fanglomerates.

Present-day drainage patterns and our seismic data indicate that coarse clastics enter the Tanganyika Rift in the following modes: (1) from the Ruzizi River at the northern end of the lake; (2) from rivers associated with platforms, such as the Malagarasi River; (3) as conglomerates and fanglomerates spilling down the border faults; and (4) from the shoaling sides of half-graben. Anywhere near these entrance points, coarse clastics dominate the stratigraphy. This is especially true in the Ruzizi basin, where our seismic data indicate that much of the fill is deltaic and prodelta clastics. In this respect, the Ruzizi basin is unique in Lake Tanganyika.

Structural controls

We now must add structure and rift geometry to the simple picture described above. The most profound effect derives from the observation that Tanganyika is a series of linked half-graben. Half-graben separated by interbasinal ridges (e.g. the geometry shown in Figs. 4 and 8) must be treated as independent depocentres until spillover occurs across the intervening ridges. This is why the acoustic stratigraphy of adjacent basins usually cannot be correlated. A case in point is the comparison of the Kigoma and Kalemie basins (Fig. 2), which are adjacent to each other but display entirely different acoustic stratigraphies above the Nyanja Event. This is shown best in line 214 of Fig. 9, which crosses both basins and the intervening ridge. The lack of acoustic continuity between the Ruzizi and Kigoma basins is also a consequence of structural isolation via an interbasinal ridge.

Although their stratigraphies may differ,

depositional processes in isolated half-graben tend to be similar. Lines 216 and 218 from the Kalemie basin (Fig. 7) are typical of these settings. Coarse clastics entering from the shoaling sides of half-graben are deposited as wedge blocks behind synthetic fault blocks. Lateral pinch-outs are common and spillover occurs between fault blocks if input outpaces subsidence along the faults. The primary facies differences between isolated half-graben relate to differences in rates of subsidence, timing of formation, and clastic sediment supply. Apparently, each of these parameters can vary dramatically from one half-graben to another (e.g. compare lines 218 and 222 in Fig. 7), although clastic supply usually overwhelms the biogenic components.

Where half-graben link together in overlapping, facing formats (Fig. 10), the interplays between geometry and deposition are extremely complex because they are dependent upon the amount of overlap, position within the overlap, and the relative rates of subsidence along the opposing border faults. We shall examine three cases which illustrate the range of possiblities.

Case 1

Case 1 is the overlap condition depicted in C–C′ of Fig. 10, with the subsidence conditions of lines 210 and 220 of Fig. 9. These conditions result in antiformal-type hinged highs flanked by narrow, elliptical sub-basins. In Tanganyika these sub-basins are always the deepest and most thickly sedimented depocentres. The modern sediments in these deep basins are organic-rich muds (Degens *et al.* 1971). The acoustic character of the sediments indicates a preponderance of shale lithologies throughout much of the depositional histories of these sub-basins.

Apparently the geometry of case 1 results in rapid subsidence and isolation of the depocentres from fluvial clastic input. This happens in spite of the fact that subsidence of the hinged highs has been retarded relative to that of the depocentres, as evidenced by offlap and divergence of reflectors away from the highs shown in lines 210 and 220 (Fig. 9). We might conclude that hinged highs have had an important influence on the patterns of deposition, almost to the present time, but they seem to have had little effect on sediment lithology. For the reasons discussed below, this may not have been the case during the very early evolution of facing half-graben geometries.

Case 2

The second case we shall examine is illustrated by line 206 (Fig. 7). Here, subsidence is occurring along a 'new' border fault near the eastern edge of the profile. Because the newer border fault faces the original border fault, a markedly asymmetric hinged high is being created. In effect, a newer half-graben is evolving on the shoaling side of a pre-existing half-graben. One depositional consequence of this geometry is the birth of a new depocentre; another is the probable loss of the coarse clastic supply to the original basin. We should note that this change in geometry is usually accompanied by unconformities.

The above case may provide important insight into the depositional histories of rift basins. Consider that the morphology shown in line 206 could occur only if the newer border fault is taking up a significant percentage of the extensional stress. The acoustic stratigraphy and fault patterns on this profile argue that activation of the newer border fault correlates with retardation or even cessation of subsidence along the older. Similar phenomena are observed elsewhere in Lake Tanganyika. In some places (e.g. lines 78 and 84 of Fig. 1) the subsidence seems to have alternated back and forth between the facing border faults. This alternation tends to create unconformities between acoustic units which thicken in opposite directions. It is likely that many of the major reflectors identified in Tanganyika owe their existence, either directly or indirectly, to the alternation of subsidence. It is probably more than a coincidence that most intrabasinal faults tend to group into generations that can be distinguished from each other on the basis of how many acoustic units they transect (note the generational groupings in Fig. 1).

This brings us to the crucial question of whether or not the conditions depicted in case 1 evolve from those illustrated in case 2 (i.e. did line 210 pass through the stage shown in line 206?). The fact that the deep acoustic stratigraphy of lines 210 and 220 does not resemble that of 206 argues in the negative. So does the fact that the hinged highs on lines 210 and 206 are not the same structure (Fig. 3). The rejoinders to these arguments are that comparisons of seismic signatures from different stratigraphic levels and ages can be very misleading, especially on time sections; that the eastern border faults of lines 210 and 206 are contiguous; and that isopach maps of the main acoustic units in the Kigoma Province suggest an alternation of depocentres through time. These isopach maps are shown in Figs. 12 and 13. The basal unit (Makara) is thickest in the southern part of the basin and was probably controlled by the eastern border fault. The upper unit (Kigoma) shows a shift of the depocentre to the N, along with its separation into sub-basins flanking the hinged high. The thicker depocentre is adjacent to

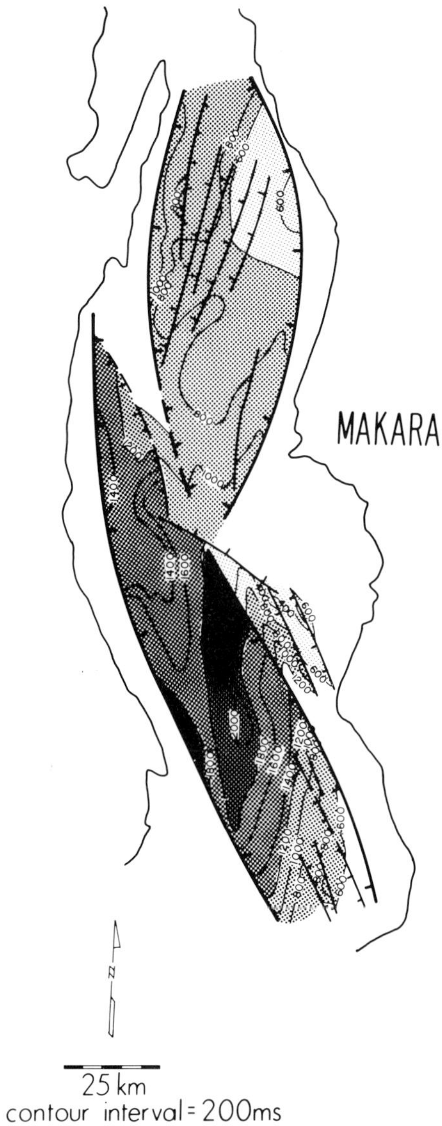

FIG. 12. Isopach map in two-way travel time for the Makara acoustic unit. This is the basal seismic sequence in the Kigoma Province. Note that the main depocentre during this interval was probably controlled by the southeastern border fault.

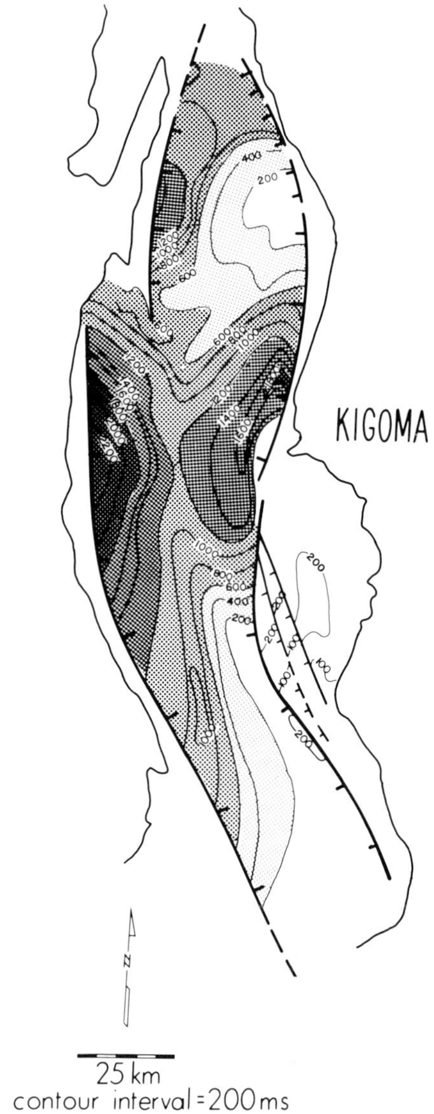

FIG. 13. Isopach map in two-way travel time for the Kigoma acoustic unit. This is the upper seismic sequence in the Kigoma Province. Note the shift of the depocentre to the N and its separation into sub-basins by the hinged high. The thicker depocentre appears to be on the W during this depositional interval.

the western border fault. At the very least, these types of depocentre shifts argue for the continuance of the subsidence shifts well into the rifting stage illustrated by line 210 (Fig. 9). If so, it is quite likely that the answer to our original question is an affirmative.

Case 3

The third case pertains to depositional processes in the regions where facing half-graben do not overlap. The morphology in these regions is usually that of a half-graben abutting a structural platform (cross-section D–D′, Fig. 10, and lines

38, 212 and 224, Fig. 11). The most important depositional facets of this case arise from the association of platforms with drainage systems (see section above on general morphology). This association creates the impression that the platforms are constructional features built into and on the rift floor by ordinary fluvial deltaic processes. This is not the case. Most Tanganyikan platforms are structural highs showing elevated basement (Fig. 11). This suggests that the drainage systems are a response to the existence of the platforms, not the other way around. This is supported by the observation that the rift flanks (or mountains) on the landward sides of platforms are relatively subdued compared to elsewhere along the rift.

One stratigraphic implication of this observation is that platforms are not deltas *per se*. Instead, they are part of the transfer systems for fluvial clastics coming into the basins proper. A case in point is the part of the Malagarasi platform shown in lines 38 and 212 (Fig. 11). At the present time this part of the platform is below lake level, and it is acting as a temporary repository for fluvial clastics flushed from the Malagarasi River. In the recent past, however, much of this platform has been subaerially exposed, as evidenced by the missing section, erosional channels, and unconformities shown in the acoustic stratigraphy of lines 38 and 212. During such low stands of the lake this platform was essentially a protrusion of the Malagarasi watershed into the Kigoma Province, and fluvial clastics and reworked sediments presumably entered the deep basins in rather direct ways. It should not be surprising that basinal sediments adjacent to this platform are much more reflective and stratified than those further away (e.g. compare the acoustic character of the basinal sediments in line 38 with those in line 210).

It is important to understand that true platforms, such as the Malagarasi, occur at predictable locations within facing half-graben geometries. This means that fluvial clastics enter facing half-graben mainly from the non-overlapping areas. Because these non-overlapping areas tend to be located near the opposing corners of the facing half-graben, it can be predicted that the input of fluvially-derived sediments to 'full-graben' should be strongly antipodal.

Conclusions

Our intention here has been to provide the stratigrapher with a conceptual model of rifting which has direct applicability to depositional processes occurring on the basinal scale. The usefulness of the model depends upon a host of factors, including the 'goodness factor' with respect to the type area (Tanganyika) and its applicability to continental rifting in general. Regarding the 'goodness factor', we can confidently state that the model is rooted entirely in empiricism—it is little more than a simplification of the structural architecture of Tanganyika. The question of global applicability is more subjective.

Individually, or in various combinations, the authors of this report have applied the Tanganyika model to the following rift systems: Malawi Rift; Kivu–Edward–Albert–West Nile system; Kenya Rift including Lake Turkana; Luangwa valley, Zambia; Kordofan, Sudan; Suez Rift; Rhinegraben; Oslo Rift; Baikal Rift; Rio Grande Rift; Triassic basins of the eastern USA; Basin and Range Province; Keweenawan Rift; South China Sea; and Viking graben. Many of these applications can be found in Reynolds (1984). Although we sometimes disagree about the specific details of the applications, the authors of this report are convinced that the Tanganyika model has broad applicability to continental rifts. The case is especially strong for those rifts which have failed prior to generation of significant widths of new lithosphere. Because space does not permit a fuller discussion here, we shall let those who have specialized in the study of these other rifts judge how well the Tanganyika model fits.

We have skirted the issue of rifting mechanisms entirely in this paper, not because the subject is uninteresting but because it is too easy to let theoretical precepts distort interpretation of empirical databases. We might note that our model implies a degree of three-dimensionality that is not explicitly evident in most models of rifting. If the Tanganyika observations prove to be as generally applicable as we suspect, then the mechanics of continental fragmentation may need some rethinking. In the same vein we ask those who study passive margins to consider the implications of Fig. 3—if this is the way that continents break apart, it is not surprising that the deep structure of passive margins seems to be so variable from place to place.

ACKNOWLEDGEMENTS: The work described in this paper was made possible by financial support, initially from NSF grant BNS 7926762, and more recently from Amoco Production Company (International), Arco Oil & Gas Company, Conoco Inc., Eason Oil Company, Société Nationale Elf Aquitaine (Production), Esso Exploration Inc., Marathon International Oil Company, Mobil Exploration & Producing Services Inc., Pecten International Petroleum, Pennzoil Exploration & Production Company, Shell Internationale Petroleum and the World Bank. We also have benefited greatly from the dialogues we have had with representatives from many of the above organizations. Special thanks go to the governments of Burundi, Kenya, Malawi, Tanzania and Zaire.

This work would not have been possible without the extraordinary field efforts at Lake Tanganyika of C. Ebinger, T. Hess, G. Howell, K. Kinkead, M. Patterson, R. Roessler and C. Schaeffer. Data processing was supervised by G. Howell and W. Schwede, with assistance from J. Buck, T. Davis and V. Heilweil. Artwork and secretarial assistance was provided by C. Adams, D. Tyler and J. Wilson.

References

BALLY, A.W. 1980. Basins and subsidence—a summary. In: BALLY, A.W., BENDER, P.L., MCGETCHIN, T.R. & WALCOTT, R.I. (eds) *Geodynamics Series 1*. American Geophysical Union, pp. 5–20.

—— 1985. Structural styles and the evolution of sedimentary basins. *Am. Assoc. Petrol. Geol., Educ. Short Course*.

DEGENS, E., VON HERZEN, R. & WONG H.K. 1971. Lake Tanganyika. *Naturwissenschaften* **58,** 229–41.

EBINGER, C.J., CROW, M.J., ROSENDAHL, B.R., LIVINGSTONE, D.A. & LEFOURNIER J. 1984. Structural evolution of Lake Malawi, Africa. *Nature* **308,** 627–9.

GIBBS, A.D. 1983. Balanced cross-section construction from seismic sections in areas of extensional tectonics. *J. Struct. Geol.* **5,** 153–60.

—— 1984. Structural evolution of extensional basin margins. *J. Geol. Soc.* **141,** 609–20.

HARDING, T.P. & LOWELL, J.D. 1979. Structural styles, their plate-tectonic habitats, and hydrocarbon traps in petroleum provinces. *Bull. Am. Assoc. Petrol. Geol.* **63,** 1016–58.

LORBER, P.M. 1984. *The Kigoma basin of Lake Tanganyika: acoustic stratigraphy and structure of an active continental rift*. Unpublished MS thesis, Duke University, Durham, NC.

PATTERSON, M.B. 1983. *Structure and acoustic stratigraphy of the Lake Tanganyika rift valley: a single-channel seismic survey of the lake north of Kalemie, Zaire*. Unpublished MSc thesis, Duke University, Durham, NC.

REYNOLDS, D.J. 1984. *Structural and dimensional repetition in continental rifts*. Unpublished MSc thesis, Duke University, Durham, NC.

—— & ROSENDAHL, B.R. 1984. Tectonic expressions of continental rifting. *Trans. Am. Geophys. Union* **65,** 1116.

ROSENDAHL, B.R. & LIVINGSTONE, D.A. 1983. Rift lakes of East Africa: new seismic data and implications for future research. *Episodes* **83(1),** 14–19.

B.R. ROSENDAHL, D.J. REYNOLDS, P.M. LORBER, C.F. BURGESS, J. MCGILL & D. SCOTT, Project PROBE, 310 Old Chemistry Building, Dept of Geology, Duke University, Durham, North Carolina 27706, USA.

J.J. LAMBIASE & S.J. DERKSEN, Marathon Oil Co., PO Box 269, Littleton, Colorado 80160, USA.

Tectonics and volcanism of the southern Kenya Rift Valley and its influence on rift sedimentation

B.H. Baker

SUMMARY: Volcanism in the northern half of the Kenya Rift Valley began at 30 Ma, and in the southern half it began at 15 Ma. In the southern rift tectonic development started with gentle downwarping followed by repeated flooding of the depression with lavas. A half-graben formed at 7 Ma by faulting on the W, succeeded by further eruptions in the rift floor. By 4 Ma a graben had formed with faulting of the flexed eastern margin of the depression. Succeeding periods of faulting and volcanism migrated inwards, creating step-fault platforms and a narrow inner graben, but in the Naivasha-Nakuru sector voluminous trachytic volcanism completely filled the rift depression at intervals between 6 and 2 Ma. The inner graben was cut by dense swarms of minor faults from 2 Ma; and in the last 0.5 Ma a series of trachytic caldera volcanoes was built axially in the inner graben. Large sedimentary basins are found in broad half-graben in the northern and southern splay-fault sectors, and are subdivided into local basins where volcanic damming has occurred. Minor basins formed on back-tilted step-fault platforms, and numerous smaller lakes existed in graben of the rift floor. In the central sector lake basins spread broadly over pyroclastic deposits between volcanic piles. Ramp and step-ramp structures locally divert drainage from the high rift flanks into rift basins. Hypersaline lakes with trona evaporites in the Natron-Magadi basin were caused by the arid climate, eruption of sodium carbonate-rich ash, and recirculation of alkaline groundwater by hot springs into closed structural basins.

Evolution of the Kenya Rift Valley during the last 30 Ma is now moderately well understood as a result of mapping and dating of volcanic units, and several accounts of its development have been given in recent years (Williams 1970, 1978; Baker *et al.* 1971, 1972, 1978; Fairhead *et al.* 1972; King & Chapman 1972; Baker & Mitchell 1976; King & Williams 1976; King 1978; Jones & Lippard 1979; Williams *et al.* 1983). This outline of the history of the southern part of the Kenya Rift gives an account of the structural style and development of a volcanically active rift, with emphasis on the factors that created sedimentary basins, using as examples a variety of Quaternary basins.

The Kenya Rift differs fundamentally from most of the Western Rift in that voluminous volcanism infilled the developing depression nearly as fast as it subsided. This resulted in the Kenya graben being characterized by shallow, ephemeral lake basins, unlike the deep, long-lived lakes with thick sediments, such as those of the Tanganyika and Malawi Rifts (Rosendahl & Livingstone 1983).

Outline of the volcanic and tectonic evolution

The rift valley in Kenya (and northern Tanzania) cuts across a local culmination in the basin and swell structure of East Africa, and extends 900 km from the Turkana depression in the N to the central Tanzania Plateau in the S (Fig. 1). The Kenya uplift is elliptical in shape, and was created by gentle upwarping of the rift shoulders accompanied by sagging of the central rift floor (Baker & Wohlenberg 1971). The uplift reached a maximum of 1.7 km (Saggerson & Baker 1965), and since the volcanic succession is at least 3 km thick in the Kamasia region (King & Chapman 1972), it can be inferred that the sub-volcanic surface now lies more than 1 km below sea-level in the central part of the rift.

The central rift consists of a graben sector 70 km wide and 450 km long, that passes laterally into widening zones of splayed step-faults in the Turkana and Natron depressions. It exhibits symmetry about its centre in the Nakuru region, where the elevation of its flanks and depression of its sub-volcanic floor are at their greatest, and where the largest volumes of volcanic rocks were deposited (Baker *et al.* 1972).

The western marginal faults of the graben branch repeatedly, and pass into antithetic splay-faults within zones 200–300 km wide. The major splay-faults generally downthrow to the E, and separate blocks with gentle westerly dips. The combined effects of greater uplift and volcanism in the rift centre leads to northerly and southerly slopes within the graben and in broad half-graben in the splay-fault zones, with the result that the largest Quaternary sedimentary basins

From Frostick, L.E. *et al.* (eds) 1986, *Sedimentation in the African Rifts*, Geological Society Special Publication No. 25 pp. 45–57.

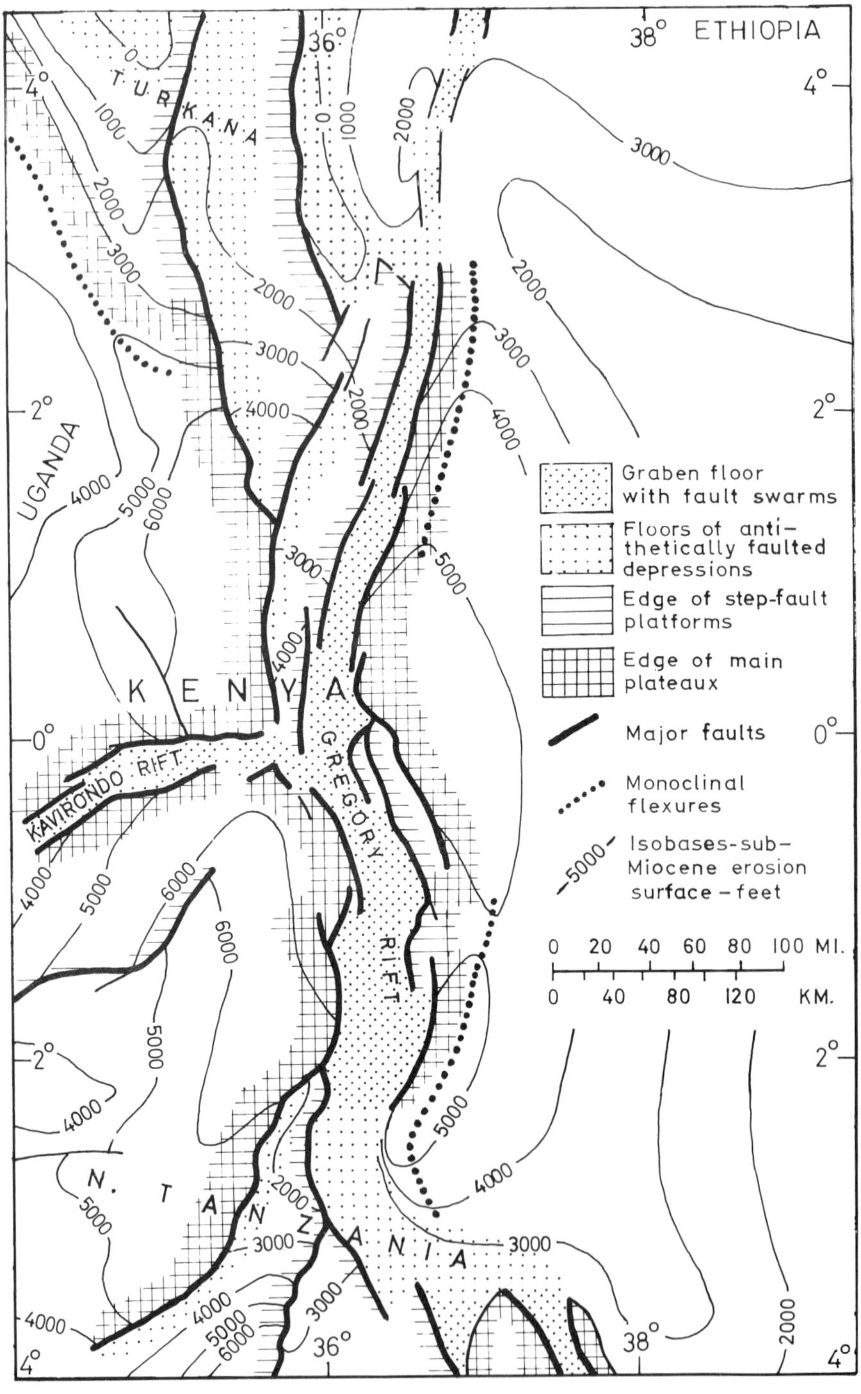

FIG. 1. Major faults of the Kenya Rift Valley, and isobases (in ft) showing the magnitude of differential uplift during rift development. Half-graben occur principally in the northern and southern sectors, step-fault platforms occur at the margins of the inner graben.

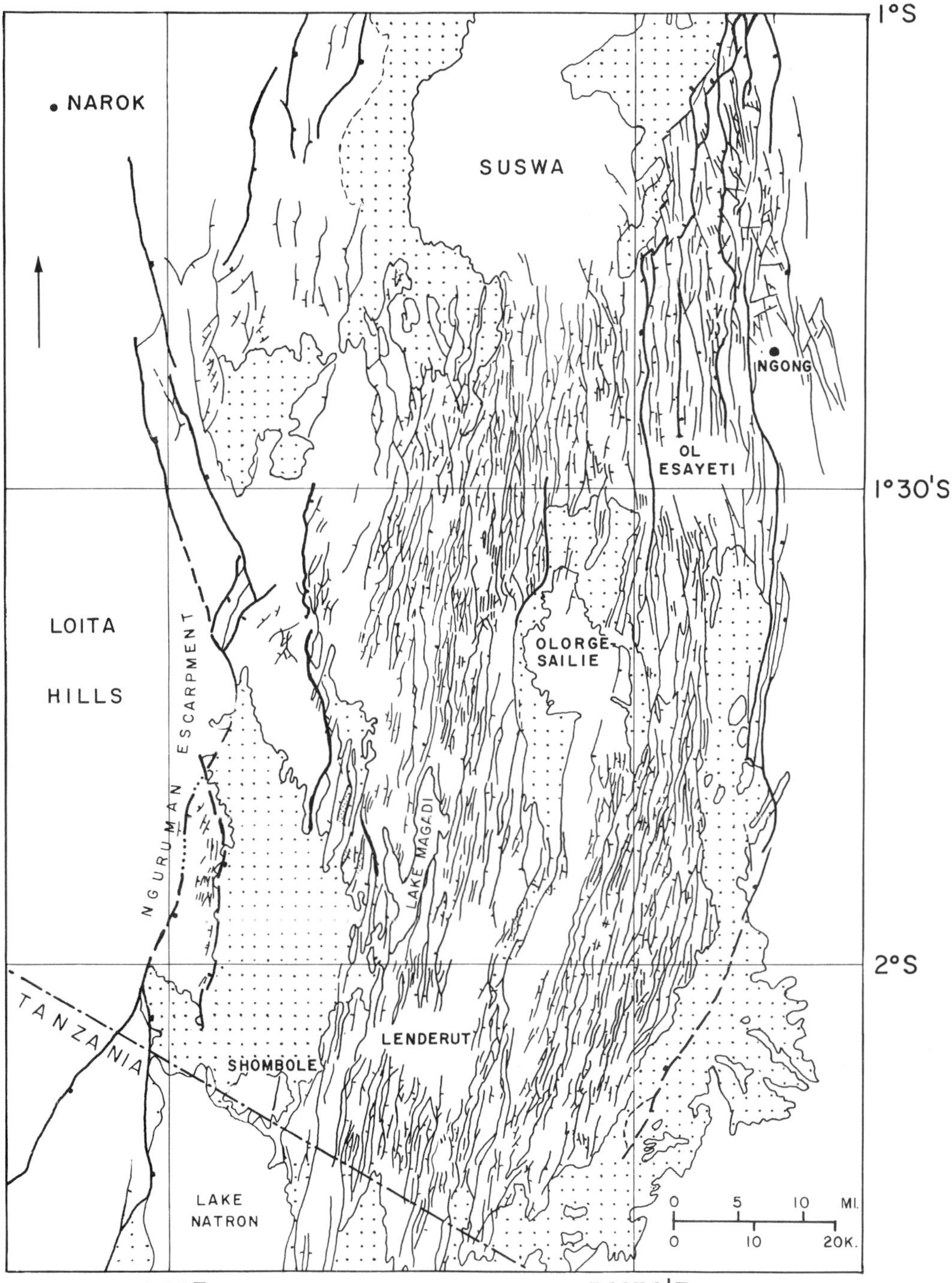

FIG. 2. Structure of the Gregory graben in southern Kenya, showing different densities (and ages) of faults on step-fault platforms and in the inner graben. Stipple represents recent sedimentary basins.

are not within the Gregory graben, but in the half-graben of the Turkana and Natron depressions.

The rift valley has developed over the last 30 Ma, during which at least 220,000 km^3 of volcanic rocks accumulated to a depth of not less than 3 km in the central sector (Williams *et al.* 1983). Individual major faults with throws of up to 2 km are known, but the cumulative displacements may be considerably larger than this. Depression of the rift zone was accomplished by broad initial downwarping in the N. This was succeeded during 16 to 7 Ma by progressive deepening of a half-graben which was faulted on its western side, and monoclinally downflexed on its eastern side. Along much of the eastern side of the rift valley monoclinal downwarping was at least as important as faulting, and it is only during the last 4 Ma that subsidence of a graben was the dominant tectonic process. The topographic expression of the rift was controlled by the opposing effects of subsidence and volcanism. During the Middle Miocene, and again in Plio-Pleistocene times, eruption of voluminous phonolite and trachytic lavas and ash-flow tuffs locally filled the rift and overflowed its flanks (Williams *et al.* 1983).

Unlike the northern half of the rift, in which volcanism dating from 30 Ma is known (Bellieni *et al.* 1981), volcanism in the southern rift began at 15 Ma (Crossley 1980; Crossley & Knight 1981). Nephelinitic lavas were erupted from low shields at the future Western Rift margin, followed by floods of phonolite which covered the future rift zone at intervals between 14 and 9 Ma. The tendency for these units to thicken towards the incipient site of the rift valley suggests that a shallow pre-rift downwarp may have existed. In the central sector of the rift a half-graben was filled by up to 2.5 km of basaltic and phonolitic lavas between 15 and 7 Ma, and some of these lavas spilled over both shoulders of the rift (Williams *et al.* 1983).

By 7 Ma the Nguruman fault had formed (Fig. 2), creating a half-graben, the northern part of which was filled by trachytic and phonolitic lavas that overflowed eastwards in the Nairobi region. The large Aberdare volcanic complex and several smaller volcanoes were built on the Eastern Rift shoulder between 6 and 3 Ma.

On the western side of the rift there were continued movements on the marginal faults accompanied by volcanism within the deepening half-graben. In the central sector (Naivasha-Gilgil) an outburst of explosive volcanism deposited a widespread series of lavas and trachytic ash-flow tuffs betwen 6 and 2 Ma, filling the central part of the rift and covering both its flanks. This was accompanied and followed by the collapse of the rift floor to form a graben for the first time.

Between 3 and 1.7 Ma several outpourings of basalt and trachyte lavas covered the rift floor, accompanied by collapse of an inner graben, leaving step-fault platforms at its sides (Fig. 2; Baker & Mitchell 1976). Several central volcanoes were built during this interval, and project above the succeeding flood lavas. Subsequent volcanic and tectonic activity was confined to the floor of the inner graben, and consisted of several phases of basaltic and trachytic volcanism, accompanied by development of swarms of closely spaced faults (Fig. 2).

During the last 0.7 Ma volcanism has been confined to building a line of salic caldera volcanoes axially in the floor of the inner graben (e.g. Suswa, Longonot, Eburru, Menengai; Figs. 3 and 4), and to localized eruptions of basalt and rhyolite. Construction of the caldera volcanoes divided the narrow inner graben into the separate sedimentary basins seen at the present day, and deposited much ash over the floor and W flank of the central rift.

While volcanism has taken place over the last 15 Ma, the structural evolution of the southern part of the rift has taken place since about 7 Ma, passing through an initial half-graben phase, and developing into a graben between 4 and 3 Ma. The zones of active volcanism and tectonism have tended to narrow with time, and the spacing between faults has decreased. Several times the central rift was completely filled with volcanic deposits, which overspilled its shoulders. Consequently the morphology of the rift was determined by the balance between subsidence and volcanism, with both becoming increasingly concentrated in a narrowing complex graben, especially in its central part.

Although the earlier structures of the central rift are largely obscured by later deposits, there is some evidence that the broad half-graben blocks of the Turkana depression are representative of the early tectonic style of the rift. This was succeeded by the formation of a graben and followed by progressive break-up of the graben floor. There has been a marked tendency for fault blocks to become narrower with time, and for tectonic depressions to become smaller and more numerous.

Characteristic tectonic terrains

The rift can be divided into tectonic terrains of contrasting age, size and type. In the splay-faulted northern and southern regions the tectonic units are antithetically-tilted large blocks 200–

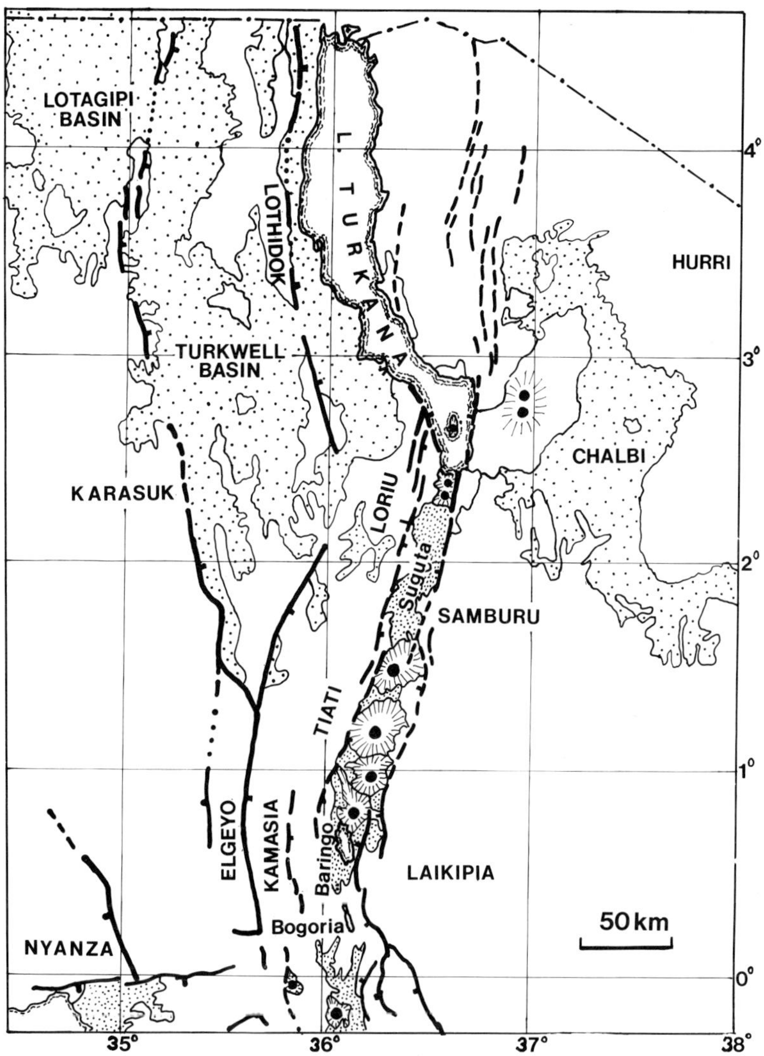

FIG. 3. Major faults and Quaternary lakes, sedimentary basins, and volcanic piles in the northern rift.

300 km long and 80–100 km wide, which slope gently away from the uplifted rift centre (Figs. 3 and 4). These structures created broad shallow basins in which fluviatile and lacustrine sediments accumulated, as is seen in the Turkana depression (Fig. 3).

The second order tectonic entity is the Gregory graben, which is 70–75 km wide and 450 km long. As the focus of faulting shifted inward, stepped ramps and step-fault platforms were formed at the rift margins (Fig. 5). Many of these sloped away from the rift axis, and served to collect drainage from the high marginal escarpments and to channel it into Lakes Turkana or Natron, leaving the central sector of the rift poorly supplied by surface water.

The third order tectonic features comprise the intensely faulted inner graben floor, where the spacing of faults averages 1.5 km (Fig. 2). The detailed structure consists of horsts, graben and step-platforms, and the features are generally so young that through-drainage has only developed where closed basins have become filled with water or sediment. Most sedimentary basins in the rift floor are small, narrow pans partly filled with fluviatile or lacustrine sediment. Larger sedimentary basins such as the Naivasha and Kedong basins are characterized by shallow ephemeral

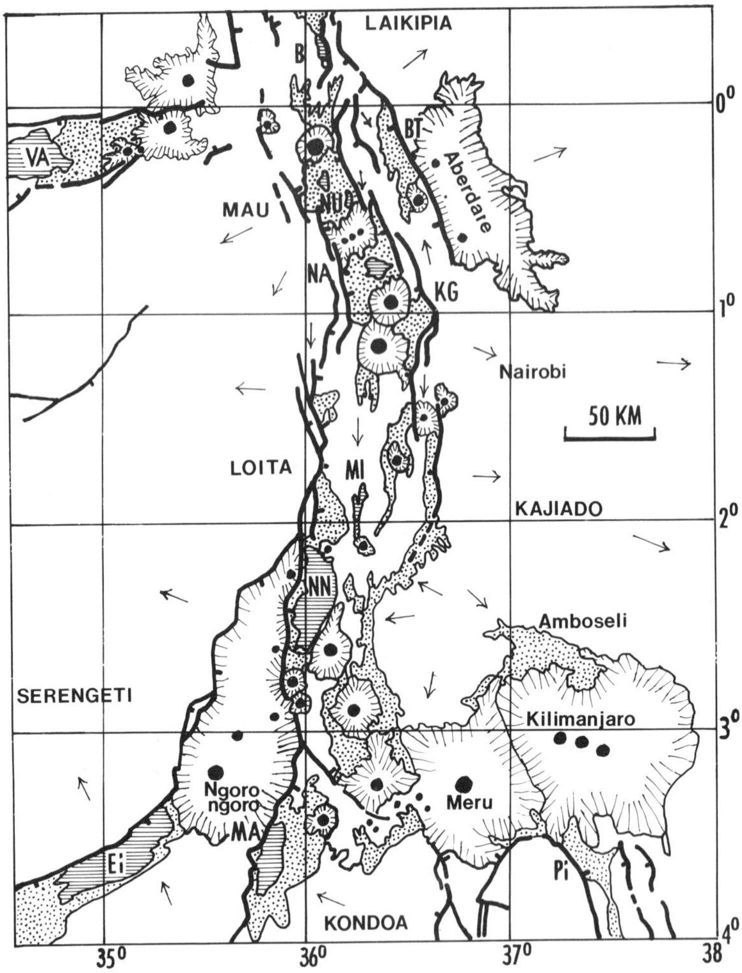

FIG. 4. Major faults and Quaternary lakes, sedimentary basins, and volcanic piles in the southern rift.

lakes where volcanoes have built volcanic dams and supplied abundant ash, that blanketed the underlying fault topography (Fig. 5).

Several characteristic structures are especially distinctive. Ramps are formed by the termination of *en echelon* major faults, such that the upthrown and downthrown sides are connected by a sloping ramp. Examples are found in the Mau fault zone (Fig. 4). Much more common are step-ramps, formed by branching of a major fault and termination of the branch fault within the rift. Examples are the Kinangop (KG, Fig. 4) and Esakut platforms (Fig. 5), which range from 7 to 23 km in width. They tend to channel drainage from marginal escarpments and the platform down the ramp onto the rift floor. Step-fault platforms are formed by sub-parallel faults, and are nearly always characterized by a gentle antithetic tilt giving slopes away from the rift valley, such as the Kinangop, Sambu and Enkorika platforms (Fig. 5).

Since the high flanks and marginal escarpments receive the greatest rainfall, step-fault platforms and ramps channel drainage into widely separated basins, leaving other regions to receive little surface water. The principal example of the influence of marginal structures is that of the course of the Uaso Ngiro River, which drains the southern slopes of the Mau, descends the Enkorika faulted ramp, and flows along the base of the Nguruman escarpment to feed Lake Natron (Fig. 5).

The main types of structural and volcanic traps that create sedimentary basins are shown in Fig. 6. Fig. 6A shows a half-graben represented by the Lake Turkana and Lake Natron basins. The Ol

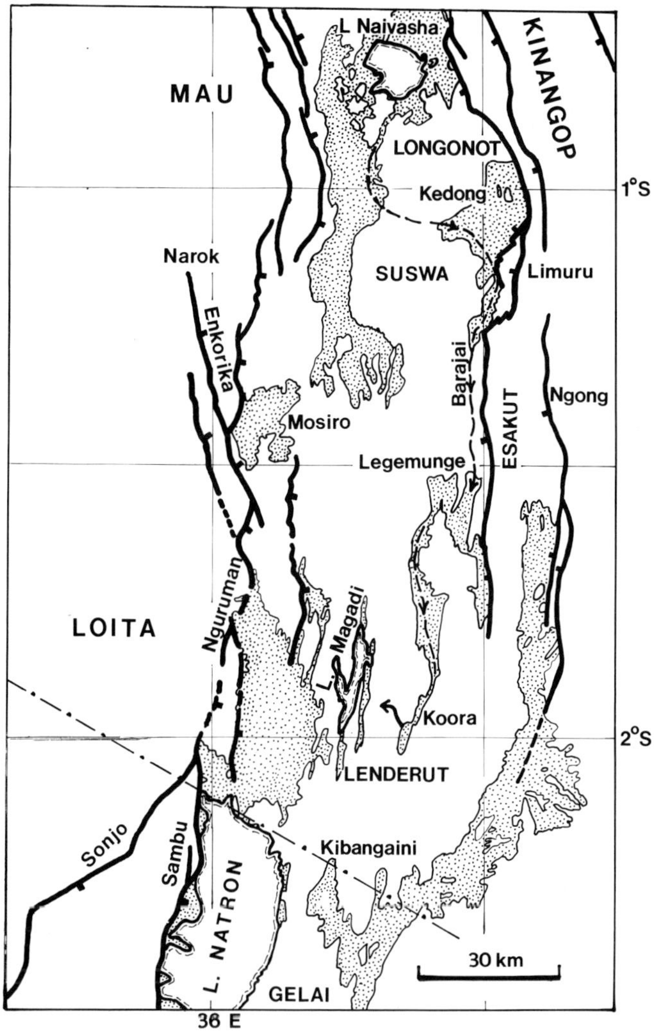

FIG. 5. Recent sedimentary basins in the southern rift (stippled areas), in relation to structural setting.

Bolossat basin on the Kinangop platform is on a tilted step-fault platform (Fig. 6B). The Naivasha and Kedong basins are in the inner graben and are dammed by volcanic piles (Fig. 6C). Numerous small graben and half-graben basins less than 2 Ma in age are found in the inner graben (Fig. 6D), and display all the structural types mentioned above on a small scale.

Examples of modern sedimentary basins

From a tectonic standpoint the larger sedimentary basins within the southern rift zone are either fault-angle basins in back-tilted step-fault platforms, basins confined in graben or half-graben with or without a volcanic dam, or piedmont basins at the base of marginal escarpments.

Fault-angle basins

The shallow, freshwater Ol Bolossat Lake rests on the Kinangop step-fault platform at the foot of the major Sattima fault (BT, Fig. 4). It owes its existence to the gentle eastward tilt of the Ol Bolossat platform, and to runoff from the well-watered Sattima escarpment and the Aberdare Range.

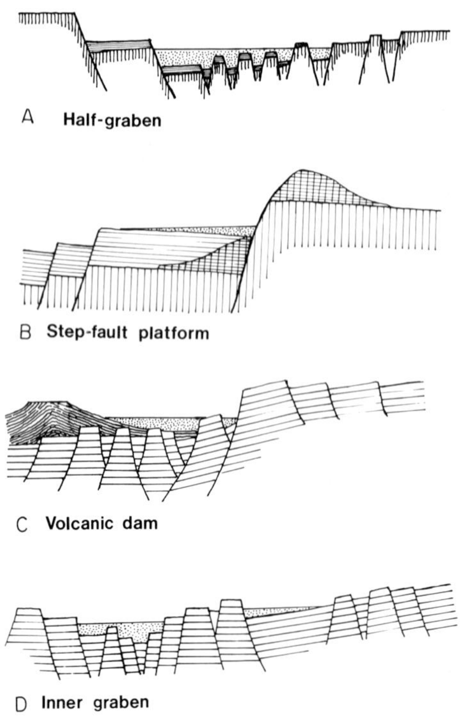

FIG. 6. Diagrammatic sections showing types of sedimentary basins in the southern rift.

Half-graben basins

Most of the major basins in Kenya and northern Tanzania are in half-graben; examples are the Lotagipi and Turkana basins in the N, and the Eyasi (Ei), Manyara (MA) and Natron (NN) basins in the S (Figs. 3 and 4). The Lake Natron basin lies at the foot of the major Ol Doinyo Sambu fault escarpment, and consists of an expanse of alluvium and mud-flats at its northern end, passing into a shallow saline lake (NN, Fig. 4, and Fig. 5). Its principal affluent is the Uaso Ngiro River, which has its headstreams on the southern slopes of the Mau Range on the high W flank of the rift valley, and is the only major river in the southern rift that discharges into the rift valley (Baker 1958; 1963; Crossley 1979).

Lake Natron is replenished by numerous saline springs, some hot, that emerge around its shores, and sustain alkaline lagoons. The lake waters are highly alkaline and contain sodium carbonate, chloride, and fluoride in solution; and the lake has a crystalline crust composed of trona, nahcolite, and traces of halite and villiaumite (NaF), overlying black unoxidized muds. The lake has not been drilled, but its size and tectonic position suggest that the basin is deep and could contain an essentially complete sedimentary record of the last 600,000 years.

The salinity of the lake is due to its being a closed basin with very high evaporation rates, and to the supply of saline groundwater from hot springs. As is the case for Lake Magadi nearby (see below), the springs represent recirculated alkaline groundwater, but there has been some debate concerning the origin of the large volume of alkaline salts. Jones *et al.* (1977) concluded that the streams and rivers carry sufficient dissolved salts to account for the salinity, but recent work on strontium and carbon isotopes of the evaporites suggests an important contribution from the active volcano Ol Doinyo Lengai at the S end of the lake (Bell *et al.* 1973). This volcano has erupted natrocarbonatite lava and ashes consisting largely of soluble sodium-potassium carbonate minerals (Dawson 1962, 1964), and the ash is readily leached to give a sodium-rich solution.

Graben lakes

The major sedimentary basins within the inner graben contain Lakes Baringo (Fig. 3; McCall *et al.* 1967), Bogoria (B, Fig. 4; Tiercelin *et al.* 1980), Nakuru (NU, Fig. 4; McCall 1967), Elmenteita, Naivasha (NA, Fig. 4; Thompson & Dodson 1963; McCall 1967), and Magadi (MI, Fig. 4; Baker 1958; Eugster 1979). These basins are on flat-lying flood lavas of the rift floor which have been cut by many faults during the last 2 Ma. The basins are therefore structurally controlled, but in the case of Lakes Nakuru and Naivasha, they were at least temporarily confined by volcanic dams consisting of Menengai volcano in the N, and the Longonot-Olkaria volcanic complex in the S (Fig. 5).

The Nakuru-Elmenteita basin is fed by streams from the Mau escarpment, but the principal surface influx to the Naivasha basin is down the Bahati ramp. Lake Naivasha is a shallow freshwater lake which has no modern outlet, although it formerly discharged to the S via the Njorowa Gorge (Thompson & Dodson 1963). It is a water-table lake which maintains its freshness by subsurface discharge into the permeable pyroclastic dam of the Longonot and Olkaria volcanic deposits. Although Lakes Nakuru and Bogoria are saline, there is no evidence that large evaporite deposits have formed as in the Magadi-Natron basin.

Several large, and many small, dried up lake basins are found in the floor of the inner graben. The Kedong basin was enclosed as a result of the growth of Suswa volcano during the last 0.4 Ma, and was fed in part by overflow from the Naivasha basin. Its deposits consist of lacustrine and alluvial reworked ash from Longonot and Suswa volcanoes (Fig. 5) (Randel 1970; Scott 1980).

The Legemunge (Olorgesailie) basin contains diatomaceous deposits *c.* 0.4 Ma old, which are now exposed as a result of minor faulting and of deep erosion by the Ol Keju Nero River (Baker 1958). The lake was shallow and slightly alkaline, and probably overflowed into the Koora graben to the S (Fig. 5).

The Kedong, Legemunge and Koora basins were traversed about 100,000 BP by a flood of extraordinary extent and force (Baker & Mitchell 1976), the effects and deposits of which can be traced from the foot of Suswa volcano to the S end of the Koora trough and into Lake Magadi, a distance of 120 km (Fig. 5). Lava flows at the foot of the NE slope of Suswa are scoured and polished by water, and the lacustrine deposits of the Kedong basin are extensively channelled and scoured along the Kedong River. The sides of the Barajai Gorge are similarly scoured, and there are terraces of coarse gravel within the gorge. Where the Kedong Gorge opens into the Ol Tepesi plain immediately E of Legemunge (Fig. 5), upstanding lava hills have been plucked and polished to a height of 30 m above the general level, and have gravel ridges on their southern sides.

The flood deposited a gravel fan at least 80 m thick in the Ol Tepesi trough, for there is no modern river there. It swept round a lava horst into the Legemunge Lake basin, depositing a succession of large exotic boulders, and cut the gorge now occupied by the Ol Keju Nero River (Baker 1958). The flood overflowed into the Koora graben, in which it deposited an expanse of feldspathic sand. At the southern end of the Koora graben a dry gorge cut in lava is the former overflow channel that discharged floodwater westwards into the Magadi basin. Crossley (1979) has suggested that thin lake sediments found at levels up to 100 m above the level of Lake Magadi may represent deposits of a lake created (or raised) by influx of the flood into the Magadi-Natron basin. This remarkable flood must have originated from the Naivasha basin, but its cause is not known, for the northern part of its course is covered by ash. It cascaded through a series of basins over a distance of 200 km and an altitude range of 1500 m.

FIG. 7. Geological map of the Magadi basin. 1 = basalt and trachyte lavas (1.65 to 0.8 Ma); 2 = Oloronga Beds (0.8 to *c.* 0.4 Ma); 3 = High Magadi Beds (> 0.1 Ma); 4 = evaporite series (*c.* 10,000 to 0 years); 5 = saline lagoons.

The Lake Magadi basin

Lake Magadi occupies a narrow complex graben in the axial part of the rift floor (Fig. 7) (Baker 1958, 1963; Eugster 1979). Volcanism in this part of the rift floor had virtually ceased by 0.8 Ma, after which formation of many small faults and rejuvenation of the Sambu fault created the basins of the Magadi-Natron region.

The oldest sediments in and near the Magadi trough are olive-green indurated silts, clays and cherts of the Oloronga Beds, which began to be deposited more than 0.78 Ma ago. They are found at the northern and southern extremities of the basin (Fig. 7). They are erionite- and chabazite-bearing clays and silts with irregular chert horizons, capped by a thick caliche layer indicating a period of desiccation. The presence of phillipsite

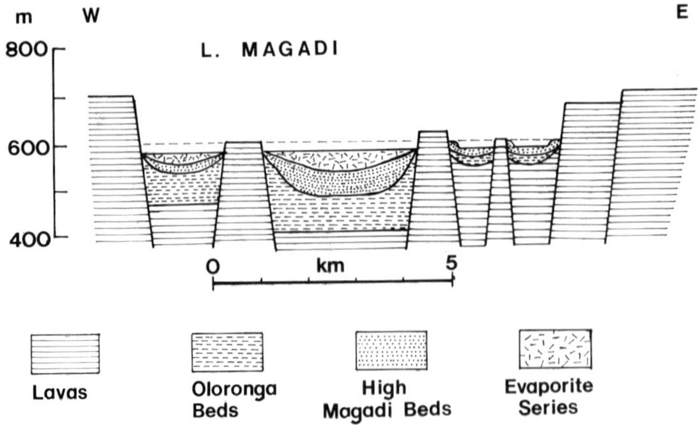

FIG. 8. Diagrammatic cross-section of deposits in the Magadi basin.

and of silicified gastropod shells suggests that the lake was weakly alkaline (Eugster 1979). The younger age limit of the beds is not known. The deposits could have been coeval with the Moinik Formation of the Sambu platform W of Lake Natron (Isaac 1967). Minor faulting of the rift floor affected the Oloronga Beds (Baker 1958), and led to deepening of the Natron and Magadi basins during the interval between 0.8 and 0.4 Ma.

Both the Natron and Magadi basins were occupied by separate alkaline lakes in which the succeeding High Magadi Beds (and High Natron Beds) were deposited. In the Magadi basin the deposits are erionite-bearing tuffs and clays containing detrital anorthoclase and amphibole, with authigenic erionite, illite, analcite, precipitated magadiite, and calcite concretions (Surdam & Eugster 1976; Eugster 1979).

The lake reached a level 13 m higher than the present lake and deposited a well-marked beach terrace. It became more alkaline, destroyed a fish population, and precipitated the sodium silicate mineral magadiite ($NaSi_7O_{13}(OH_3)3H_2O$) at a brine–epilimnion interface in an annually stratified lake (Eugster 1967; Hay 1968). Increasing desiccation lowered the lake level to at least 92 m below the maximum, and caused trona crystallization to begin, thus developing the evaporite series which is still accumulating. The magadiite layers deposited in the lake convert to chert by loss of sodium, as is seen where High Magadi Beds are exposed near the present surface. Leaching by water has resulted in transitions from plastic magadiite to chert within one metre (Eugster 1967). A diagrammatic section of the contents of the Magadi basin is shown in Fig. 8, which is based on the results from drill cores (Baker 1958; Eugster 1979).

The alkalinity of the basin water is maintained by recirculation of alkaline groundwater by hot springs, which represent a mixture of lake brine with a small component of infiltrated meteoric water. It is estimated that *c.* 75% of the saline input to the lake is recycled through the underlying groundwater (Fig. 9). Of the dissolved elements entering the basin, Ca and Mg were precipitated early as carbonates, K was adsorbed by clays, Si was precipitated as magadiite or by diatoms, and sulphate was removed by bacterial reduction, leaving Na and Cl to accumulate until trona began to crystallize.

The evaporite series consists of alternating trona beds and black anoxic muds, containing trona, minor nahcolite and villiaumite (NaF), with a sodium carbonate and chloride-rich brine. Sodium carbonate input from hot springs is estimated at 2.5×10^6 t a^{-1}. The trona crust is bordered by saline lagoons fed by hot springs, in which silica gels are deposited. The mud flats around the lake shore are locally covered by a gravel consisting of calcite oncolites and chert

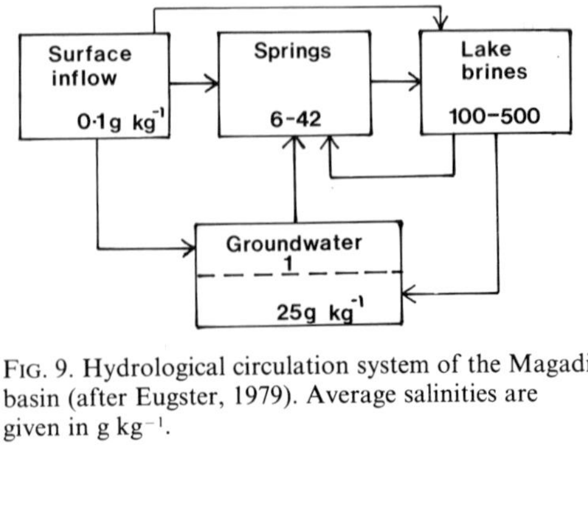

FIG. 9. Hydrological circulation system of the Magadi basin (after Eugster, 1979). Average salinities are given in g kg^{-1}.

fragments. Both the lagoons and the pools of brine in the lake contain large amounts of varicoloured bacteria and cyanobacteria which should give rise to muds high in organic carbon, but the carbon contents have not been determined.

Lakes Natron and Magadi are examples of hypersaline lakes in closed basins in the lowest and driest part of the rift valley. The large quantity of trona that has accumulated in them is due to recent natrocarbonatite ash eruptions and the recirculation of alkaline groundwater by hot springs.

Several rift valley lakes in Kenya are alkaline to some degree, but Lakes Magadi and Natron, and to a lesser degree Lake Bogoria, represent a strongly saline environment that depends on the chemical decomposition of alkaline igneous rocks and the recirculation of groundwater by hot springs (Eugster 1970). The peralkaline pyroclastic deposits that were widely erupted in the rift valley floor contain soluble sodium metasilicate, and their decomposition provides most of the sodium in saline lakes such as Nakuru and Bogoria. But the very large volume of trona evaporites in the Natron and Magadi basins, and the inferred presence of an extensive saline groundwater body, suggest that the millions of tons of trona evaporites of the Magadi and Natron basins are due to leaching of natrocarbonatite ash from Ol Doinyo Lengai volcano. A characteristic of lake waters and sediments of these lakes is their abnormally high cerium content.

The effects of natrocarbonatite volcanism on sedimentation are difficult to recognize because of the soluble nature of much of the evidence. Recognition that fossiliferous lacustrine limestones of the Koru Beds in the Nyanza Rift are calcified natrocarbonatite tuffs (Deans & Roberts 1984), and the occurrence of similar deposits at the Kaiserstuhl volcano in the Rhine graben (Keller 1981), suggest that many more such occurrences will be found in volcanic rifts. One effect of the deposition of natrocarbonatite tuffs and development of a hyperalkaline lacustrine environment, is the emergence of stratified meromictic lakes with sediments high in organic carbon.

Other deposits at the margins of rifts are of piedmont-type, having accumulated in fans below fault escarpments. Such deposits are exposed as interbeds in volcanic sequences of step-fault platforms adjacent to marginal faults. Another environment of such deposits is at the eastern margin of the southern rift E of Lenderut (Fig. 5), where the eastern marginal fault passes into a gentle monoclinal flexure. Here accumulation of lavas in the rift floor blocked drainage from higher ground to the E, forming a zone of sandy piedmont fan deposits.

Controls of rift sedimentation

The sedimentary environment of the southern Kenya Rift is characterized by the dominant influence of nearly continuous tectonic movements and volcanism, which created a multitude of small, short-lived basins in the main graben, and larger, longer-lived basins in the broad half-graben at the northern and southern extremities of the rift. Within the Eastern Rift there are only two examples of major rivers that drain a substantial area of the rift flank and enter the rift itself. In both cases drainage from the flanks is channelled into the rift along ramp structures. The catchment areas of most lacustrine basins are within the rift, most of the water being supplied by streams descending high-standing marginal escarpments. The courses of major rivers, such as the Uaso Ngiro which feeds Lake Natron, and the Kerio River which feeds Lake Turkana, are channelled N or S along fault-angle depressions.

The tendency for sedimentary basins to fragment due to subsequent faulting, results in complex stratigraphic relations and rapid variation of sediment type in response to changes in topography and hydrology. Growth-faults and rapid lateral variations in facies and thickness are common. The relative effects of fault displacements, erosion and volcanism have varied greatly in time and space, and have caused frequent changes in the sedimentary environment.

The influence of volcanism is principally to divert or dam surface drainage, thereby tending to multiply the number of basins within a tectonic depression. Pyroclastic activity overloads streams and supplies an abundance of sediment, which tends to smooth the topography and create larger, shallow basins. Leaching of peralkaline ash can lead to mildly alkaline waters, and leaching of natrocarbonatite ash can lead to hyperalkaline environments. Combined with hydrothermal activity and high evaporation, the latter results in trona deposition and the formation of stratified lakes with high organic content.

Interpretation of rift sedimentation cannot be carried out successfully without careful consideration of the tectonic and volcanic evolution. Equally, careful study of the sediments in a rift basin can lead to a more complete understanding of its tectonic and volcanic history. There is a need to combine tectonic, geomorphological and sedimentological studies to quantify rates of fault displacement, erosion and deposition of sediment. If seismic reflection profiles like those that have been obtained from the largest rift lakes (Rosendahl & Livingstone 1983) can be obtained from smaller lakes, this would add greatly to knowledge of their sedimentary and tectonic history.

References

BAKER, B.H. 1958. Geology of the Magadi area. *Rep. Geol. Surv. Kenya* **42,** 82 pp.

—— 1963. Geology of the area south of Magadi. *Rep. Geol. Surv. Kenya* **61,** 27 pp.

——, CROSSLEY, R. & GOLES, G.G. 1978. Tectonic and magmatic evolution of the southern part of the Kenya rift valley. In: NEUMANN, E.R. & RAMBERG, I.B. (eds) *Petrology and Geochemistry of Continental Rifts.* D. Reidel Publishing Co., Holland, pp. 29–50.

—— & MITCHELL, J.G. 1976. Volcanic stratigraphy and geochronology of the Kedong–Olorgesailie area and the evolution of the south Kenya rift valley. *J. Geol. Soc.* **132,** 467–84.

——, WILLIAMS, L.A.J., MILLER, T.A. & FITCH, F.J. 1971. Sequence and geochronology of the Kenya rift volcanics. *Tectonophysics* **11,** 191–215.

——, ——, —— & MOHR, P.A. 1972. Geology of the eastern rift system of Africa. *Spec. Pap. Geol. Soc. Am.* **136,** 67 pp.

—— & WOHLENBERG, J. 1971. Structure and evolution of the Kenya rift valley. *Nature* **229,** 538–42.

BELL, K., DAWSON, J.B. & FARQUHAR, R.M. 1973. Strontium isotope studies of alkalic rocks: the active carbonatite volcano Ol Doinyo Lengai, Tanzania. *Bull. Geol. Soc. Am.* **84,** 1019–30.

BELLIENI, G., JUSTIN VICENTIN, E., ZANETTIN, B., PICIRILLO, E.M. 1981. Oligocene transitional tholeiitic magmatism in northern Turkana (Kenya): comparison with coeval Ethiopian volcanism. *Bull. Volc.* **44,** 411–27.

CROSSLEY, R. 1979. The Cenozoic stratigraphy and structure of the western part of the rift valley in southern Kenya. *J. Geol. Soc.* **136,** 393–405.

—— 1980. Structure and volcanism in the S. Kenya rift. In: *Geodynamic Evolution of the Afro-Arabic Rift System.* Acad. Nazionale dei Lincei, Rome, pp. 89–98.

—— & KNIGHT, R.M. 1981. Volcanism in the western part of the rift valley in southern Kenya. *Bull. Volc.* **44,** 117–28.

DAWSON, J.B. 1962. The geology of Ol Doinyo Lengai. *Bull. Volc.* **24,** 349–87.

—— 1964. Carbonatitic volcanic ashes in northern Tanzania. *Bull. Volc.* **27,** 1–11.

DEANS, T. & ROBERTS, B. 1984. Carbonatite tuffs and lava clasts in Tinderet foothills, western Kenya: a study of calcified natrocarbonatites. *J. Geol. Soc.* **141,** 563–80.

EUGSTER, H.P. 1967. Hydrous sodium silicates from Lake Magadi, Kenya: precursors of bedded chert. *Science* **157,** 1177–80.

—— 1970. Chemistry and origin of the brines of Lake Magadi, Kenya. *Spec. Pap. Mineral. Soc. Am.* **3,** 215–35.

—— 1979. Lake Magadi, Kenya, and its precursors. In: NISSENBAUM, A. (ed) *Hypersaline Brines and Evaporitic Environments.* Elsevier, Amsterdam, pp. 195–232.

FAIRHEAD, J.D., MITCHELL, J.G. & WILLIAMS, L.A.J. 1972. New K/Ar determinations on rift volcanics of S. Kenya and their bearing on the age of rift faulting. *Nature* **238,** 66–9.

HAY, R.L. 1968. Chert and its sodium-silicate precursors in sodium-carbonate lakes of East Africa. *Contrib. Mineral. Petrol.* **17,** 255–74.

ISAAC, G.L. 1967. The stratigraphy of the Peninj Group—early Middle Pleistocene formations west of Lake Natron, Tanzania. In: BISHOP, W.W. & CLARK, J.D. (eds) *Background to Evolution in Africa.* University of Chicago Press, pp. 221–28.

JONES, B.F., EUGSTER, H.P. & RETTIG, S.L. 1977. Hydrochemistry of the Lake Magadi basin, Kenya. *Geochim. Cosmochim. Acta* **41,** 53–72.

JONES, W.B. & LIPPARD, S.J. 1979. New age determinations and the geology of the Kenya rift—Kavirondo rift junction, W. Kenya. *J. Geol. Soc.* **136,** 693–704.

KELLER, J. 1981. Carbonatitic volcanism in the Kaiserstuhl alkaline complex: evidence for highly fluid carbonatitic melts at the earth's surface. *J. Volcanol. Geoth. Res.* **9,** 423–31.

KING, B.C. 1978. Structural and volcanic evolution of the Gregory Rift Valley. In: BISHOP, W.W. (ed) *Geological Background to Fossil Man.* Scottish Academic Press, Edinburgh, pp. 29–54.

—— & CHAPMAN, G.R. 1972. Volcanism of the Kenya rift valley. *Phil. Trans. R. Soc. Lond.* **271A,** 185–208.

—— & WILLIAMS, L.A.J. 1976. The East African Rift System. In: DRAKE, C.L. (ed) *Geodynamics: Progress and Prospects.* American Geophysical Union, Washington, pp. 63–74.

MCCALL, G.J.H. 1967. Geology of the Nakuru–Thompson's Falls–Lake Hannington area. *Rep. Geol. Surv. Kenya.* **78,** 122 pp.

——, BAKER, B.H. & WALSH, J. 1967. Late Tertiary and Quaternary sediments of the Kenya rift valley. In: BISHOP, W.W. & CLARK, J.D. (eds) *Background to Evolution in Africa.* University of Chicago Press, pp. 191–220.

RANDEL, R.P. 1970. *Geological Map of the Suswa Area.* Mines & Geological Dept., Nairobi, Kenya.

ROSENDAHL, B.R. & LIVINGSTONE, D.A. 1983. Rift lakes of East Africa: new seismic data and implications for future research. *Episodes* **1,** 14–19.

SAGGERSON, E.P. & BAKER, B.H. 1965. Post-Jurassic erosion surfaces in eastern Kenya and their deformation in relation to rift structure. *J. Geol. Soc.* **121,** 51–72.

SCOTT, S.C. 1980. The geology of Longonot volcano, central Kenya: a question of volumes. *Phil. Trans. R. Soc. Lond.* **296,** 437–65.

SURDAM, R.C. & EUGSTER, H.P. 1976. Mineral reactions in the sedimentary deposits of the Lake

Magadi region, Kenya. *Bull. Geol. Soc. Am.* **87,** 1739–52.

Thompson, A.O. & Dodson, R.G. 1963. Geology of the Naivasha area. *Rep. Geol. Surv. Kenya* **55,** 80 pp.

Tiercelin, J.J., Le Fournier, J., Herbin, J.P. & Richert, J.P. 1980. Continental rifts: modern sedimentation, tectonic and volcanic control. Example from the Bogoria-Baringo graben, Gregory Rift, Kenya. In: *Geodynamic Evolution of the Afro-Arabic Rift System*, Acad. Nazionale dei Lincei, Rome, pp. 143–63.

Williams, L.A.J. 1970. The volcanics of the Gregory Rift Valley, East Africa. *Bull. Volc.* **34,** 439–65.

—— 1978. The volcanological development of the Kenya Rift. In: Neumann, E.R. & Ramberg, I.B. (eds) *Petrology and Geochemistry of Continental Rifts*. D. Reidel Publishing Co., Holland, pp. 101–21.

——, MacDonald, R. & Leat, P.T. 1983. Magmatic and structural evolution of the central part of the Kenya Rift. In: *Proceedings of the Regional Seminar on Geothermal Energy in Eastern and Southern Africa*. UNESCO–USAID, pp. 61–7.

B.H. Baker, Center for Volcanology, University of Oregon, Eugene, Oregon 97403, USA.

Relationships between major structures, salic volcanism and sedimentation in the Kenya Rift from the equator northwards to Lake Turkana

L.A.J. Williams & G.R. Chapman

SUMMARY: At the equator, the Kenya (Gregory) Rift Valley is a half-graben, faulted on the eastern side. It passes northwards into a deep graben with a double western margin, and then into what is essentially a downwarped depression. The deep graben sector (0°15′–1°00′N) is virtually devoid of late Miocene–Quaternary trachytic central volcanoes: their counterparts in areas of strong faulting are salic flood lavas. Moreover, the scarcity of pyroclastics in early–late Miocene 'plateau' phonolite sequences of the deep graben suggests that the thick tabular lava units are some distance from source. The phonolites exposed on the western side of the rift (Kamasia-Uasin Gishu) probably came from eruptive centres beyond the deep graben, whereas on the eastern side (Laikipia) a substantial part of the sequence could well belong to a lava shield on the plateau. In contrast to intra-volcanic pyroclastic sediments that occur predominantly N of latitude 1°N, the formations in the Kamasia region consist of waterlaid deposits intercalated with tabular lava flows. This spectacular development of waterlaid sediments is attributable to the absence of major eruptive centres and the existence of tectonic-geomorphic sediment traps in the deep graben sector of the rift.

The northern part of the Kenya Rift offers more rewarding areas for study of Tertiary events than the southern section (see Baker, this volume) where much of the early history can only be inferred from examination of attenuated and incomplete sequences exposed along the rift margins. Most of the ground between the equator and latitude 2°N has been mapped in detail by the East African Geological Research Unit, and it is now possible to identify major structural features and to investigate their influences on styles of salic volcanism and patterns of sedimentation.

Structure

From the equator to latitude 2°30′N, the NNE-trending Kenya Rift Valley is about 100 km across—much wider than in southern Kenya. Lying E of the centre and within this broad depression, is an inner trough occupied by Quaternary volcanics and sediments. It maintains a remarkably constant width of 20–30 km despite the fact that the floor descends from an altitude of some 1500 m at the equator to less than 300 m near Lake Turkana.

Some major structures swing to a NW–SE alignment at the southern end of the region, and the rift adopts this trend S of the equator (Fig. 1). The elbow coincides with the junction between the Kenya Rift and the WSW-trending Nyanza (Kavirondo) trough.

The eastern margin of the rift valley is well defined S of the latitude of Lake Turkana, but the western boundary becomes increasingly diffuse N of 1°15′N where the Elgeyo fault bifurcates (Figs. 1 and 2). Its direct continuation soon loses all topographic expression, but major scarps (Lomut, Sekerr and Turkwel faults) branch off with a general NNW trend towards the Uganda border, framing the Turkana depression which widens towards the Ethiopian and Sudan borders, where it is at least 300 km across. The Turkana depression is bounded to the W by a major flexure along the Uganda border, but the eastern margin is poorly defined. The Tertiary formations of the Turkana depression have a regional westward dip except near the Uganda border. They are cut by antithetic faults with NNW–NNE trends, and the structural pattern exerts strong control on the orientation of the northern half of Lake Turkana.

The Turkana depression is seen as an essentially pre-Miocene feature, cut across at its southern end by the later Tertiary and Quaternary structures of the Kenya Rift. At 2°30′N, the western boundary of the rift is taken to be defined by a change in the regional dip direction in Miocene basalts (from NW to dipping towards the present trough) and by the appearance of faults with the present rift trend. Both criteria place the boundary as a strike continuation of the Elgeyo fault.

The essential structural features of the rift between the equator and the southern end of Lake Turkana (Fig. 2) are conveniently summar-

From FROSTICK, L.E. *et al.* (eds) 1986, *Sedimentation in the African Rifts*, Geological Society Special Publication No. 25, pp. 59–74.

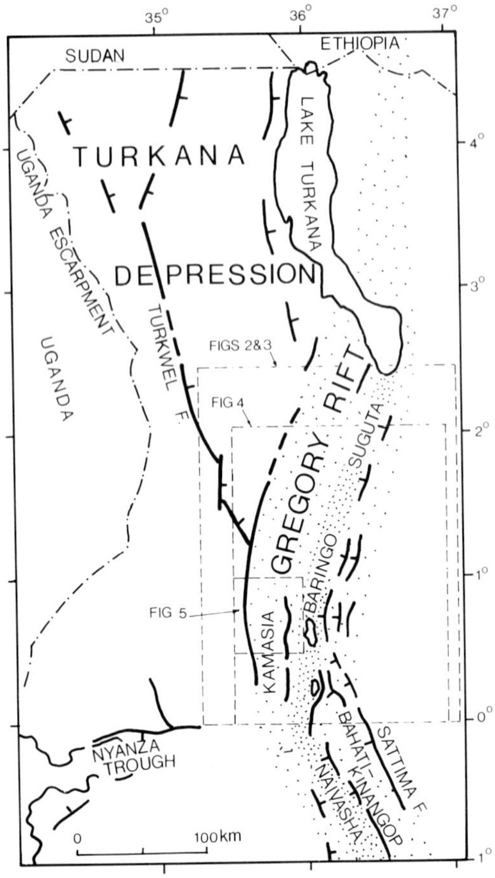

FIG. 1. Outline structural map of NW Kenya showing the Kenya (Gregory) Rift Valley (stippled; fine stipple represents inner troughs), the Turkana depression, and the areas covered by Figs. 2–5.

ized by dividing this region into five sub-areas, which are treated from S to N, stepping from well-studied domains into terrain covered only by rapid reconnaissance mapping.

Equator–0°15′N

The rift is defined on the western side by a gentle flexure, and to the E by a step-fault platform; it is thus a half-graben. The downwarp at the western margin has come to be known as the 'equator monocline' (Jones & Lippard 1979), but this name is somewhat misleading in view of the meridional trend of the hinge-line. The step-fault structure on the opposite side of the rift is a continuation of the Bahati–Kinangop platform S of the equator (Fig. 1). The Marmanet faults represent an extension of the NW-trending Sattima fault, but show complications not apparent to the S. The throw on the central section of the Marmanet fault must be at least 1000 m (Griffiths 1977). The Bogoria fault, defining the margin of the inner trough of the rift, has a total throw exceeding 1150 m, with Quaternary displacements up to 900 m.

The inner trough at this latitude is a unique feature having a near-central graben (Molo graben), flanked to the E by numerous antithetic faults which lower the floor to a narrow depression occupied by Lake Bogoria (altitude about 960 m).

0°15′–1°00′N

In this area, the rift is a deep graben with a double western boundary formed by the Elgeyo fault (maximum throw at least 3000 m) and the Kamasia faults (total displacement up to 4000 m). Between these faults lies the tilted Kamasia block in which basement rocks are overlain by 3000 m of Miocene volcanics and sediments (Chapman *et al.* 1978). Repeated movements on the major faults have been demonstrated. These date back to the early Miocene, but the most important displacements occurred in the late Miocene (7 Ma) and in Plio-Pleistocene times (4–0.5 Ma). Between the Kamasia faults and the inner trough, an anticline (Kaparaina 'arch') exposes mainly late Miocene basalts and trachytes, and Plio-Pleistocene sediments. The essential structure of the Kamasia Range (Tugen Hills) and its eastern foothills is thus a major asymmetric arch dislocated by large faults just to the W of its axis (Chapman *et al.* 1978).

The eastern margin of the rift at this latitude consists of back-tilted blocks and ramps formed by *en echelon* antithetic faults of the Laikipia escarpment. Basement rocks are not exposed by the faults, so that total displacements cannot be estimated with precision. Carney (1972) suggests throws of 1000–1500 m on some of the more important faults, and recognized three major episodes of faulting. The first probably occurred at about 13 Ma, and locally horsts were produced by downthrows to the E. The second episode was about 7–6 Ma, and was broadly equivalent to the late Miocene event in the Kamasia area. The third episode (Pleistocene) led to displacements of up to 900 m in the floor of the rift as well as rejuvenation of the Laikipia faults.

The Quaternary faulting defined an inner trough narrower than the present spread of Pleistocene–Holocene volcanics and sediments, which thus cover a broad step-fault platform at the base of the Laikipia escarpment. On the opposite side of the inner trough, the Quaternary formations rest on Tertiary basalts, trachytes and sediments in the eastern limb of the Kaparaina arch.

1°00′–1°30′N

The rift continues to show strong structural asymmetry in E–W profiles, but major faults give way progressively to steep flexures so that the deep graben of the previous sub-area is replaced northwards by what is essentially a downwarped depression with complications on the western side.

The Elgeyo fault forms an imposing western boundary where it truncates the high-standing basement mass of the Cherangani Range which rises to 3200 m. The fault then bifurcates, and its direct continuation (the Kula fault) now has no topographic expression. Few outcrops of Tertiary rocks are preserved on the upthrow side, but minimum displacements of 250–400 m can be inferred locally. Initial movements on the Elgeyo and Kula faults probably occurred in the early Miocene, but rejuvenation in later Tertiary and Quaternary times was confined to the Elgeyo section.

The tilted Kamasia block passes northwards in this area into an asymmetric anticline. Basement rocks are exposed in the core to give a NNW-trending inlier, within which the foliation in gneisses tends to be parallel to the regional orientation of the rift. The axis of the anticline is thus oblique to both the basement and the main rift structures. It is tempting to regard this anticline ('basement arch') as a feature controlled by some pre-rift or early rift structure, possibly one associated with the Turkana depression. A further hint of the influence of such a structure is seen at the northern end of the Kamasia block where some major faults swing to a NNW trend. The Kaparaina 'arch', which separates the Kamasia block from the inner trough, passes across the lineament and can be traced northwards for some distance before it gives way to a pattern of consistent eastward dips. Some of the dips here have a depositional component where the volcanics have a broad shield-like form, but a major Pliocene trachyte volcano (Ribkwo) rising from the western side of the inner trough was tilted 5–6° towards the present rift floor (Webb & Weaver 1975).

Along the eastern margin of the rift the tilted block regime of the Laikipia escarpment extends into this sub-area but it soon gives way to a steep flexure termed the Samburu monocline (Shackleton 1946). The hinge is well defined by an abrupt change in the attitude of Miocene phonolites. These lavas are flat-lying on the plateau, but dip at an angle of up to 45° into the rift along the escarpment. There are local complications in the dip patterns displayed by overlying pyroclastics and by dissected trachytic centres flanking the inner trough: nevertheless, the dominant structure is a downwarp, and faults become progressively less important northwards.

The broad low-angle trachytic shield of Silali volcano (late Quaternary) occupies much of the inner trough in this region. It is surmounted by an impressive oval caldera 7.5×5 km across, with the long axis lying normal to the regional trend of the rift (Williams *et al.* 1984). Another large caldera volcano (Emuruangogolak) straddles latitude 1°30′N. The eastern flanks are severely faulted, and trachytes tentatively assigned to the early shield rest on more steeply dipping Tertiary formations beyond the structural confines of the inner trough, which is here only about 20 km across.

1°30′–2°00′N

The Kula fault, defining the western margin of the rift, probably dies out northwards across this area, though some topographic feature must have controlled the distribution of early Miocene basalts and underlying sediments. The possibility that the fault passes northwards into a pre-volcanic flexure cannot be dismissed. A lobe of basement extending into the volcanic outcrop marks the axial region of an anticline having a nearly meridional trend. As in the somewhat similar 'basement arch' to the S, the extent to which metamorphic rocks were deformed in the axial regions of these Cenozoic anticlines is not clear.

Southeast of the basement arch several faults with a total throw of at least 750 m step down the Miocene basalts and trachytes (Weaver 1973), but the overall structure is a flexure which reveals progressively younger formations towards the inner trough (Suguta valley). The steepest dips (*c.* 30°) are in Quaternary volcanics along the western margin of the inner trough (Truckle 1977), so the pattern is not simply one of uniform tilting nor of decreasing severity of deformation towards the most recently active zone.

On the eastern side of the Suguta valley, the rift margin is a steep flexure affecting Miocene volcanics and sediments; this is a continuation of the Samburu monocline (Baker 1963). Near latitude 2°N, Pliocene trachytes and basalts rest unconformably on the downwarped Miocene formations. The trachytes are flat-lying in eastern outcrops, but westwards they dip beneath the Quaternary deposits of the Suguta valley. The axis of the flexure thus shifted westwards with time.

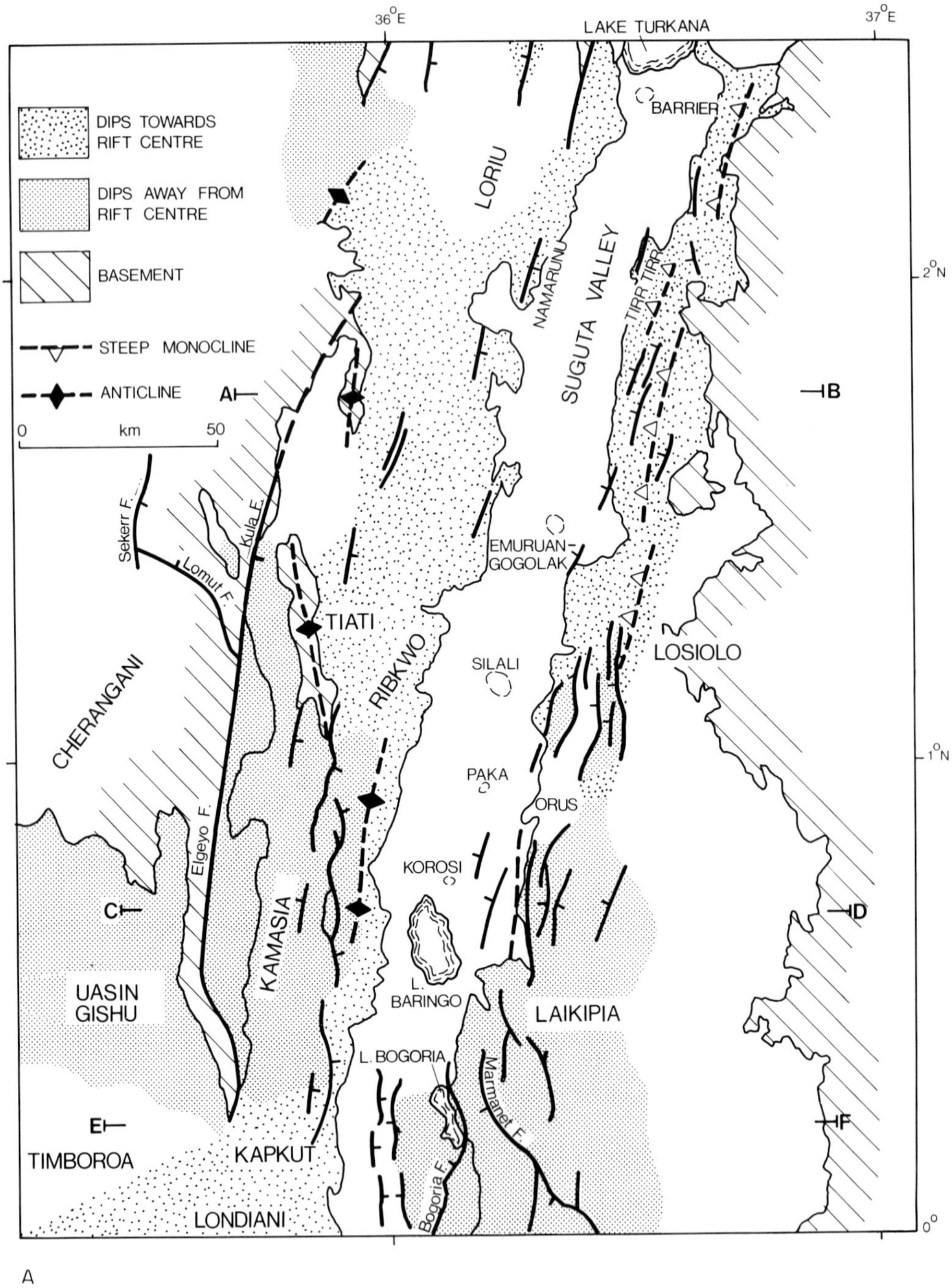

FIG. 2.A. Structural and locality map of the Kenya Rift between the equator and the southern end of Lake Turkana. Unornamented areas represent Quaternary volcanics and sediments of the inner trough, and near-horizontal Tertiary strata.

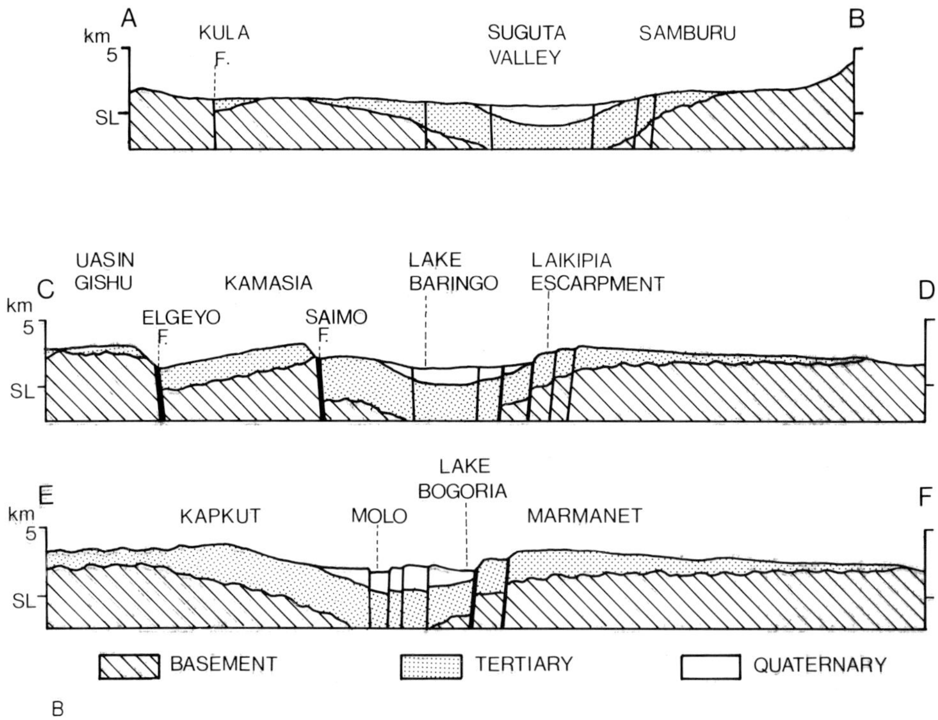

FIG. 2.B. Diagrammatic sections (vertical exaggeration *c.* ×3) illustrate the northern downwarped depression (A–B), the deep graben sector (C–D), and the equatorial half-graben (E–F).

2°00′–2°30′N

This area has only been covered by rapid reconnaissance mapping, and structural data are sparse. The Tertiary sequence on the western side of the rift is much thinner here than in areas to the S. Basement rocks outcrop on the shores of Lake Turkana along the fringe of the inner trough, and they are again exposed in fault-scarps beneath Miocene(?) basalts of the Loriu Plateau. Rhemtulla (1970) reported basement exposures on the western flanks of Loriu, which he considered to be a tilted horst-like block.

Salic volcanism

A full discussion of the volcanic stratigraphy of the region is beyond the scope of this article, and attention is focused here on broad patterns of salic volcanism in order to demonstrate some fundamental longstanding differences in the style of activity in contrasting structural regimes. Many important sedimentary formations are intercalated in sequences of phonolites and trachytes. Some of the sediments are waterlaid deposits; others consist predominantly of primary pyroclastics. A full understanding of the sedimentary environments depends heavily, therefore, on knowledge of the nature and distribution of sources of salic volcanics, as well as on an appreciation of the structural setting.

Early attempts (e.g. Williams 1970) to distinguish, on a regional scale, flood volcanics (produced by fissure and/or multicentre eruptions) from those associated with major central volcanoes understandably led to gross oversimplification. Detailed studies have shown, for instance, that broad shield-like source areas are frequently identifiable in regions of both basic and salic flood volcanism. Such major centres are characterized by thickening of flank lava sequences into complex areas having substantial contents of pyroclastics and numerous intrusive bodies. Some examples have been convincingly demonstrated within the voluminous Miocene 'plateau' phonolites (Lippard 1973). Nevertheless, there remain certain lava formations within which no major central complexes have been identified. Some fall readily into the multicentre category; others bridge the gap between multicentre and shield categories, in having local concentrations of vents, and a related thickening of the formation; yet others have yielded few

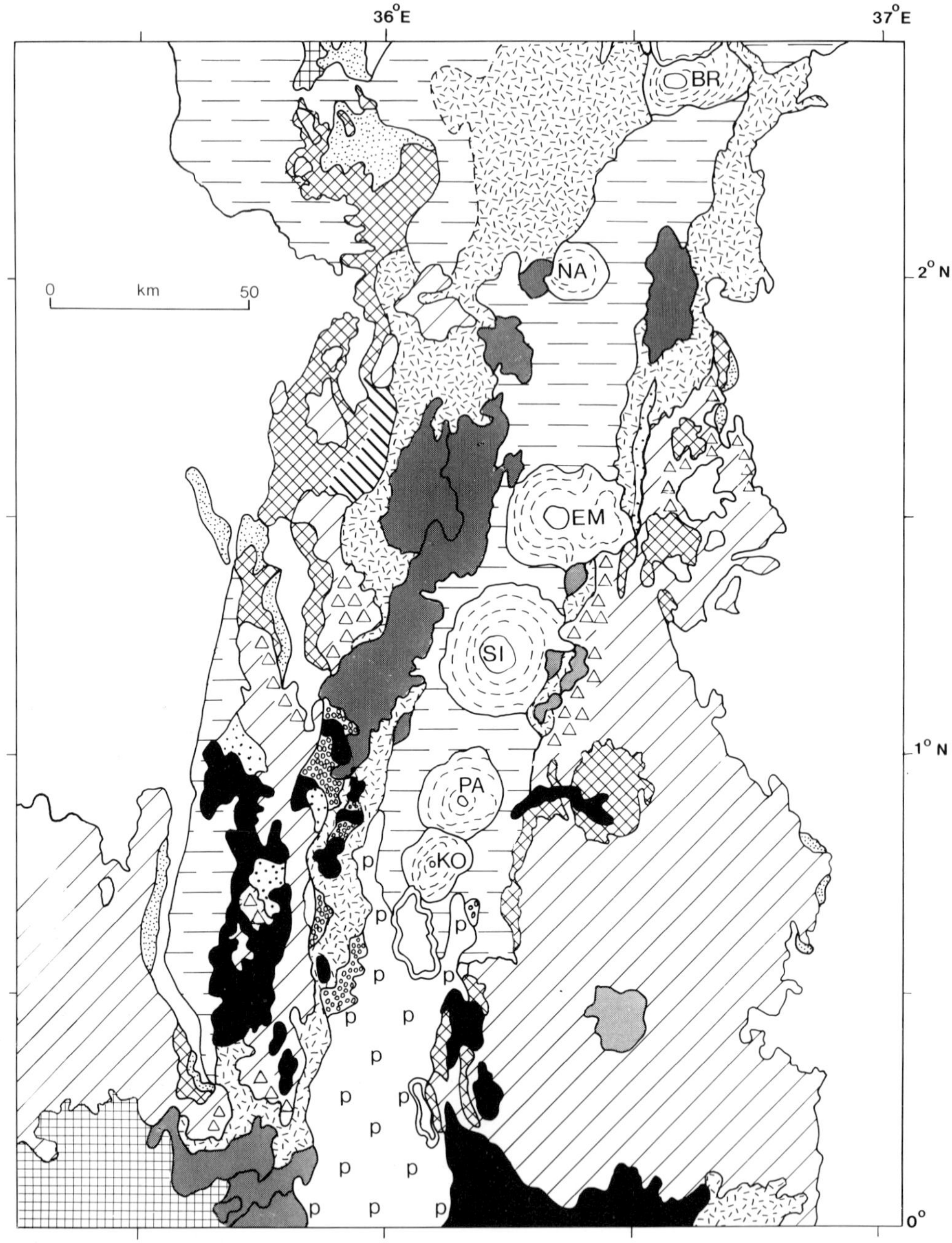

FIG. 3. Simplified geological map of the region covered by Fig. 2.

indications of any eruptive sources. All such examples are still conveniently described as flood lavas.

The salic volcanics between the equator and Lake Turkana (Fig. 3) fall into the following three groups, each representing widespread activity separated by major tectonic events: (1) Quaternary trachytes and trachyphonolites; (2)

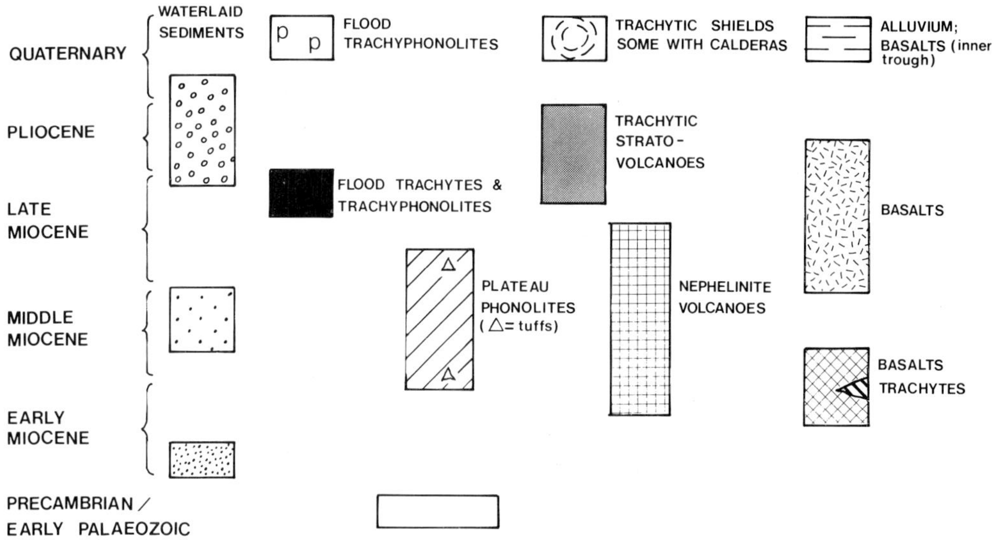

FIG. 3. Continued.

late Miocene–Pliocene trachytes and trachyphonolites; and (3) early–late Miocene 'plateau' phonolites.

The two trachytic groups are subdivided (Fig. 3) to distinguish late Tertiary strato-volcanoes from flood lavas, and Quaternary trachytic shields (commonly surmounted by calderas) from flood trachyphonolites.

Miocene 'plateau' phonolites

The thickest (1600 m) and most complete (16–8 Ma) exposed succession of Miocene phonolites (Tugen Hills Group) is found in the Kamasia Range, W of Lake Baringo (Chapman *et al.* 1978). A direct correlation of a unit in the middle of the Tugen Hills Group with a basal flow in the relatively thin lava sequence of the Uasin Gishu Plateau, shows convincingly that phonolites overtopped the western flank of the rift depression in the Middle Miocene. No detailed correlations can be made between members of the Tugen Hills Group and the various phonolite formations exposed along the eastern side of the rift, though the latter are clearly the time equivalents of at least part of the Kamasia sequence (Chapman & Brook 1978; King 1978; Williams & Truckle 1980). Some reviews give the impression that all the phonolites on the eastern plateaus came from concealed sources within the rift, and that overtopping of eastern and western rift shoulders occurred simultaneously. Such notions are not supported by firm evidence, and they fail to take into account stratigraphic complexity on the eastern side of the rift (Fig. 4).

The early Middle Miocene Sidekh flows (16–14 Ma) of Kamasia were confined to an early rift depression, and no representatives occur on the western (Uasin Gishu) plateau. The Sidekh phonolites thicken northwards towards Tiati (Fig. 4) where thick pyroclastics are accompanied by plugs and dykes. The Tiati centre was one of three proposed by Lippard (1973) who considered the plateau successions to be the flanks of large low-angle shield volcanoes some 70–80 km across. At the time of Lippard's review, little detailed work had been carried out on the opposite side of the rift where all the phonolites on the plateau were thought to belong to the Rumuruti Formation (12–10 Ma).

Later studies by Golden (1978) showed that the oldest phonolites involved in the Samburu monocline date back to 15 Ma; they are thus the time equivalents of the Sidekh phonolites of the Kamasia. Golden assigned these early Miocene phonolitic lavas of the eastern rift margin to a formation named Losiolo (see King 1978; Williams & Truckle 1980), but Shackleton's (1946) use of the term Losiolo phonolites has priority. It was applied to post-Rumuruti lavas, and relationships have been confirmed by recent mapping on the plateau (Samburu-Marsabit Project, Bri-

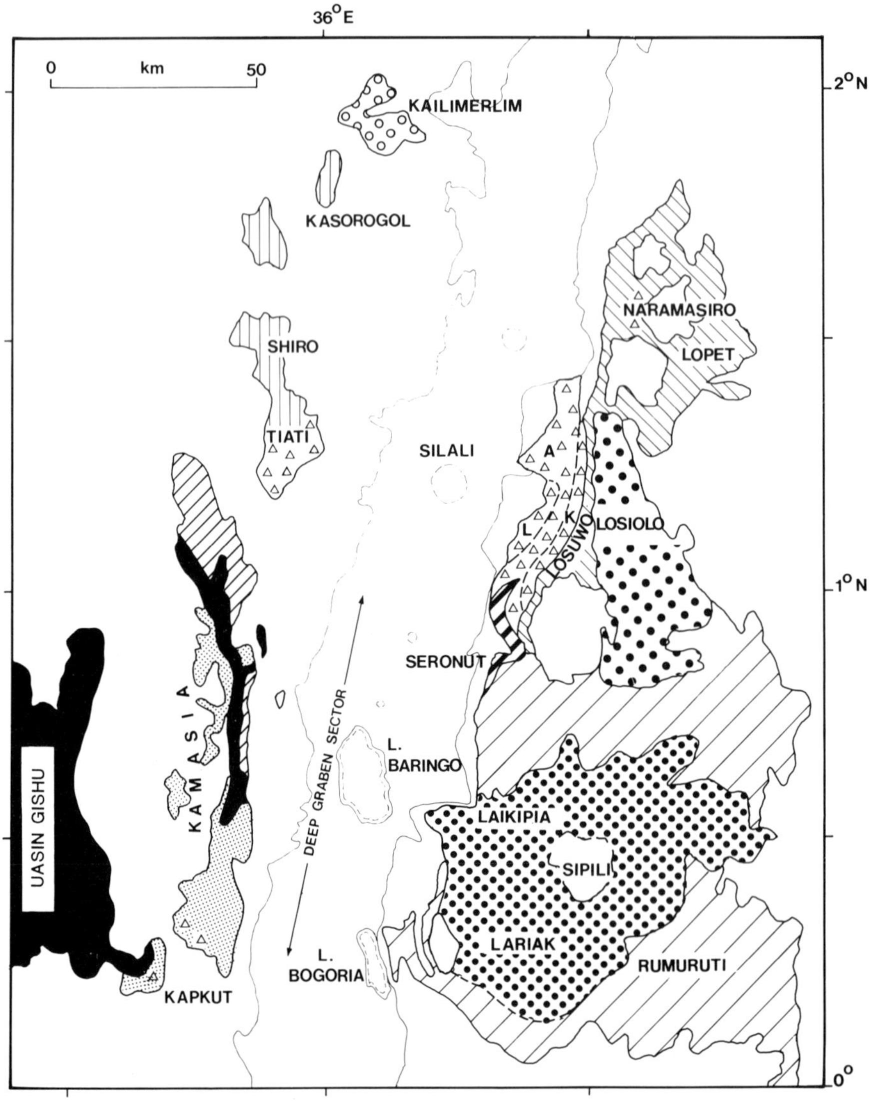
36°E
2°N
1°N
0°
0 km 50
KAILIMERLIM
KASOROGOL
SHIRO
TIATI
SILALI
NARAMASIRO
LOPET
A
K
L
LOSUWO
LOSIOLO
SERONUT
DEEP GRABEN SECTOR
L. BARINGO
KAMASIA
UASIN GISHU
LAIKIPIA
SIPILI
LARIAK
RUMURUTI
L. BOGORIA
KAPKUT

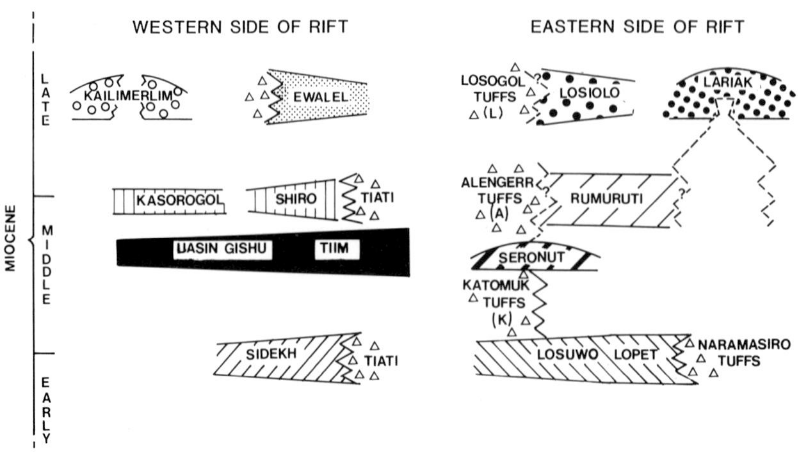
WESTERN SIDE OF RIFT
EASTERN SIDE OF RIFT
MIOCENE
LATE
MIDDLE
EARLY
KAILIMERLIM
EWALEL
LOSOGOL TUFFS (L)
LOSIOLO
LARIAK
KASOROGOL
SHIRO
TIATI
ALENGERR TUFFS (A)
RUMURUTI
UASIN GISHU
TIIM
SERONUT
KATOMUK TUFFS (K)
SIDEKH
TIATI
LOSUWO
LOPET
NARAMASIRO TUFFS

tish Geological Survey). Golden's Losiolo formation is here renamed Losuwo. Though as yet unconfirmed by isotopic dating, the Losuwo lavas are probably equivalent to what are now termed the Lopet phonolites (BGS team, pers. comm., 1984) which form the plateau NE of Losiolo (Fig. 4). Further N, the phonolites have substantial intercalations of pyroclastics (Baker 1963), indicating a major source area in the vicinity of Naramasiro.

The Losuwo phonolitic lavas are overlain by thick pyroclastics (Katomuk and Alengerr tuffs) along the downwarped rift margin. Ignimbrites are prominent, and there are several important intercalated lava units; one is a mugearite-trachyte-phonolite shield (Seronut formation; *c.* 13 Ma). The bulk of the pyroclastics accumulated during the Middle Miocene, and they are broadly equivalent in age to those of the Tiati pile on the opposite side of the rift where activity continued after the eruption of the Sidekh phonolites. These outcrops of tuffs on both sides of the inner trough point to a much larger area of lower Middle Miocene eruptive centres than suggested by Lippard. The term 'proto-Silali' is perhaps appropriate to describe the active region, because that unique Quaternary volcano (Williams *et al.* 1984) is situated more or less centrally within the inferred spread of Miocene pyroclastics.

Lippard (1973) concluded that the thick Tiim phonolites (*c.* 14–12 Ma) of the Kamasia Range, and their attenuated lateral equivalents on the Uasin Gishu plateau, came from a concealed source beneath the Baringo sector of the inner trough. The Tiim phonolites thin northwards from an area of maximum development at the latitude of Lake Baringo, so that derivation from proto-Silali centres must be considered unlikely. A source area at the latitude of Lake Bogoria is a distinct possibility, and it would be in keeping with flow directions in the Uasin Gishu phonolites.

On the eastern side of the rift there are no widespread lava formations equivalent to the Tiim phonolites; only the Seronut shield is of comparable age. Flooding of the eastern plateau by the Rumuruti phonolites (12–10 Ma) occurred later than overtopping of the western margin of the depression by the Uasin Gishu flows. Three formations, separated by unconformities related to faulting, have been recognized in the lavas assigned to a Rumuruti group (Carney 1972; Griffiths 1977). In this account, only the lower (Ngelesha) and middle (Uaso Narok) units are described as Rumuruti phonolites. The youngest formation (Marmanet) is distinguished in Fig. 4 to show its approximate distribution; it is provisionally renamed Lariak because the flows may not extend SW as far as the Marmanet escarpment. The sources of the phonolites of the Laikipia Plateau remain speculative, but indications of westward and northward thinning of units tend to favour derivation from vents on the plateau, rather than eastward flow from central complexes concealed in the floor of the rift. The Lariak phonolites seem to have the form of a broad shield centred on the Sipili area, where a possible source is obscured by later Tertiary trachytes. The Rumuruti phonolites underlying this Lariak shield could well have come from the same centre. This is considered more likely than derivation from the proto-Silali region to the N, despite the fact that the Alengerr tuffs there are the time equivalents of the Rumuruti flows.

The youngest plateau phonolites in the Kamasia sequence constitute the Ewalel formation (*c.* 9–8 Ma). Increasing contents of pyroclastics in southern outcrops indicate an approach to a source now concealed by late Miocene trachytes at Kapkut (Lippard 1973). The undated Losiolo phonolites on the eastern plateau are possibly time equivalents of the Ewalel lavas. If the Losiolo flows came from a within-rift source (proto-Silali), there must have been a rapid transition from lavas to pyroclastics (e.g. Losogol Tuffs) across the width of the present downwarp. The Lariak phonolites of southern Laikipia may be somewhat older than the Losiolo formation, though there are similarities in the degree of dissection. Specimens from the extreme fringes of the Lariak outcrop yielded dates indistinguishable from those obtained from the underlying Rumuruti phonolites.

Miocene phonolitic and trachytic activity occurred to the W of the inner trough at about 2°N where a foyaite plug is exposed at Kailimerlim volcano (Truckle 1977). Dates from the complex are not reconcilable with the established stratigraphy, which indicates a late Miocene age for the centre. The Kailimerlim volcanics and scattered microfoyaites and phonolites to the NW are tentatively regarded as the time equivalents of the Ewalel formation of the Kamasia Range. An uppermost Middle Miocene age is provisionally

FIG. 4. (Opposite page.) Map showing the distribution of exposed representatives of the Miocene 'plateau' phonolites. The Lariak, Losiolo, Kailimerlim, Kasorogol and Lopet phonolites remain undated or have given questionable results.

assigned to the Shiro (Tiriomim) phonolites on the northern flanks of Tiati, and the Kasorogol phonolites between Tiati and Kailimerlim.

It will be apparent from this discussion that the sources of the plateau phonolites remain largely speculative. Nevertheless, the evidence currently available suggests that most of the flows came from centres N and S of the deep graben sector of the rift, and that the eastern plateau differs from that on the W in having shields or source areas outside the structural confines of the rift. This in turn explains why difficulties have arisen in attempts to correlate phonolite sequences of the Kamasia and Laikipia regions.

Late Miocene–Pliocene trachytic strato-volcanoes

Tilted coalescing trachyte shield volcanoes (6–2 Ma) form the western margin of the inner trough between latitudes 1° and 2°N (Webb & Weaver 1975). Major trachytic centres of comparable age (Kapkut, 7 Ma; Londiani, 3 Ma) occur between the equator and 0°15′N, but there is a substantial gap in the chain along the deep graben sector of the rift (Fig. 3).

Representatives of this group of strato-volcanoes on the eastern side of the rift are the Pliocene Tirr Tirr trachytes (Baker 1963; Dodson 1963) straddling 2°N, and the dissected remnants of late Miocene centres (Lopogno and Nakali trachytes; Golden 1978) at the margin of the inner trough near Silali. The undated, but probably late Miocene, Sipili trachytes (McCall 1967) on the Laikipia Plateau at 0°30′N do not constitute a discrete strato-volcano, and the flows should perhaps be regarded as a late (conceivably post-caldera) phase of the Lariak phonolite shield (Fig. 4).

It is probably no coincidence that the trachytic strato-volcanoes are well developed on both sides of the inner trough within one of the postulated source areas (proto-Silali) for the plateau phonolites.

Late Miocene flood trachytes and trachyphonolites

The trachytes occupying the gap in the western chain of strato-volcanoes are late Miocene flood lavas (Kabarnet formation, *c.* 7 Ma; Chapman *et al.* 1978) exposed on the dip-slope of the Kamasia block and in inliers along the eastern foothills. Their equivalents, in terms of age and mode of eruption, on the eastern side of the rift are the Thomson's Falls trachyphonolites (6.5 Ma; Baker *et al.* 1971) and the predominantly phonolitic trachytes of the Tasokwan formation (6.5–7.5 Ma; Griffiths 1977). These two formations are combined on Fig. 3. The southernmost outcrops of the Tasokwan have yet to be studied in detail, but a date of *c.* 6 Ma from the Sabukia phonolitic trachytes (Williams *et al.* 1983), just S of the equator, strongly suggests that these are Tasokwan lavas rather than Rumuruti phonolites as proposed by McCall (1967).

All these lavas were emplaced after major faulting of the plateau phonolites at *c.* 7 Ma. On the eastern side of the rift, the late Miocene flood lavas occur within both graben and half-graben sectors; significantly, they are restricted to the faulted side of the latter. The Pliocene Orus trachytes, just S of 1°N in the deep graben sector, are part of a sheet that flooded the rift floor (Golden 1978), but the source has not been identified.

Quaternary trachytic centres and flood lavas

North of latitude 1°N, the low-angle trachytic shields of three late Quaternary caldera volcanoes (Silali, Emuruangogolak and the 'Barrier') occupy nearly the full width of the inner trough (Fig. 3). Together with a fourth centre (Namarunu) that has no recognizable caldera, they form a chain overlapping, but extending beyond, the belt of late Tertiary strato-volcanoes mentioned above.

Two centres exist in the deep graben sector of the rift, but the shields tend to be smaller than those to the N (Williams *et al.* 1984). Moreover, the summit collapse on one (Paka) barely qualifies as a caldera, and the other (Korosi) has an explosion crater but lacks evidence of any subsidence.

Southwards from Korosi, the inner trough is occupied by flood lavas (Hannington trachyphonolites). The northern and southern limits of these flows coincide approximately with the spreads of late Miocene flood trachytes and phonolitic trachytes (Kabarnet and Tasokwan formations) exposed outside the central trough.

Sediments

Two main groups of sediments are conveniently distinguished for the purpose of this summary: (1) sub-volcanic deposits of basement provenance; and (2) intra-volcanic sediments embracing a wide range of lithologies but generally lacking basement clasts. The former provide critical evidence of early tectonic events in the region, though they are sparsely fossiliferous and their ages remain speculative. In discussing the latter, emphasis is placed on formations of the Kamasia area, in the deep graben sector of the rift, where a unique tectono-volcanic setting influenced patterns of sedimentation from the early Miocene onwards.

Only the sediments belonging to predominantly waterlaid formations are shown in Fig. 3. Broad distinctions are made between early Miocene, Middle Miocene and late Miocene–Pleistocene units, to illustrate an inward contraction with time of the sedimentary basins. The youngest Quaternary deposits are intimately associated with the volcanics of the inner trough, and they defy representation on the scale of the diagram.

Sub-volcanic sediments

Arkosic sandstones, silts, gravels and conglomerates exposed along the western margin of the Gregory Rift northwards from 1°N are often collectively referred to as Turkana Grits. However, this term (Murray-Hughes 1933) should be reserved for a formation in the Turkana depression where overlying basalts have given Upper Oligocene dates (Bellieni *et al.* 1981). The Turkana Grits of the type area could well be of early Tertiary or late Mesozoic age, whereas their alleged equivalents in the Gregory Rift are probably not older than Miocene.

The restriction of volcanic fragments to the uppermost part of a thick (up to 450 m) clastic wedge on the downthrow side of the Kula fault shows that the main displacements occurred before the onset of volcanism (Weaver 1973). The sediments thin progressively southwards along the margin of a major basement inlier. No natural breaks were found in sequences having 'Turkana Grits' overlain by tuffs, and the base of the early Miocene Kapchererat basaltic formation (17–16 Ma) was arbitrarily defined to include the more tuffaceous members (Webb 1971). At the southern end of the inlier, thin arkosic sandstones occur locally at the base of a succession of tuffaceous mudstones, but the entire sequence was assigned to a 'lower sedimentary member' of the Kapchererat formation (McClenaghan 1971).

The mainly basement-derived sediments of the Kimwarer and Tambach formations exposed in the Elgeyo escarpment were deposited in separate basins on a surface of considerable relief. At the southern end of the escarpment, fluviatile deposits in the Kimwarer sequence pass upwards into basaltic tuffs, which are overlain conformably by basic volcanics of the early Miocene Elgeyo formation (16–15 Ma), the base of which is arbitrarily defined. Further N, the fluviatile and lacustrine Tambach beds are somewhat younger than the Kimwarer sediments, and have thin interbedded phonolitic/trachytic lavas and tuffs. The succession is capped by the Middle Miocene Uasin Gishu phonolites (14–12 Ma). No major scarp existed on the line of the Elgeyo fault at the time of accumulation of the Kimwarer and Tambach sediments. Thus, despite a possible overlap in the ages of sedimentation in the Elgeyo and Kula regions, there were substantial differences in the tectonic environments.

In Kamasia, the Kamego formation (Fig. 5) crops out locally beneath the thick phonolite and sediment sequence of the Tugen Hills Group. It consists of quartzites, sandstones and siltstones resting on an irregular surface in hornblende-gneisses with up to 100 m of topography. These rocks have yielded no macrofossils, but examination of pollen has shown that the formation is not of Karroo or Mesozoic age (Martyn 1969). Perhaps the most significant feature of these enigmatic rocks is the fact that they dip eastwards into the rift, and so are separated by a pronounced angular unconformity from the Tugen Hills Group which dips to the W along the length of the Saimo fault. The attitude of the Kamego suggests that the early 'rift' feature along the Kamasia at this latitude may have been a downwarp similar to that which has persisted on this line to the N of 1°N.

Intra-volcanic sediments

Sediments interbedded with the Tertiary lavas tend to occur in two typical associations, one embracing pyroclastic sediments, and the other waterlaid sediments intercalated with tabular lava flows.

The primary air-fall tuffs, lapilli tuffs, and agglomerates appear in massive accumulations, commonly punctuated by crystalline intrusive bodies of various types. This association is exemplified by: the Middle Miocene Lelgrong tuffs of the Tiati massif (Webb 1971); the Katomuk and Alengerr tuffs of the Middle-late Miocene sequence to the E of Silali volcano (Golden 1978); and the late Miocene–Pliocene trachyte volcanoes such as Ribkwo (Webb & Weaver 1975). All these examples are located to the N of 1°N, in the 'proto-Silali' sector of the rift (Figs. 3 and 4). The association is not widely developed in the southern part of the region, although a southward increase in pyroclastic content in the youngest of the plateau phonolite formations of Kamasia has been noted (Lippard 1973), and tuffs also occur at Kapkut volcano.

Virtually all these pyroclastic sediments are trachytic or phonolitic in composition, and in the past they have frequently been categorized as 'volcanics' and distinguished from the fossilifer-

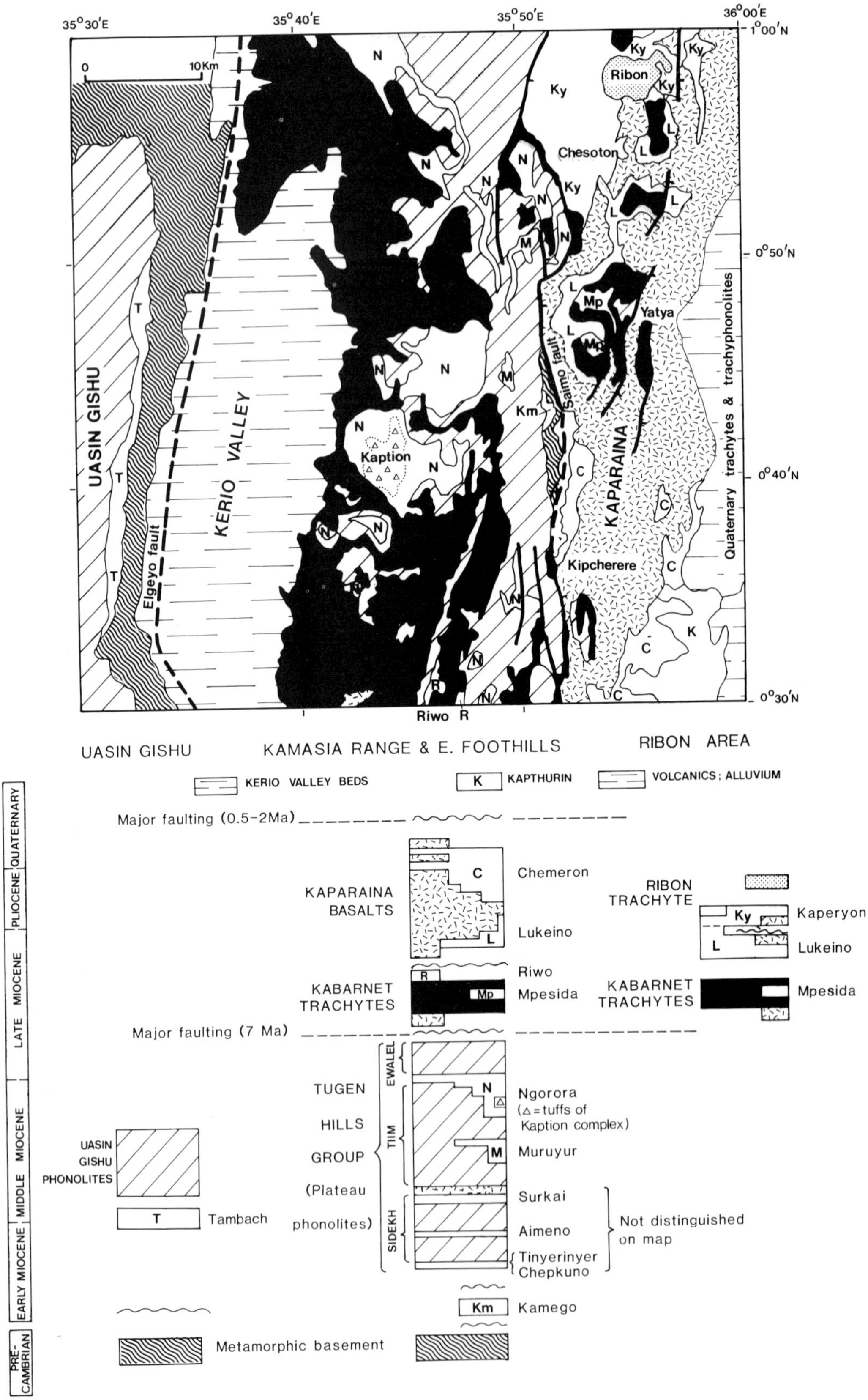

35°30'E
35°40'E
35°50'E
36°00'E
1°00'N
0°50'N
0°40'N
0°30'N
0
10Km
N
Ky
Ribon
Chesoton
L
Mp
Yatya
Saimo fault
Km
M
Kaption
UASIN GISHU
KERIO VALLEY
Elgeyo fault
T
KAPARAINA
Kipcherere
C
K
Quaternary trachytes & trachyphonolites
R
Riwo R
UASIN GISHU
KAMASIA RANGE & E. FOOTHILLS
RIBON AREA
KERIO VALLEY BEDS
KAPTHURIN
VOLCANICS; ALLUVIUM
QUATERNARY
PLIOCENE
LATE MIOCENE
MIDDLE MIOCENE
EARLY MIOCENE
PRE-CAMBRIAN
Major faulting (0.5–2Ma)
KAPARAINA BASALTS
Chemeron
Lukeino
RIBON TRACHYTE
Kaperyon
Lukeino
Riwo
KABARNET TRACHYTES
Mpesida
KABARNET TRACHYTES
Mpesida
Major faulting (7 Ma)
EWALEL
TIIM
SIDEKH
TUGEN HILLS GROUP (Plateau phonolites)
Ngorora
(△ = tuffs of Kaption complex)
Muruyur
UASIN GISHU PHONOLITES
Surkai
Aimeno
Tinyerinyer
Chepkuno
Not distinguished on map
Tambach
Kamego
Metamorphic basement

ous waterlaid sediments of the second association. This distinction is sometimes useful but, apart from the basal beds of certain formations (e.g. Lukeino and Chemeron shown in Fig. 5), the waterlaid sediments contain virtually no lithologies which can be demonstrated to be other than pyroclastic in origin. Even the numerous 'diatomite' horizons appear to have a significant tuffaceous component, though few of these rocks have been studied in detail. Well-sorted conglomerates in the Muruyur and Ngorora formations of Kamasia, once thought to be epiclastic, are more likely to be composed of volcanic blocks transported, rounded and sorted by water. Analogues may be seen in the water-courses draining the Quaternary volcanoes of the inner trough, and study of those young volcanoes is an essential prerequisite for the interpretation of the lithologies of the older waterlaid sediments.

The lava and waterlaid sediment association is typical of the northern part of the Kamasia Range (Fig. 5), where the volcanic components are thick tabular units of the plateau phonolites far removed from source areas, extensive flood trachytes, and fissure basalts. Pyroclastic centres are not totally absent from the Kamasia succession. Examples occur in the Ngorora Formation at Kaption (Bishop & Pickford 1975) and at one locality in the Lukeino Formation (Pickford 1975), but these are local and atypical phenomena in a predominantly waterlaid sequence. The Kaparaina basalts contain a linear eruptive centre in the Kaparaina Range (Martyn 1969), but the volumes of primary pyroclastics are insignificant compared to those of the lavas.

The spectacular development of waterlaid sediments in Kamasia appears to have been caused equally by the absence of major eruptive centres and the existence of tectono-geomorphic sediment traps. The inner trough of the rift supplies, and probably not by coincidence, an exact parallel in the volcanic 'quiet zone' containing Lake Baringo that lies to the S of a chain of Quaternary volcanoes.

Individual waterlaid formations are described in detail elsewhere in this volume, and only a brief outline is provided here. Relationships between sedimentary and volcanic units of the Kamasia area are summarized in Fig. 5.

The sediments (*c.* 16–14 Ma) intercalated in the lower part of the Tugen Hills Group are possibly the equivalents of the Tambach beds of the Elgeyo escarpment, showing as they do a transition upwards from arenaceous lithologies with basement clasts to thin-bedded argillites.

The Muruyur (*c.* 13–12 Ma) and Ngorora formations (Bishop & Chapman 1970; Bishop *et al.* 1971; Bishop & Pickford 1975) are similar in their range of constituent lithologies, but the former is more restricted both in facies and in present (exposed) extent. Both formations contain a large component of coarse reworked pyroclastic material. The few sedimentological studies carried out so far on the Ngorora suggest that sediment transport in conglomerates was from directions between N and E. To the N stands the Tiati massif (early-Middle Miocene volcanics), but no attempt has yet been made to match Ngorora pebbles with rocks cropping out in that area. Petroliferous calcarenites have been observed in the Ngorora, S of Kaption (Martyn 1969; Chapman 1971).

The Ngorora shows better than any other unit a close correlation between sediment thickness and tectonic relief, and it is clear that the pattern of sedimentation was being affected by the growing Kamasia faults. The formation thickens steadily westwards from the crest of Kamasia into the Kerio valley, and it is likely that part of the strong negative gravity anomaly in that area (Swain *et al.* 1981) is caused by a very great thickness of Ngorora and older sediments now concealed on the downthrow side of the Elgeyo fault.

The Mpesida beds (*c.* 7–6.5 Ma) of the Kabarnet trachytes formation are almost entirely restricted to the E of the Kamasia faults. Despite the intervention of the first major faulting episode at 7 Ma, with its consequent pronounced unconformity beneath the Kabarnet trachytes, the Mpesida beds more closely resemble the Ngorora and the Muruyur than the other, younger, sediments of the eastern foothills. The sources of the Mpesida are not known, but a local area of primary air-fall pumice tuff occurs near Yatya. The Riwo beds (Martyn 1969) of the Kabarnet, which outcrop at latitude 0°30′N, are lithologically similar to the Mpesida beds.

The lithological contrast between the Mpesida beds and the Lukeino Formation (Bishop *et al.* 1971; Pickford 1975) is due chiefly to the change in the chemistry of the associated eruptive rocks. The Lukeino, above its basal horizons of trachytic weathering products (Martyn 1969), is strongly influenced by the initial (and possibly later) eruptions of the Kaparaina basalts. The

FIG. 5. (Opposite page.) Map of the northern Kamasia Range (Tugen Hills) showing the main volcanic and waterlaid sedimentary formations.

Lukeino has usually been regarded as underlying the Kaparaina basalts, but the present authors suspect a degree of coeval accumulation of the two units, following Martyn (1969) in this respect.

The Chemeron formation (Martyn 1967; McCall *et al.* 1967; Bishop *et al.* 1971) overlies the Kaparaina. Above the epiclastic and palaeontologically contentious basal horizons, Chemeron lithologies suggest a renewal of sediment (tuff) contribution from trachytic eruptions, but no obvious source can be identified. The late Miocene–Pliocene trachyte strato-volcanoes to the N are in general slightly older than the bulk of the Chemeron, which seems to have continued to accumulate until *c.* 2 Ma (Chapman & Brook 1978).

A unit which unquestionably originated in Ribkwo strato-volcano (Fig. 3) is the upper ('yellow tuff') member of the Kaperyon formation (Bishop *et al.* 1971; Chapman 1971; Chapman & Brook 1978). The Kaperyon has been dismembered by M. Pickford (unpub. data), who has reallocated its parts to the Lukeino, below, and the Chemeron, above. The latter designation depends on fossil evidence, and a single marker horizon that can perhaps be traced into the Kipcherere outcrops of the Chemeron. Since there is a degree of uncertainty (Martyn 1969) about the relationship of these outcrops to the Chemeron of the type area, the present authors do not favour the use of the name 'Chemeron' for the erstwhile upper Kaperyon, and suggest that further work is required to clarify the sedimentation history of the northern Kamasia scarp-foot.

During the Quaternary, a major episode of faulting caused a fundamental shift in the sedimentation pattern of this section of the rift. The Kamasia scarp-foot basins of the Lukeino, Chemeron and Kaperyon were relatively strongly deformed, and ceased to be sediment traps as the main Kamasia faults moved. On the opposite side of the Kaparaina arch, the Chemeron of the eastern outcrops was tilted eastwards towards the central trough, an effect also shown by the eastern flank of Ribkwo volcano. The eastern margin of the central trough became defined by a series of *en echelon* antithetic faults, and the regional centre of accumulation shifted emphatically to its present position. The later Chemeron of the eastern outcrops, particularly if linked (Pickford *et al.* 1983) with the Chesowanja sediments E of Lake Baringo (Bishop *et al.* 1978), may have anticipated this pattern in the late Pliocene.

Rejuvenated relief on the Kamasia Range resulted in rapid erosion and deposition on the rift floor of the Kapthurin formation (McCall *et al.* 1967). This is the first formation since those at the base of the Tugen Hills Group to contain evidence of basement erosion, testifying to the magnitude of the Quaternary movements on the Saimo fault. The Kapthurin resembles none of the earlier sediment formations, and represents the beginning of a pattern of erosion and sedimentation that continues to the present. Deeper and totally concealed horizons of 'alluvium' in the Kerio valley are almost certainly as old as the Kapthurin. The latter should perhaps most logically be regarded as 'early alluvium', separated as it is by the consequences of relatively minor tectonic events from the present alluvium of the Baringo shore.

Conclusions

A deep graben sector of the rift (0°15′–1°00′N) passes northwards into what is essentially a downwarped trough, and southwards into a half-graben faulted on the eastern side. The deep graben displays marked asymmetry resulting from exceptionally large displacements on two fault systems in the W, and a more dispersed pattern of *en echelon* faulting to the E. It was first clearly defined by major faulting in the late Miocene (7 Ma), and the early rift structure in this region was evidently a downwarp similar to that which persisted to the N.

Late Miocene–Quaternary trachytes and trachyphonolites provide the most convincing examples of relationships between structural setting and the style of salic volcanism. Trachytic central volcanoes tended to develop along sections of the rift defined by major flexures, whereas areas of strong faulting (the deep graben and the eastern side of the equatorial half-graben) were characterized by eruptions of salic flood lavas.

The transition zone from deep graben to the northern downwarped trough remained a long-active site of pyroclastic activity, both phonolitic and trachytic, from early Miocene times onwards. This 'proto-Silali' region was undoubtedly one of the major source areas for the voluminous early–late Miocene plateau phonolites, but the precise distribution of eruptive centres elsewhere remains controversial. Nevertheless, there is general agreement that the thick phonolite sequence of Kamasia, on the western side of the present deep graben, is made up of tabular lava units some distance from source. Overtopping of the western rift margin by Middle Miocene phonolites is undisputed, but corresponding overspill on the eastern side of the rift, though commonly envisaged, has never been conclusively demonstrated. Currently available evidence is more consistent with the eruption of at least some Middle-late Miocene phonolites on the eastern plateau, rather than wholesale derivation from within-rift sources.

Two associations are recognized in Tertiary intra-volcanic sediments. Pyroclastic sediments occur mainly N of 1°N, whereas waterlaid deposits intercalated with tabular lava flows are spectacular features of Kamasia, reflecting the absence of major eruptive centres and the existence of tectono-geomorphic sediment traps in the deep graben sector of the rift.

One factor controlling the character of both salic volcanism and sedimentation could well have been the rate of extension of different structural regimes.

ACKNOWLEDGEMENTS: Much of the information presented in this paper comes from the records of the East African Geological Research Unit which was supported by the Overseas Development Administration, the Natural Environment Research Council, and the Kenyan Ministry of Natural Resources. The authors are indebted to the late Professor B.C. King and all members of the unit for advice and stimulating discussions over the years. Assistance from the team led by the late Professor W.W. Bishop is gratefully acknowledged. The authors are grateful also to Professor B.H. Baker for his helpful comments on this paper, and to Jane Rushton who drew the figures. GRC publishes by permission of the Director, British Geological Survey.

References

BAKER, B.H. 1963. Geology of the Baragoi area. *Rep. Geol. Surv. Kenya* **53,** 74.

——, WILLIAMS, L.A.J., MILLER, J.A. & FITCH, F.J. 1971. Sequence and geochronology of the Kenya rift volcanics. *Tectonophysics* **11,** 191–215.

BELLIENI, G., JUSTIN VISENTIN, E., ZANETTIN, B., PICCIRILLO, E.M., RADICATI, D.I., BROZOLO, F. & RITA, F. 1981. Oligocene transitional tholeiitic magmatism in northern Turkana (Kenya): comparison with coeval Ethiopian volcanism. *Bull. Volc.* **44,** 411–27.

BISHOP, W.W. & CHAPMAN, G.R. 1970. Early Pliocene sediments and fossils from the northern Kenya Rift Valley. *Nature* **226,** 914–18.

—— & PICKFORD, M.H. 1975. Geology, fauna and palaeoenvironments of the Ngorora Formation, Kenya Rift Valley. *Nature* **254,** 185–92.

——, CHAPMAN, G.R., HILL, A. & MILLER, J.A. 1971. Succession of Cainozoic vertebrate assemblages from the northern Kenya Rift Valley. *Nature* **233,** 389–94.

——, HILL, A. & PICKFORD, M. 1978. Chesowanja: a revised geological interpretation. In: BISHOP, W.W. (ed) *Geological Background to Fossil Man.* Scottish Academic Press, Edinburgh, pp. 309–27.

CARNEY, J.N. 1972. *The geology of the area to the east of Lake Baringo, Rift Valley Province, Kenya.* Unpublished PhD thesis, University of London.

CHAPMAN, G.R. 1971. *The geological evolution of the northern Kamasia Hills, Baringo District, Kenya.* Unpublished PhD thesis, University of London.

—— & BROOK, M. 1978. Chronostratigraphy of the Baringo Basin, Kenya. In: BISHOP, W.W. (ed) *Geological Background to Fossil Man.* Scottish Academic Press, Edinburgh, pp. 207–33.

——, LIPPARD, S.J. & MARTYN, J.E. 1978. The stratigraphy and structure of the Kamasia Range, Kenya Rift Valley. *J. Geol. Soc.* **135,** 265–81.

DODSON, R.G. 1963. Geology of the South Horr area. *Rep. Geol. Surv. Kenya* **60,** 53.

GOLDEN, M. 1978. *The geology of the area east of Silali, Rift Valley Province, Kenya.* Unpublished PhD thesis, University of London.

GRIFFITHS, P.S. 1977. *The geology of the area around Lake Hannington and the Perkerra River, Rift Valley Province, Kenya.* Unpublished PhD thesis, University of London.

JONES, W.B. & LIPPARD, S.J. 1979. New age determinations and the geology of the Kenya Rift–Kavirondo Rift junction, W. Kenya. *J. Geol. Soc.* **136,** 693–704.

KING, B.C. 1978. Structural and volcanic evolution of the Gregory Rift Valley. In: BISHOP, W.W. (ed) *Geological Background to Fossil Man.* Scottish Academic Press, Edinburgh, pp. 29–54.

LIPPARD, S.J. 1973. The petrology of phonolites from the Kenya Rift. *Lithos* **6,** 217–34.

MARTYN, J.E. 1967. Pleistocene deposits and new fossil localities in Kenya. *Nature* **215,** 476–80.

1969. *The geological history of the country between Lake Baringo and the Kerio River, Baringo District, Kenya.* Unpublished PhD thesis, University of London.

MCCALL, G.J.H. 1967. Geology of the Nakuru–Thomson's Falls–Lake Hannington area. *Rep. Geol. Surv. Kenya* **78,** 121.

——, BAKER, B.H. & WALSH, J. 1967. Late Tertiary and Quaternary sediments of the Kenya Rift Valley. In: BISHOP, W.W. & CLARK, J.D. (eds) *Background to Evolution in Africa.* University of Chicago Press, pp. 191–220.

MCCLENAGHAN, M.P. 1971. *The geology of the Ribkwo area, Baringo District, Kenya.* Unpublished PhD thesis, University of London.

MURRAY-HUGHES, R. 1933. Notes on the geological succession, tectonics and economic geology of the western half of Kenya Colony. *Rep. Geol. Surv. Kenya* **3,** 8.

PICKFORD, M. 1975. Late Miocene sediments and fossils from the northern Kenya Rift Valley. *Nature* **256,** 279–84.

——, JOHANSON, D.C., LOVEJOY, C.D., WHITE, T.D. & ARONSON, J.L. 1983. A hominoid humeral fragment from the Pliocene of Kenya. *Am. J. Phys. Anthrop.* **60,** 337–46.

RHEMTULLA, S. 1970. The South Turkana expedition

scientific papers III. A geological reconnaissance of South Turkana. *Geogr. J.* **136,** 61–73.

SHACKLETON, R.M. 1946. Geology of the country between Nanyuki and Maralal. *Rep. Geol. Surv. Kenya* **11,** 54.

SWAIN, C.J., KHAN, M.A., WILTON, T.J., MAGUIRE, P.K.H. & GRIFFITHS, D.H. 1981. Seismic and gravity surveys in the Lake Baringo–Tungen Hills area, Kenya Rift Valley. *J. Geol. Soc.* **138,** 93–102.

TRUCKLE, P.H. 1977. *The geology of the area to the south of Lokori, South Turkana, Kenya.* Unpublished PhD thesis, University of Leicester.

WEAVER, S.D. 1973. *The geology of the Nasaken area of the Rift Valley, South Turkana, Kenya.* Unpublished PhD thesis, University of London.

WEBB, P.K. 1971. *The geology of the Tiati Hills, Rift Valley Province, Kenya.* Unpublished PhD thesis, University of London.

—— & WEAVER, S.D. 1975. Trachyte shield volcanoes: a new volcanic form from South Turkana, Kenya. *Bull. Volc.* **39,** 1–19.

WILLIAMS, L.A.J. 1970. The volcanics of the Gregory Rift Valley, East Africa. *Bull. Volc.* **34,** 439–65.

——, MACDONALD, R. & CHAPMAN, G.R. 1984. Late Quaternary caldera volcanoes of the Kenya Rift Valley. *J. Geophys. Res.* **89,** 8553–70.

——, —— & LEAT, P.T. 1983. Magmatic and structural evolution of the central part of the Kenya Rift. In: *Proceedings of the Regional Seminar on Geothermal Energy in Eastern and Southern Africa.* UNESCO/USAID, Nairobi, pp. 61–7.

—— & TRUCKLE, P.H. 1980. Volcanic sequences in the Kenya rift. In: *Geodynamic Evolution of the Afro-Arabic Rift System.* Acad Nazionale dei Lincei, Rome, pp. 123–32.

L.A.J. WILLIAMS, Dept of Environmental Sciences, University of Lancaster, Lancaster LA1 4YQ, UK.

G.R. CHAPMAN, British Geological Survey, 154 Clerkenwell Road, London EC1R 5DU, UK.

Depth to metamorphic basement in the Koobi Fora region from seismic and gravity data

C.J. Swain, N.J. Skinner & M.A. Khan

SUMMARY: A detailed 2.3 km long seismic refraction line near Koobi Fora shows velocities increasing continuously with depth, apart from a 100 m thick low-velocity layer whose top lies at a depth varying from 100 to 250 m. This is thought to be a lacustrine sequence. Total thickness of rocks overlying basement is 1200–1300 m. Interpretations of four shorter profiles in this area are also presented, one of which indicates basement at only 40 m depth. These data are used to control the interpretation of part of the Bouguer gravity map of the region in terms of an upfaulted basement block partly coinciding with the Kokoi horst.

During 1978, seismic refraction profiles were shot at five locations near Koobi Fora (Fig. 1) to investigate the Plio-Pleistocene sediments and volcanics. This area has been intensively studied by a multidisciplinary team in the Koobi Fora Research Project led by R.E.F. Leakey and G.Ll. Isaac (Coppens *et al.* 1976) and has been a rich source of vertebrate fossils, including hominids, and archaeological artefacts. Specific objectives of the seismic experiment were to establish depths of basement and volcanic horizons in order to provide a control for interpreting the gravity data (collected by the University of Leicester in 1977). In addition seismic velocities can be used to identify sedimentary formations.

Field procedure

The seismic investigations were carried out using a 12-channel seismograph with paper recorder and 10 Hz vertical geophones with a 15 m separation, giving spread lengths of 165 m. Shots of up to a few kg of gelignite were fired in augered holes 1–3 m deep and, where possible, in the lake. Adjacent spreads were laid out end to end (with a common geophone) to make longer profiles. Shots were fired at both ends of the profiles and off-end (in line) to probe deeper. Shots were also fired into individual 165 m spreads to give shallow information.

The gravity measurements were made with a LaCoste Romberg gravimeter G-16, while heights were measured by barometric leap-frog traversing with an estimated accuracy of ± 3 m relative to lake level in July 1977 (364 m above sea-level). Stations were established along tracks at 1 km intervals and were marked on the 1970 air photos. The positions were established on 1:100,000 maps after they had been transferred from the 1970 photos to the old photos on which these maps were based. This gave an estimated relative accuracy of ± 100 m, but an absolute accuracy no better than ± 500 m. Terrain corrections are negligible for most stations. Bouguer anomalies are estimated to be accurate to ± 1 mGal, and the Bouger anomaly map is shown in Fig. 2.

Seismic results

The records obtained at sites 1 to 4 (Fig. 1) could be interpreted adequately by standard methods, based on models consisting of homogeneous layers separated by plane (or slightly irregular) interfaces. However, some of the records (Fig. 3 being typical) showed abnormally rapid attenuation of the first arrivals, which is characteristic of refractions from a thin embedded high-speed layer such as a basalt flow. Such flows are seen in the areas of exposed Cenozoic volcanics (Fig. 1). The thickness of a thin high-speed layer can be estimated roughly from the rate of attenuation (Schmoller 1982). The layer below the high speed one is a low-velocity layer (LVL) which does not give first arrivals and therefore its velocity must be assumed.

The site 5 data indicated a continuous velocity—depth variation as well as a LVL. Their interpretation required techniques more common in crustal seismology than in small scale work. Basement refractions were observed by firing a series of off-end shots into a single geophone spread: cross-over points were accurately located, however, thereby allowing reliable depth estimates to be made.

Seismic interpretation

The interpretations of most of the data from the first four sites were reasonably straightforward and are summarized in Fig. 4. Attention is drawn

From FROSTICK, L.E. *et al.* (eds) 1986, *Sedimentation in the African Rifts*, Geological Society Special Publication No. 25, pp. 75–84.

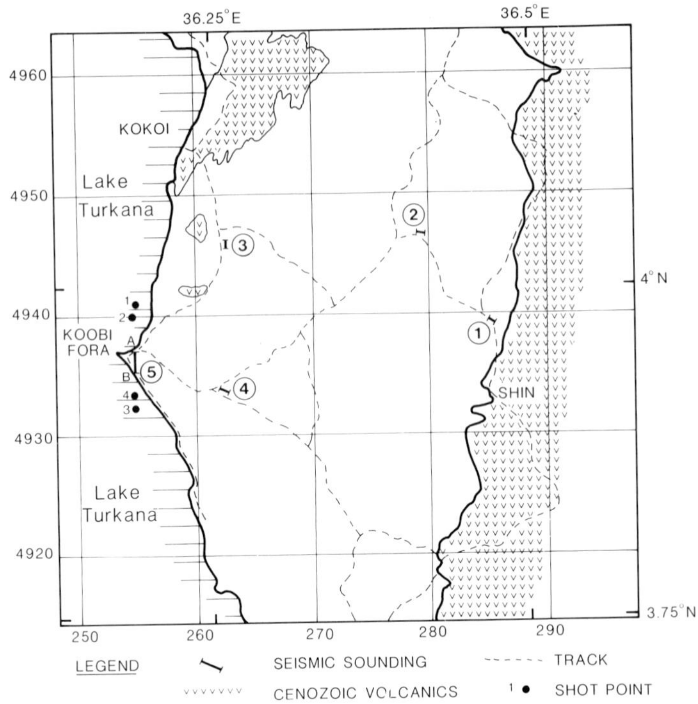

FIG. 1. Map of the Koobi Fora region showing sites of seismic profiles. Both grid and geographic co-ordinates are given, the former being 10 km squares of the East African grid used on the Survey of Kenya maps.

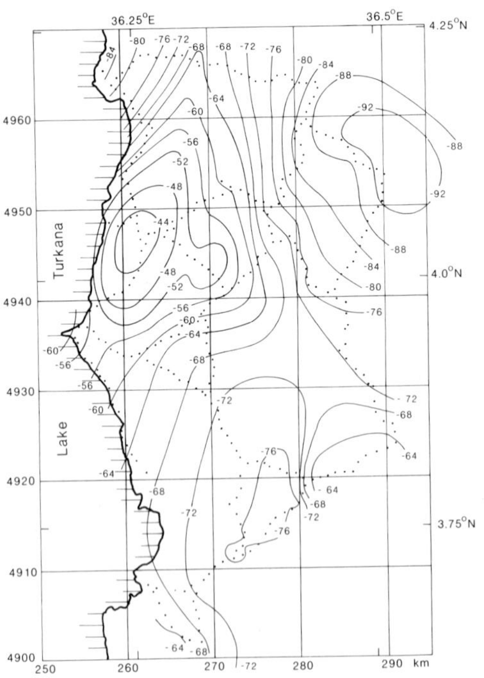

FIG. 2. Bouguer anomaly map of the Koobi Fora region, showing gravity stations and contours in mGal.

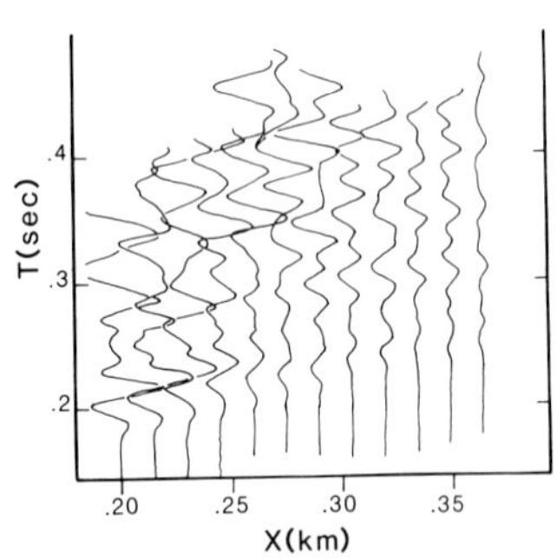

FIG. 3. Record section for site 1, showing the rapid attenuation of first arrivals from the supposed lava refractor.

to the thin high-velocity layers at sites 1 and 2 which are almost certainly basalt flows. Site 3 is important as it appears to have shallow metamorphic basement. Not only is the velocity typical of basement but the first arrival amplitudes decrease as the reciprocal of the square of the shot–geophone distance (Fig. 5), implying head waves from a thick homogeneous non-attenuating layer (O'Brien 1967). This contrasts with the refracted arrivals obtained at site 1 (Fig. 3) and site 2. The

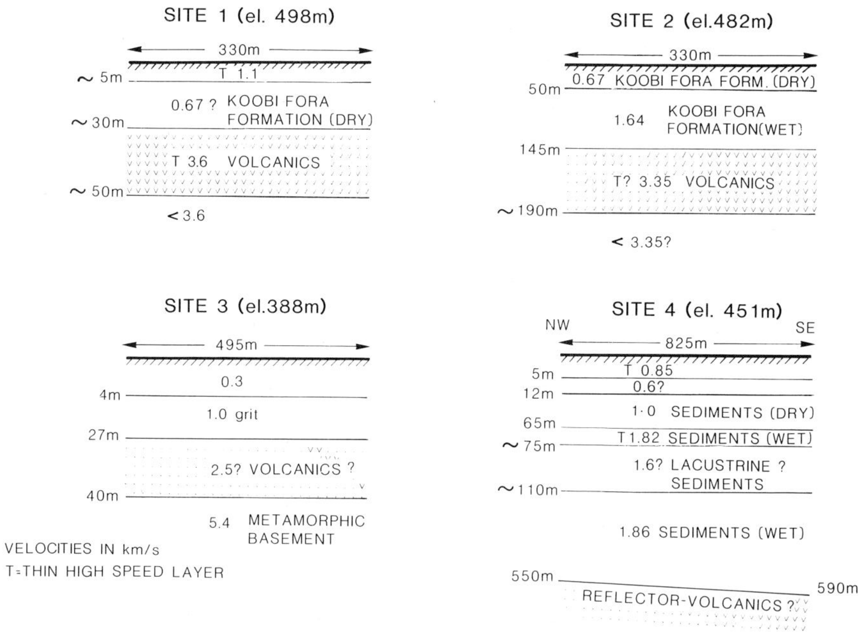

FIG. 4. Interpretations of seismic soundings at sites 1–4.

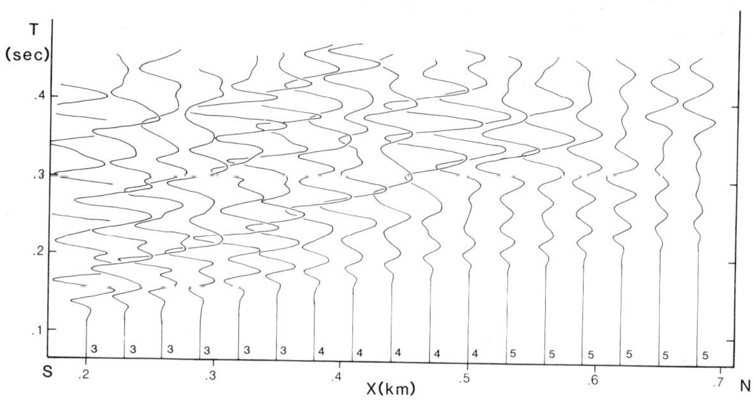

FIG. 5. Record section for site 3. Numbers by traces are relative charge sizes. Compare the attenuation rate of the first arrivals with that for site 1 (Fig. 3).

third layer is a 'hidden layer', too thin (relative to its depth) to give first arrivals.

The interpretation of site 4 is tentative. There seem to be at least two thin high-speed layers, which prevent accurate interpretation. However, there are two aspects that are not affected by details of the layering. Firstly, no high-velocity early arrivals, which could be basement refractions, were seen, even for a 2 km off-end shot (which gave phase velocities near 2 km s^{-1}) so the minimum depth to a 5.5 km s^{-1} horizontal refractor is 800 m. Secondly, a persistant reflector was observed from both end shots and interpreted as a plane dipping interface. A T^2/X^2 method gave a mean velocity of 1.77 km s^{-1} above the reflector, which agrees well with the velocities given in Fig. 4.

As noted above, the site 5 results were quite interesting, so their interpretation will be described in more detail. Fig. 6 is a reduced time

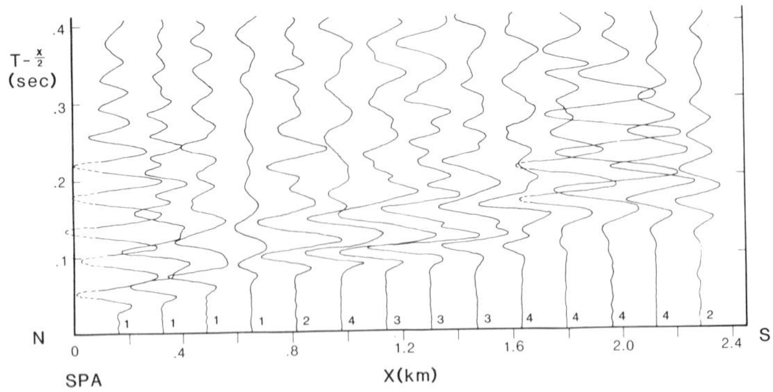

FIG. 6. Record section for site 5, with reduced travel times, for SP A. Numbers by traces are relative charge sizes. Note the large amplitude first arrivals in the range 0.8–1.3 km, their rapid attenuation beyond 1.3 km, and the ringing at X=0.15 and 1.95 km from SP A.

section for shot point (SP) A showing every twelfth recorded trace. Delays in the near surface 0.4 km s^{-1} layer have been removed. Note the large-amplitude first arrival from 0.8 to 1.3 km and its rapid demise (4 db/wavelength) beyond 1.3 km. SP B gave broadly similar results though the amplitude peak occurs at shorter offsets (see Fig. 8). The large amplitudes can best be explained as focusing of seismic energy by predominantly vertical velocity gradients, and the rapid attenuation as a shadow zone due to a velocity inversion (i.e. a LVL). Note also the 'ringing' that occurs throughout the profile, but most strongly near X=0.15 and 1.95 km from SP A. We think the ringing is due to first order multiple reflections of the upcoming waves between the base of the 0.4 km s^{-1} layer and the free surface (O'Brien 1957). Since the thicknesses of this (3–8 m) are known from the delay times, they were used to calculate the dominant frequencies of such multiples, and the results agreed well with observations. Quantitative analyses of both travel times and amplitudes of first arrivals for SP A and SP B were carried out.

Travel times

The travel time graphs (Fig. 7) could be approximated by straight line segments, but such an interpretation, implying head waves and homo-

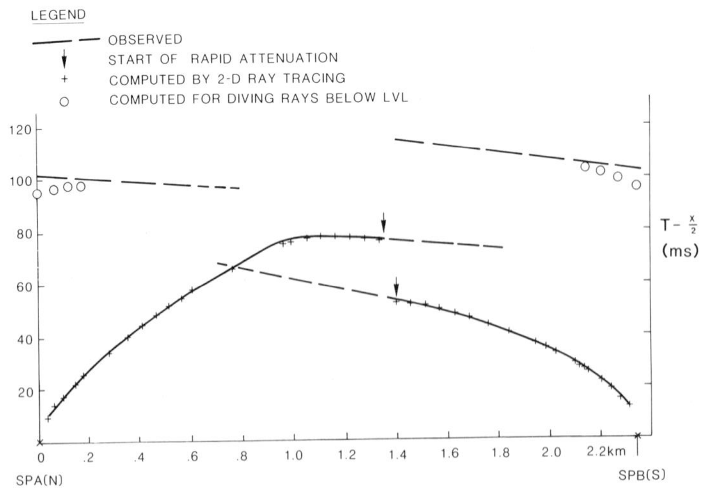

FIG. 7. Reduced time–distance plots for first arrivals at site 5. Smooth curves have been fitted to the corrected data and V(z) calculated by suitable integration. Crosses (and circles) were computed by ray tracing through a 2-D model. Circles are for a strong velocity gradient below the LVL (see text).

geneous layers, conflicts with the amplitude data. Therefore smooth curves were fitted to the data and the vertical velocity distribution, V(z), calculated from each curve using the WHB integral (Grant & West 1965). The two shot points do not yield identical V(z), so V must vary (slightly) with x as well as with z. However, as Kennett (1976a) pointed out, a one dimensional (1-D) velocity model derived in the presence of small lateral velocity variations applies approximately along the locus of turning points of the rays, so a 2-D model was constructed by plotting the two 1-D sections along these loci and assuming straight isovelocity lines. Travel times were calculated for this model using a simple ray tracing program (Will 1976). A fit within 1 ms was achieved by steepening the dips slightly. A LVL was added and its top adjusted to match the observed sharp cut-off of the first arrivals at 1.35 km (SP A) and 0.95 km (SP B). The resulting model is incorporated in Fig. 10. By adding a base to the LVL and a velocity gradient below it, an attempt was made to match the arrival times near the ends of the profile (*c.* 100 ms reduced time: Figs. 6 and 7). A thickness of *c.* 100 m, for a velocity of 1.9 km s^{-1}, was required for the LVL. However, as a strong gradient (1s^{-1}) was required to produce diving ray arrivals before the end of the profile (whereas the observed arrivals have a nearly constant apparent velocity near 2 km s^{-1}) these arrivals are more likely to be interference head waves from just below the lower boundary of the LVL, which do not require a strong gradient (Kennett 1977). Such arrivals cannot be modelled by ray tracing.

Amplitudes

Theoretical amplitudes of first arrivals from the 1-D models derived (as above) for SP A and B were calculated using a reflectivity program (Fuchs & Muller 1971; Kennett 1976b). The results, shown in Fig. 8, are in broad agreement with the observations, and discrepancies can be ascribed to lateral variations and anelastic attenuation. The main purpose of this exercise was to estimate the velocity (1.9 km s^{-1}) of the LVL, which cannot be deduced from travel time data but which governs the rate of attenuation in the shadow zones.

Reflections

A reflection, similar to the one at site 4, was observed over a range of > 700 m and interpreted in terms of a plane dipping interface (see Fig. 10). The amplitudes are quite different for the two shot points, which could perhaps be explained by a basalt flow of small but variable thickness.

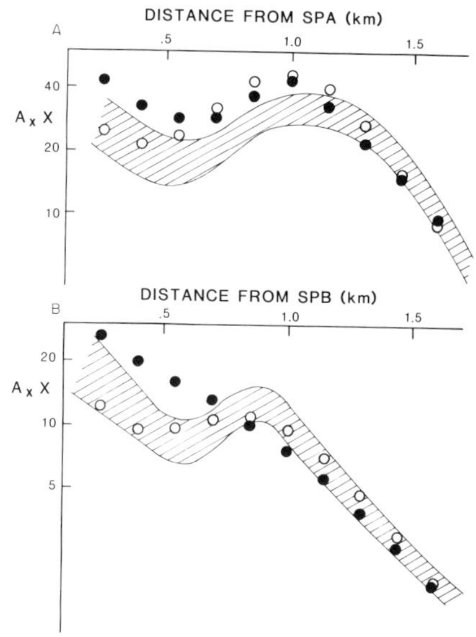

FIG. 8. Normalized amplitude vs. distance for SP A and SP B (site 5). The shaded area contains 95% of the data. Open circles are for perfectly elastic propagation and black circles for Q = 100.

Basement refractions

As no high-speed early arrivals were observed from the end shots, a series of untimed off-end shots was fired, in the lake, into the northernmost 165 m geophone spread (SP 1–4, Fig. 1). Weak first arrivals with apparent velocities near 5 km s^{-1} were observed from SP 1, 3 and 4, followed closely by a strong second event and then several further strong arrivals all with phase velocities of 2–2.5 km s^{-1} (Fig. 9). Because the first arrivals were so weak in these cases, relative to the second arrivals, these shots were repeated with ten times the charge, giving first arrivals that could be accurately timed, but overloading later arrivals. The first arrival has almost certainly been refracted by the basement and the second is the diving wave from the overlying sequence. The strong later arrivals are probably multiples or bubble pulses resulting from shooting in deep water.

The amplitudes of the first arrivals from SP 1 were about 1/7 those of the second arrivals, compared to a ratio of 1/20 for SP 3 and 4. Positions of the cross-over points were determined to be 250 m S of SP A for SP 4 and 150 m N of SP A for SP 1, but the resulting cross-over distance estimates (3.35 km from SP 4, 3.75 km from SP 1) are subject to the ±200 m uncertainty in the shot positions.

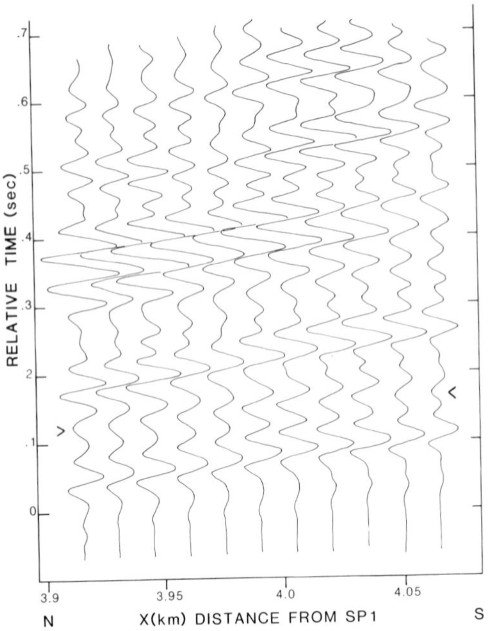

FIG. 9. Record section for SP 1, (site 5). Note the small high-speed first arrival. This shot was repeated with 10 × charge size. Note also the possible basement reflection (marked ˆ) and the strong later events that are probably related to shooting in moderately deep water.

The average cross-over distance and observed mean second arrival velocity (2.2 km s^{-1}) can be modelled by a 1-D velocity variation that increases uniformly from 2 km s^{-1} at 300 m depth to 2.6 km s^{-1} at 1300 m where it increases abruptly to 5 km s^{-1}. This velocity variation is similar to Faust's (see Dobrin 1976) for Cenozoic sediments where compaction is the major factor governing V(z). If a uniform dip is allowed then the observations indicate depths of 1150 m near SP 4 and 1450 m near SP 1. Synthetic seismograms generated from the 1-D model reproduce the relative amplitudes of the first and second arrivals observed for SP 1. They also show a very strong post-critical reflection about 0.2 s after the first arrival. There is in fact an arrival of about the expected apparent velocity 0.15 s after the first arrival on some of the records (e.g. Fig. 9), but it is rather overwhelmed by the strong water-related events mentioned. A possible explanation for the relative weakness of the reflected energy is that it is scattered by rough basement topography, perhaps due to faulting.

A 2-D velocity section for site 5, including the features discussed above, some of which are fairly tentative, is given in Fig. 10.

Geological interpretation of the seismic models

The identification of the high velocity (5–5.4 km s^{-1}) refractors at sites 3 and 5 as metamorphic basement rocks (rather than volcanics) seems fairly clear. Sites 1 and 2, close to the basin margin where the thick Cenozoic volcanic sequence of basalts and ignimbrites is exposed, gave refractor velocities of 3.35 and 3.6 km s^{-1} and the attenuation characteristic of lava flows. Such velocities

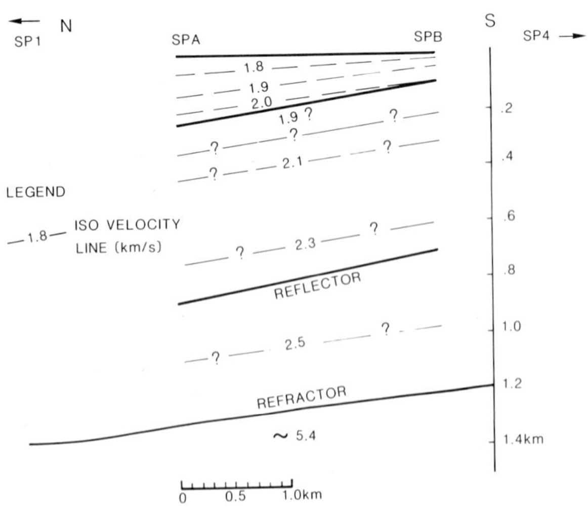

FIG. 10. Interpreted 2-D velocity section for site 5.

are typical for *in situ* measurements of Cenozoic extrusive rocks from Kenya and elsewhere (e.g. Swain *et al.* 1981; Pankratz & Ackermann 1982), though higher values occur. For example, a (previously unreported) value of 5.5 km s^{-1} was obtained for a Uasin Gishu phonolite flow, but the arrivals also attenuated rapidly. Allunuation of the arrivals is perhaps the more reliable characteristic of such flows.

At site 3 only a thin layer of possible volcanic rock is present (whose velocity was not well determined) although there are volcanic outcrops nearby (Fig. 1). At sites 4 and 5 the prominent reflections may well be from volcanic layers which are presumably quite thin as they do not seem to increase the mean velocities of the sequence above basement, to judge by the phase velocities (2–2.5 km s^{-1}) observed at the largest offsets.

It is well established that velocity generally increases with grain size in marine sediments (e.g. Hamdi & Taylor Smith 1982), so the LVL at site 5 can probably be identified with one of the lacustrine sequences of siltstones and claystones that are described by Vondra & Bowen (1978) and summarized by Brown *et al.* (1978). One of these occurs just below the KBS Tuff with a thickness of > 100 m (Cerling & Brown 1982), and another consists of the Guomde Formation with a mapped thickness of > 37 m, although the latter has not apparently been mapped near Koobi Fora. In view of the dip inferred from the seismic (Fig. 10) and gravity (Fig. 11) data and the position of the Koobi Fora Tuff outcrop 1 km SE of SP B (Fig. 11), it seems likely that this LVL lies above the Koobi Fora Tuff and hence is the Guomde.

Extrapolation of basement depths using gravity data

The gravity dataset (Fig. 2) is unfortunately rather sparse, as measurements were confined to the motorable tracks; quantitative interpretation has been attempted of the better covered area, NE of Koobi Fora. The Bouguer anomaly map shows large variations over a region that is topographically quite flat: these variations must be caused either by lateral changes of density or by changes in depth to a density contrast, or both. We consider the principal cause to be variations in basement depth: not only is there an obvious correlation between the gravity data and the seismic results at sites 3 and 5, but aeromagnetic data from a recent survey indicated that the gravity low in the NE corner of the map (Fig. 2) marks the end of a palaeo-rift, beneath the

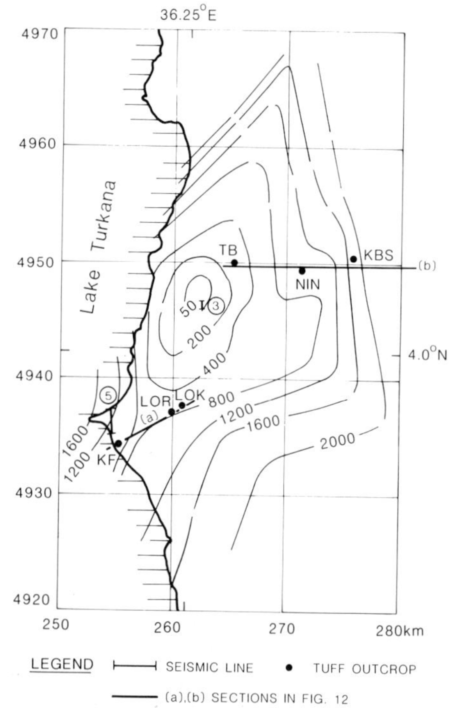

FIG. 11. Basement depth in m below lake level (364 m a.s.l. in 1977), based on the gravity data and assuming a uniform density contrast.

Cenozoic volcanics, which depresses the metamorphic basement to a depth of several thousand metres. However, we do not feel it is justified to extrapolate basement depths too far from the seismic control, so we have limited the interpretation to the area shown in Fig. 11, and even this may be extrapolating too far.

In making this interpretation we have assumed a single interface between two homogeneous media so that there would be a unique solution for the configuration of the basement surface, and for the density contrast, if the gravity field were known accurately everywhere. The assumption of a single interface is, of course, geologically a gross simplification: we know, for example, that volcanic rocks, probably with a higher density than sediments, outcrop in the Kokoi and may well occur within or beneath the sedimentary sequence elsewhere. However, the seismic evidence is that, with the exception of some thin layers of higher velocity (probably lava flows), velocities within the supra-basement rocks do not cover a wide range. Thus we feel that a model incorporating just one density contrast is geophysically justified, particularly in view of the sparseness of both the gravity data and the seismic control. An additional point in favour of the model is that although only two control depths were used in the inversion of the gravity data, namely at seismic site 3 and at the N end of the long profile (site 5), the depth predicted by the gravity model at the S end (Fig. 11) agrees with the seismic interpretation (Fig. 10).

The model shown in Fig. 11 satisfies the gravity data above the −70 mGal contour (Fig. 2) with an RMS error of *c.* 1.5 mGal, for a density contrast of 0.40 g cm^{-3}. If the basement density is assumed to be 2.6 g cm^{-3} then the average sediment density obtained is 2.2 g cm^{-3}, which is typical for rocks of this type and age. Taking a background of −70 mGal and a model base at a depth of 2000 m means that basement topography outside the 2000 m contour was ignored, and this may lead to some error in the model where the dips are >10° or so, i.e. around the N and NE edges. The model shown is based on polygonal contours whose vertices were joined to make a solid 3-D body, with triangular facets, for which Okabe's (1979) method was used to calculate the exact gravity anomaly.

Geological interpretation of the gravity model

Some 65 km S of Koobi Fora there occurs a rather larger gravity maximum over the eroded syenite plug of El Moiti mountain (Swain 1976). This anomaly may be caused by a basic magma chamber from which the syenite (and associated trachytes) differentiated. It is associated with a major magnetic anomaly, whereas the Koobi Fora gravity anomaly has only a very slight magnetic expression. It is unlikely that this latter anomaly has an intrusive source.

Comparison of Figs. 1 and 2 shows that there is no correlation between the Bouguer anomaly and the outline of the Cenozoic sedimentary basin. This is presumably because average densities for volcanics and sediments hardly differ. There is a general correlation between the gravity high (Fig. 2) and the Kokoi horst (Fig. 1), although the former culminates S of the Kokoi near subsidiary volcanic outcrops. There is also a tendency for the older sedimentary formations to outcrop in this area. Although at present the geological maps of the area (Findlater 1978; Vondra & Bowen 1978) are unreliable because of erroneous correlations between the tuff outcrops used as marker horizons, Cerling & Brown (1982) have presented a revised stratigraphical section and a map of type localities where marker horizons have been positively identified. Some of these type localities are shown in Fig. 11. It is possible to construct rough stratigraphical cross-sections from this data, and this is done in Fig. 12 where the 'geophysical basement' has been added. In drawing each section we have assumed uniform thicknesses of the formations. However, this assumption cannot

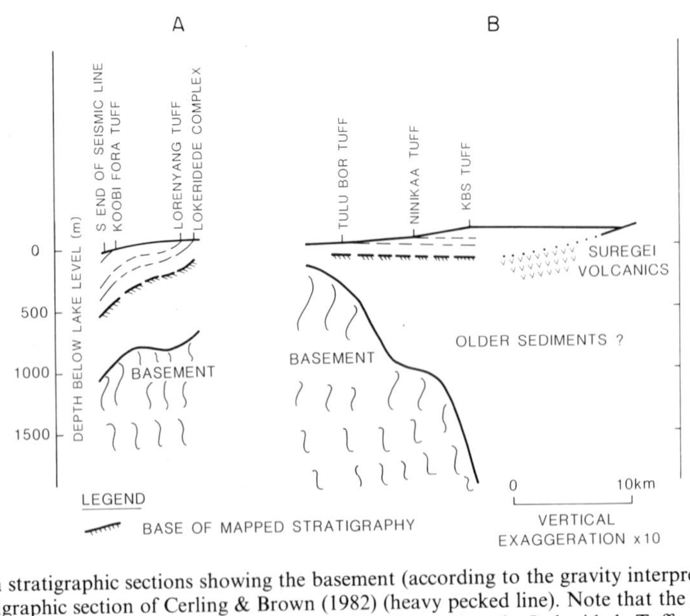

FIG. 12. Cartoon stratigraphic sections showing the basement (according to the gravity interpretation) and the base of the stratigraphic section of Cerling & Brown (1982) (heavy pecked line). Note that the KBS Tuff occurs *c.* 30 m above the Lorenyang Tuff, and the Tulu Bor Tuff *c.* 80 m below the Lokeridede Tuff.

be applied over the basin as a whole, as was shown by Cerling & Brown (1982). For example, their Fig. 3 shows *c*. 300 m of section below the Koobi Fora Tuff in area 102 (just E of Koobi Fora), but only some 50 m between the Okote Tuff (time equivalent to the Koobi Fora Tuff) and the Tulu Bor Tuff in areas 129–131 (40 km NE of Koobi Fora). Much of this difference they ascribe to erosion. Thus the formation thicknesses, particularly in Fig. 12B, may be somewhat in error; we do not believe this will affect our conclusions, which centre on the great thickness of unexposed section between the base of the Cenozoic sediments and the metamorphic basement according to the geophysics.

In Fig. 12A there is good general agreement between the configuration of the 'geophysical basement' and the stratigraphic cross-section. The > 500 m of material between the Suregei Tuff and the basement may include a rather thick sequence of ignimbrites (like those of Sibilot to the S) having a density similar to that of the sediments (F.H. Brown, pers. comm.), as well as older sediments. It is quite possible that the 700–800 m deep reflection on the Koobi Fora seismic line represents a basalt flow near the top of such a sequence.

In Fig. 12B there is no correlation between the basement and the Cenozoic stratigraphy. As mentioned earlier, the gravity low in the NE corner of Fig. 2 is believed to be due to a deep Mesozoic sediment-filled rift, underlying the Cenozoic volcanics. If this is so, then the lack of correlation is hardly surprising, as the gravity variations along this profile would be largely due to density contrasts between basement and Mesozoic sediments and within the latter. The form of the basement profile may be an indication of step faulting, with the steepest gradient representing a major rift fault. Using a larger and more realistic density than 2.2 g cm^{-3} for these older sediments would result in still greater basement relief.

Conclusions

It seems evident, from our rather limited geophysical experiments, that over much of the NE Lake Turkana region the Cenozoic sediments form merely a veneer on the total sediment/volcanic section above the metamorphic basement. It is probably the variations in thickness of Mesozoic rocks that are responsible for much of the relief in the Bouguer anomaly map. However, the basement seems to be very shallow in one area: this could quite easily be checked by, for example, an electrical sounding. Close to Lake Turkana both the depth to basement and the thickness of Cenozoic sediments appear to increase rapidly towards the lake, and it would be interesting to see what happens beneath the lake itself. This data may already be available (Reynolds & Rosendahl 1984).

ACKNOWLEDGEMENTS: We thank R.E. Leakey for logistical support; N. Bhatt, J.M. Ndombi, A.G. Ferguson and D. Barker for assistance in the field; F.H. Brown, D. Livingstone and B.L.N. Kennett for helpful correspondence; A. Cheetham for reducing the gravity data and Mrs Susan Swain for drafting the original figures. The fieldwork and computing were supported by the Universities of Leicester and Nairobi, and the seismic equipment generously lent by the Mines and Geological Department, Government of Kenya.

References

Brown, F.H., Howell, F.C. & Eck, G.G. 1978. Observations on problems of correlation of late Cenozoic hominid-bearing formations in the North Lake Turkana Basin. In: Bishop, W.W. (ed) *Geological Background to Fossil Man*. Scottish Academic Press, Edinburgh, pp. 473–98.

Cerling, T.E. & Brown, F.H. 1982. Tuffaceous marker horizons in the Koobi Fora region and the Lower Omo Valley. *Nature* **299,** 216–21.

Coppens, Y., Howell, F.C., Isaac, G.Ll. & Leakey, R.E. (eds) 1976. *Earliest Man and Environments in the Lake Rudolf Basin*. University of Chicago Press, Chicago.

Dobrin, M.B. 1976. *Introduction to Geophysical Prospecting*, 3rd edn. McGraw Hill, New York.

Findlater, I.C. 1978. Isochronous surfaces within the Plio-Pleistocene sediments east of Lake Turkana. In: Bishop, W.W. (ed) *Geological Background to Fossil Man*. Scottish Academic Press, Edinburgh, pp. 415–20.

Fuchs, K. & Muller, G. 1971. Computation of synthetic seismograms with the reflectivity method and comparison with observations. *Geophys. J. R. Astr. Soc.* **23,** 417–33.

Grant, F.S. & West, G.F. 1965. *Interpretation Theory in Applied Geophysics*. McGraw Hill, New York.

Hamdi, F. & Taylor Smith, D. 1982. The influence of permeability on compressional wave velocity in marine sediments. *Geophys. Prosp.* **30,** 622–40.

Kennett, B.L.N. 1976a. The influence of lateral variations on seismic refraction interpretation. *Pure Appl. Geophys.* **119,** 647–52.

—— 1976b. The effects of attenuation on seismograms. *Bull. Seism. Soc. Am.* **65,** 1643–51.

—— 1977. Towards a more detailed picture of the oceanic crust and mantle. *Mar. Geophys. Res.* **3,** 7–42.

O'Brien, P.N.S. 1957. Multiply reflected refractions in a shallow layer. *Geophys. Prosp.* **5,** 371–80.

O'Brien, P.N.S. 1967. The use of amplitudes in seismic refraction survey. In: Musgrave, A.W. (ed) *Seismic Refraction Prospecting*. S.E.G., Tulsa.

Okabe, M. 1979. Analytical expressions for gravity anomalies due to homogeneous polyhedral bodies and translations into magnetic anomalies. *Geophysics* **44,** 730–41.

Pankratz, L.W. & Ackermann, H.D. 1982. Structure along the northwest edge of the Snake River Plain interpreted from seismic refraction. *J. Geophys. Res.* **87,** 2676–82.

Reynolds, D.J. & Rosendahl, B.R. 1984. Tectonic expression of continental rifting (Abstract). *EOS* **65,** 1116.

Schmoller, R. 1982. Some aspects of handling velocity inversion and hidden layer problems in seismic refraction work. *Geophys. Prosp.* **30,** 735–51.

Swain, C.J. 1976. Some gravity results from the northern Gregory Rift. In: Pilger, A. & Rossler, A. (eds) *Afar Between Continental and Oceanic Rifting*. Schweizerbart'sche, Stuttgart, pp. 120–5.

——, Khan, M.A., Wilton, T.J., Maguire, P.K.H. & Griffiths, D.H. 1981. Seismic and gravity surveys in the Lake Baringo–Tugen Hills area, Kenya Rift Valley. *J. Geol. Soc.* **138,** 93–102.

Vondra, C.F. & Bowen, B.E. 1978. Stratigraphy, sedimentary facies and palaeoenvironments, East Lake Turkana, Kenya. In: Bishop, W.W. (ed) *Geological Background to Fossil Man*. Scottish Academic Press, Edinburgh, pp. 395–414.

Will, M. 1976. Calculation of travel times and ray paths for lateral inhomogeneous media. In: Geise, P., Prodehl, C. & Stein, A. (eds) *Explosion Seismology in Central Europe*. Springer-Verlag, New York.

C.J. Swain, Dept of Physics, University of Zimbabwe, PO Box MP 167, Harare, Zimbabwe.

N.J. Skinner, Dept of Physics, University of Botswana, Private Bag 0022, Gaborone, Botswana.

M.A. Khan, Dept of Geology, University of Leicester, Leicester LE1 7RH, UK.

Volcano-tectonic control on sedimentation in the Koobi Fora sedimentary basin, Lake Turkana

R.T. Watkins

SUMMARY: It has previously been considered that the sedimentary rocks of the Koobi Fora basin, NE of Lake Turkana, lie disconformably upon older volcanic strata, local volcanism and sedimentation being separated in time. Volcanic and sediment relationships observed at the boundary of the basin reveal that major Pliocene tholeiitic flood volcanism took place after the initiation of proto-Lake Turkana. Significant lacustrine and fluviatile deposits lie beneath lavas in the upland margin of the northern part of the basin. These are lithologically similar, and may be laterally equivalent, to strata outcropping along the northern edge of the basin and on the southeastern flank of the Kokoi upland. The fine-grained lacustrine rocks formed during an extended period of slow sedimentation. The age of Lake Turkana, therefore, may be significantly greater than 4.3 Ma, which is recorded from the lowermost tuff bed within the Koobi Fora basin (McDougall 1985). Further lacustrine sediments form thin but laterally persistent intercalations between flood lavas. Concurrence of large scale subaerial eruptions and lacustrine sedimentation is explained by a model of volcano-tumescence and deflation. The early Pliocene lake was relatively shallow and of greater extent than present-day Lake Turkana. Swellings of the land surface preceding volcanic eruptions E of the lake produced significant withdrawals of the shoreline westwards. Subsidence, following lava outpourings, allowed transgression of the lake towards the former shorelines.

The Koobi Fora basin is an embayment of sedimentary exposures on the northeastern shores of Lake Turkana, northern Kenya, bordered to the E by a predominantly volcanic upland 'margin' (Fig. 1). The relationships between sediments and volcanics at the basin boundary provide important information on the early history of the lake.

Vondra & Bowen (1978) considered the 325 m of Pliocene–Pleistocene sedimentary strata exposed within the basin to lie with simple disconformity upon older Miocene and Pliocene volcanics. According to these authors, the earliest sediments, those assigned to the lower Kubi Algi Formation, were characterized by pebble conglomerate units. Such coarsely clastic deposits were consistent with sedimentation in a half-graben developing rapidly through faulting to the W, and monoclinal flexuring to the E. Findlater (1976a, pers. comm.) cited deep weathering of alkali basalt lavas beneath the sedimentary cover along the northern boundary of the basin as indicating a significant volcanic hiatus prior to the birth of Lake Turkana. This view was supported by whole rock K-Ar isotopic ages from two of the lavas of 11.9 ± 0.5 Ma and 14.5 ± 1.4 Ma (Fitch *et al.* 1974; Fitch & Miller 1976) (these and all following isotopic ages are corrected where necessary using the decay constants of Steiger & Jäger 1977). Further S along the basin margin, rocks forming the mountains of Shin and Kubi Algi (see Fig. 1) were also found to have minimum Miocene ages of 12.1 ± 0.9 Ma and 7.7 ± 0.8 Ma, respectively (Fitch *et al.* 1974).

Miocene rocks, however, form only a part of the exposures in the marginal uplands which are otherwise capped by little deformed, nearly flat-lying, tholeiitic basalt flood lavas. Preliminary K-Ar dating of these lavas (Fitch *et al.* 1974) gave Pliocene ages of *c.* 3.9 Ma. Since tuffaceous beds preserved in the southern part of the Koobi Fora basin were provisionally dated at around 4.0 and 4.6 Ma, it was apparent that some lava–sediment interdigitation might occur along the basin's southern edge. Both Bowen (1974) and Findlater (1976a, b) considered sedimentation to have commenced progressively later northwards. An $^{40}Ar/^{39}Ar$ age of *c.* 4.3 Ma for a tholeiitic basalt lava of the Harr Formation capping the Suregei Plateau (Fig. 2) (Hooker & Watkins 1980) appeared to confirm the temporal separation of volcanism and sedimentation in the N, where the lowest marker bed within the basin succession was the Suregei 'Tuff' (Bowen & Vondra 1973) which was assigned a stratigraphical level above that of the 4.0 Ma Hasuma Tuff (Findlater 1978).

Brown & Cerling (1982) employed chemical criteria for correlating the tuffaceous marker beds throughout the Koobi Fora basin and concluded that there was a need for major revision of the stratigraphy. One result of this is the re-assignment of the coarse, conglomeratic sediments at the base of the Kubi Algi Formation to a much higher stratigraphical level. According to Brown

From FROSTICK, L.E. *et al.* (eds) 1986, *Sedimentation in the African Rifts*, Geological Society Special Publication No. 25, pp. 85–95.

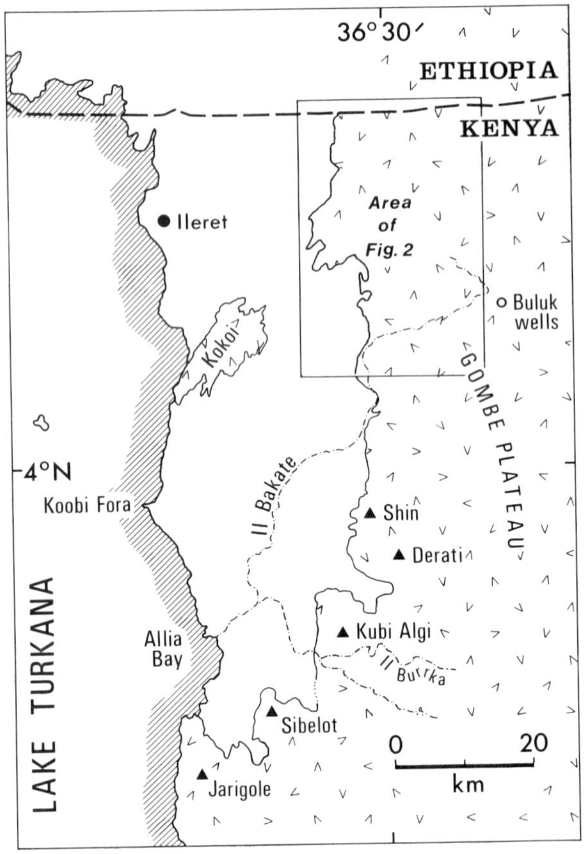

FIG. 1. Location of the Koobi Fora basin (unshaded) on the northeastern shores of Lake Turkana. The basin is taken as the entire embayment of sediments including the area N of the Kokoi upland, sometimes termed the Ileret basin. Short sections of the boundary of the volcanic margin (ornamented) N of Shin and Jarigole mountains are after Vondra & Bowen (1978).

& Cerling, the oldest basin sediments thus become a series of laminated claystones and fine sandstones overlain by the Suregei 'Tuff' (actually a series of recrystallized diatomites), which they renamed the Suregei Complex. The fine-grained lacustrine sequence, lying upon weathered Miocene alkali basalt lavas, constitutes much of the northern part of the basin boundary (Fig. 2), and also outcrops on the southeastern flank of the Kokoi upland (Fig. 1) (Findlater 1976a).

Field relations

Despite a general lack of relief, much of the 200 km of volcanics–sediment interface studied is well exposed. Particularly informative sections were observed in the walls of river valleys entering the basin and in some shallow excavations. Whilst these confirmed the existence of a major 'angular' unconformity where basin strata overlie volcanics of Miocene age, more complex relationships were evident in those segments of the boundary formed by Pliocene flood lavas. The relationships between the flood lavas and the earliest lake sediments is the subject of the remaining part of this paper, of which two aspects are of particular importance:

1 Significant volumes of lacustrine and fluviatile sediment of Pliocene age demonstrably predate the eruption of tholeiitic flood basalt. In the absence of a formal nomenclature, these rocks will herein be termed 'sub-flood lava sediments'.

2 Lacustrine sediments and flood lavas are interbedded in many exposures around the basin boundary, the lava flows having poured out over 'dry' lake floors.

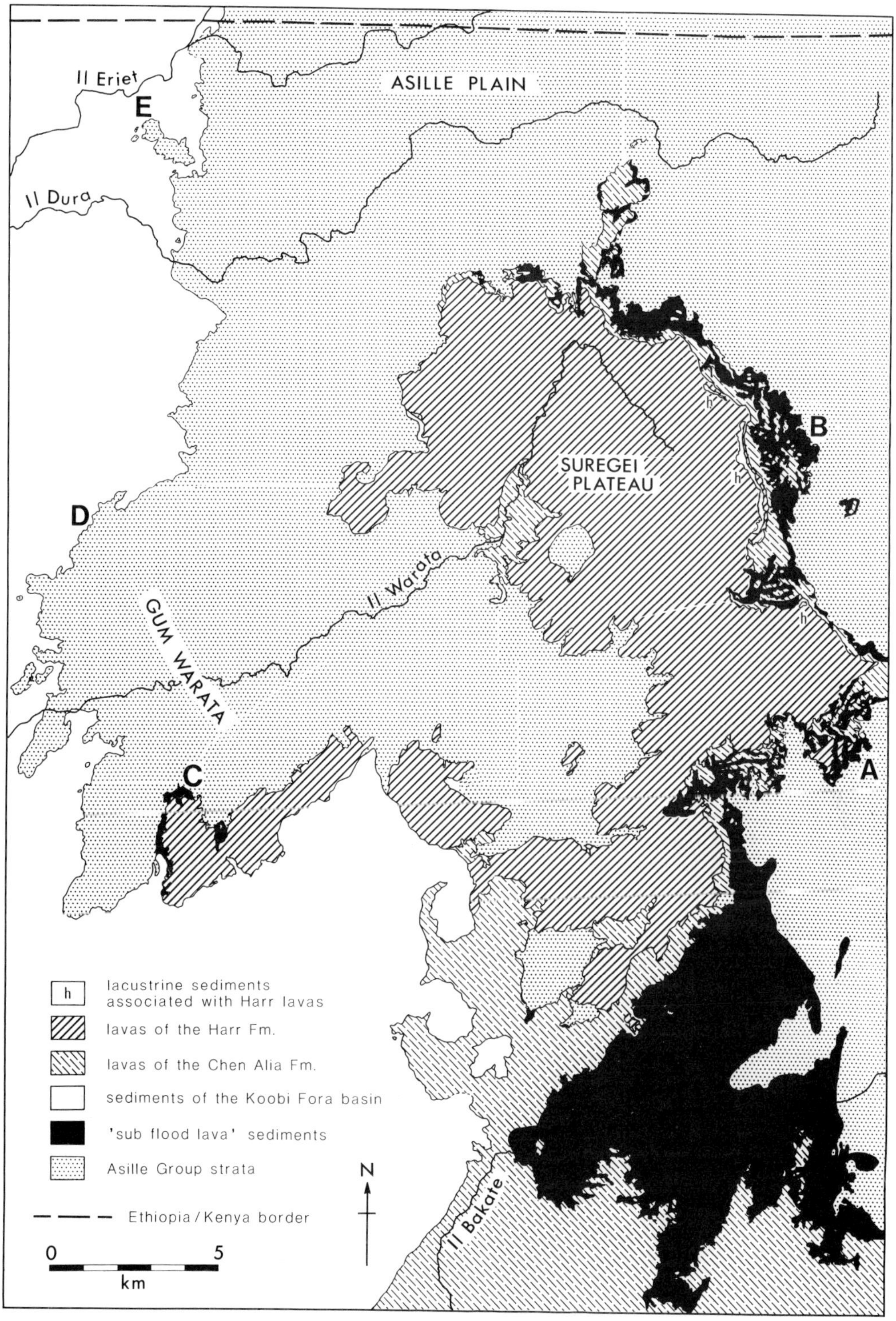

FIG. 2. Map of the Suregei–Asille district and the upper Bakate valley, showing the extent of sub-flood lava Pliocene sediments in the northern segment of the upland basin margin. A–E refer to section locations used in Fig. 3.

Sub-flood lava sediments

On the northeastern side of the sedimentary basin, the Il Bakate River has cut its valley through thick tholeiitic flood basalts which are left capping the Suregei and Gombe Plateaus (Figs. 1 and 2). In the resulting 'Bakate Gap', there is an area of some 5 km^2 floored by almost flat-lying sediments. These lie unconformably upon Miocene strata which have a regional dip of *c*. 6° to the WSW. They further contrast with the older strata in being unaffected by major faulting and remaining undeformed across the axes of two sharp monoclines. They are clearly older than the flood tholeiites since they are seen to pass directly beneath the lava sheets. The latter are chilled to a dark glass and rucked into an 'upside down' pahoehoe structure at their base. Fragments of the thin chill margin, a centimetre or so thick and partly oxidized to a bright orange colour, are a useful indicator of the base of the flows where found in surface lag.

In their lower part, the sediments consist mainly of claystones and fine siltstones with a number of conspicuous pale diatomite beds. The diatomites are largely recrystallized and frequently porcellanous in character. This lacustrine sequence, containing large fish vertebrae, crocodile teeth and bones, and fragments of turtle carapace, generally lies directly upon a slightly irregular surface cut into the Miocene rocks, and only locally are there basal conglomerates. The lake sediments are succeeded, upon a sharply erosional boundary, by coarser-grained fluviatile or fluvio-deltaic rocks bearing a terrestrial vertebrate fauna. Well-sorted, cross-bedded beach sandstones and shallow water molluscan sandstones and limestones are also present. A similar sequence of sediments of equal thickness (40 m) is exposed in the Gum Warata Hills and to the NE of the Suregei Plateau where greater relief has facilitated the recording of stratigraphical sections (Fig. 3).

The complete sedimentary sequence thus appears to have formed in, and around, a lake extending well to the E of the present Koobi Fora basin. The vertebrate fauna from the non-lacustrine beds is Pliocene in age. Individual fossils, such as a horn core of *Tragelephas nakuae*, are morphologically akin to specimens recovered from similar lithologies in the Lower Member of the Koobi Fora Formation within the Koobi Fora basin (J.M. Harris, pers. comm.). The assemblage, however, is presently insufficient to enable precise biostratigraphical correlation.

Similar relationships of lavas and sediments to those seen in the Suregei-Asille district are found in exposures on the southeastern flank of the Kokoi upland. Streams cutting back into this basalt-capped SW–NE ridge have revealed a series of claystones, siltstones and diatomaceous beds. These lacustrine sediments directly overlie the same deeply weathered alkali basalt lavas of the Nakwele Formation (Watkins 1983) that crop out on the basin margin to the NE. The sediments are overlain by tholeiitic basalt lavas which have given a K-Ar isotopic age of 3.72 ± 0.36 Ma (Fitch *et al.* 1974, 1985). A substantially older age of 6.57 ± 0.35 Ma has been obtained from a further tholeiitic lava flow on the Kokoi, but the relationship of this lava to the sediments is not clear.

Sediment intercalations in flood lavas

In addition to those sediments already described that pre-date flood volcanism, others were deposited subsequently during the period of tholeiitic eruptions. These are preserved as thin, yet often laterally persistent, units interbedded with the lavas. In the Suregei-Asille district, a thin (generally less than 2 m) but widespread unit of bioclastic limestone and sandstone, the Warata Formation (Watkins & Williamson 1980), outcrops between major lava floods. A significant micrite component in many of the Warata molluscan limestones and the total absence from them of bassomatophoran pulmonates, led Watkins & Williamson (1980) to interpret them as having formed through the reworking of shell material some distance from the shoreline, although possibly in part as relatively shallow water shell-pavements. The overlying lava of the Harr Formation (Hooker & Watkins 1980), however, flowed over dry lake sediments. A further thin lacustrine unit, including a brick-red stained diatomite bed, outcrops between flood lavas on the eastern rim of the Suregei Plateau (Watkins 1983) where, once again, the overlying lavas show no evidence of having come in contact with lake water. Indeed, throughout the outcrop of flood tholeiites, including those lying upon the sub-flood lava sediments, there are only two known occurrences of pillow lavas. At one of these, in the Gum Warata Hills, discrete pillows are preserved within a structureless, strongly lithified siltstone suggesting that lava flowed into very shallow water or, more likely, merely advanced over wet sediment which was fluidized by the liberation of steam. Elsewhere, local concentrations of calcite veining at the bases of lava flows probably further indicate the mobilization of connate water in the sediment substratum. East of Kubi Algi, there is a limited exposure of pillow lavas with accompanying hyaloclastites.

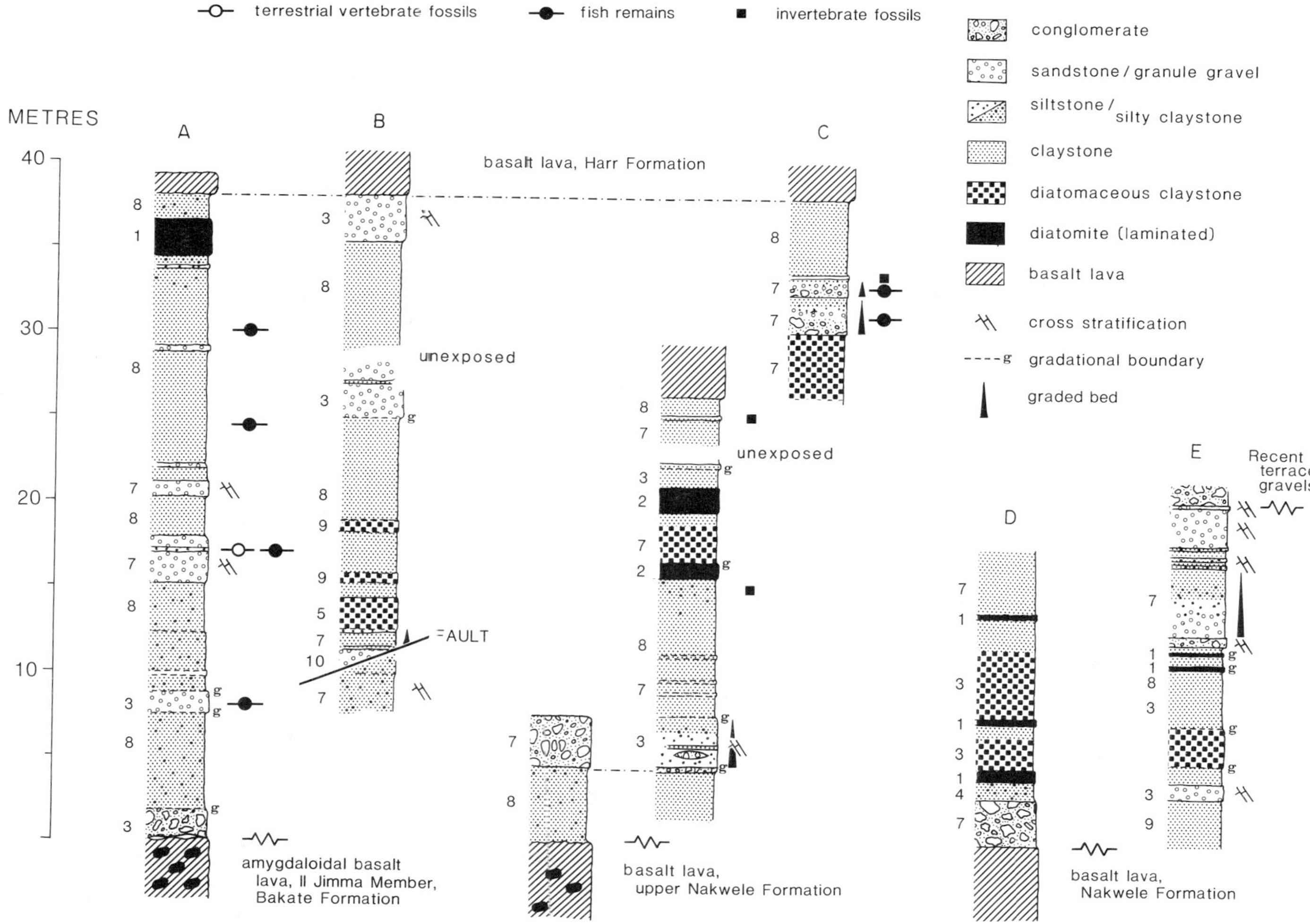

FIG. 3. Stratigraphical sections showing comparison between sequences of Pliocene sediments demonstrably underlying flood lavas in the Suregei–Asille district (A–C) and those exposed on the northeastern edge of the Koobi Fora basin (D–E). Approximate colour of lithologies (Geological Society of America 1975): 1 = white; 2 = pinkish grey; 3 = yellowish grey; 4 = olive grey; 5 = greenish yellow; 6 = greyish yellow green; 7 = yellowish brown; 8 = pale to moderate brown; 9 = very pale orange; 10 = yellowish orange. (The sections are illustrated in greater detail in Watkins 1983.)

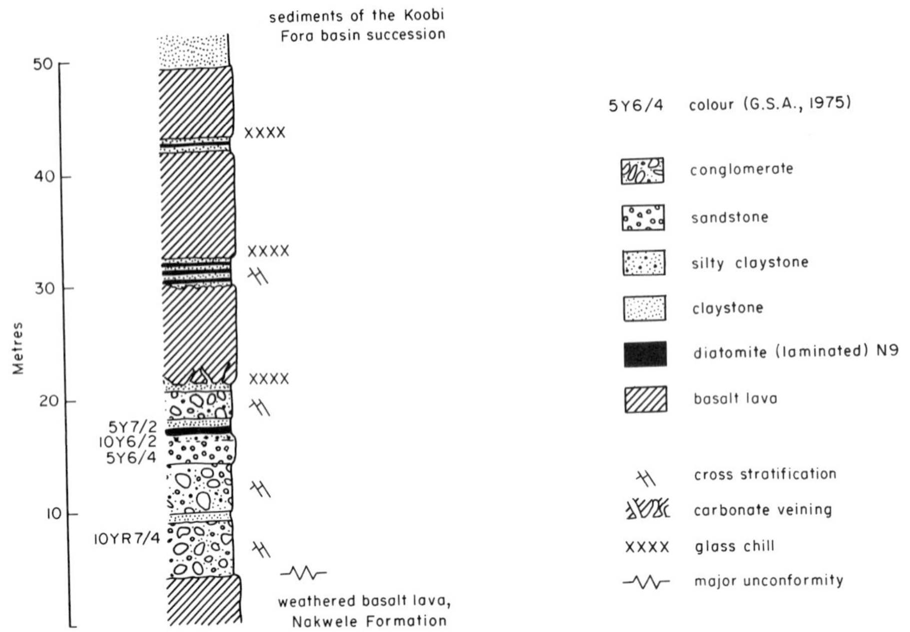

FIG. 4. Stratigraphical section recorded at a river cliff on the S side of the Il Burrka valley, showing the intercalation of subaerial tholeiitic basalt lavas and lacustrine sediments. Breccias near the base of the section are of fluviatile origin.

Interdigitation of lacustrine sediments and subaerial lavas is most convincingly illustrated in a steep-sided valley immediately S of the Kubi Algi Mountain. Here the Il Burrka River has cut through Pliocene strata of the basin margin to expose Miocene lavas (Nakwele Formation) on the valley floor. The Pliocene sequence comprises three subaerial tholeiitic lava flows with sediment intercalations (Fig. 4). Laminated diatomite and claystone occur towards the top of a sequence of breccias and sandstones that was laid down before the eruption of the first flood lava, and similar fine-grained lacustrine sediments were deposited in intervals between subsequent lava outpourings. Minor amounts of coarser clastic material in the intercalations may have been deposited during transgression and regression of a lake; but there is little evidence of erosion in the form of channelling or other disruptions of the laminations towards the top of each sedimentary division. There is similarly little erosion of the lava flows. Highly irregular, vesiculated and glassy surfaces of individual flow-units are frequently preserved beneath sediments, despite the inherent instability of basaltic glass even under semi-arid conditions such as have prevailed in the recent geological past. Molluscs can be found cemented in their life positions within larger gas cavities, whilst wave erosion has not destroyed relatively delicate structures on the surfaces of the flows.

Discussion

Stratigraphical correlation

The tholeiitic basalts form generally extensive flows, up to 50 m or more in thickness, and there are probably only a small number of flows represented in the exposures at the rim of the Koobi Fora basin. Presence of sedimentary intercalations may enable future mapping of individual flows. However, the lavas display uniform petrography and chemical composition making the recognition of individual flows by these criteria impossible. They have proved to be unsuitable for precise K-Ar and $^{40}Ar/^{39}Ar$ dating since they contain a glass mesostasis and variable amounts of secondary alteration products. Available ages suggest that the magmatism extended between *c.* 6–3.5 Ma (Fitch *et al.* 1985; Hooker & Watkins 1980; I. McDougall, unpub. data). They offer little hope of future geochronometric discrimination of individual flows. Broadly equivalent ages have been obtained from lavas of the Il Burrka section (Fig. 4), and lavas outcropping in the upper Il Bakate valley (I. McDougall, unpub. data). It is therefore tentatively presumed that the oldest sedimentary units are the sub-flood lava sediments in the northern segment of the basin margin.

Laminated lacustrine clays and diatomaceous rocks pre-dating the flood lavas formed through slow sedimentation; the subdued relief and semi-arid climate restricted production of coarse clastic deposits. After subsequent mantling of large tracts to the E of the lake by flood lavas, the shoreline adjoined a largely featureless and barren landscape similar to that bordering the southern part of the lake today (Cohen *et al.*, this volume). Sediment input from this terrain was sufficiently slow to enable periodic accumulation of reasonably pure diatomite, as well as diatom-rich clays. As a consequence, the fine-grained deposits occurring beneath and between the flood lavas may extend back the history of lacustrine sedimentation considerably beyond the 4.3 Ma K-Ar age recently recorded from the lowermost tuff bed within the Koobi Fora basin (McDougall 1985). They may pre-date the succession within the Koobi Fora basin; alternatively, they may be, wholly or in part, stratigraphically equivalent to basin exposures.

The similarity between sub-flood lava sediments in the basin margin and those on the southeastern flank of the Kokoi has already been mentioned. Both sequences lack basal conglomerates, are comprised predominantly of claystones passing up into sandstones, and contain highly conspicuous, pale-coloured diatomaceous beds. The strata flanking the Kokoi were considered by both Bowen (1974) and Findlater (1976a) to belong to the Kubi Algi and Lower Member, Koobi Fora Formations and to contain the Suregei (Tuff) Complex. The same sequence was recorded along the northern periphery of the basin and is compared with the sub-flood lava rocks in Fig. 3. The possibility of lateral correlation between the sediments is particularly apparent on the southern side of Gum Warata, where 40 m of near flat-lying lacustrine strata overlain by a tholeiitic lava flow (Fig. 3, section C) crop out in a river valley only 1500 m N of exposures of similar strata at the edge of the basin. Indeed, poorly exposed claystones and siltstones capped by lava form the walls of a stream valley entering the basin in this region, and must abut basin rocks if they are not, in fact, laterally continuous with them (Fig. 2) (Watkins 1983). The sediments are only very gently dipping and no faulting or erosion surface is evident to account for a vertical displacement.

For lateral correlation to exist, the stratigraphical horizon of the flood lavas must lie within the basin sedimentary succession. This has been discounted in the past by workers in the basin since no remnants of basalt lava sheets, besides that capping the Kokoi, have been recorded within the area of the basin. The great thicknesses of individual lava flows at the basin rim were thought to indicate their former extension much further westwards. This, however, may not have been the case. Despite their large volume and evident fluidity, the lavas advanced behind steep flow-fronts. Flow geometry is well seen in the vicinity of Sibelot Mountain. Broad nosed flow-units encroaching on the basin from the E terminate abruptly, one flow-front being particularly well exposed at Karsa waterhole. If the flood lavas extended little beyond their present outcrops, there might be little or no evidence of unconformity in the basin succession, a lava eruption representing an extremely brief period of time.

Lava–sediment relationships

Despite the considerable extent of lake sediments in the early Pliocene, overlying lavas exposed in the margin of the Koobi Fora basin are everywhere subaerial, there being no large scale disruptions indicative of the interaction of molten lava and lake water. Interbedding of subaerial lavas and lacustrine, often diatomaceous, sediments is seen elsewhere in the Eastern Rift, for example, within the Kaparaina Formation (Pickford 1974) and the Nachola Formation (M. Pickford, pers. comm.) in Kenya. It has generally been explained through climatic influences and the commonly ephemeral nature of rift valley lakes.

Flood lavas of the Koobi Fora basin margin were erupted over a time span of no more than 2.5 Ma. During this brief period, lavas advanced over dry lake floors on at least three, and probably more, occasions. (Difficulty of correlation between individual exposures showing intercalated lavas and sediments makes the number of such intercalations uncertain.) It appears from isotopic dating studies (Fitch *et al.* 1985; I. McDougall, unpub. data) that some sequential flows, with lacustrine units between them, are closely spaced in time. Thus for a climatic control to be effective, not only would changes of climate have to be frequent but lava eruptions would have to have occurred fortuitously during periods of increased aridity. The unlikelihood of such coincidence is increased by evidence that the periods of non-lacustrine conditions were brief. The very fine-grained sediments would have been unlithified and little consolidated on emergence from the lake. Preservation of thin, and laterally persistent, units of this material testifies to the relatively rapid superimposition of the lavas. Some evidence has also been presented to suggest that submergence of the lavas on each occasion was similarly rapid. Dominance of fine-grained laminated rocks within the sedimentary intercalations reflects how quickly low energy lacustrine conditions were restored during each transgression.

Minor sediments occurring within successions of flood lavas elsewhere have been envisaged as forming in localized playa lakes existing briefly in shallow depressions on the extensive flat-lying lava sheets. However, diatomites and claystones lying between lavas in the Il Burrka section (Fig. 4) are 2.6 m and 1.4 m thick, respectively. Dewatering and compaction beneath encumbent lavas must have considerably reduced the original thickness of the fine-grained sediment. Such accumulations require a water body of reasonably long duration (R.B. Owen, pers. comm.). Furthermore, similar, possibly laterally equivalent, sediments can be traced close to precipitous lava flow fronts in the southern part of the basin margin, an improbable location for water ponding. The Warata Formation (Watkins & Williamson 1980) and the earlier sub-flood lava sediments were certainly associated with substantive lakes. Preliminary SEM studies of the recrystallized diatomites from the upper Bakate valley and Gum Warata indicate that the dominant microfossils may be the cylindrical forms most often associated with deeper water accumulations (Gasse *et al.* 1983; R.B. Owen, pers. comm.).

It is unlikely that the highly dynamic control of lake morphology operative during the early Pliocene could have been brought about by either climatic changes or regional tectonism. The contemporaneity of major lava eruptions and sedimentation suggests that a volcano-tectonic control may have been responsible for the repeated variations in lake level. The tholeiite lavas represent a very considerable volume of magma. An estimated 15,000 km^2 E of Lake Turkana is presently covered by the lavas to an average depth of *c.* 200 m. The fluid lavas were extruded rapidly, at least initially from fissure eruptions, E of Lake Turkana. Volcanism consisted of a small number of floods, each involving a large amount of magma. The major flood of the Harr Formation, for example, has a preserved volume estimated at 6 km^3 with probably a similar volume of lava having been removed by erosion. The approach to the surface of such voluminous batches of magma would inevitably have caused a swelling of the land centred in the vicinity of the forthcoming eruption.

Proto-Lake Turkana, at the time of the Pliocene eruptions, lay upon the post-Miocene erosion surface in a depression formed by gentle downwarping. The first Pliocene sediments largely infilled irregularities of relief remaining at the end of a volcanic hiatus which lasted *c.* 5 Ma (Fitch *et al.* 1985), although they were subsequently etched by stream erosion before being covered by lavas. Unconstrained by rift faults and flexuring, the lake would have been less elongate and probably far more extensive in area than its present-day counterpart. Even a small degree of volcanic tumescence could have effected dramatic changes in the extent of such a relatively shallow lake.

A volcano-tectonic model

A model of volcano-tectonic control on the geometry of proto-Lake Turkana is illustrated in Fig. 5. It employs vertical movements of the land surface associated with major volcanic events to the E of the lake to explain recurrent and rapid lateral movements of the eastern shoreline. Outpouring of lava occurs shortly after tumescence, the lava flows advancing over a recently emerged portion of the lake floor. Deflation of the land surface after cessation of eruptive activity allows a readvance of the lake. The extensive lavas maintain and enhance the flat terrain. Once the lake level surmounts the fronts of the newly erupted lava flows, eastward transgression is swift. With re-establishment of the shoreline well to the E of the present edge of the Koobi Fora basin, deposition of coarse clastic material in the latter region is minimized.

The effectiveness of the model relies upon an overall increase in water level relative to the accumulating thickness of lava, enabling the lake on each occasion to surmount the flow fronts. The lava would have advanced over only a small part of the former lake basin. Post-eruption deflation of this area may have reduced the topographic height of the lava pile. Compaction of the underlying sediments and subsidence under the weight of accumulating lava may have further reduced the need for a rise in water level. Independent factors, such as long term climatic change or constriction of the lake within a basin subsiding on the flanks of the rift axis to the E, may have played a role.

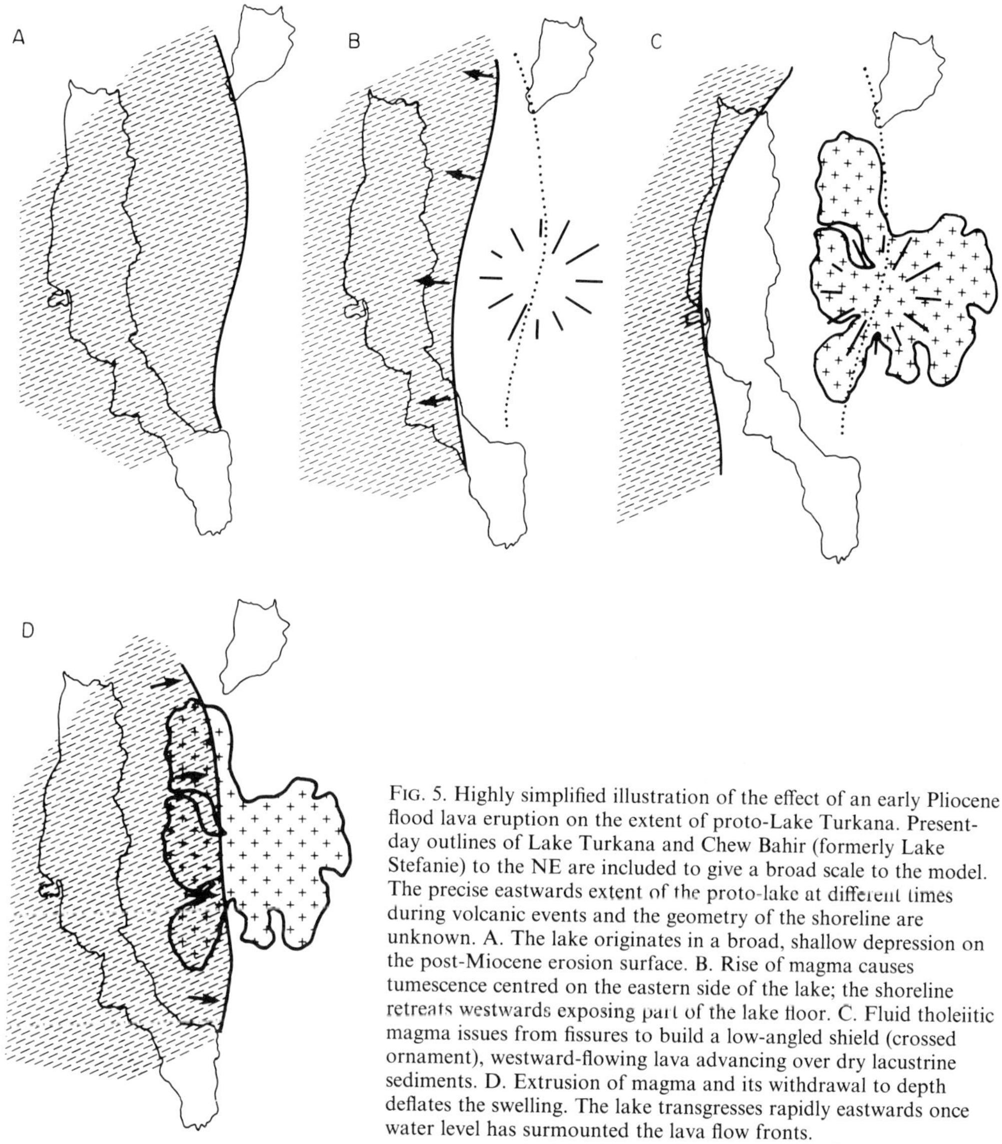

FIG. 5. Highly simplified illustration of the effect of an early Pliocene flood lava eruption on the extent of proto-Lake Turkana. Present-day outlines of Lake Turkana and Chew Bahir (formerly Lake Stefanie) to the NE are included to give a broad scale to the model. The precise eastwards extent of the proto-lake at different times during volcanic events and the geometry of the shoreline are unknown. A. The lake originates in a broad, shallow depression on the post-Miocene erosion surface. B. Rise of magma causes tumescence centred on the eastern side of the lake; the shoreline retreats westwards exposing part of the lake floor. C. Fluid tholeiitic magma issues from fissures to build a low-angled shield (crossed ornament), westward-flowing lava advancing over dry lacustrine sediments. D. Extrusion of magma and its withdrawal to depth deflates the swelling. The lake transgresses rapidly eastwards once water level has surmounted the lava flow fronts.

Conclusions

Study of the volcanics–sediment interface at the boundary of the Koobi Fora basin shows there to have been significant temporal overlap of basaltic volcanism and lake sedimentation. Frequent intercalation of subaerial lava and lacustrine sediments presents major problems in interpreting the early history of Lake Turkana.

Sedimentary strata pre-dating flood lava in the basin margin and on the Kokoi upland may have formed during a period of lake history prior to that represented by the rocks of the Koobi Fora basin. It is considered more probable that the sediments are, at least in part, laterally equivalent to basin strata, and accumulated within a lake extending much further eastwards than the present basin.

Fine-grained deposits, including diatomites, at the base of the Koobi Fora basin succession accumulated slowly and may represent a considerable period of time before the deposition of the oldest isotopically dated marker bed, the Moiti Tuff (Brown & Cerling 1982; McDougall 1985). Sedimentation within the basin did not necessarily commence any later in the N than in the S. Correlation between sediments on the periphery of the basin and those of the upland margin is made difficult by the present lack of knowledge of sedimentary strata outcropping towards the outside of the basin, away from areas of prime palaeontological and archaeological interest. Precise isotopic dating of the glassy lavas has proved most difficult, and age determinations on associated sediments are impossible owing to the absence of suitable rock types. It will, therefore, fall upon biostratigraphy to provide firm correlation. Further collection and evaluation of the sparse vertebrate fauna within the upper part of the sub-flood lava sedimentary sequence is most desirable. Microfossil studies could be useful in correlation. They might also elucidate the depth of water in which the sediments accumulated and hence the scale of lake level changes. Unfortunately, such studies may be precluded by the recrystallization which has occurred in diatomites both from the basin (Brown & Cerling 1982) and the margin.

Major floods of tholeiitic basalt lava have advanced on a minimum of three occasions to cover areas of lake sediment previously drained of water. Volcano-tectonic movements provide a viable explanation of changes in lake morphology during the early Pliocene. With a magnitude that could override climatic and regional tectonic influences, they were of sufficiently short duration to account for the rapid and recurring migration of the shoreline. During the last 3.5 Ma or so, since the last flood eruption, climate and rift tectonism have dominated the evolution of Lake Turkana. If volcanically induced tumescence were to recur E of the lake today, it would have a far less profound effect than formerly, when regional relief was subdued.

Major volcanic eruptions taking place during the development of other segments of the Eastern Rift must have produced surface tumescence. It would be unlikely if this had not influenced the pattern of sedimentation of other lakes. Any effect upon lake morphology would be brief, its extent greatly dependent upon the nature, particularly the depth, of the basin. The shallow character of the Lake Turkana basin during the early Pliocene provided conditions under which volcano-tectonism could substantially modify lake morphology and exert a dynamic control on the position of the shorelines.

ACKNOWLEDGEMENTS: Fieldwork in Kenya was supported by NERC. I would like to thank colleagues on the Koobi Fora Research Project for their useful discussions in the field, and, in particular, Dr J.M. Harris for his comments on the stratigraphical implications of the vertebrate fossils. I am also grateful to Dr A.R. Macgregor for his helpful criticism of the draft manuscript, and to Dr P. Bowden and Richard Greenwood who also suggested how it might be improved.

References

Bowen, B.E. 1974. *The geology of the Upper Cenozoic sediments in the East Rudolf embayment of the Lake Rudolf basin, Kenya.* Unpublished PhD thesis, Iowa State University.

—— & Vondra, C.F. 1973. Stratigraphical relationships of the Plio-Pleistocene deposits, East Rudolf, Kenya. *Nature* **242,** 391–93.

Brown, F.H. & Cerling, T.E. 1982. Stratigraphical significance of the Tulu Bor Tuff of the Koobi Fora Formation. *Nature* **299,** 212–19.

Findlater, I.C. 1976a. *Stratigraphic analysis and palaeoenvironmental interpretation of a Plio/Pleistocene sedimentary basin east of Lake Turkana.* Unpublished PhD thesis, University of London.

—— 1976b. Tuffs and the recognition of isochronous mapping units in the East Rudolf succession. In: Coppens, Y. *et al.* (eds) *Earliest Man and Environments in the Lake Rudolf Basin.* University of Chicago Press, Chicago and London, pp. 94–104.

—— 1978. Isochronous surfaces within the Plio-Pleistocene sediments east of Lake Turkana. In: Bishop, W.W. (ed) *Geological Background to Fossil Man.* Scottish Academic Press, Edinburgh, pp. 415–20.

Fitch, F.J., Findlater, I.C., Watkins, R.T. & Miller, J.A. 1974. Dating of the rock succession containing fossil hominids at East Rudolf, Kenya. *Nature* **251,** 213–15.

——, Hooker, P.J., Miller, J.A., Mitchell, J.G. & Watkins, R.T. 1985. Reconnaissance potassium-argon geochronology of the Suregei–Asille district, northern Kenya. *Geol. Mag.* **122,** 609–22.

—— & Miller, J.A. 1976. Conventional potassium-argon and argon-40/argon-39 dating of volcanic rocks from East Rudolf. In: Coppens, Y. *et al.* (eds) *Earliest Man and Environments in the Lake Rudolf Basin.* University of Chicago Press, Chicago and London, pp. 123–47.

Gasse, F., Talling, J.F. & Kilham, P. 1983. Diatom assemblages in East Africa: classification, distribution and ecology. *Rev. Hydrobiol. Trop.* **16,** 3–34.

Geological Society of America. 1975. *Rock-color chart.* Geol. Soc. of American, Boulder, Colorado.

Hooker, P.J. & Watkins, R.T. 1980. The nature and age of the youngest basalts of the Suregei–Asille district, northern Kenya. *Proc. 8th Panafrican Congress of Prehistory and Quaternary Studies,* Nairobi 1977, pp. 100–2.

McDougall, I. 1985. K-Ar and $^{40}Ar/^{39}Ar$ dating of the hominid-bearing Plio-Pleistocene sequence at Koobi Fora, Lake Turkana, northern Kenya. *Bull. Geol. Soc. Am.* **96,** 159–75.

PICKFORD, M. 1974. *Stratigraphy and palaeoecology of five late Cainozoic formations in the Kenya Rift Valley.* Unpublished PhD thesis, University of London.

STEIGER, R.H. & JÄGER, E. 1977. Subcommission on geochronology: convention on the use of decay constants in geo and cosmochronology. *Earth Planet. Sci. Lett.* **36,** 359–62.

VONDRA, C.F. & BOWEN, B.E. 1978. Stratigraphy, sedimentary facies and palaeoenvironments, East Lake Turkana, Kenya. In: BISHOP, W.W. (ed) *Geological Background to Fossil Man.* Scottish Academic Press, Edinburgh, pp. 395–414.

WATKINS, R.T. 1983. *The geology of the Suregei–Asille district and the upper Bakate Valley, northern Kenya.* Unpublished PhD thesis, University of London.

—— & WILLIAMSON P.J. 1980. The Warata Formation, northern Kenya: its character and palaeoenvironmental significance. *Proc. 8th Panafrican Congress of Prehistory and Quaternary Studies,* Nairobi 1977, pp. 107–8.

R.T. WATKINS, Dept of Geology, University of Cape Town, Rondebosch, Cape Town, South Africa.

Siliciclastic, Chemical, Pedogenic and Organic Sediments in Contemporary Rift Environments

Slope processes, sediment derivation and landform evolution in a rift valley basin, northern Kenya

I. Reid & L.E. Frostick

SUMMARY: The Plio-Pleistocene sediments of the Lake Turkana basin are the product of episodic deposition interspersed with long periods of reworking. Their stratigraphy is complex and includes a wide range of lithotypes from fine lake clays to extensive coarse gravel sheets. The current phase of landscape evolution is dominated by fluvial incision and slope retreat under semi-arid conditions. Rates of sediment derivation are high, with measured values of up to 1600 t km^{-2} a^{-1}, and a catalogue of evidence of surface processes (rainsplash pillars, mobile soil mantles, etc.) illustrates the erosional dynamism of an area that currently receives a rainfall of only 300 mm a^{-1}. Detailed process studies reveal the ten-fold decline in infiltration capacity of silty regoliths that results from impaction of the surface by the first fall of a rain season. They also explain the rapid onset of overland flow which only accounts for between 2 and 7% of stormwater but which ravages the soil, causing erosion that ranges from 0.2 to 0.5 kg m^{-2} a^{-1}. Shallow gradient slopes are shown to export fine material ($D_{50}=0.2$ mm) and to concentrate coarse fragments >4 mm as a surface lag. But studies on steeper gradient slopes where the regolith is typically a talluvium, and where processes are gravity-dominated, sliding or rolling movements indicate a direct relationship between particle size and downslope movement towards the stream system. Despite the variable importance of different processes on different type-slopes, a clearcut relationship emerges between slope gradient and erosion rate that allows us to model slope retreat and landscape evolution. A volumetric analysis gives an age to the present basin landform of 0.11 Ma. Analysis of sequential links in the sediment transport train permits us to match clastic materials deposited in the accumulation zones of rift valley basins (riverine floodplain, lake delta, beach) with those of the erosional source areas.

The Plio-Pleistocene sediments of the Lake Turkana basin are the product of depositional episodes interspersed with long periods of reworking. They occupy a trough thought by Baker *et al.* (1972) and Cerling & Powers (1977) to be a northward extension of the Gregory Rift, which has largely been abandoned in favour of a line passing N from the Suguta valley through Kino Sogo to Chew Bahir (Fig. 1). In the Koobi Fora region, the eastern boundary of the Turkana depression finds topographic expression in the basalt scarp of Suregei-Gombe. It is through this scarp that most of the sediments of the Koobi Fora embayment were brought by a substantial river draining the southern Ethiopian highlands. This was the proto-Sagan (Grove *et al.* 1975; Vondra & Bowen 1978; Frostick & Reid 1980). Climatic changes in the Lower and Middle Pleistocene have meant an increasing aridity in the region (Bonnefille 1979) and there is no evidence that the river has filled Chew Bahir (Lake Stefanie) to overspill into Lake Turkana since the Upper Pleistocene, although shortlived Holocene discharges may have occurred. The physiographic legacy of this ancient drainage is the Bakate Corridor (Fig. 1), currently a wind-gap large enough to be clearly visible on Landsat imagery.

Such wide-ranging changes in river discharge have been characteristic throughout the Plio-Pleistocene. Butzer *et al.* (1972), reviewing the evidence for higher lake levels in East Africa, observed that the sedimentary succession at Olduvai indicates comparatively short episodes of humid conditions that punctuate a drier climatic regime. In the Koobi Fora embayment, the long periods of subaerial erosion and sedimentary reworking are marked by tuffs that clothe now-buried landscapes (Findlater 1978). Indeed, Cerling & Brown (1982), in a detailed comparison of the sediments of the Lower Omo valley N of present-day Lake Turkana and those of Koobi Fora, suggest a substantial unconformity in the Lower Member of the Koobi Fora Formation representing *c.* 0.4 Ma of subaerial erosion.

The current phase of landscape evolution is dominated by the incision of local ephemeral streams and by slope retreat under semi-arid conditions, characterized by an annual rainfall of *c.* 300 mm and an annual potential evaporation of *c.* 3000 mm. Vegetation cover is scanty and consists of widely spaced *Acacia* and *Commi-*

From FROSTICK, L.E. *et al.* (eds) 1986, *Sedimentation in the African Rifts*, Geological Society Special Publication No. 25, pp. 99–111.

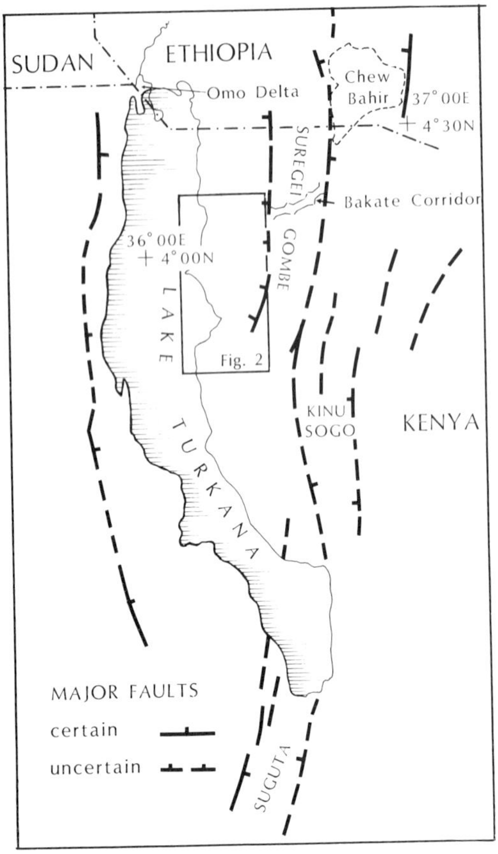

FIG. 1. Map of the Lake Turkana basin and adjacent terrain.

phora. Intervening bare ground sees rapid germination of grasses during the extremely brief rain seasons, though this provides little extra protection for the ground.

In order to assess the rates of physiographic change in these drier periods, a number of experiments have been conducted to evaluate the erosion of sediment from the valley slopes of a number of drainage basins. The results have then been used cautiously to derive an estimate of the length of time required to evolve the present landscape, and to indicate the volume of sediment that has been reworked under these conditions and redeposited at greater distance from the bounding fault scarp.

Geology

Important palaeoanthropological finds in the area E of Lake Turkana (Leakey & Leakey 1978) have encouraged intensive investigation of the stratigraphy (Findlater 1978; Vondra & Bowen 1978; Cerling & Brown 1982; Watkins, this volume). The earliest exposed sediments of the Koobi Fora embayment, those of the Koobi Algi Formation, are dated radiometrically as older than 3.8 Ma (Fig. 2). Above these are the Pleistocene Koobi Fora and Guomde Formations and the Holocene Galana Boi Beds. The deposits range in character from pebble conglomerates, through fluvial and littoral sands and silty sands, to lacustrine clays. They are intercalated with widespread tuffs which act as isochronous marker horizons, while palaeosols suggest various stages of calcrete formation. The sediments are variably lithified, and tensile strength comes from cementation rather than from consolidation, associated with overburden pressure. However, there has been considerable minor faulting and flexuring associated with adjustments of the Turkana Rift boundary faults and with downwarping of the sedimentary basin. As a result, local dips can be comparatively high (>30°).

Scarp retreat

Minor arching of the Koobi Fora basin in the late Pleistocene (post-Silbo tuff dated as 0.8 Ma by Cerling & Brown 1982) produced a consequent drainage pattern that dissects the Koobi Fora plateau gravel (which is post-Silbo tuff and pre-Galana Boi, where the Galana Boi are sediments of a lake transgression which reached a maximum at 9500 BP, Butzer *et al.* 1972). This is a deposit that represents the last major overspill of Chew Bahir through the Bakate Corridor (Frostick & Reid 1980). Strike-etching has left a number of scarps of *c.* 80–100 m, typified by the Koobi Fora escarpment and the Karari ridge (Fig. 3). These are heavily dissected by small steep-sided valleys (Fig. 4) which form the fingertips of ephemeral stream networks that expand at the expense of the scarps, in a fashion similar to badlands in other semi-arid regions (e.g. Shumm 1964).

The slopes of the scarp valleys are typically clothed with a coarse regolith. Because gradients are generally steep—everywhere greater than 26°, but generally close to the angle of rest common for loose granular material, 30°—this regolith is a mobile talluvium (a mixture of coarse talus size particles and finer colluvium). Detailed accounts of the processes transporting sediment downslope towards the basal streams are presented elsewhere (Frostick & Reid 1982, 1983). A summary of annual movements of the talluvium's coarser constituents (>10 mm diameter) is given in Fig. 5, and is derived from a three year tracing of

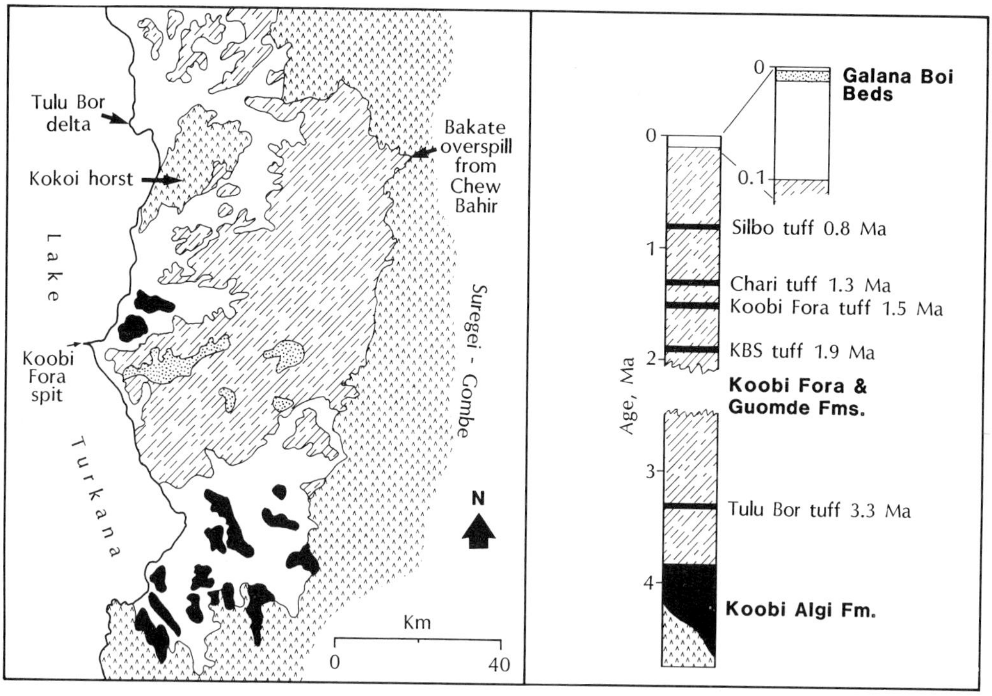

FIG. 2. Geology of the Koobi Fora sedimentary basin E of Lake Turkana (after Vondra & Bowen 1978; Cerling & Brown 1982).

marked particles introduced to a slope typical of the basins' escarpments. As can be seen, the majority of particles that move are displaced comparatively short distances, with 45% moving less than 0.5 m a^{-1}. But some particles travel a long way, with individuals moving in excess of the experimental slope's length of 15 m and entering the stream in a single year.

The least-squares curves of mass movement given in Fig. 5B indicate that there is a tendency for smaller particle displacements in the summit portion of these slopes compared with mid and basal portions. This is because the gathering of overland flow during intense rainstorms leads to its concentration in small rills and chutes away from the slope summit. The addition of an undermining process in the chutes where the finer particles are etched out by flowing water, and the ubiquitous damage caused by rainsplash, encourages slightly greater particle travel rates.

This interaction of processes complicates the mechanics of erosion. It can be seen from Fig. 6A that gravity movements on simple planar slopes unaffected by chutes or rills are functionally related to particle mass. This is to be expected from a consideration of shear force alone which, *ceteris paribus*, is 37 times larger for a 50 mm diameter particle than for one of 15 mm. Where the undermining of grains in chutes occurs (Fig. 6B), the relationship between particle mass and particle velocity is complicated by these additional processes. Fortunately for our attempts to derive a predictive model of erosion, chutes, though important, affect only a small fraction of the typical scarp valley slope.

This experimental work provides a very reproducible estimate of net erosion for this part of the drainage basin. Annual export of sediment to the stream was found to be 2.9, 2.8 and 1.2 kg m^{-2} in consecutive years, with an average rate of 2.3 kg m^{-2} a^{-1}. This is remarkably consistent considering the erratic nature of rainfall in such a region.

Having evaluated process rates in the steeper headwater zones of the basin we now turn our attention to the pediments left by scarp retreat.

Dissection of the pediments

Pediments make up more than half the landscape. They are not planar and are heavily dissected (Fig. 7), in keeping with the high drainage density typical of semi-arid terrain (Frostick *et al.* 1983), and in response to changes in base level of Lake Turkana (see Fig. 8).

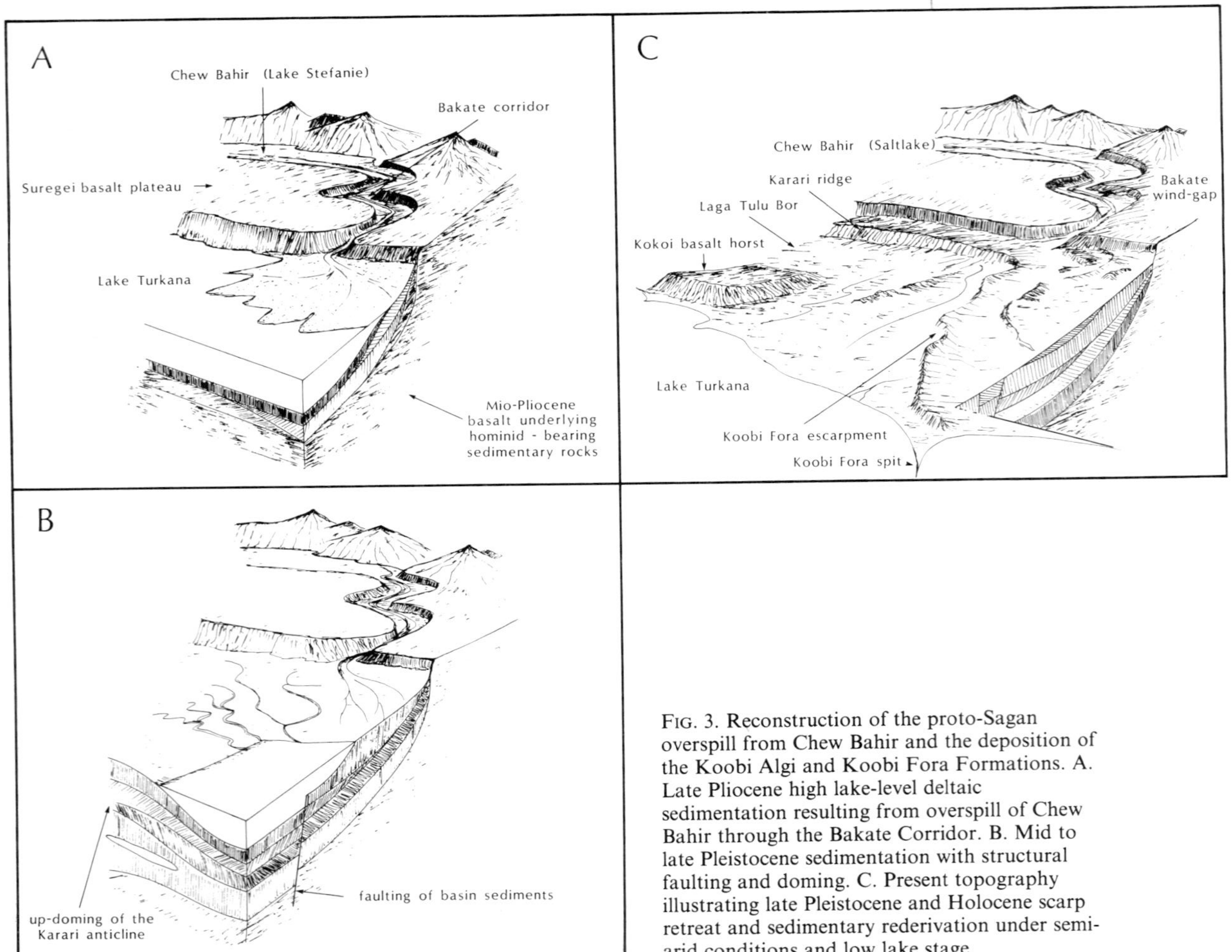

FIG. 3. Reconstruction of the proto-Sagan overspill from Chew Bahir and the deposition of the Koobi Algi and Koobi Fora Formations. A. Late Pliocene high lake-level deltaic sedimentation resulting from overspill of Chew Bahir through the Bakate Corridor. B. Mid to late Pleistocene sedimentation with structural faulting and doming. C. Present topography illustrating late Pleistocene and Holocene scarp retreat and sedimentary rederivation under semi-arid conditions and low lake stage.

FIG. 4. Incised valleys of the Koobi Fora escarpment showing steep slopes clothed with talluvium.

Slopes are generally much less steep than those of the scarp valleys, with modal classes at 15° and 4°. The nature of the regolith surface depends upon the character of the country rock, but is generally less coarse than that on the steeper slopes of the escarpment. This is a pattern similar to that remarked on by Kirkby & Kirkby (1974) for slopes of semi-arid Arizona. Typical median particle size of the pediment soils ranges from 0.46 to 8.6 mm, compared with 26 mm for the talluvium of the scarp slopes (Fig. 9).

The abundance of rainsplash pillars (Fig. 10) on the slopes of the dissected pediment testifies to the importance of rainsplash and overland flow in moulding this part of the landscape. Tests with a constant-head ring infiltrometer reveal an interesting difference in response to rainfall of the two major textural soil types. This has considerable importance for short and long-term erosion in an environment where the surface is largely bare of vegetation at the time rain falls during the two short wet seasons. Sandy and silty soils are shown to have similar infiltration curves before the onset of the rain season (Fig. 11, 29 March). But after the first rainfall (Fig. 11, 31 March) the surface crust of the silty soils—disrupted during the preceding dry season by desiccation cracks—is re-established and final infiltration rates (the infiltration characteristic) are reduced to only 10% of the pre-rain season value (Fig. 11 inset). In contrast, the maintenance of an unclogged open structure at the surface of sandy soils leaves these soils able to accept water at similar rates as the season progresses. Silty soils are, therefore, much more vulnerable to erosion since, *ceteris paribus*, they are likely to generate substantially more overland flow than sandy soils on equivalent slopes (Reid & Frostick 1984).

The picture is, however, complicated by other variables. Principal among these are slope declivity and slope length. In order to determine their effect and to establish rates of erosion, troughs were cut into selected slopes that possess well-defined upper boundaries and these were used to measure runoff and sediment yield during the March–April rain season of 1979. Seventeen determinations of response to natural rainfall were made on three slopes as the season progressed. With no apparent variability in soil between slopes, the runoff percentages are given in Fig. 12 and reveal that the discharge of overland flow decreases with slope length, but increases with slope angle. These values of percentage water yield are similar to those found during a sprinkler experiment on slopes in the Negev Desert (Yair & Klein 1973), but no clearcut relationship between percentage runoff and slope declivity was found at that location.

Despite the low runoff percentages soil erosion is active on slopes of all lengths and declivities. But the entrainment process is selective: the median diameter of the eroded surface wash-material ranges between 0.11 and 0.26 mm, while that of the residual surface material ranges between 0.46 and 8.6 mm (Fig. 9). On these shallow slopes, it is the coarse fragments which remain as a dispersed lag gravel; material greater than 4 mm is almost stationary. This is in direct contrast to the gravitationally more dynamic environment of the steep slopes of the escarpments, where the median diameter of the mobile talluvium is 26 mm, and where there appears to be no distinction between the size of moved and unmoved clasts.

Erosion rates on the slopes of the pediment zone are remarkably high considering that they are the product of a highly variable number of rainfalls per year which normally does not exceed ten. Sediment losses from these zones of the basin range from 0.19 to 0.49 kg m^{-2} a^{-1}.

General model of slope retreat

By combining the average erosion rates in both the escarpment and the dissected pediment zones, a remarkably clearcut relationship between slope

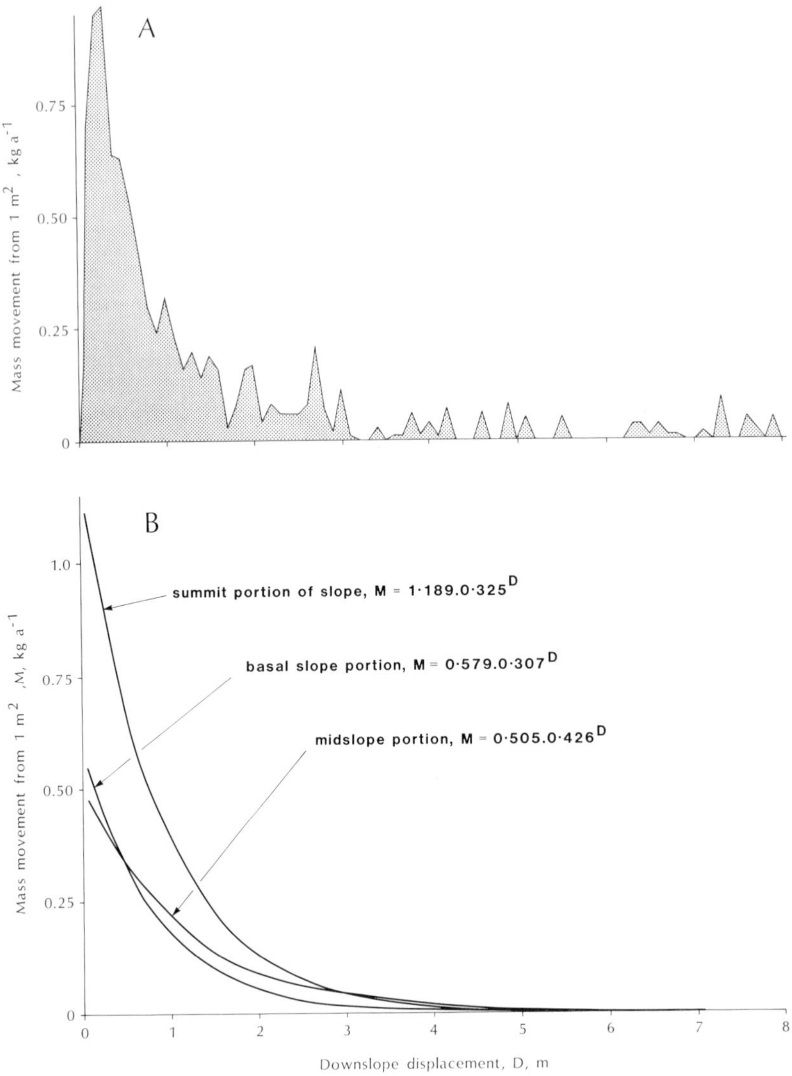

FIG. 5. A. Averaged annual pattern of displacement of talluvium on a 30° scarp valley slope. B. Least-squares curves describing displacement of talluvium from unit area at three positions on a 30° slope.

gradient and slope retreat emerges (Fig. 13). (Slope retreat is defined as back-wearing orthogonal to the slope surface, and is calculated using average bulk densities pertinent to each experimental site.)

The values compare favourably with those established for other semi-arid environments: for example, Yair & Gerson (1974) obtain rates of between 1 and 2 mm a^{-1} for 28° slopes in southern Israel, a value close to the Koobi Fora value of 2 mm a^{-1} for a 30° slope. This accordance, together with the good correlation between slope gradient and erosion on slopes of differing regolith character ($r=0.931$, Fig. 13) encourages a presumption that the relations could be applied throughout the sedimentary basin as a general model of slope retreat.

Evolution of the Koobi Fora landscape

The proto-Sagan, issuing from the Bakate Corridor as an overspill of Chew Bahir, laid down an extensive plateau gravel (Frostick & Reid 1980).

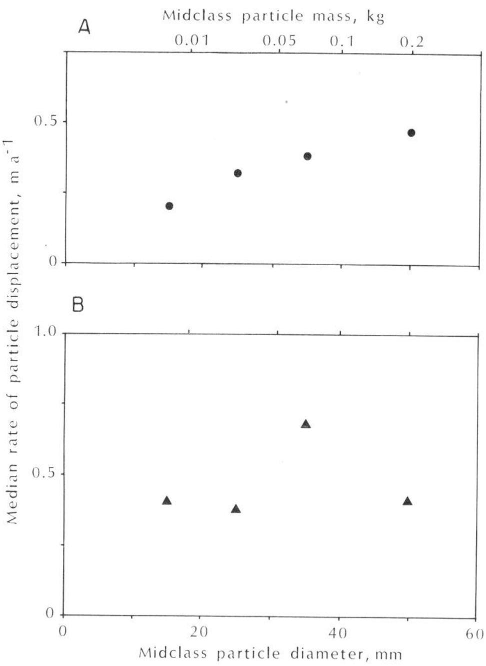

FIG. 6. Relationship between particle mass or size and particle velocity on a 30° scarp valley slope for: A. gravity dominated planar slope forms; and B. portions of slope where overland flow concentrates in chutes and rills.

FIG. 7. View of pediplain left by retreating scarp and dissected by high drainage density stream networks. The strike stream (ephemeral) can just be seen as a line of larger trees running right to left in the background. View from scarp face.

This was its last major contribution to the Turkana trough. The plateau gravel achieves a maximum measured thickness of 6 m and sits unconformably on a surface that is post-Silbo Tuff (which itself is dated as 0.8 Ma by Cerling & Brown 1982). In fact, the plateau gravel is the highest deposit of the basin with parts of it untouched by the Holocene high lake stand of +80 m dated at 9500 BP. It can be envisaged as a near-horizontal sedimentary blanket laid down by a braided stream discharging to a lake at a level of *c.* +100 m (e.g. Fig. 3B).

This level has been taken as a base line from which the present incised topography has developed. Low lake stage and subaerial conditions are presumed to have prevailed for a major part of the period elapsing since deposition of the gravel. This is entirely reasonable, since the onset of the last major high latitude glaciation (the Würm) would have been accompanied by lowered temperatures and greater aridity in equatorial latitudes (Adamson *et al.* 1980). Only minor interludes of high lake stage would have interrupted the process (see Fig. 8), and the last of these, the Galana Boi, can be shown to have left little mark on the landscape, a mere hiccup in the long-term denudation of the basin. We are also making a presumption about the persistence of a semi-desert vegetation association, with its limited influence on the rainfall–runoff process. There is, in fact, very little palynological information about the way climatic change has affected the regional vegetation. That which does exist (Bonnefille 1979; Vincens *et al.*, this volume) confirms the general picture of an arid landscape similar to that of today.

The erosion rates that have been established by experiment on a wide range of slopes were evaluated in the light of the comments of Douglas (1967) and Dunne (1979) that few sites can be free of man's influence in having accelerated the processes of erosion. The exclusion of nomads from the National Park which encloses the experimental sites, the accordance of established local rates of erosion with those reported for other undisturbed semi-arid areas, together with the fact that the Koobi Fora rates are substantially less (only 8%) than those reported by Leopold *et al.* (1966) for a New Mexico location which is suffering accelerated erosion under the influence of mankind, has encouraged us to apply the Uniformitarian Principle in estimating the length of time needed for evolution of the present topography.

In order to do this, the valley slopes of a number of catchments were placed in broad gradient classes, and each class assigned an appropriate erosion rate derived from the least-

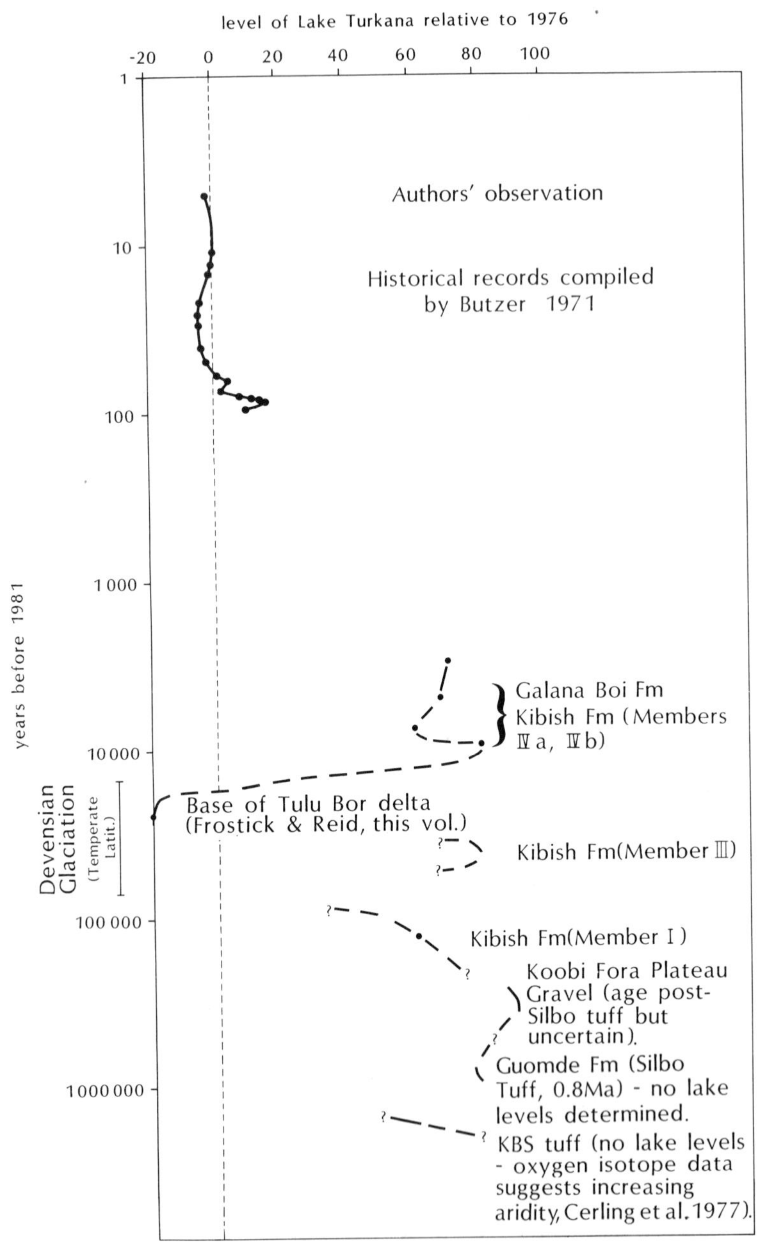

FIG. 8. Mid-Pleistocene to Recent levels of Lake Turkana. (Information compiled from Butzer 1971; Butzer *et al.* 1972; Cerling *et al.* 1977; Frostick & Reid 1980.)

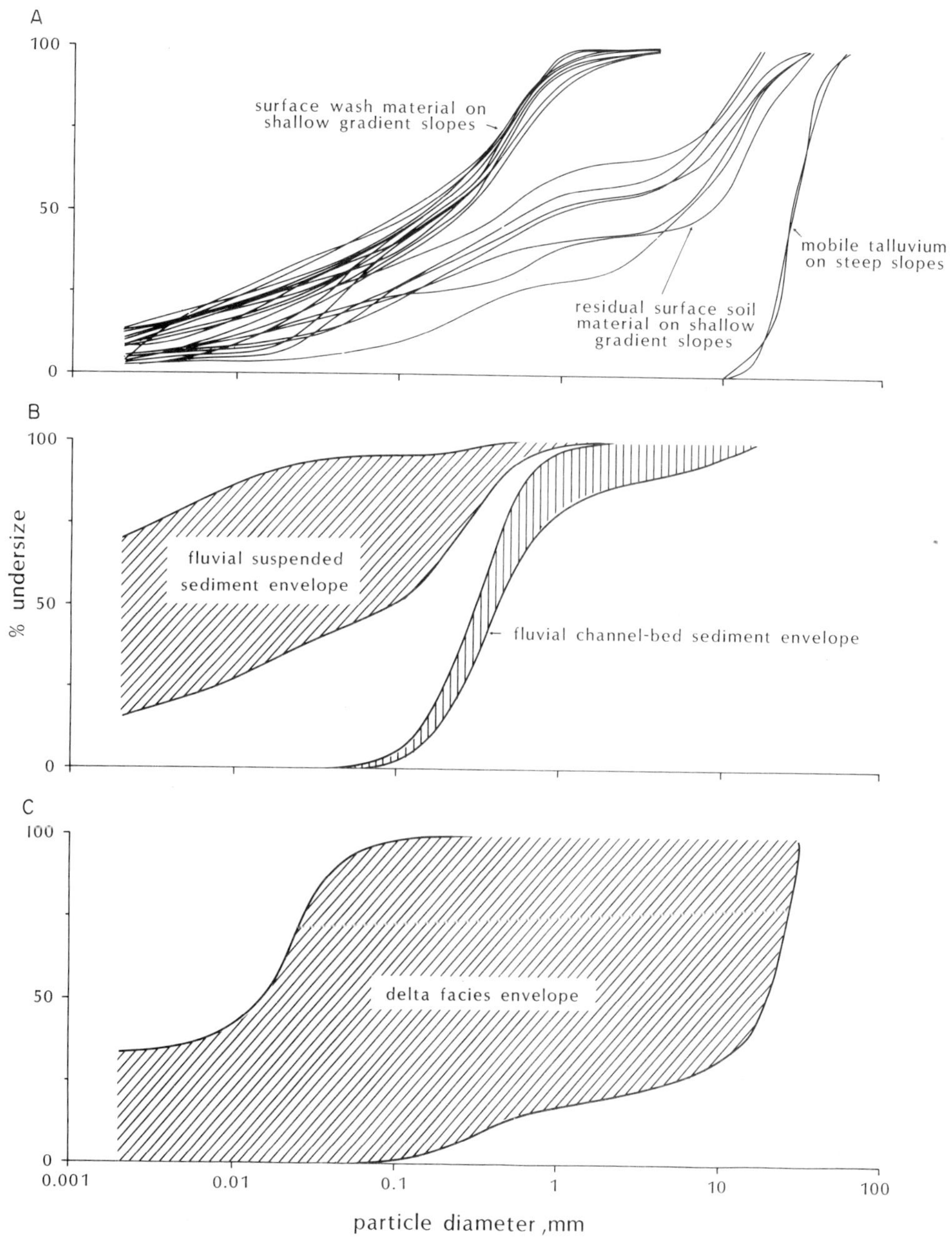

FIG. 9. Size distribution of the arid zone sediment train. A. Sediment source and slope transport: residual and mobile regoliths in the scarp and pediment zones of the landscape. B. River transport: sandy river bed sediments and flash-flood suspended sediment. C. Delta deposits.

FIG. 10. Rainsplash pillar supporting sandstone flag. The pedestal represents about 30 cm of soil lost through erosion. Incipient pedestals can be seen developing under adjacent flags.

squares relationship between gradient and slope retreat given in Fig. 13. Summing for each drainage basin provides estimates of average erosion rates in defined landscape units (i.e. river catchments) having varying proportions of steep (30°), intermediate (15°) and shallow (4°) slopes. The range is 850–1600 t km^{-2} a^{-1}, or, using average measured bulk densities, 630–1180 m^3 km^{-2} a^{-1}. These are very similar to those found in other semi-arid environments.

Taking the Koobi Fora Plateau gravel surface (+100 m) as a more or less horizontal base line, and estimating the present-day hypsometric curve of the whole basin with the aid of aerial photographs and form-line maps, gives a crude measure of the amount of sediment removed by erosion. Using the unit rates established for various individual drainage basins, this translates into a period of between 112,000 and 60,000 years for the evolution of the present landscape. The larger figure is of the right order of magnitude given that the plateau gravel is of uncertain age, but known to be younger than the 0.8 Ma Silbo Tuff. It also conveniently spans the period of the last high-latitude glaciation (Fig. 8).

Conclusions

Since downfaulting of the Turkana trough in the mid to late Pliocene, episodes of sedimentation in the Koobi Fora embayment have been separated by long periods of subaerial erosion and rederivation. The present topography represents a late stage in one of the erosional phases (Fig. 14) in which retreating scarps form a smaller fraction of the landscape than the dissected pediments they leave behind.

Rates of erosion are high despite the low frequency of rainfalls. The poor ground cover given by the xeric vegetation encourages crusting of the soil surface by rainbeat, and this in turn leads to a higher incidence of overland flow and high drainage densities. The tentacles of expanding stream systems consume the escarpments from within, by riddling the leading edge with a network of steep-sided valleys (Fig. 14). Using average values of drainage density and established rates of slope erosion, this converts to an effective scarp retreat of 0.2 m a^{-1}. A pediment is left behind. This is dismantled at rates that may be ten times lower than those of the scarp because of the shallower nature of local slopes and the different combination of surface processes.

With caution, contemporary process rates can be used to estimate the length of time taken for the present landscape of the Koobi Fora basin to evolve. A very plausible answer of 0.11 Ma is obtained. In this period, a large volume of sediment has been distributed in a downstream direction to form the modern deltaic and near-shore lacustrine depositional wedges (see Fig. 9). This rederivation is an important process in the development of rift margin sediments.

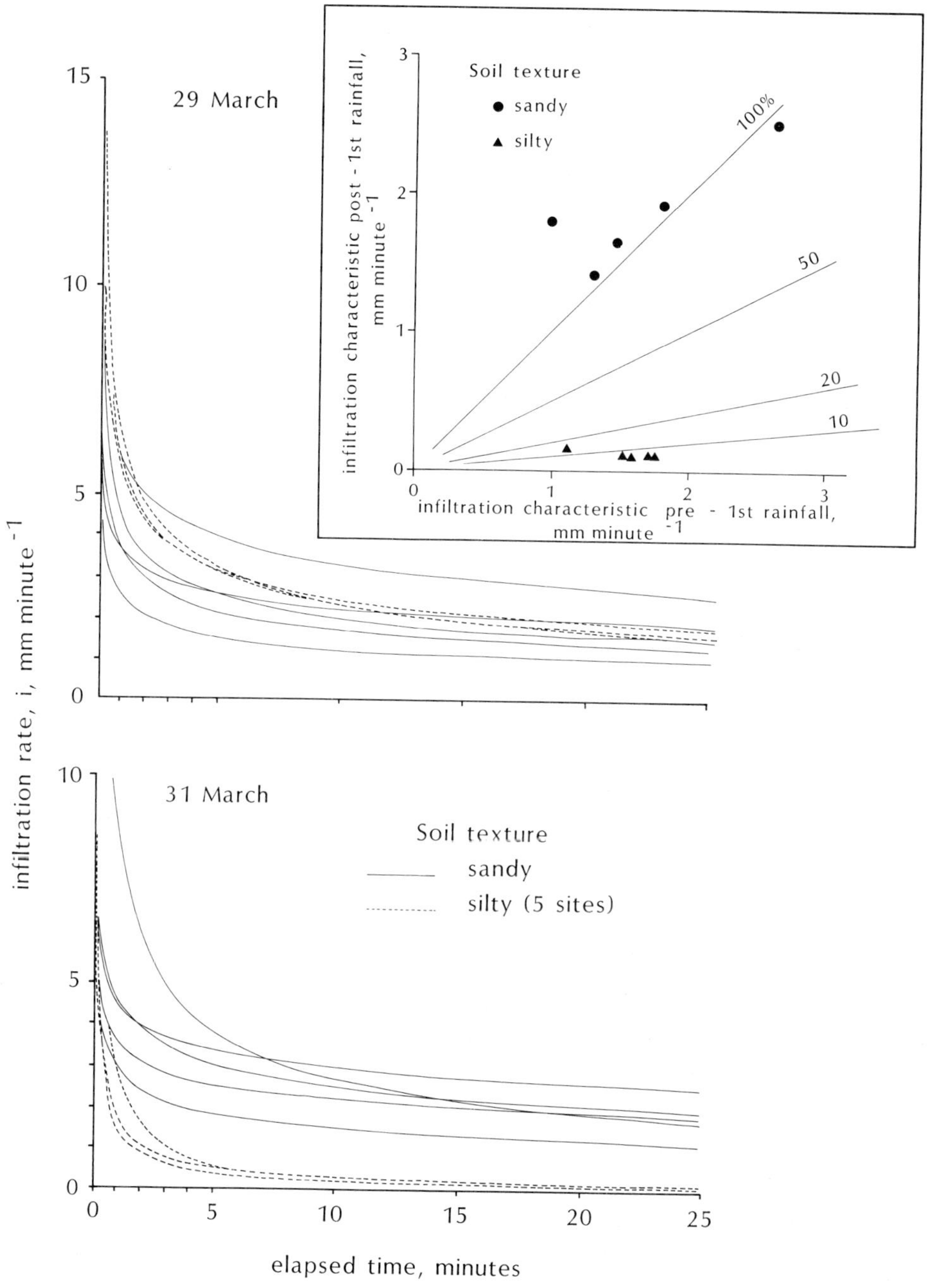

FIG. 11. Infiltration curves for sandy and silty soils of the dissected pediment zone before (29 March) and immediately after (31 March) the first rainfall of the short wet season. The inset illustrates the changes in infiltration characteristic—the final level of infiltration which approximates the saturated hydraulic conductivity of the soil—and reveals no change for sandy soils but a dramatic effect of surface crusting in silts.

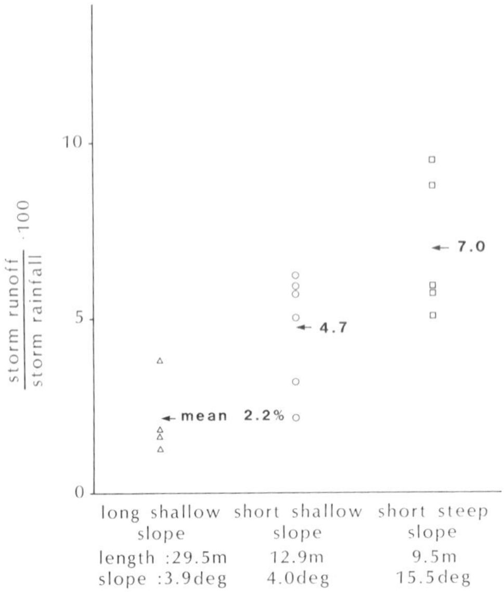

FIG. 12. Overland flow as percentage natural rainfall for three storms on three experimental plots in the dissected pediment zone.

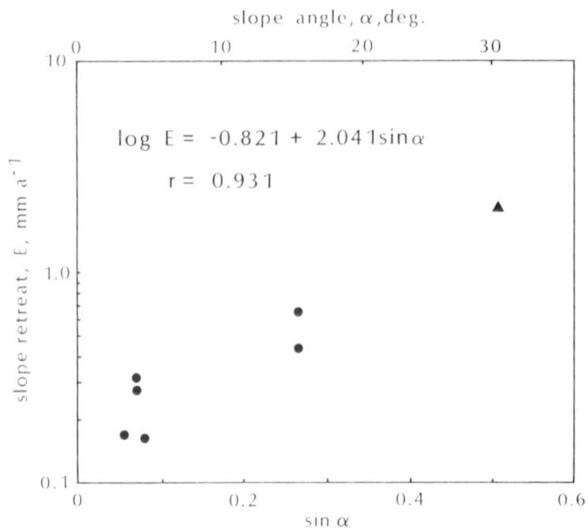

FIG. 13. Summary diagram of slope retreat as a function of slope gradient for a variety of regolith types in both escarpment and pediment landscape zones.

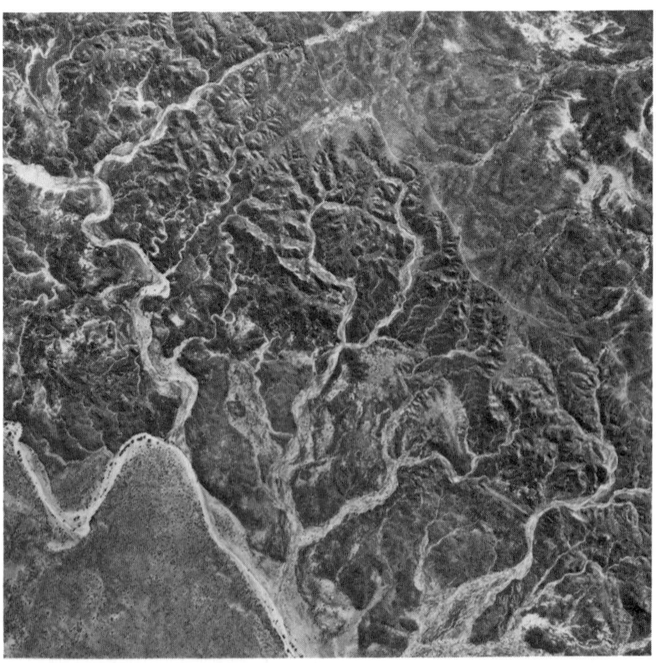

FIG. 14. Aerial photograph of the extending stream networks that consume the leading edge of the Koobi Fora escarpment. Part of the pediment can be seen in the lower part of the photograph between the riddled scarp and the ephemeral strike stream. (Photo by Hunting Surveys Ltd.)

The examination of present-day landscape evolution provides a yardstick by which to judge events more distant in time and represented by disconformity in a stratigraphic sequence. It allows us to assess the likely scale of reworking that has taken place during semi-arid subaerial phases of rift basin history.

ACKNOWLEDGEMENTS: We are grateful for the encouragement and logistical help given by Richard Leakey and Frank Fitch during our many field expeditions. Musaa Kyeva, Cosmas Mbondo and Kumo Ngandi were our valued field assistants. The work was funded by the Natural Environment Research Council.

References

ADAMSON, D.A., GASSE, F., STREET, A.F. & WILLIAMS, M.A.J. 1980. Late Quaternary history of the Nile. *Nature* **287,** 50–5.

BAKER, B.H., MOHR, P.A. & WILLIAMS, L.A.J. 1972. Geology of the Eastern Rift System of Africa. *Spec. Pap. Geol. Soc. Am.* **136,** 67pp.

BONNEFILLE, R. 1979. Méthode palynologique et reconstitutions paléoclimatiques au Cénozoïque dans le Rift Est Africain. *Bull. Soc. Géol. Fr.* **21,** 331–42.

BUTZER, K.W. 1971. Recent history of an Ethiopian delta. *Univ. Chicago, Res. Pap. Dept. Geogr.* **136,** 184pp.

——, ISAAC, G.L., RICHARDSON, J.L. & WASHBOURN-KAMAU, C. 1972. Radiocarbon dating of East African lake levels. *Science* **175,** 1069–75.

CERLING, T.E. & BROWN, F.H. 1982. Tuffaceous marker horizons in the Koobi Fora region and the Lower Omo Valley. *Nature* **299,** 216–21.

——, HAY, R.L. & O'NEIL, J.R. 1977. Isotopic evidence for dramatic climatic changes in East Africa during the Pleistocene. *Nature* **267,** 137–8.

—— & POWERS, D.W. 1977. Paleorifting between the Gregory and Ethiopian Rifts. *Geology* **5,** 441–4.

DOUGLAS, I. 1967. Man, vegetation and the sediment yield of rivers. *Nature* **215,** 925–8.

DUNNE, T. 1979. Sediment yield and land use in tropical catchments. *J. Hydrol.* **42,** 281–300.

FINDLATER, I.C. 1978. Isochronous surfaces within the Plio-Pleistocene sediments east of Lake Turkana. In: BISHOP, W.W. (ed) *Geological Background to Fossil Man.* Scottish Academic Press, Edinburgh, pp. 415–20.

FROSTICK, L.E. & REID, I. 1980. Sorting mechanisms in coarse-grained alluvial sediments: fresh evidence from a basalt plateau gravel, Kenya. *J. Geol. Soc.* **137,** 431–41.

—— & —— 1982. Talluvial processes, mass wasting and slope evolution in arid environments. *Z. Geomorph. Suppl. Bd.* **44,** 53–67.

—— & —— 1983. Taphonomic significance of subaerial transport of vertebrate fossils on steep semi-arid slopes. *Lethaia* **16,** 157–64.

——, —— & LAYMAN, J.T. 1983. Changing size-distribution of suspended sediment in arid-zone flash floods. *Spec. Pub. Int. Assoc. Sedimentol.* **6,** 97–106.

GROVE, A.T., STREET, F.A. & GOUDIE, A.S. 1975. Former lake levels and climatic change in the Rift Valley of Southern Ethiopia. *Geogr. J.* **141,** 177–94.

KIRKBY, A. & KIRKBY, M.J. 1974. Surface wash at the semi-arid break in slope. *Z. Geomorph. Suppl. Bd.* **21,** 151–76.

LEAKEY, M.G. & LEAKEY, R.E. (eds) 1978. *Koobi Fora Research Project. 1. The Fossil Hominids and an Introduction to Their Context—1968–74.* Clarendon Press, Oxford, 191 pp.

LEOPOLD, L.B., EMMETT, W.W. & MYRICK, R.M. 1966. Channel and hillslope processes in a semi-arid area, New Mexico. *Prof. Pap. US Geol. Surv.* **353-G,** 193–253.

REID, I. & FROSTICK, L.E. 1984. Rivers that rarely run. *Geogr. Mag.* **56,** 178–83.

SCHUMM, S.A. 1964. Seasonal variations of erosion rates and processes on hillslopes in Western Colorado. *Z. Geomorph. Suppl. Bd.* **5,** 215–38.

VONDRA, C.F. & BOWEN, B.E. 1978. Stratigraphy, sedimentary facies and palaeoenvironments, East Lake Turkana, Kenya. In: BISHOP, W.W. (ed) *Geological Background to Fossil Man.* Scottish Academic Press, Edinburgh, pp. 395–414.

YAIR, A. & GERSON, R. 1974. Mode and rate of escarpment retreat in an extremely arid environment. *Z. Geomorph. Suppl. Bd.* **21,** 202–15.

—— & KLEIN, M. 1973. The influence of surface properties on flow and erosion processes on debris covered slopes in an arid area. *Catena* **1,** 1–18.

I. REID Dept of Geography, Birkbeck College, University of London, London WC1E 7HX, UK.

L. E. FROSTICK, Dept of Geology, Royal Holloway & Bedford New College, Egham Hill, Egham, Surrey TW20 0EX, UK.

Evolution and sedimentary character of lake deltas fed by ephemeral rivers in the Turkana basin, northern Kenya

L.E. Frostick & I. Reid

SUMMARY: The growth of deltas in Lake Turkana, northern Kenya, is shown to be controlled by the erratic supply of sediment, and the short- and long-term fluctuations in lake level brought about by climatic change and tectonic activity. The sparse and infrequent rainfall of this arid region dictates that all rivers draining to the lake, except the Omo, are ephemeral. Accretion along much of the lake margin is therefore discontinuous, interrupting extended periods during which littoral sediments are reworked by wave action and aeolian processes. Although fluvial activity is infrequent (rivers may run for a total of less than 20 hours per year) the sediment load is high, in common with other arid environments (measured sediment yields reach 1600 t km^{-2} a^{-1}) and delta construction is rapid. Numerous seismic profiles of the Tulu Bor delta—a typical example—reveal a Holocene fill that reaches 20 m. They also show changes in position and gradient of palaeo-channels at successive levels in the sediment wedge as the system has responded to rising lake levels. Older sediments of the delta have undergone early stages of diagenesis and can be differentiated easily from the younger fill. Sediments of the present delta have been sectioned to provide a catalogue of microstratigraphy at type-sites that is useful in interpreting older, Plio-Pleistocene sediments of the Turkana basin. Upper and lower distributary, overbank levée, and interbedded littoral and fluvial environments are identified by specific associations of primary structures and grain-size characteristics. Grain size is also used to trace source areas in the hinterland and so establish sediment transport pathways.

Delta sediments accumulate where river waters disgorge into a larger water body. Development is often rapid, and the character of the material varies widely, not only between deltas but also within a single depositional system. There is considerable interest in the economic potential of these sediments, since the delta plain represents an environment that is both favourable to the formation of coal (e.g. Ferm 1970; Collinson & Banks 1975) and produces facies variations which eventually provide excellent traps for oil and gas (e.g. Fisher *et al.* 1969; Rainwater 1975). This is reflected in a wealth of published information on both the formation of recent deltas (e.g. Sykes 1937; McKee 1939; Coleman & Gagliano 1964; Brown 1969; Oomkens 1974; Wright & Coleman 1974; Coleman & Wright 1975; Khaidarov 1975; Wright 1977) and on the characteristics of ancient equivalents (notably Taylor 1963; Greensmith 1965; Smith 1975; Elliott 1975, 1976; Cherven 1978; Retallack & Ryburn 1982).

The study of deltas dates back to Gilbert's (1885) pioneering research. Later work tended to concentrate on case studies which were then used to establish those features common to all deltas (Fisk 1960; Coleman *et al.* 1964). Wright & Coleman (1973) and Coleman & Wright (1975) were amongst the first to acknowledge and highlight inherent variability. Using field and remote-sensing data, they contrasted the morphology and deposits of 50 modern deltas throughout the world, explaining differences in facies and sedimentary sequences in terms of differences in hinterland, in the river drainage net, and in the receiving basin character.

Studies to date have concentrated on delta development in marine basins and at the mouths of perennial streams. Little attention has been given to lake deltas. Notable exceptions are the work of Axelsson (1967) in Sweden, Brown (1969) in Australia, Butzer (1971) in Kenya, and Hyne *et al.* (1979) in Venezuela. No attention has been given to deltas developing in continental rift lakes and fed by ephemeral streams, although the economic importance of these sediments is now being appreciated. An estimated 10% of the world's known oil reserves are in continental rift basins, where sandy delta-channel and beach sediments abut and overlie organic-rich lake deposits, and are sealed in by the superimposed clays and silts of overbank and lagoon environments. Since these accumulations are buried rapidly in response to crustal downwarping, they are potentially ideal oil and gas reservoirs. But the tectonic instability associated with such lake basins encourages a complex delta plain sedimentation that is not normally experienced in a marine situation. Short- and long-term fluctua-

From Frostick, L.E. *et al.* (eds) 1986, *Sedimentation in the African Rifts*, Geological Society Special Publication No. 25, pp. 113–25.

tions in the rate of basin downwarping lead to changes in lake level that shift shorelines and require adjustments of contributing rivers. This is compounded by seasonal rainfall patterns and by longer-term changes in water balance brought about by climatic changes, both of which influence lake level. These changes alter sedimentation patterns, especially at the delta fringe.

The semi-arid climate that characterizes a large part of the area traversed by the East African Rift System also dictates the magnitude and frequency of sediment-carrying river floods. Although a few deltas may be fed by large perennial streams with marked flood seasons (e.g. Omo, Lake Turkana; Butzer 1971), many are built by smaller, ephemeral systems which flow for only a few hours each year (Frostick *et al.* 1983). These ephemeral flows supply vast quantities of sediment over very brief periods and are interspersed with prolonged intervals during which reworking and redistribution take place. Little is known about such deltas. This study focuses attention on them by examining the three-dimensional morphology and alluvial architecture of a representative example on the shores of Lake Turkana in Kenya's Northern Province.

Location

The Turkana sedimentary basin

Turkana is the most northern of a chain of lakes which punctuate the Kenyan section of the East African Rift Valley (Fig. 1). The present rift bypasses much of the lake, cutting eastwards via the Kino Sogo fault belt to ephemeral Chew Bahir and the Galla lakes of Ethiopia. The N–S trending Turkana basin lies on a westward-tilted fault block which began downwarping in the early Miocene (Cerling & Powers 1977).

The sedimentary fill charts the evolution of the lake from the Pliocene to the present. For much of this time Turkana has been an internal drainage basin, although overflow towards the Nile system in a northwestward direction has occurred in both recent and more distant times (Adamson *et al.* 1980). It has also received water from Chew Bahir during wetter periods when the Ethiopian lakes overflowed through the Bakate Corridor (Frostick & Reid 1980; Reid & Frostick, this volume).

Water balance and circulation of Lake Turkana

Lake Turkana is 250 km long and 15–30 km wide, with mean and maximum water depths of 35 m and 125 m respectively. Rainfall in the area is very variable but averages 300 mm a^{-1}. River inputs to the lake more or less equal evaporative losses, which are estimated at 2844 mm a^{-1} (Frostick & Reid 1982). This demonstrates the importance of the contribution made by the perennial flows of the River Omo, which has its source in the Ethiopian Highlands and drains to the northern shore of the lake (Fig. 1). Nevertheless the monthly change in balance between river discharge and evaporation produces an annual rise and fall in lake level of about one metre.

Apart from the Omo, drainage is dominated by ephemeral streams. The largest of these is the Turkwel which debouches into the SW corner of the lake (Fig. 1). This is followed in size by the Kerio. Other ephemeral streams are much smaller, but are numerous, and cut into the sedimentary and volcanic fill of the basin (Reid & Frostick, this volume).

Internal drainage combined with evaporation produce a saline lake, and solute concentrations currently average 0.25%. But there is little chemical and thermal stratification, largely due to a combination of relatively shallow water depths and strong wind–wave turbulence (Yuretich 1979). The prevailing southeasterly winds dictate an anticlockwise circulation of surface water in the northern half of the lake and a northwards drift in its southern portion.

The deltas of Lake Turkana

The extent to which deltas are able to build, and the character of their deposits, depend upon a number of interacting factors. Among these are: the quantity and composition of the river's particulate load; the frequency and magnitude of flood events; the strength, direction and variability of shoreline currents at the point of debouchment; local rates of subsidence; off-shore gradient; and the rise and fall of base level.

The largest delta currently building into Lake Turkana is that of the River Omo, the present subaerial morphology of which appears to assume a classic, river-dominated birds-foot form similar to that of the Mississippi delta (Coleman & Gagliano 1964). This is a little misleading since Butzer (1971) has shown that the shoreline owes as much to the submergence of interdistributary zones during a recent rise in lake level as it does to levée building and extension of the distributary mouth. Similar inundation has modified other deltas shown in Fig. 1. Indeed the Bura Hasuma is particularly affected since it is not only subject to changes in lake level but also occupies an area that has been locally downwarped for much of the Pleistocene. Its delta plain is consequently dominated by swamp.

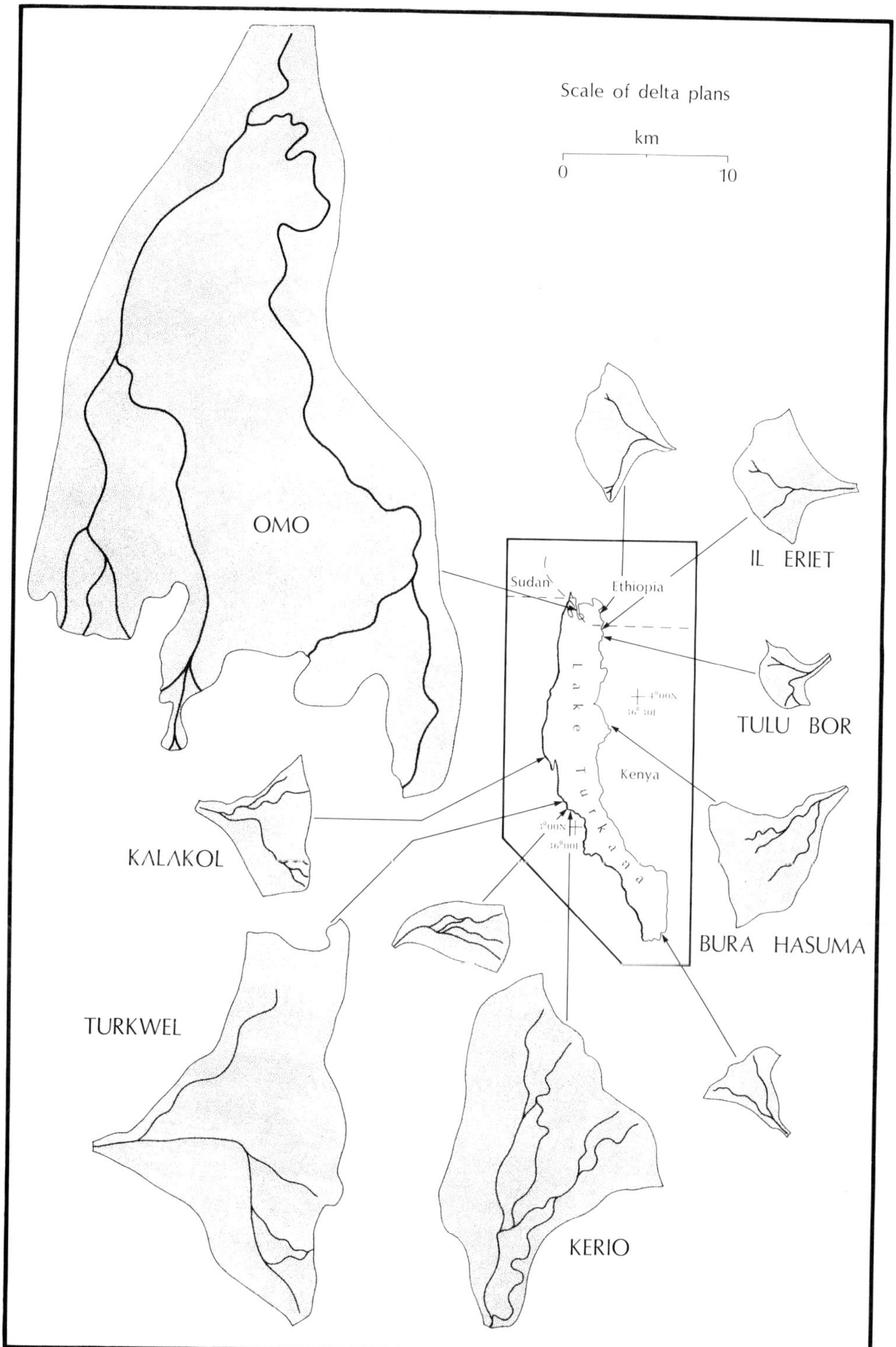

FIG. 1. Location and planimetric maps of the major deltas building into Lake Turkana, northern Kenya.

Longshore movement has comparatively little effect on the shape of large deltas, and most are approximately symmetrical about their river axes (Fig. 1). Contrary to this general pattern, the Turkwel shows a degree of asymmetry. In this case a spit and barrier beach complex is building northwards in sympathy with local lake circulation and wave activity.

Experimental programme

A detailed investigation of the Tulu Bor delta (Fig. 1) was carried out during three field seasons, 1976–78. Aerial photographs were used to delimit the river catchment (Fig. 2A) and the extent of the subaerial delta (Fig. 2B). These were modified in the light of field observations.

Nine traverse lines were established on the delta plain and accurately surveyed for surface elevation (Fig. 2B). Shallow seismic profiles were established along each traverse at *c.* 125 m intervals using a portable hammer seismograph. The results were processed using standard geophysical techniques to give local thicknesses of the alluvial fill and depths to major discontinuities in the sedimentary sequence.

A series of trenches were cut into the delta plain sediments along the main distributary (Fig. 2B, sites A–D). Careful recording of the microstratigraphy at each trench provided the data necessary to establish typical patterns for lateral facies changes, and to determine the nature of stratification, primary structures and so on, which are characteristic of different zones in the delta. Some of these zones are subject to littoral reworking, others are the product of unidirectional river processes alone. Samples were extracted from vertical sections in the side walls of each trench and returned to the laboratory for detailed textural analysis.

Results

The Tulu Bor River and delta

The Tulu Bor River has a catchment area of 147 km^2 (Fig. 2A), comprising effusive and pyroclastic volcanic rocks, and sediments, all of which range in age from the Miocene to the Holocene. It drains to the northeastern shore of Lake Turkana and is an ephemeral stream flowing, if at all, during the March–April and October–November rain seasons.

When floods do occur they are spectacular, achieving peak discharge within 30 minutes of the flood bore (Frostick *et al.* 1983). Flow is supercritical ($Fr > 1$) throughout much of the flood wave and, as a result, the river is capable of carrying large quantities of detritus. Suspended sediment concentrations exceed 15 g l^{-1} (Frostick *et al.* 1983) and bedload transport rates, though unmeasured, must be considerable since shear velocity is high but the flow remains shallow (Reid & Frostick 1985a). The slopes of the catchment yield sediment at rates of between 850 and 1600 t km^{-2} a^{-1} (Reid & Frostick, this volume) most of which is deposited in, or passes through, the delta.

The Tulu Bor River water is more dense than the saline lake waters, although this density contrast changes during the course of a flood as the concentration of suspended solids waxes and wanes. A river sediment plume spreads out over the lake bottom which, in this area, has a comparatively shallow gradient of 0.005 for several kilometres offshore. Bottom friction is therefore the dominant factor controlling offshore deposition of material, at least until the flood wave is in its final stages. At this point, buoyancy becomes an important consideration in the dispersal of the clay-size material that is still suspended in the flow. As a result of these processes, grain size diminishes rapidly offshore, as illustrated for another ephemeral stream delta S of the Tulu Bor where maximum particle diameter falls at a rate of 21 mm per 100 m over the first 0.5 km (Frostick & Reid 1979).

The subaerial Tulu Bor delta plain is crossed by two major distributaries (Fig. 2B). The most southerly of these is currently the more active. A fall in lake level since 1970 has caused the channel to cut into its sediments to a depth of *c.* 1.5 m and a new promontory is being built where it enters the lake. The subaerial delta plain covers an area of *c.* 7 km^2. Sands deposited in the delta are generally quartz-rich, with an average feldspar content of 25%, and also contain a wide range of accessory heavy minerals. The finer-grained sediments contain much comminuted quartz and mica. The clay fraction has a characteristic montmorillonite to kaolinite ratio of between 1.5 and 2.0 (Yuretich 1979).

Geometry of the alluvial fill

The results of the seismic survey illustrate effectively the response of this delta system to long-term changes in lake level. Major cut-and-fill sequences can be distinguished on the basis of seismic velocities. (These were shown to fall into two main classes and to vary with degree of lithification after conducting a series of tests at sites carefully selected for the exposure of different facies.)

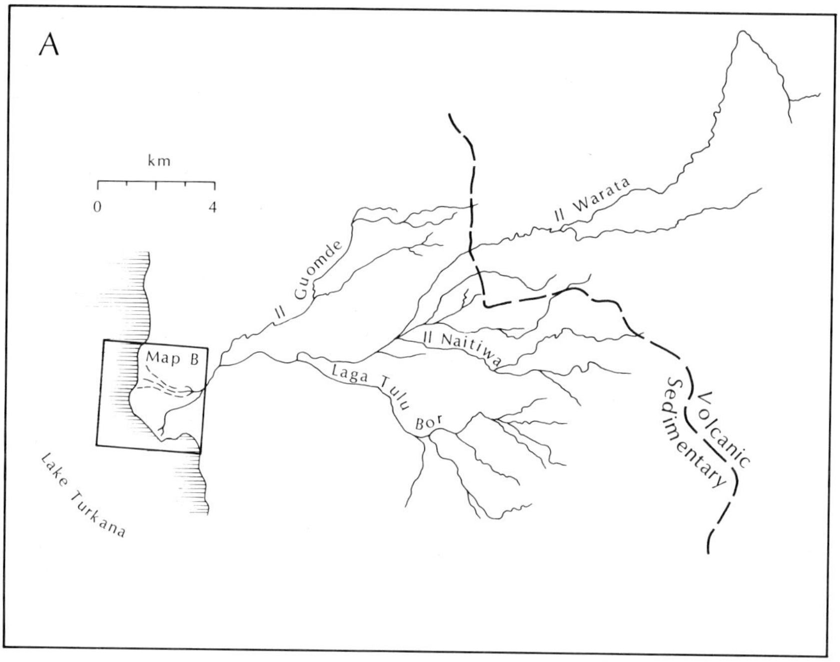

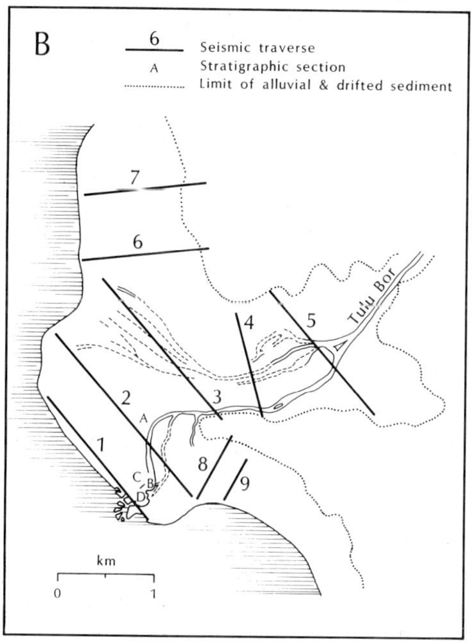

FIG. 2. A. River catchment and drainage net of the Laga Tulu Bor and Il Guomde. B. Planimetric map of the Tulu Bor delta showing major distributaries and the positions of the nine seismic traverse lines.

Although minor localized changes in delta stratigraphy are evident in the seismic profiles, two major discontinuities are traced throughout the alluvial fill. The first of these forms the base of the delta and is the level to which the proto-Tulu Bor and its major right-bank tributary, the proto-Il Guomde, had cut into the Plio-Pleistocene sediments of the Koobi Fora and Guomde Formations (Fig. 3A). The second lies immediately below the less consolidated sediment of the most recent phase of deposition (Fig. 3B).

The similar nature of the pre-delta and present-day topography in the vicinity of the delta plain is striking. The only difference is that, in the older landscape, the Il Guomde and Laga Tulu Bor were almost undoubtedly separate streams, not yet joined until some point further W, and now covered by the transgressive lake and delta deposits. Gradients derived from least-squares regression for each of these channels range narrowly between 0.0048 and 0.0051 (Fig. 3D). These are consistent with drainage towards a lake shore at some distance and at a much lower level than the present one, and are values very similar to those found in the middle reaches of the present river systems in the area. The gradients of the present distributaries are much lower (at 0.0027 and 0.0030) and are not significantly different from that of the channel at the base of the most recent major cut-and-fill (0.0023, Fig. 3D). All three are

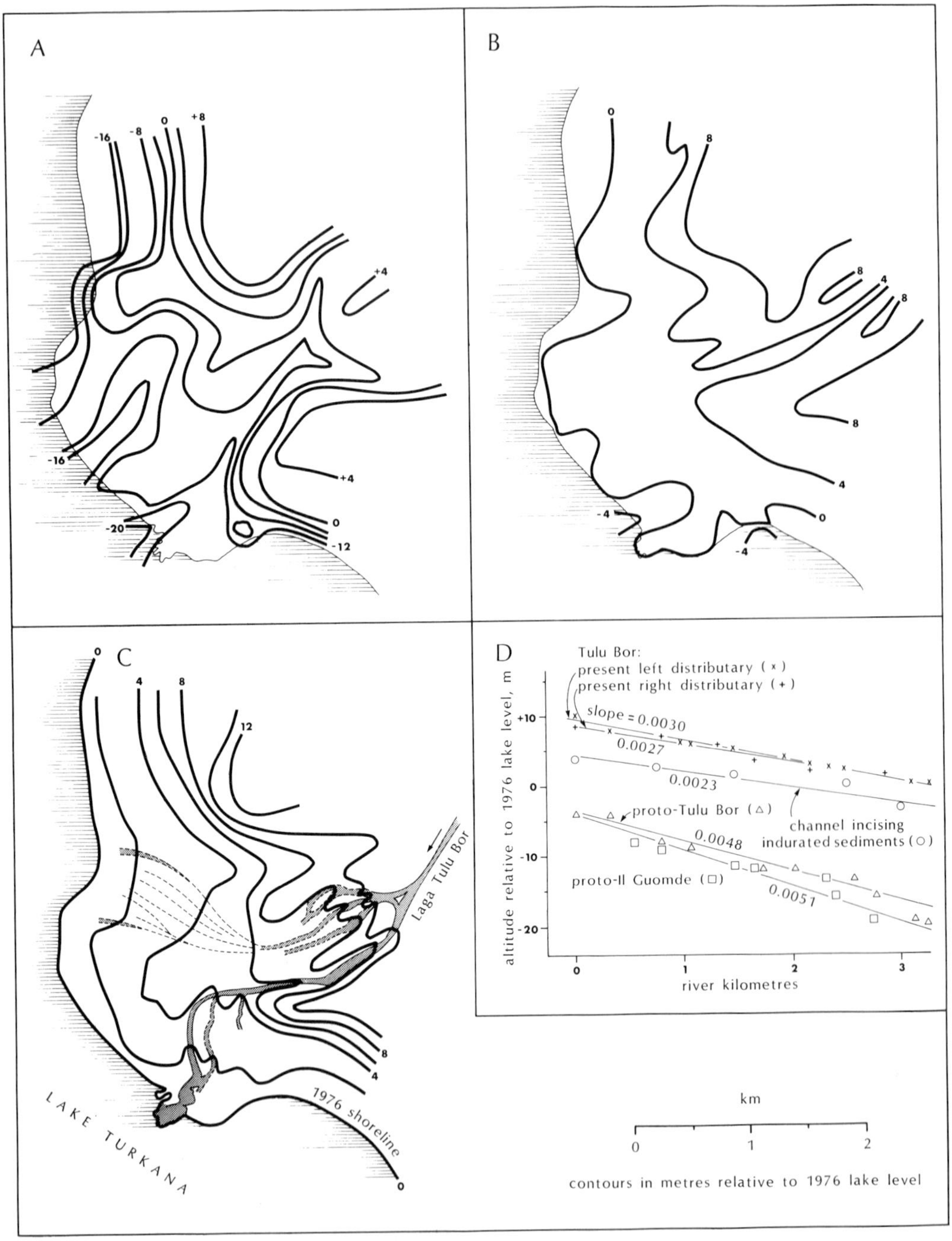

FIG. 3. Geometry of the alluvial fill of the Tulu Bor delta as revealed by seismic profiles. A. Pre-delta topography. B. Top of indurated sediments. C. Present topography. D. Gradients of present and ancient channels.

consistent with river gradients expected of delta plains close to base level.

It is difficult to derive a date for the various horizons within the delta wedge. Direct observations suggest that the fill post-dates the Guomde Formation (0.8 Ma), since sediments of this age are incised by the delta. However, we can estimate the length of time over which the sediments have been accumulated. The volume of material making up the subaerial part of the delta is established from the seismic data as 1.0095×10^9 m^3. Offshore extrapolation of the wedge to the break of slope that is assumed to be the base of the delta front, indicates that there is a further 4.154×10^8 m^3 giving a total of 1.425×10^9 m^3. At an average bulk density obtained for the sediment of 1.6 t m^{-3}, this equals 2.28×10^9 t. By using the rates of sediment production established in long-term slope erosion experiments within the Koobi Fora basin (Reid & Frostick, this volume), it is possible to calculate that the fill has taken a minimum of between 9000 and 17,000 years to accumulate. Because there would be relatively low rates of deposition at the present delta site during the high lake stages that are known to have occurred between 9500 and 3000 BP (see Reid & Frostick, this volume), the development of the pre-delta topography can be tentatively placed within the period of the Devensian glaciation of high latitudes. This is a time of known East African low lake levels (Adamson *et al.* 1980).

A derivation of delta age in this manner makes the assumption that the sediment delivery ratio of the contributing drainage basin has been close to 100%. This is not unreasonable considering the relatively small size of the catchment area, the reasonably long span of time involved, and the fact that there are no obvious significant sediment sinks between the headwaters and the delta. The other major assumption that is made is that all the material delivered by floodwaters will be deposited within the delta wedge. This does not allow for the transport offshore of clay particles, and Frostick *et al.* (1983) show that 24% of the suspended load of flash floods is of this calibre. However, our observations of the sediment plumes of ephemeral streams discharging into the lake suggest that bottom friction and loss of momentum limit penetration to less than 500 m. If this is general, then most of the clays will flocculate in the nearshore zone and will settle onto the delta front. This would account for the substantial deposits of mud at, and close to, the shoreline (Frostick & Reid 1979). With both these assumptions in mind, the estimated time span of delta development becomes a minimum.

An isopach map of the delta-fill indicates that there is some considerable lateral variation in thickness (Fig. 4). Maximum values of over 20 m are located along the axes of the distributary systems, notably close to the lake shore. These would be sites where coarser, more permeable channel sediments occur in ribbon-like lenses and it is these that would eventually provide the traps for migrating fluids, including hydrocarbons.

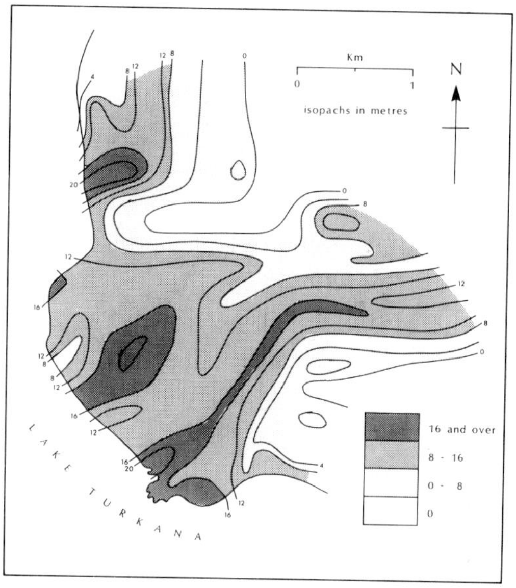

FIG. 4. Isopach map of the alluvial fill of the Tula Bor delta.

Microstratigraphy of the subaerial delta

The microstratigraphy of four trenches cut through the subaerial delta sediments amply demonstrates the complexity of sequence and facies generated within these rift lake deltas (Fig. 5), and helps explain some of the variations encountered within delta plain deposits of the underlying Koobi Fora Formation (Behrensmeyer 1975; Vondra & Bowen 1978). Upper distributary deposits are dominated by channel sands and gravels (Fig. 5A). Coarser, more massive beds up to 0.3 m thick alternate with thinner bedded, plant-rich silts characteristic of overbank environments (Fig. 6). These form a typical 0.4 m thick deposit capping the sequence and actively accumulating as part of the present levée deposit (Fig. 5A). Small troughs that constitute part of the primary structure of these fine silts bear witness to the velocity and transporting capacity of flood spills.

This sequence is made more complex because of the shifting nature of the delta distributaries. Scour-and-fill during brief floods intersperses

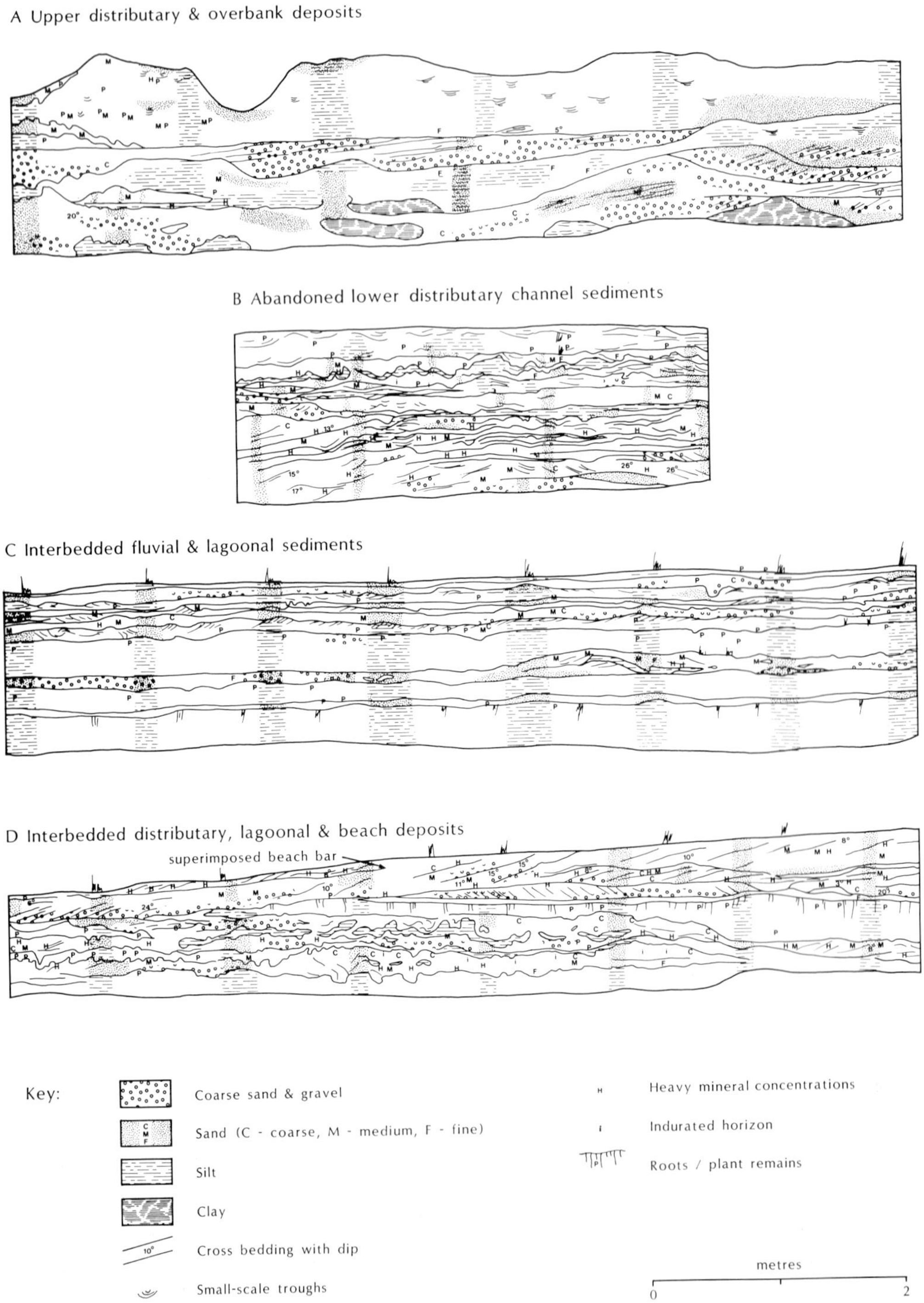

FIG. 5. Microstratigraphy of sections cut into subaerial delta deposits. The sections are arranged such that there is progression from section A which is nearest the delta head, through to section D which is close to the lake shore.

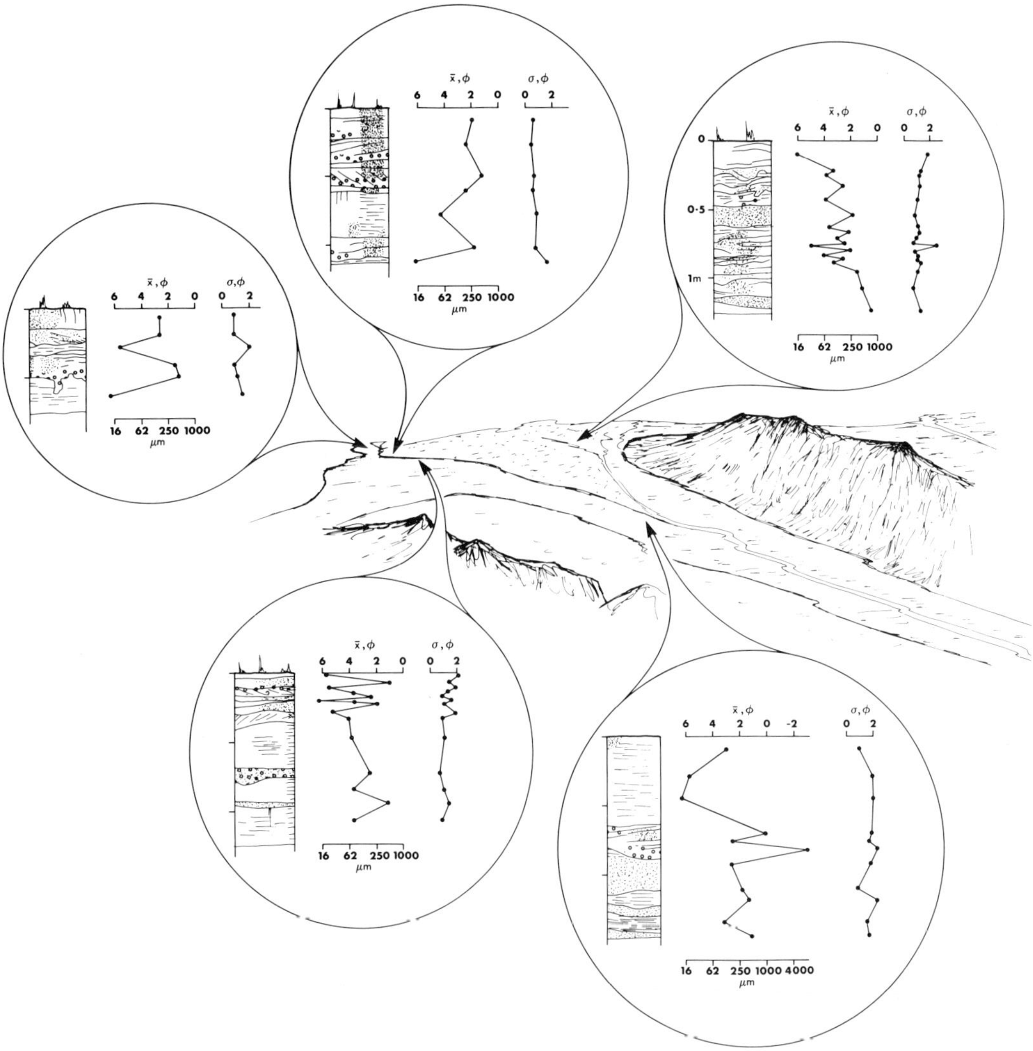

FIG. 6. Summary diagram of textural variations in different parts of the subaerial delta plain.

with prolonged periods of aeolian reworking which redistributes material, depositing it especially in topographic lows such as river channels. These may be sufficiently blocked to divert the waters of succeeding floods and produce the interdigitation of channel and overbank material in this rapidly aggrading system. Similar sequences are common within the Upper Member of the Koobi Fora Formation (Vondra & Bowen 1978).

Further down-delta the reduction in channel gradient and distribution of flow in several channels means that sand and silts dominate the bed material. These accumulate in thinner beds at the sites of the lower distributary channels (Fig. 5B). Flow here is unidirectional and the depositional environment relatively uncomplicated.

Closer to the lake shore the influence of the lake becomes dominant and the deposits become more complex in response (Fig. 5C and D). Seasonal inundation of the delta plain and reworking of deposits by wind waves adds to the effect of sporadic river flows to produce interbedded sediments of contrasting size, sorting and structure. A small rise in lake level submerges the low-lying fringes of the delta plain, creating extensive lagoons and marshes. The fine sediment and plant debris which accumulates in such areas becomes

FIG. 7. A. Bioturbation of lagoon silts and clays as a result of trampling by wild and domesticated animals.

FIG. 7. B. Detail of bioturbation: a single hoofprint.

interbedded with coarser distributary deposits (Figs. 5C and 6). Thicker sequences of lagoon and marsh fine sediments (e.g. seen at the base of Fig. 5C) must represent more persistent periods of relatively high lake-stand. These sediments are similar to those described from the Koobi Fora Formation (Behrensmeyer 1975) which contain abundant vertebrate fossils.

Of considerable interest is the disruption of bedding recorded low in Figs. 5D and 7A. This is bioturbation caused by trampling of the wet sediments at the very edge of the lake by wild and domesticated animals that come to drink. In places it is possible to distinguish individual hoof or footprints preserved in the clay-rich sediments (Fig. 7B). Similar structures have been found in Quaternary delta sediments (Laporte & Behrensmeyer 1980), and have been useful in resolving, among other things, the controversy surrounding the question of bipedalism of our early ancestors when discovered in these ancient sediments (Behrensmeyer & Laporte 1981).

Sediment at the shoreline is inevitably subject to wave activity and this adds yet further complexity to the microstratigraphy (Fig. 5D). Beach bars are formed by bidirectional currents. They may be superimposed on other deposits of different origin, and may be abandoned by a regressive lake. In the sedimentary record these beach deposits are recognizable as well-sorted, cross-bedded sands (Fig. 6, top insert; Reid & Frostick 1985b).

Sedimentary environments

The Tulu Bor delta can be taken as a genotype both for other modern deltas in similar topographical situations, and for Plio-Pleistocene equivalents that have contributed to the sedimentary accumulation of the Turkana basin. It differs from the classic, Gilbert-type lake delta in that the effect of tectonic activity in the rift adds to that of seasonal rainfall and long-term climatic fluctua-

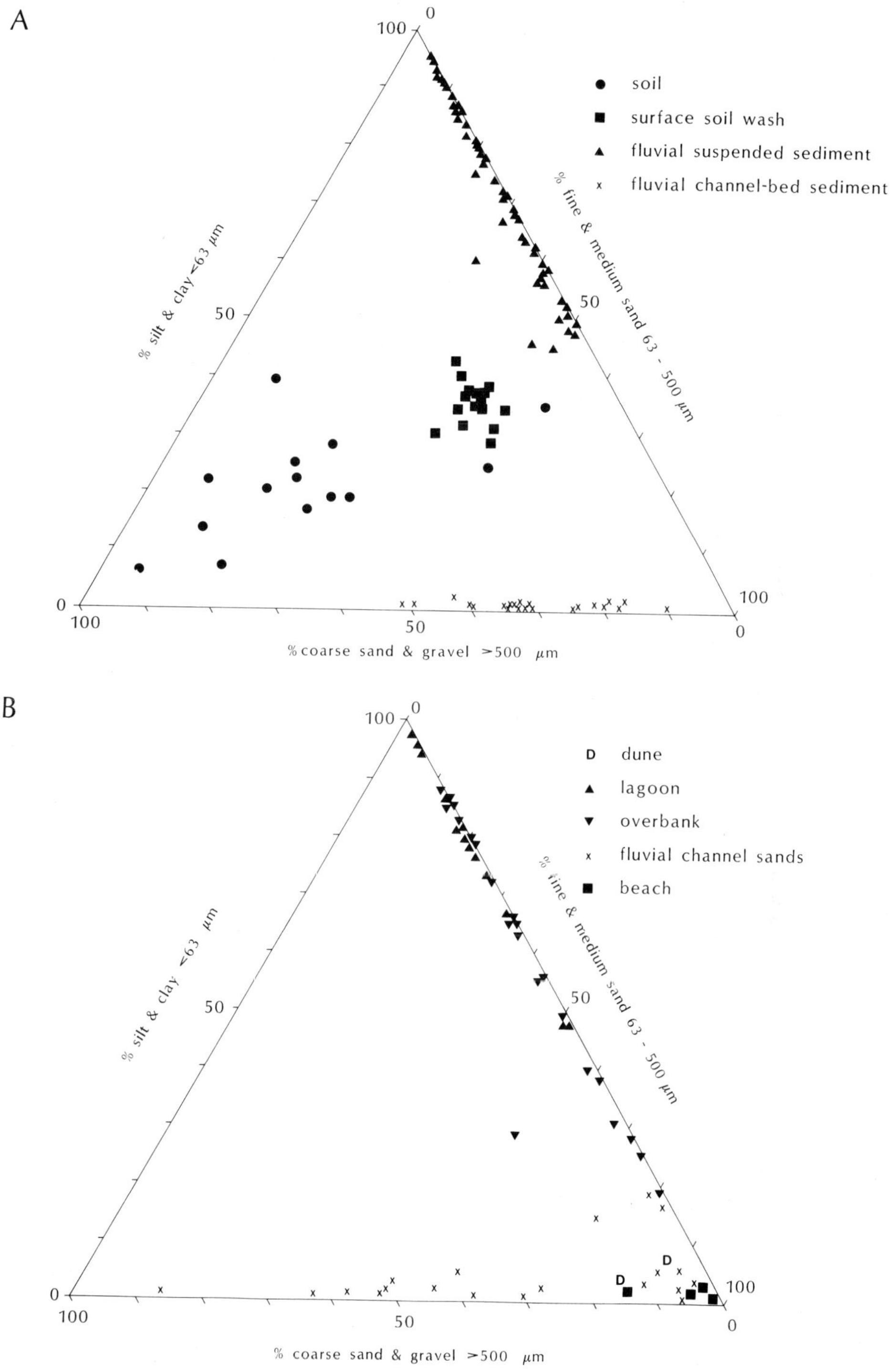

FIG. 8. Source, transport and deposition of sediments within the Tulu Bor system differentiated on the basis of size distribution.

tions, to dictate wide and rapid fluctuations in base level. These require considerable adaptation by the contributing river. This is especially important in complicating patterns of sedimentation since it brings about rapid fluctuations in facies and sedimentary sequence.

In this, as in all deltas (Coleman & Wright 1975), environments with contrasting hydraulic regimes are juxtaposed. High current velocity channels are flanked by interdistributary areas that accommodate shallow overbank flows; still-water lagoons are separated from wave-washed beaches by migrating beach bars, and so on. The grain-size distributions of deposits in each environment reflect prevailing hydraulic conditions, and sedimentary environments can be distinguished from each other using size parameters (Fig. 8B). Coarse sands and gravels occur only in beach and channel environments and are reworked into knob dunes. Lagoon and overbank material is composed almost exclusively of finer sands, silts and clays.

Textural data can be used to locate both source and means of transport for sediment deposited in the delta. All particles originate in the rudimentary soils developed on the slopes within the drainage basin (Fig. 8A). Initial movement in the form of slope wash selects finer material which moves rapidly into the channel system and travels in suspension during flood flows. Larger particles move more slowly and tend to be concentrated as a lag deposit on slope surfaces. When they reach the river they dominate the bed material, travelling largely as bedload. They make up the bulk of the distributary channel, beach and dune sediments in the delta. Suspended sediment is flushed rapidly downstream. Much of the finer clay material escapes to be deposited on the delta slope or in deeper parts of the lake. The remainder settles in the quiet lagoon and overbank environments (Fig. 8).

Conclusions

Despite the sporadic supply of sediment, subsidence and changing lake levels have encouraged the accumulation of almost 2.5×10^9 t of sediment at the Tulu Bor delta. If present-day sediment production rates are extrapolated they give the delta a history in the order of 15–20,000 years and an average accumulation rate over the delta plain and pro-delta shelf of 1 mm a^{-1}. This compares with long-term rates estimated by Brown *et al.* (1978) for the Koobi Fora basin during the Pleistocene of 0.09 mm a^{-1}, during part of which time a large, sometimes perennial, river with a large hinterland (the proto-Sagan) was issuing through the Bakate Corridor.

The incised proto-Tulu Bor and proto-Il Guomde are easily identified by seismic study, and their gradients and depth clearly indicate an ancient lake level well below the present one. The sediments are at their thickest above these channels.

Microstratigraphical and textural analysis allows distinct segregation of facies into four type-environments that are linearly related to distance from either the delta head or the delta fringe. Sedimentary complexity increases towards the delta fringe, where the influence of lake level rise and fall is reflected in the interdigitation of lagoon fines and coarser channel and beach deposits (see Fig. 6). Sedimentary patterns established for the Tulu Bor delta provide an important interpretative tool for the stratigrapher dealing with sedimentation in continental rift basins in semi-arid environments.

ACKNOWLEDGEMENTS: The work was financed by NERC grant number GR3/2897. We gratefully acknowledge the help of R.E. Leakey and Kamoya Kimeu whose enthusiasm encouraged this project. We also thank all of our colleagues associated with the Koobi Fora Research Project for profitable discussions of field problems.

References

ADAMSON, D.A., GASSE, F., STREET, F.A. & WILLIAMS, M.A.J. 1980. Late Quaternary history of the Nile. *Nature* **287,** 50–5.

AXELSSON, V. 1967. The Laiture delta. A study of deltaic morphology and processes. *Geogr. Annl.* **49A,** 1–127.

BEHRENSMEYER, A.K. 1975. The taphonomy and palaeoecology of Plio-Pleistocene vertebrate assemblages, east of Lake Rudolf, Kenya. *Bull. Mus. Comp. Zool.* **146,** 473–578.

—— & LAPORTE, L.F. 1981. Footprints of a Pleistocene hominid in northern Kenya. *Nature* **289,** 167–9.

BROWN, B. 1969. Complex delta—Macquarie rivulet, Lake Illawarra. *Aust. Geogr.* **11,** 87–9.

BROWN, F.H., HOWELL, F.C. & ECK, G.G. 1978. Observations on problems of correlation of late Cenozoic hominid-bearing formations in the North Lake Turkana Basin. In: BISHOP, W.W. (ed) *Geological Background to Fossil Man.* Scottish Academic Press, Edinburgh, pp. 473–98.

BUTZER, K.W. 1971. Recent history of an Ethiopian delta. *Univ. Chicago, Res. Pap. Dept. Geogr.* **136,** 184 pp.

CERLING, T.T. & POWERS, D.W. 1977. Palaeorifting between the Gregory and Ethiopian rifts. *Geology* **5,** 441–4.

CHERVEN, V.B. 1978. Fluvial and deltaic facies in the Sentinel Butte Formation, Central Williston Basin. *J. Sed. Petrol.* **48,** 159–70.

COLEMAN, J.M. & GAGLIANO, S.M. 1964. Cyclic sedimentation in the Mississippi River deltaic plain. *Trans. Gulf Coast Assoc. Geol. Soc.* **14,** 67–80.

——, —— & Webb, J.E. 1964. Minor sedimentary structures in a prograding distributary. *Mar. Geol.* **1,** 140–258.

—— & Wright, L.D. 1975. Modern river deltas: variability of processes and sand bodies. In: Broussard, M.L. (ed) *Deltas—Models for Exploration.* Houston Geol. Soc., 99–149.

Collinson, J.D. & Banks, N.C. 1975. The Haslingden Flats (Namurian G_1) of south-east Lancashire: bar finger sands in the Pennine basin. *Proc. Yorks. Geol. Soc.* **40,** 431–58.

Elliott, T. 1975. The sedimentary history of a delta lobe from a Yoredale (Carboniferous) cyclothem. *Proc. Yorks. Geol. Soc.* **40,** 505–36.

—— 1976. Upper carboniferous sedimentary cycles produced by river-dominated elongate deltas. *J. Geol. Soc.* **132,** 199–208.

Ferm, J.C. 1970. Albegheny deltaic deposits. In: Morgan, J.P. & Shaver, R.H. (eds) *Deltaic Sedimentation Modern and Ancient.* Spec. Publ. Soc. Econ. Palaeontol. Mineral. No. 15, Tulsa, pp. 246–55.

Fisher, W.L., Brown, L.F., Scott, A.J. & McGowen, J.H. 1969. Delta systems in the exploration for oil and gas. *Res. Coll. Univ. Texas Bur. Econ. Geol.* 78 pp.

Fisk, H.N. 1960. Recent Mississippi river sedimentation and peat accumulation. *Congr. Avan. Etudes Stratigraph. Géol. Carbonifere Compte Rend.* **4,** 187–99.

Frostick, L.E. & Reid, I. 1979. Drainage net control of sedimentary parameters in sand-bed ephemeral streams. In: Pitty, A.F. (ed) *Geographical Approaches to Fluvial Processes.* Geo Books, Norwich, pp. 173–201.

—— &—— 1980. Sorting mechanisms in coarse-grained alluvial sediments: fresh evidence from a basalt plateau gravel, Kenya. *J. Geol. Soc.* **137,** 431–41.

—— & —— 1982. Talluvial processes, mass wasting and slope evolution in arid environments. *Z. Geomorph. Suppl. Bd.* **44,** 53–67.

——, —— & Layman, J.T. 1983. Changing size distribution of suspended sediment in arid-zone flash floods. *Spec. Publ. Int. Assoc. Sedimentol.* **6,** 97–106.

Gilbert, G.K. 1885. The topographic features of lake shores. *Ann. Rept. US Geol. Surv.* **5,** 104–8.

Greensmith, J.T. 1965. Calciferous Sandstone Series sedimentation at the eastern end of the Midland Valley of Scotland. *J. Sed. Petrol.* **35,** 223–42.

Hyne, N.J., Cooper, W.A. & Dickey, P.A. 1979. Stratigraphy of intermontane, lacustrine delta, Catatumbo river, Lake Maracaibo, Venezuela. *Bull. Am. Ass. Petrol. Geol.* **63,** 2042–57.

Khaidarov, R.M. 1975. Experience and results of studying the delta dynamics of rivers with a high sediment content (exemplified by the delta of the Ili river). *Vodriye Resursy* **1,** 142–67.

Laporte, L.F. & Behrensmeyer, A.K. 1980. Tracks and substrate reworking by terrestrial vertebrates in Quaternary sediments of Kenya. *J. Sed. Petrol.* **50,** 1337–46.

McKee, E.D. 1939. Some types of bedding in the Colorado river delta. *J. Geol.* **47,** 44–81.

Oomkens, E. 1974. Lithofacies relations in the Late Quaternary Niger delta complex. *Sedimentology* **21,** 195–222.

Rainwater, E.H. 1975. Petroleum in deltaic sediments. In: Broussard, M.L. (ed) *Deltas—Models for Exploration.* Houston Geol. Soc., pp. 3–11.

Reid, I. & Frostick, L.E. 1985a. The incidence and nature of bedload transport during flood flows in course-grained alluvial channels. *Earth Surf. Proc. Landforms* **10,** 33–44.

—— & —— 1985b. Beach orientation, bar morphology and the concentration of metalliferous placer deposits: a case study, Lake Turkana, N. Kenya. *J. Geol. Soc.* **142,** 837–48.

Retallack, G.J. & Ryburn, R.J. 1982. Middle Triassic deltaic deposits in Long Gully, near Otematata, North Otago, New Zealand. *J. R. Soc. New Zeal.* **12,** 207–27.

Smith, A.E. 1975. Ancient deltas: comparison maps. In: Broussard, M.L. (ed) *Deltas—Models for Exploration.* Houston Geol. Soc., pp. 529–55.

Sykes, G.G. 1937. The Colorado delta. *Carnegie Inst. Washington Publ.* **460,** 193pp.

Taylor, J.H. 1963. Sedimentary features of an ancient deltaic complex; the Wealden rocks of south-eastern England. *Sedimentology* **2,** 2–28.

Vondra, C.F. & Bowen, B.E. 1978. Stratigraphy, sedimentary facies and palaeo-environments, East Lake Turkana, Kenya. In: Bishop, W.W. (ed) *Geological Background to Fossil Man.* Scottish Academic Press, Edinburgh, pp. 395–414.

Wright, L.D. 1977. Sediment transport and deposition at river mouths: a synthesis. *Bull. Geol. Soc. Am.* **88,** 857–68.

—— & Coleman, J.M. 1973. Variations in morphology of major river deltas as functions of ocean wave and river discharge regimes. *Bull. Am. Assoc. Petrol. Geol.* **57,** 370–98.

—— &—— 1974. Mississippi river mouth processes: effluent dynamics and morphologic development. *J. Geol.* **82,** 751–78.

Yuretich, R.F. 1979. Moden sediments and sedimentary processes in Lake Rudolf (Lake Turkana) Eastern Rift Valley, Kenya. *Sedimentology* **26,** 313–31.

L.E. Frostick, Dept of Geology, Royal Holloway & Bedford New College, Egham Hill, Egham, Surrey TW20 0EX, UK.

I. Reid, Dept of Geography, Birkbeck College, University of London, London WC1E 7HX, UK.

The distribution of coarse-grained sediments in modern Lake Turkana, Kenya: implications for clastic sedimentation models of rift lakes

A.S. Cohen, D.S. Ferguson, P.M. Gram, S.L. Hubler & K.W. Sims

SUMMARY: The nearshore environment of Lake Turkana, Kenya was studied to determine how coarse clastic sediments are being distributed in the modern lake. Four lake margin study localities were examined in detail: (1) an area of basaltic colluvial headlands; (2) an area of sand, sandy silt and silty mud substrates; (3) an area of highly variable substrate conditions; and (4) an area of alternating basaltic headlands and gravel/shingle terraces. Based upon quantitative observations in these areas and qualitative observations elsewhere in the lake, we believe that two interacting factors have the greatest influence on the distribution of nearshore clastics. The first is a sediment distribution system which is primarily dependent on localized supply, via rockfall and intermittent streams. The second is the presence of physiographic barriers, principally faulted and eroded volcanic flows, which border on local coarse sediment delivery systems and isolate them from adjacent systems. Volcanic barriers may operate at various scales in regulating the delivery systems of clastic sediments in rift lakes, from localized facies disruption, as observed in Lake Turkana, to the isolation of entire sedimentary basins. At the largest levels the presence or absence of volcanic activity may have an important bearing on the controversy of hydrocarbon accumulation in rift lakes.

Patterns of sedimentation in rift lake systems have traditionally been investigated within the context of palaeoclimatic, palaeohydrologic or palaeotectonic models, each of which has led to important advances in our understanding of sedimentary processes in the lacustrine environment. In this paper we will take a somewhat different approach to the question of facies geometry regulation, based upon preliminary analyses of marginal lacustrine sediments from Lake Turkana, Kenya.

We wish to emphasize from the outset the dual nature of this paper. The first section will consider the results obtained to date from a grain size and mineralogical analysis of the nearshore, modern sediments of Lake Turkana. We will attempt to demonstrate, through an analysis of sediment distribution, how localized geomorphic features (both subaerial and subaqueous) which are primarily of volcanic origin have exerted a major influence over the geometry of nearshore clastic sediments in the lake.

In the second section of this paper, we will generalize from the Turkana example, to develop the basic elements of a process-response model of volcanogenic controls on lacustrine, rift sedimentation. Because the model has some important theoretical *and* applied consequences (especially with regard to the accumulation of hydrocarbons), we will also include in our discussion some suggestions as to how the model might be tested.

Study location and field methods

Lake Turkana (Lake Rudolf in literature prior to 1976) lies in northernmost Kenya, partially within the Kenyan segment of the Gregory Rift. It has received considerable scientific attention due to the extensive Plio-Pleistocene fauna recovered from along its northern and eastern shores (Coppens *et al.* 1976), which in turn has stimulated much interest in the region's stratigraphical history.

Considerably less attention has been paid by geologists to the modern lake. Yuretich (1979) described the sedimentology of offshore regions throughout the lake basin, with emphasis on fine clastic distribution, clay mineralogy and sedimentary geochemistry. The reader is also referred to Yuretich (1976) and Yuretich & Cerling (1983) for chemical data on modern Lake Turkana. Hopson (1983) has published an extensive account of the fisheries biology of Lake Turkana, which contains a wealth of environmental information of interest to geologists. Butzer (1971) has discussed the modern environments associated with the Omo River delta, at the extreme N end of the lake. Cohen (1982) has described the nearshore environments of Lake Turkana, with particular emphasis on faunal–sediment interrelationships.

Lake Turkana is fed by one perennial stream (the Omo River), two quasiperennial streams (the

From Frostick, L.E. *et al.* (eds) 1986, *Sedimentation in the African Rifts*, Geological Society Special Publication No. 25, pp. 127–39.

Kerio and Turkwel Rivers), and numerous seasonal streams and springs. These streams, along with local aeolian sources and colluvial rockfall, provide pathways for sediment input into the lake. Frostick & Reid (1979, 1980) have previously discussed fluvial sediment transport near Lake Turkana.

During 1978 and 1979, A.C. conducted an extensive dredging programme of modern marginal and deep water lacustrine sediments throughout Lake Turkana. Ekman dredge samples were taken from 331 stations (Fig. 1). One hundred and twenty-four of these localities lie within four primary lake-margin (less than 10 km from shore) study areas discussed below. Sediment sub-samples representing surficial bedding surfaces from these stations were subjected to quantitative (dry sieve/pipette) grain size and mineralogical analysis.

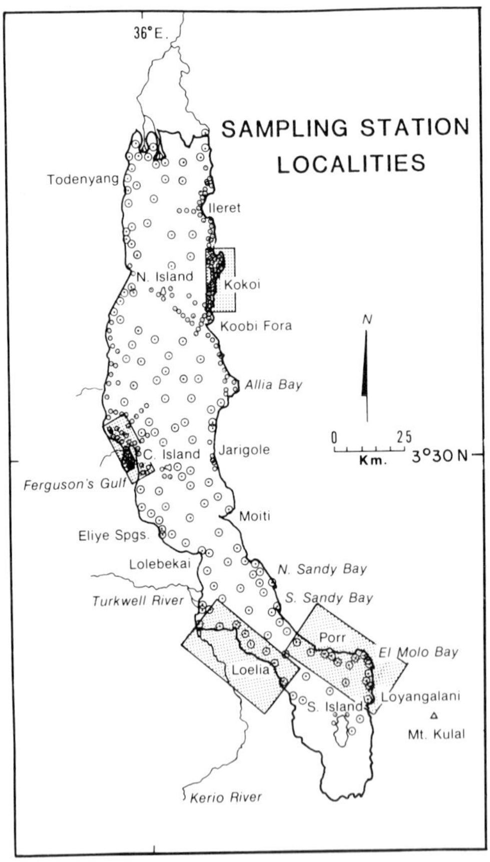

FIG. 1. Sediment sampling station localities; 331 stations recorded from 1978 and 1979 surveys. Where several stations are in very close proximity, a single dot indicates all of them. The Omo River, discussed in the text, enters the lake from the N. The four inset rectangles are the detailed study areas discussed in the text and shown at larger scale in Figs. 2–5.

Analysis of environment–substrate interrelationships at Lake Turkana has shown that five major substrate classes exist in the marginal lacustrine environment. If observed in the rock record, they would appear as distinctive lithofacies associations (or in the fifth class, erosion surfaces). A detailed description of these substrate types and their associated environments, and a lake margin substrate map of Lake Turkana, are given in Cohen (1982). Only a brief description of each type is given here, to clarify our discussion for the reader.

Marshland substrates

Vegetated, organo-sedimentary substrates are conspicuously rare in Lake Turkana, in contrast to many other East African lakes. They occur in the Omo delta region, and to a lesser degree around the Kerio-Turkwel delta, and in south-central Allia Bay. Generally, the heavy vegetation cover of these areas resulted in a low relief, and a low mechanical energy shoreline, due to wave baffling.

Silty mud substrates

This substrate type occurs in most protected embayments along the western shore, as well as on the eastern shore, N of Allia Bay. It is associated with extremely low relief, so that such regions are variously inundated or exposed over large areas during the flood and low water seasons respectively.

Sand and sandy silt substrates

Sand substrates are distributed primarily in lengthy strips along the western shore, from Loelia almost to Todenyang, unbroken except for stretches around the Kerio-Turkwel delta and Ferguson's Gulf. Sandy substrates are found less commonly along the more irregular eastern shore. Vigorous longshore drift in these areas is responsible for continuous sediment transport parallel to the shoreline.

Rocky gravel and shingle substrates

This substrate type is most common on the southeastern side of the lake and around the volcanic islands. It is typically expressed as regions of sedimentation offshore from cuspate beaches, which are themselves interspersed with colluvial headlands.

Continuous rocky colluvial headlands

Extensive cliff-faced shorelines occur primarily in the southernmost parts of Lake Turkana, and more sporadically near Kokoi, between Jarigole and Moiti, and between Porr and El Molo. The substrates in these regions are either Precambrian basement rocks (predominantly in the SW) or basaltic lava flows. This substrate differs from previously mentioned ones, in being primarily an erosional environment rather than a depositional one.

Due to the great heterogeneity of nearshore clastic substrates, Lake Turkana is an ideal locality for investigating how various mechanisms of, and barriers to, sediment transport affect lacustrine sediment distribution. We have chosen to contrast depositional styles between areas where volcanic landform barriers to sedimentation are abundant (predominantly along the eastern shore) and those areas where they are essentially absent (mostly in the NW). This distinction is a particularly important area for investigation, because most rift valley lakes are *either* volcanically dominated basins (hereafter referred to as VDBs) *or* non-volcanically dominated basins (NDBs). Turkana is almost unique among the African Rift lakes in having substantial proportions of its coastline of both volcanic and non-volcanic character. In one lake therefore, it is possible to contrast the microscale facies association and geometries produced by both situations.

Substrate geometries

To date, we have concentrated our analysis on four sets of samples from four regions of lake shore, which taken as a group, are representative of substrate variation in the marginal lacustrine environment of Lake Turkana. These regions, shown in Fig. 1, are: (1) the large inlet immediately N of Kokoi, on the NE side of the lake; (2) the region around and immediately to the N of Ferguson's Gulf; (3) the region surrounding Loelia, on the SW shore; and (4) the region from immediately N of Porr to Loyangalani, on the SE shore.

For each region sets of transects running both parallel and perpendicular to the shoreline were analysed for grain size and composition trends. Shoreline parallel transects were run *c*. 200 m offshore. Between three and five shoreline perpendicular transects were conducted for each study area, with sample spacings of *c*. 0.5 km. Samples were taken within the swash zone (here defined as the zone between the breaking waves and the highest wave wash during non-storm intervals) with a spacing interval of *c*. 1 km. These data were then compared with local physiography and drainage patterns, all of which are summarized in Table 1.

Kokoi area (Fig. 2)

This area consists of a 6 km reach of continuous, Mio-Pliocene basaltic, colluvial headlands in the southern part of the inlet. Here there are gravel and shingle substrates in some places. Sand and sandy silt substrates occur along the shoreline both N and S of the headland (shown by the angled pattern in the isopleths on Fig. 2A).

The more consistent patterns of offshore sand belts (coarser than 3ϕ as measured normal to the shoreline) occur at the N and S ends of the study area (away from the rocky headlands) where sands with mean size coarser than 3ϕ occur in a strip from 0.5–2 km wide. The widths of the medium-coarse sand belts (coarser than 3ϕ) along the rocky headlands are highly variable (even within small areas) and are generally much narrower, ranging from 0.1–1.5 km. At the transition between the two shoreline types the sand body geometries are wedge-shaped, while the remaining regions have strip-like sand bodies which seem to conform to the shape of the coastline.

Sorting variations in the nearshore sands are shown in Fig. 2C. The values given for each strip of shoreline indicate the range of the graphic standard deviation values as defined by Friedman & Sanders (1978). The poorest sorting is observed around the stream channel mouths (particularly near the Tulu Bor, but also near some small streams entering at the S), and along the rocky headlands. Sorting improves with distance from both of these types of features parallel to shoreline.

Poorer sorting in the vicinity of the seasonal streams is easily explained by variations in the streams' flow regime throughout the year, which produces an admixture of grain sizes on the stream deltas. These become progressively better sorted as the sands are carried away from the deltas.

Poorer sorting along the headlands, similarly, results from an admixture of colluvial, volcanogenic particles with grains derived from the surrounding stream sources. Longshore current flow and turbulence in the nearshore environment is apparently incompetent to carry much sand across the deep water divide formed by the headland.

Sand fraction compositions are summarized in

TABLE 1. *Comparison of physical characteristics of the four principal study areas*

Characteristics	Kokoi	Ferguson's Gulf	Loelia	Porr
Major Drainages (size of individual drainage basin(s) in km^2)	Laga Tulu Bor (400)	Kalikol River (2200) Kalorukungole (200), shown entering Ferg. Gulf from S (not labelled on map)	Kerio River (14,000) Other drainages all < 150 km^2	All small (< 300 km^2)
Sediment provenance within drainages	Laga Tulu Bor: Pliocene–Holocene fluvio-lacustrine sediments and volcanics Small drainages: Mio-Pliocene basalts and alluvium	Kalikol-pε met. Tert. basalts, Miocene–Holocene fluvial sediments Kalorukungole: Tert. basalts, Pleist.–Holocene alluvium Small drainages: Tert. basalts and alluvium	Kerio-pε gneiss, Tert. basalts, minor Miocene–Pleist. fluvio-lacustrine sediments. Small drainages: Tert. basalts	pε schists and gneisses, late Mesozoic–Miocene fluvio-lacustrine sed., Tert.–Quaternary basalts
Shoreline substrates	Basaltic colluvial headlands in central area; sand and sandy silts to N & S	Sands S of Kalikol delta; silts N of Kalikol delta	Continuous volc. headlands S of Loelia. Sand and sandy silts N of Loelia. Marshlands and silty muds around Kerio delta	Alternating basaltic headlands and gravel/shingle terraces. Sands and sandy silts in protected embayments
Bottom topography along shoreline	Sharp dropoff along headlands; gentler to N & S	Conformable with shoreline to depth of 10 m	Mirrors shoreline substrate variations; dropoff > 50 m within 2 km of shore; relief and slope decline rapidly N of headlands	Steep dropoffs to > 40 m, some small terraces
Prevailing nearshore surface current direction(s)	S during Omo flood (July–November), N during remainder of the year	N and S (converging on spit) during Omo flood, S during remainder of the year	From SE throughout the year	From SE throughout the year
Volcanic headland(s) present along shoreline?	Yes	No	Yes	Yes
Mean grain size data for sand belts	Swash zone range: 1.49–2.82ϕ, decreases away from Laga Tula Bor (1.5–1.95ϕ mean at mouth, 1.88–2.21ϕ 2 km from mouth, N & S). Size also decreases away from headlands: 1.75ϕ mean at 0.25 km from S end of headland, 2.28ϕ mean at 1 km S of headland, 2.82ϕ mean at 2 km N of headland. Grain size changes erratic going offshore	Fine sands occur in narrow irregular zones following bathymetric contours, coarse sands (coarser than 0ϕ) are rare	Extremely variable with silts (5.19ϕ mean) at Kerio River mouth and S to prominent spit, then rapid coarsening to coarse sands and gravels (-2ϕ mean) near Loelia	Pattern follows bathymetry, extremely variable, with medium sands (coarser than 2ϕ mean) in El Molo Bay and extreme N of study area. Pebbly mudstones elsewhere

Sorting variations in sand belts	Poorest near Laga Tulu Bor delta, small streams in S and along headlands	Moderate throughout study area. Possibly sorting improves between S end of study area (mean s.d. = 1.43ϕ) and N end of major sand spit (mean s.d. = 0.92ϕ)	Very poor near headlands (up to 2.69ϕ s.d.) improves to N (0.85 mean s.d. near Kerio River in silts)	Very poor near headlands (1.72–2.46ϕ s.d. around Porr, 1.97–2.02ϕ s.d. at N end of El Molo Bay). Sorting improves slightly (1.37–1.87ϕ s.d.) in inter-headland area, and greatly (0.69–1.07ϕ s.d.) in extreme N (away from headlands)
Composition variations in sand belts	Avg. composition from throughout study area: quartz (85%)—subangular and pitted: volc. rock fragments (12.5%)—subangular; met. rock fragments (2.2%)—subangular; feldspars (all types) (1.0%)—subangular; minor biogenic (ostracode, gastropod) calcite and sed. aggregates	Avg. composition from throughout study area: quartz (70%)—subangular: volc. rock fragments (20%)—subround; biogenic calcite (4%); sed. rock fragments (3%); feldspars (all types) (2%); met. rock fragments (< 1%). Quartz component increases from 12% at southern end of spit to 97% at northern end of spit, while volc. rock fragments decrease from 55% to 3% over same ground	Very variable. Volc. rock fragments decrease to N from 70% 10 km S of Loelia, to 22% 10 km N of Loelia, to 15% at Kerio delta. Quartz shows opposite trend (14%, 37% and 53% resp.). Avg. values for remaining components throughout area; biogenic calcite (11%); sed. aggregates (7%) and feldspars (mostly k-spar) (4%)—all highly variable. Met. rock fragments locally abundant in S	Avg. composition from throughout study area: volc. rock fragments 41%—subrounded; quartz 36%—subangular; sed. rock fragments and sed. aggregates make up most of remainder; met. rock fragments locally abundant. Volc. rock fragments much more abundant near volc. headlands and N end of El Molo Bay (39–75% range) and less abundant in inter-headland area and region W of Porr (12–17%), while quartz varies inversely (9–37% and 15–70% resp.)

Avg. = average; met. = metamorphic; Pleist. = Pleistocene; pϵ = Precambrian; sed. = sedimentary; Tert. = Tertiary; volc. = volcanic.

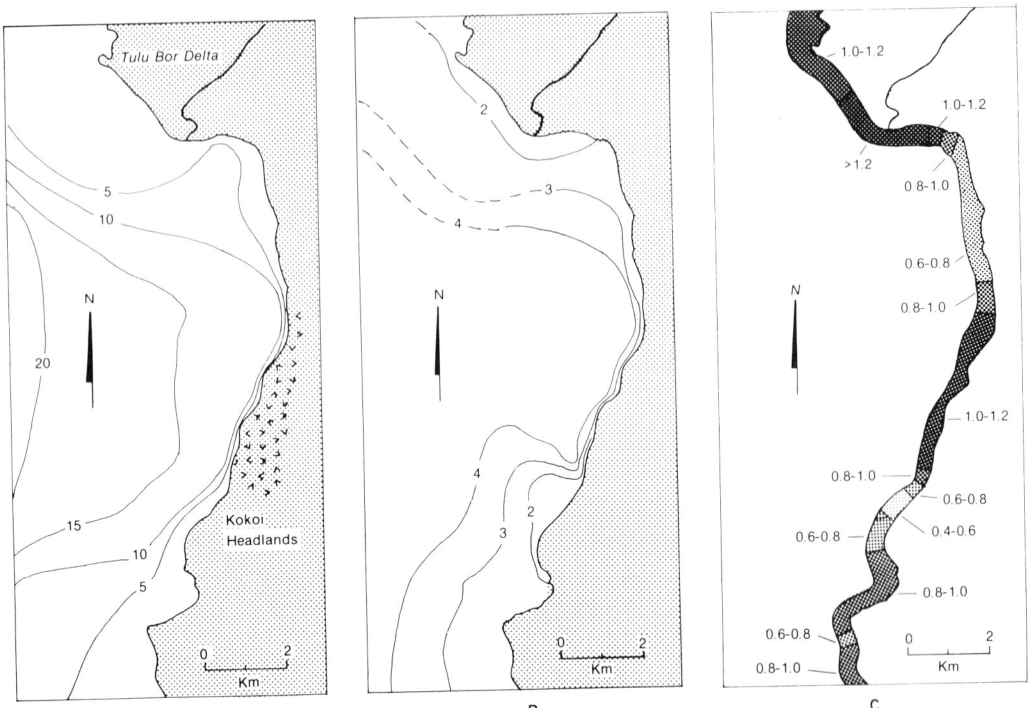

FIG. 2. Kokoi study area described in text. A. Local physiography and bathymetry. B. Mean grain size (numbers are ϕ units). C. Size sorting (graphic standard deviation, in ϕ units) of all measured nearshore sands.

Table 1. Fieldwork has revealed that the absence of abundant volcanogenic sand, even in the immediate proximity of volcanic outcrops, is a function of the weathering pattern of shoreline volcanics. Volcanogenic cobbles and gravel-sized fragments are frequently encountered in nearshore muddy matrices, but apparently these rocks weather quickly to clays. This idea is supported by the higher frequency of volcanogenic grains among the smallest sand-sized fraction (3–4ϕ), relative to the coarser sand grains.

We tentatively suggest that the sand bodies of this study area are locally derived from the drainages and colluvium in, and immediately adjacent to, the study area. Erratic, localized variations in grain size and mineralogy argue strongly for the Tulu Bor and other smaller nearby drainages as the principal sources for the sands. We suspect that the abundant quartz is derived from upland Plio-Pleistocene sediments which in turn are derived from a metamorphic source terrain; we are currently conducting grain mount studies to test this idea.

Ferguson's Gulf area (Fig. 3)

This study area centres around one of the largest embayments on the lake, Ferguson's Gulf, which covers *c.* 22 km^2. The gulf displays low physiographic relief, reaching its greatest depths (*c.* 4 m) along its eastern side.

The prominent spit along the eastern side of the gulf is composed primarily of aeolian and reworked aeolian sands, which continue for *c.* 10 km S of the study area. North of the gulf, drainages are again localized, running over alluvium and basalts. No continuous or intermittent volcanic headland shorelines occur within the study area.

Bottom topography within the lake (Fig. 3A) conforms fairly well with shoreline shape out to a depth of 10 m. This depth is attained within 1 km of the shore everywhere except around the mouth of the gulf, where the Kalikol River delta has produced an area of subdued relief. The distal portion of this delta appears on the bathymetric map as the steepened slope, NE of the delta, at a depth of 5–10 m. The steepened slope immediately to the E of the northern tip of the large spit may be the distal portion of an older, inactive lobe of the delta, over which aeolian sands subsequently prograded. Lobate shapes in bottom contours S of this point also suggest inactive delta fronts.

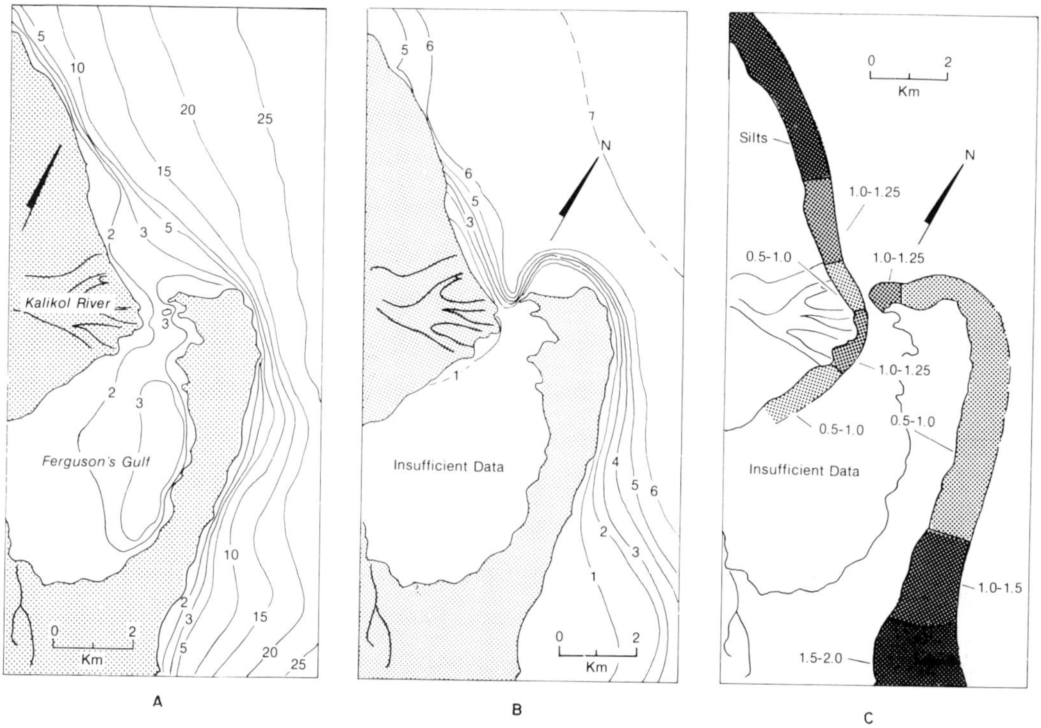

FIG. 3. Ferguson's Gulf study area described in text. A. Local physiography and bathymetry. B. Mean grain size (numbers are ϕ units). C. Size sorting (graphic standard deviation, in ϕ units) of nearshore sands.

By comparison with similar features along the western shore that are in various stages of growth, emergence, deflation and submergence (Cohen 1982), we interpret Ferguson's Gulf as being an inter-dune deflationary sag, which formed during a period of lowered lake level.

Grain size and dispersion variations are somewhat difficult to interpret within this study area. Fine sands are developed in narrow, irregular zones which roughly follow bathymetric contours (Fig. 3B). Most of the coastal sand in this area (particularly on the large spit) is currently subaerial. At the N end of the study area and within much of Ferguson's Gulf, patchily vegetated, silty mud substrates dominate the shore zone. At the S end of the study area the sand belt widens considerably to a maximum of *c.* 4 km (at the abandoned delta lobe front). South of the study area the sand belt extends subaqueously for *c.* 7 km, it is 0–2 km wide, and narrows gradually to the S.

Moderate sorting occurs in most of the study area sand bodies (Fig. 3C and Table 1). A possible dispersion trend can be seen in the increased sorting going N from the abandoned delta front at the S end of the study area, towards a large area of moderately sorted, mineralogically very mature sands at the N end of the major sand spit. Quartz contents of up to 97% occur in some samples, which, if lithified, would be considered quartz arenites.

The mineralogical and textural data are consistent with the interpretation, previously suggested, that extensive reworking has occurred among the sands on the Ferguson's Gulf spit. The exceptionally high quartz frequencies argue for a multicycle-sediment origin. Sorting is not consistent with a strictly aeolian origin, but does agree with a mixed aeolian-fluvial origin. This is consistent with the physiographic argument advanced earlier, which envisages a migrating beach dune/fluvio-deltaic border in the Ferguson's Gulf area. The ultimate source of this sand is almost certainly the Kalikol River drainage (no other drainages exist in the area which could supply large amounts of quartz), although presently very little sand appears to be moving out of the immediate vicinity of the delta (note the silts N of the present delta).

This would also support the notion that the

submerged sand body is a relict of a period of lower lake-stand. During that time the Kalikol delta would have been much less vegetated than at present, allowing for the dispersal of large quantities of sand into the lake proper. Subsequent lacustrine reworking (the spit area is presently a zone of surface current convergence, according to Yuretich 1976) and an admixture of quartzose, aeolian sand, would account for the modern observable sand textures and compositions. It appears that sand does not extend much N of the Kalikol delta today due to the modern longshore drift direction (from the N in that area). Since the other two proposed, abandoned delta fronts both lie S of the modern one, it is not surprising that the subaqueous portion of the sand body pinches out abruptly N of the modern delta, but tapers gradually to the S (pinching out 7 km S of the study area).

Loelia area (Fig. 4)

This area extends for *c.* 45 km along the south-western side of the lake. The offshore sand belt in the Loelia study area is narrow along the volcanic headlands and then widens dramatically just N of them. To the N of Loelia, the sand wedge attains a maximum offshore width of *c.* 6 km. Near the prominent spit this wedge narrows again.

Sorting is poorest in nearshore sands along the volcanic headlands (Fig. 4C and Table 1). North of Loelia sorting appears to improve considerably in nearshore areas, although offshore sands are still quite poorly sorted. At the Kerio River outlet silts show remarkably good sorting.

Two aspects of sediment distribution patterns in the study area are noteworthy. The general absence of coarse clastics around the Kerio River at first appears anomalous. Sediments at the mouth of the Omo river are similarly fine. We suspect that most of the coarse clastic bedload is being trapped in the proximal parts of the Kerio delta, and is not delivered into the lake. Extensive marshlands on the delta, also present at the Omo, may be responsible for the incompetence of the Kerio to deliver significant quantities of sand into the lake. In this respect it is important to note that ephemeral streams without a marshland in the proximal delta *do* deliver large quantities of sand into the lake (see Kokoi and Porr area descriptions).

The other important observation in this study

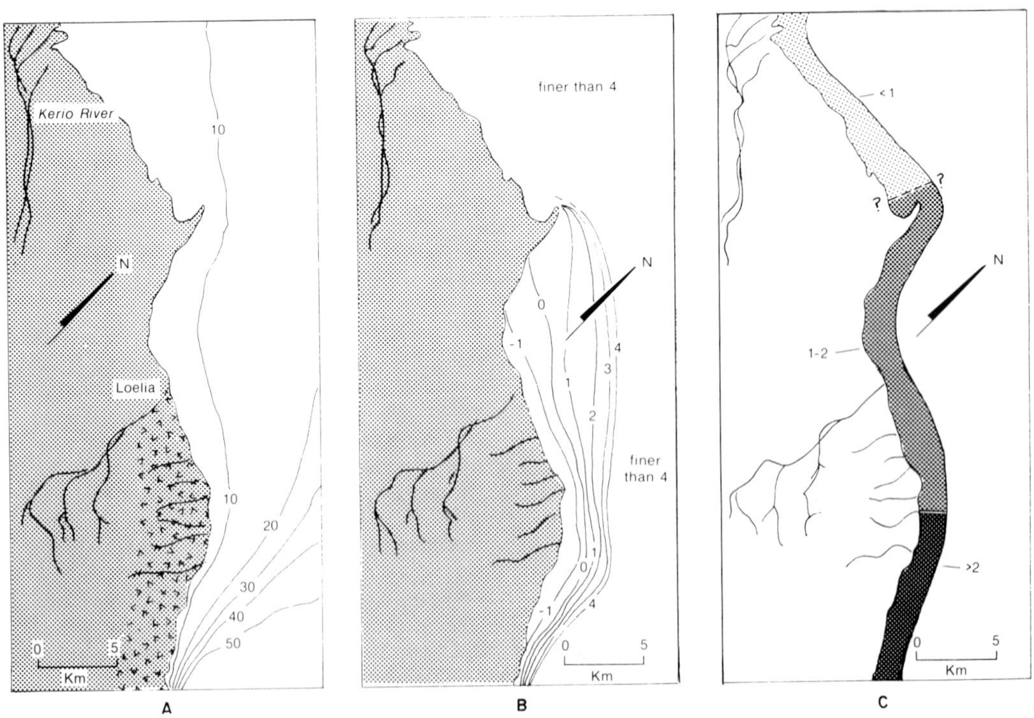

FIG. 4. Loelia study area described in text. A. Local physiography and bathymetry. B. Mean grain size (numbers are ϕ units). C. Size sorting (graphic standard deviation, in ϕ units) of nearshore sands.

area is the unusual width of the sand body located around Loelia. The fact that the percentages of quartz and volcanic rock fragments vary independently (see Table 1) suggests that the two major components of this sand body are independently derived. The volcanic grains are almost certainly derived from local headland sources (directly or via small local drainages). This is clear from both their diminishing percentage contribution away from the headlands, and the similarity in petrology between headland volcanics and clasts (both are predominantly olivine basalts). Occasional metamorphic grains in the southernmost part of the study area are probably derived from outcrops of Precambrian basement nearby.

The quartz has several possible origins. It may be aeolian, given the large rock and sand desert found immediately to the W of this portion of the lake. Alternatively, it may be a relict sand, deposited by a large river system (presumably the Kerio) during a lower lake-stand. This idea is consistent with a reduction in marshland (and therefore sand retention) during a dry phase at some point in the lake's recent history (Cohen 1982). At present we do not have sufficient evidence to favour one hypothesis over the other.

Porr area (Fig. 5)

The Porr study area lies on the southeastern side of the lake, around the prominent dog-leg (Fig. 1). The absence of any major drainages in the area has prevented large scale shoreline progradation, and as a result shoreline relief is high throughout the study area. The primary source of the relief along both the N–S and E–W shorelines are two major normal faults.

Variations in mean grain size (Fig. 5B) follow the bathymetry fairly closely. A sand belt of up to 1 km wide (in plan view), has accumulated NW of El Molo Bay, while a second, much larger one, occurs W of Porr. This second belt tapers from *c*. 5 km wide, 12 km W of Porr, to less than 100 m at the Porr Point headland. Immediately to the E of this point sands and gravels are very localized, occurring only in the swash zone and immediately offshore. Basalt granules of up to 1 cm occur commonly embedded in the nearshore silts of this region.

Sorting is very poor throughout most of the sand bodies. The poorest sorting is encountered around the Porr headland and in the N part of El Molo Bay. Sorting improves in the inter-head-

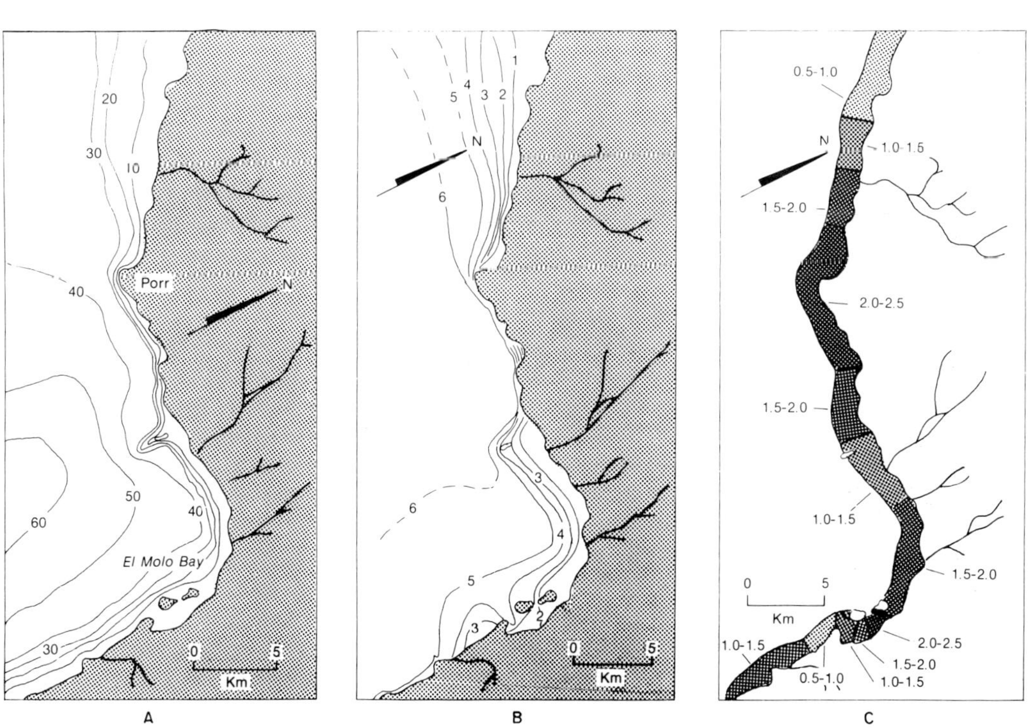

FIG. 5. Porr study area described in text. A. Local physiography and bathymetry. B. Mean grain size (numbers are ϕ units). C. Size sorting (graphic standard deviation, in ϕ units) of nearshore sands.

land areas, SE of El Molo Bay and in the extreme NW parts of the study area.

Sand belt geometries and compositions clearly reflect the interaction of highly localized drainages with volcanic headland barriers to sediment dispersal. The percentages of quartz and metamorphic rock fragments are associated with nearby metamorphic-sedimentary source terrains. These components do not appear to be readily transported either around Porr or past El Molo Bay.

Analysis of textural variations shows that a large sand wedge at the NW end of the study area abuts upon the Porr headland, becoming progressively narrower and more poorly sorted as it approaches that point. Data from the sand body E of Porr are more ambiguous, but at least on the W end of this second sand body similar relationships can be seen. The Porr headland appears to represent a nearly impassable boundary to sediment movement from the S and E, given the extremely narrow (or absent) coarse clastic zone along its shoreline. At the same time, it appears to be contributing sediment locally to the nearby impounded sands and offshore muds, imparting an unusually poorly sorted texture to the sediments in its vicinity.

Models of volcanogenic regulation on lacustrine sedimentation

Based upon the observations discussed above and work by A.C. discussed elsewhere (Cohen 1982), we believe that two interacting factors currently have the greatest influence on the distribution of nearshore, coarse clastics in Lake Turkana. These are:

A sediment distribution system that is primarily dependent on localized supply via rockfall and intermittent streams

The Omo River, considered a major source of fine-grained sediments (Yuretich 1979), does not, in fact, function as a major source of sand away from its delta front. The main source is local. The absence of sands in the distal portions of the Omo delta and around its margins, suggests that its sandy bedload is trapped in the more proximal reaches of the delta, probably through continued subsidence. This hypothesis receives support from the thick Plio-Pleistocene section found in the proximal portion of the delta (de Heinzelin *et al.* 1976; Brown *et al.* 1978). Similarly the other major streams entering the lake, the Kerio and Turkwel, provide sources of sand for perhaps no more than 20–30 km N and S of their deltas.

Physiographic barriers, principally faulted and eroded volcanic flows, which impinge on local coarse sediment delivery systems and isolate them from adjacent systems

It is clear that tectonic and sedimentary barriers to sediment transport exist in Lake Turkana. We wish to emphasize that barriers of a volcanic nature appear to be of greater importance, but also note that the Ferguson's Gulf spit is a good example of a sedimentary barrier. The variations in sediment textures and compositions observed around high-relief volcanic headlands suggests that relatively little coarse-grained sediment is transported around such barriers by longshore drift. Clearly the effectiveness of a given headland as a barrier is in part a function of its size, its vertical relief, and the competence of local currents to circumvent it.

Given the importance of volcanism in the Lake Turkana basin we believe it is useful to generalize a variety of hypothetical models for the impact of volcanism on rift lake sedimentation. Both Yuretich (1976, p. 232) and Logatchev (1978) have previously discussed the concepts, and we will elaborate on them here. We have identified processes operating at three different scales by which patterns of sedimentation may be, in part, regulated by volcanic activity, and suggest ways in which these hypothetical relationships might be tested. We believe that a detailed examination of these and other potential effects is critical to an understanding of the differences in physiography and sedimentation between volcanically dominated and non-volcanically dominated rift basins.

Microscale effects—facies interruption and dislocation

A comparison of localized, lateral lake facies variations in the immediate vicinity of lava flows with nearby control areas where volcanics are absent (or a comparison with NDB lakes) can be expected to reveal both nearshore and offshore facies differences. Following the extrusion of a lava flow, both nearshore and offshore sediments will be deflected initially lakewards by up to several kilometres as local currents pile sediments up against the barrier. This deflection will become less marked as the headland erodes or as the shoreline progrades. Where volcanic obstacles are absent, no deflection of coarse-grained facies will be observed and nearshore substrates will possess horizontally strip-like geometries away from influent streams.

Finer-grained facies of the offshore mud blanket in VDBs should display very rapid lateral transitions from volcanically-derived colluvium in the immediate vicinity of the volcanic landforms to offshore silts and clays with little intervening sand deposition. Poor sorting would be characteristic of the narrow transition zone. For NDBs (or in VDBs away from volcanic headlands), grain size transitions from nearshore to offshore could be expected to be more gradual, and sorting in the transition zones would be better.

In a stratigraphic section, rapid vertical facies changes are a predicted consequence of lateral facies interruption by volcanic flows. Expected transitions of increased frequency include mud–pebbly mud and mud–pebbly sand. In exposed sections of VDB lake margin sediments or in cores, facies transitions should occur with a greater vertical frequency in the immediate vicinity of flows than away from them.

Mesoscale effects—sediment ponding and formation of subsidiary 'perched' basins

Volcanic barrier development is an important mechanism for the formation of subsidiary basins, either marginal to, or in sequence with, the principal axial basins of the rift valley. Truckle (1976) has suggested such a mechanism for the isolation of the Suguta trough (located S of present-day Lake Turkana) from the Turkana basin. Similar, although less drastic, basin isolation is currently being effected by the three major volcanic islands of Lake Turkana, as evidenced by their localized controls over bathymetry. Tiercelin *et al.* (1980a) have noted the importance of volcanics in the development of marginal basins in the Afar.

Other, tectonic, mechanisms of sedimentary basin division in rift systems have been discussed by Rosendahl & Livingstone (1983), with suggested basinal dimensions of *c.* 20–40 km by 50–70 km. In basins smaller than this, and in the absence of volcanism, underlying basement or sedimentary ramps might allow direct sediment flow into the central part of the basin, with little long-term sediment accumulation on the lake margin. This will be particularly important when sedimentation rates can keep pace with local fault growth, maintaining continuous downsloping ramps for sediment movement towards the centre of the lake basin. Sediment flow into the central basin under these conditions would include the movement of coarser grain sizes as grain and turbidity flows.

Development of nearshore and offshore volcanic barriers will result in sediment ponding, particularly of coarser-grained sediments. Coarse-grained sediments which do occur offshore would be primarily derived from localized volcanic weathering. If the volcanic barrier rises above local base level across an entire segment of the rift basin, a complete isolation of sedimentation systems would result, as is the case with the Suguta trough in southern Turkana. At Lake Turkana, given the extremely localized nature of its sediment delivery system, such isolation could produce adjacent and contemporaneous stratigraphic packages with very different lithologies.

This model may be very useful in understanding what might otherwise be a rather puzzling situation. There appears to be a major difference between the number and extent of marginal lacustrine sequences of Tertiary or Quaternary age preserved around NDBs and VDBs. With a few notable exceptions (the Chiwondo sequence for example), the major NDB lakes of the African Rift System (Malawi and Tanganyika) are devoid of marginal deposits, while such sequences are very abundant in the VDB lakes of Kenya. This observation may be more comprehensible if we accept the idea that in NDB lakes, the lake margins may be only temporary holding areas for sediment in transit to more central portions of the basin. In VDBs, by way of contrast, large amounts of clastic material, particularly the coarser-grained fractions, may become trapped by volcanics in 'perched' basins along the lake margins, leading to their higher preservation potential as compared with NDBs. Such a model is strongly supported by the work of Watkins (in this volume) on volcanic controls on late Cenozoic sedimentation in the Koobi Fora basin. Watkins has proposed that volcanic tumescence of local physiographic features in the Koobi Fora basin resulted in the localized entrapment of sedimentary packages, similar to the method proposed here.

This model produces two testable stratigraphic predictions. The first of these, that VDBs should possess more lake margin loci of sedimentation than NDBs, appears to be the case, as discussed above. It must be noted however, that the geology of the regions around most of the VDBs (particularly in Kenya) is much better known than around the large NDBs of the Western Rift. Until this situation changes, the possibility of sampling bias cannot be entirely ruled out.

The second testable prediction of the mesoscale model is that coarse-grained sediments, originating as turbidity and grain flows, should be more abundant in deep water NDB settings than in similar settings in VDBs. While sediment ponding and current deflection may often lead to turbidity underflows in marine environments, there they are accompanied by, in this case, much greater relief than is observed in most VDBs. Deeper NDBs are more conducive environments for long-distance gravity transport of sediments. Marine offshore sands are abundantly interbedded with muds, often with sheet-like or shoestring geometries (LeBlanc 1972). In VDBs, we would predict that where volcanic barriers are abundant, gravity flows down steep sedimentation ramps would be relatively uncommon, and mostly restricted to deltaic fronts. The testing of this prediction must await a detailed seismic and core analysis of deep water sediments from both VDB and NDB settings.

Macroscale effects—volcanic infilling of rift basins

The largest scale sedimentation effects which can be postulated as being affected by volcanism revolve around the total or near total infilling of a rift segment by volcanics; a situation which could develop when the rate of volcanic extrusion is greater or equal to the rate of rift propagation. While this group of effects does not strictly revolve around coarse clastic sedimentation, we have included it here because we believe its consequences are very important, particularly for an understanding of the origin of hydrocarbon accumulations in rift settings.

In VDBs where extrusion rates are high relative to the rate of rift extension, the topographic surface relief of the rift in question will be limited, severely constraining the capacity of the rift segment to act as a water and sediment catchment. Rift lakes under such situations can be expected to be very shallow.

In contrast, when volcanism is absent, or rates of volcanic activity are low relative to the rate of rifting, relief on the rift walls can be much greater, allowing water and sediments to act as the primary rift infillings. In this case, extremely deep rift lakes can develop in the presence of moist climatic conditions.

The volcanic infilling model suggests that VDBs should contain relatively thin sections of predominantly shallow water lacustrine, marginal lacustrine and fluvial sediments, interrupted by massive volcanic sequences, such as in the Lake Turkana basin. In rifts formed under temperate climatic conditions, the sediments of VDBs can be expected to be well oxidized due to seasonal water column overturn. Exceptions to this generalization might occur in hyperalkaline or hypersaline lakes, where ectogenic meromixis could develop. Under tropical and subtropical conditions, due to shallow water stratification effects, rift sediments in VDBs would alternate between being dominantly oxidized and being locally reduced and organic rich (eg. Lakes Baringo and Bogoria, discussed by Tiercelin *et al.* 1980b).

Large NDBs formed under tropical and subtropical conditions should contain great thicknesses of reduced, organic rich, deep water muds (with occasional interfingered sands as described earlier). Robbins (1983) has discussed the implications of rift lake variability for hydrocarbon accumulation. NDB lakes which develop under cooler climatic conditions, where seasonal overturn prevails, would also form very thick stratigraphic sequences of profundal lacustrine muds, but these would be oxidized or neutral, with relatively low organic carbon contents. A corollary of these predictions is that for ancient rift sequences, hydrocarbon formation and accumulation potential is greatest in NDBs which formed in a tropical environment. Formation potential will be high in such areas due to the large accumulation of organic carbon, while hydrocarbon reservoir formation may be facilitated by the presence of interfingering, deep water, shoestring sands, which have considerable reservoir potential.

These predictions could be tested through an analysis of worldwide modern and ancient rift lake systems, with particular attention paid to (1) the stratigraphic/seismic stratigraphic relationship between rift volcanism (or lack of volcanism) and total basin fill characteristics; and (2) the degree to which tropical NDBs (and ancient NDBs which formed in tropical environments) correlate with known hydrocarbon occurrences. Due to its potential significance as a petroleum exploration tool, if correct, we feel that this last set of predictions deserves immediate attention.

Conclusions

In this paper we have attempted to demonstrate the significance of volcanic barrier formation with respect to the distribution patterns of coarse clastic sediments. We have chosen Lake Turkana for our analysis primarily because of data availability, but also because it provides a good location to investigate the microscale differences between sediment transport in areas where volcanic rocks are present and those where they are absent. We are aware that the application of generalizations drawn from one study area in the East African Rift to most or all rift lakes is dangerous, but we have presented our model here more to generate discussion about the potential impact of volcanism upon sedimentation than to provide definitive solutions. Unlike the importance of climate and tectonism on sedimentation in the rift lakes, volcanism has received relatively little attention. We hope that this contribution will serve to stimulate future discussion in what seems to be an undeservedly neglected aspect of the study of rift lakes.

ACKNOWLEDGEMENTS: We would like to thank Carol Erickson for secretarial assistance, Chris Wray for drafting and Nancy Dickinson and Karen Higgins for field assistance. Debra Gevirtzman and Eric Leonard read the manuscript and made useful suggestions for improving it. We would also like to acknowledge NSF grant No. EAR-77-2349 (University of California-Santa Cruz) and Leo Laporte and Kay Behrensmeyer for field support. Laboratory support was financed by a Colorado College Natural Sciences Divisional Research Grant.

References

BROWN, F.H., HOWELL, F.C. & ECK, G.G. 1978. Observations on problems of correlation of late Cenozoic hominid bearing formations in the North Lake Turkana Basin. In: BISHOP, W.W. (ed) *Geological Background to Fossil Man*. Scottish Academic Press, Edinburgh, pp. 473–98.

BUTZER, K.W. 1971. Recent history of an Ethiopian delta. *Univ. Chicago, Res. Pap. Dept. Geogr.* **136,** 184 pp.

COHEN, A.S. 1982. *Ecological and paleoecological aspects of the rift valley lakes of East Africa.* Unpublished PhD thesis University of California-Davis, 314 pp.

COPPENS, Y., HOWELL, F.C., ISAAC, G.L. & LEAKEY, R.E.F. (eds) 1976. *Earliest Man and Environments in the Lake Rudolf Basin.* University of Chicago Press, Chicago, 615 pp.

DE HEINZELIN, J., HAESAERTS, P. & HOWELL, F.C. 1976. Plio-Pleistocene formations of the lower Omo basin with particular reference to the Shungura Formation. In: COPPENS, Y. *et al.* (eds) *Earliest Man and Environments in the Rudolf Basin.* University of Chicago Press, Chicago, pp. 24–9.

FRIEDMAN, G.M. & SANDERS, J.E. 1978. *Principles of Sedimentology.* John Wiley & Sons, New York, 792 pp.

FROSTICK, L.E. & REID, I. 1979. Drainage net control of sedimentary parameters in sand-bed ephemeral streams. In: PITTY, A.F. (ed) *Geographical Approach to Fluvial Processes.* Geo Books, Norwich, pp. 173–201.

—— & —— 1980. Sorting mechanisms in coarse-grained alluvial sediments: fresh evidence from a basalt plateau gravel, Kenya. *J. Geol. Soc.* **137,** 431–41.

HOPSON, A.J. (ed) 1983. *A Report of the Findings of the Lake Turkana Project*, Vols. 1–6. University of Stirling Press, 1900 pp.

LEBLANC, R.J. 1972. Geometry of sandstone reservoir bodies. *Mem. Am. Assoc. Petrol. Geol.* **18,** 133–90.

LOGATCHEV, N.A. 1978. Main features of evolution and magmatism of continental rift zones in the Cenozoic. In: RAMBERG, I.B. & NEUMANN, E.R. (eds) *Tectonics and Geophysics of Continental Rifts*, Vol. 2. Proceedings of the NATO Advanced Study Institute with emphasis on the Permian Oslo Rift. D. Reidel Publishing Co., Dordecht, Holland, pp. 351–66.

ROBBINS, E.I. 1983. Accumulation of fossil fuels and metallic minerals in active and ancient rift lakes. *Tectonophysics* **94,** 633–58.

ROSENDAHL, B.R. & LIVINGSTONE, D.A. 1983. Rift lakes of East Africa. New seismic data and implications for future research. *Episodes* **83–1,** 14–19.

TIERCELIN, J.J., TAIEB, M. & FAURE, H. 1980a. Continental sedimentary basins and volcano-tectonic evolution of the Afar Rift. In: *Geodynamic Evolution of the Afro-Arabian Rift System.* Acad. Nazionale dei Lincei, Rome, pp. 491–504.

——, LEFOURNIER, J., HERBIN, J. & RICHERT, J. 1980b. Continental rifts: modern sedimentation, tectonic and volcanic control. In: *Geodynamic Evolution of the Afro-Arabian Rift System.* Acad. Nazionale dei Lincei, Rome, pp. 143–63.

TRUCKLE, P.H. 1976. Geology and Late Cainozoic lake sediments of the Suguta Trough, Kenya. *Nature* **263,** 380–3.

YURETICH, R.F. 1976. *Sedimentology, geochemistry and geological significance of modern sediments in Lake Rudolf (Lake Turkana), Eastern Rift Valley, Kenya.* Unpublished PhD thesis, Princeton University, 305 pp.

—— 1979. Modern sediments and sedimentary processes in Lake Rudolf (Lake Turkana) Eastern Rift Valley, Kenya. *Sedimentology* **26,** 313–31.

—— & CERLING, T.E. 1983. Hydrogeochemistry of Lake Turkana, Kenya: mass balance and mineral reactions in an alkaline lake. *Geochim. Cosmochim. Acta* **47,** 1099–109.

A.S. COHEN, D.S. FERGUSON, P.M. GRAM, S.L. HUBLER & K.W. SIMS, Dept of Geology, The Colorado College, Colorado Springs, Colorado 80903, USA.

Controls on the composition of modern sediments, Lake Turkana, Kenya

R.F. Yuretich

SUMMARY: Source rock composition plays a primary role in controlling the sediment composition in modern Lake Turkana, but interactions between sediment and lake water both during and after deposition may serve to modify this distribution. The Omo river drainage basin, which comprises about 58% of the Lake Turkana catchment area, is underlain primarily by Tertiary mafic volcanics exposed to a humid, tropical climate. In contrast, the smaller Kerio-Turkwel River drainage basin (28% of the total lake catchment) is extremely arid, and bedrock exposures are dominated by Precambrian gneisses and schists. Sediments delivered by the Omo system reflect the intense weathering of the area; fine grain sizes dominate and consist of kaolinite, along with iron oxides and resistate minerals with high Al and Ti. The Kerio-Turkwel sediments are generally coarser (more silt and fine sand) and contain appreciable quartz, feldspar, smectite and illite. The chemistry and mineralogy of the bottom sediments near the river mouths reflect these differences, but compositional changes in deep water deposits may be caused by reactions with the alkaline and slightly saline (2500 ppm TDS) lake water. The sediments here are very fine grained ($71 \pm 9\%$ clay) and, despite an apparently small influx of smectite into the lake, Mg-rich smectite is the principal clay. Hydrodynamic segregation of smectite from coarser materials could explain this trend, but geochemical calculations suggest that the reaction of weathered aluminosilicates with Mg^{2+}, HCO_3^- and SiO_2 in the lake water may also produce diagenetic smectite. Calcite is also precipitated inorganically, particularly in the more distal parts of the lake. Hydrogeochemical balances predict other mineral–water reactions to remove dissolved species, but direct evidence for these is yet to be documented.

Lake Turkana, in the northwestern extremity of the Kenya Rift Valley (Fig. 1), is the largest lake in the East African Rift System. This large size, coupled with substantial areal and volumetric changes during the latest Cainozoic, has produced an abundance of exposed lucustrine sediments documenting the evolution of the basin. These are currently the subject of considerable research interest from the viewpoints of palaeoclimate and organic evolution (Williamson 1981; Owen *et al.* 1982; Vincens 1982). Successful interpretation of these palaeolacustrine deposits requires an understanding of the principal controls on the sediment composition in modern Lake Turkana. In this paper, I elaborate upon some of my earlier results obtained from studying the present lake (Yuretich 1979, 1982; Yuretich & Cerling 1983) and discuss some additional data concerning geochemistry and mineralogy of the lake sediments.

Depositional framework and sediment sources

Lake Turkana occupies a topographic low within the eastern rift, often called the 'Turkana Depression'. Consequently, the basin receives runoff and detrital sediment from a wide geographic area. Average sedimentation rates are very high; estimates derived from ^{210}Pb and palaeomagnetic studies (C. Barton, unpub. data), burial of interstitial chloride (Yuretich & Cerling 1983) and calcium removal from the lake (Yuretich 1979), all produce modern sedimentation rates of *c.* 0.2 gm cm^2 a^{-1}, which translates into about 1 cm a^{-1} when corrected for sediment water content. Sedimentation rates are extraordinarily high because (1) the tectonic setting of Lake Turkana makes it a 'sediment sump' so that drainage from a wide area (131,000 km^2) is received by the lake; and (2) the climate in the source uplands is humid subtropical to tropical (McCall 1964; Butzer 1971) which produces abundant runoff.

The source areas for Lake Turkana sediments are dominated by volcanic rocks (Fig. 2). On the western side of the lake, most of the uplands are remnants of Tertiary volcanism, containing soda-rich mineral assemblages, with scattered outcrops of Precambrian basement. The weathered products of these rocks are delivered to Lake Turkana by numerous small, intermittent streams which constitute the primary control along the western margin of the lake. The headwaters of the Kerio and Turkwel Rivers, which empty into the central lake region, are dominated by extensive outcrops of Precambrian basement. Most of these

From FROSTICK, L.E. *et al.* (eds) 1986, *Sedimentation in the African Rifts*, Geological Society Special Publication No. 25, pp. 141–52.

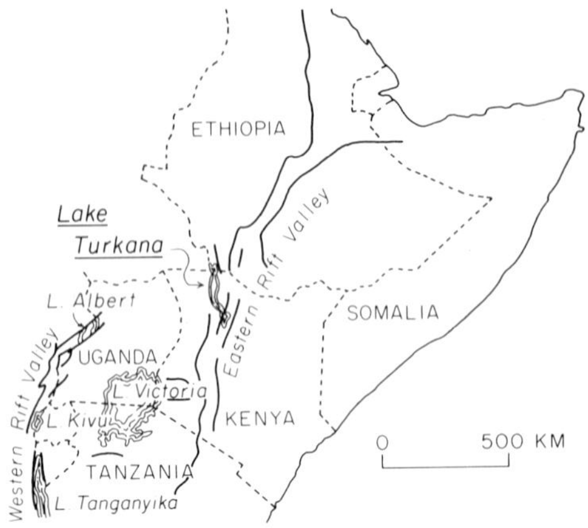

FIG. 1. Location map of Lake Turkana in the East African Rift System.

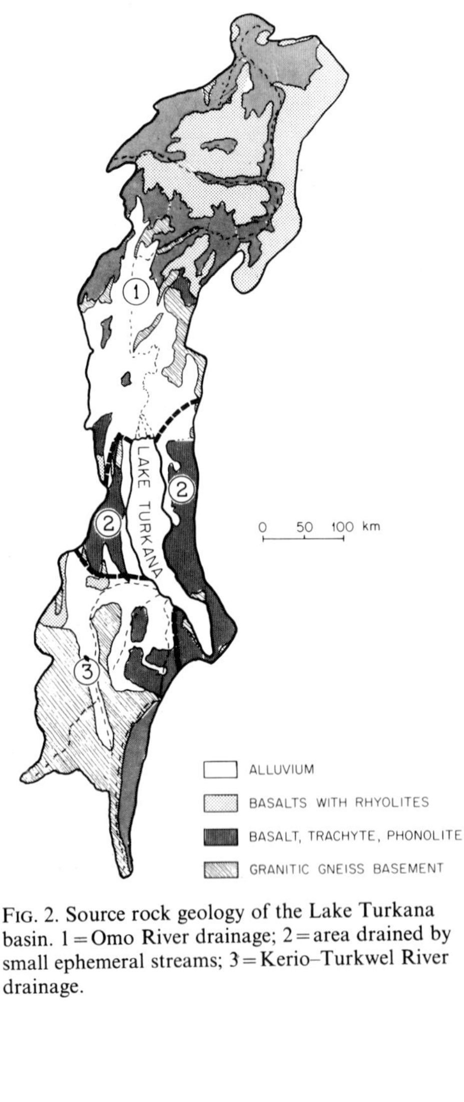

FIG. 2. Source rock geology of the Lake Turkana basin. 1 = Omo River drainage; 2 = area drained by small ephemeral streams; 3 = Kerio–Turkwel River drainage.

TABLE 1. *Lake Turkana catchment areas*

Region	Area (km^2)*
Omo basin	74,000
Kerio–Turkwel	39,400
Ephemeral rivers	9,900
Lake surface	7,500
Total	130,860

* After Ferguson & Harbott (1982).

are biotite gneisses, hornblende gneisses, migmatites and plagioclase amphibolites of the Upper Proterozoic Turbo-Kitale Group (Pallister 1971). The catchment of the Omo River is the largest in the lake drainage basin (Table 1). Tertiary to Pleistocene volcanic rocks constitute the primary lithologies (Mohr 1971). The Omo River is the only perennial water input to the lake and provides some 80–90% of the total annual water budget to the lake (Butzer 1971; Yuretich & Cerling 1983).

Sediment characteristics: shore zone

The differences in the source area geology and climate in the various drainage systems produce distinctive characteristics in the detrital sediments delivered to the lake near the points of influx.

Texture

The dominantly fine-grained nature of the sediments in modern Lake Turkana has been reported elsewhere (Yuretich 1979). Grain size decreases, as expected, towards the deeper parts of the lake basin (Fig. 3), however, other textural characteristics serve to distinguish between sediment sources. Deltaic regions are characterized by concordant medians and modes in the silt-size range (Fig. 4A); the graphic standard deviation (Folk 1974), a measure of sorting, has consistent values near 1.5. On an absolute scale, this number is fairly high, reflecting poor sorting, but these sediments are comparatively well sorted for Lake Turkana. Samples taken from the centre of the lake possess a tail of extremely fine-grained material (Fig. 4B), although the overall sorting is about the same. Marginal environments other than deltaic regions reflect the variability of sources and sedimentary processes. Sorting is noticeably poorer here, although cumulative curves exhibit the same general pattern as for the

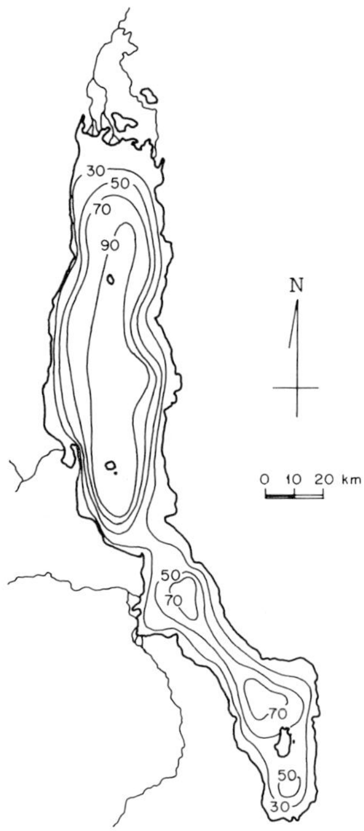

FIG. 3. Isopleths of grain size distribution within Lake Turkana. Isopleths are in percent clay size (<2 μm) fraction.

deltaic sediments (Fig. 4C). Such textural distinctions should also be applicable to the interpretation of palaeoenvironments for the older lake deposits.

Mineralogy

The different lithologies in the source areas are the principal controls on the mineral composition of the sediments in nearshore, shallow water areas. In the volcanically dominated plateaux of Ethiopia where the Omo River has its source, the climate is mesothermal humid, and deep soils are developed (Butzer 1971). Smectites are the dominant clay minerals in the vertisols developed on valley floors underlain by basalt (Semmel 1964; Butzer 1971). However, ferrisols (laterites) are often found on hillslopes and interfluves in montane valleys. Kaolinite dominates in these latter soils.

Butzer (1971) reports finding smectite in the alluvial deposits of the Omo River, yet X-ray analyses I have conducted on both Omo River suspended sediments and shallow subaqueous sediments of the Omo delta reveal only a subdued and broad peak in the smectite range (Fig. 5). Using unadjusted peak height analyses, the smectite:kaolinite ratio in these samples is less than one. In addition, these sediments are lacking an illite component. The broad 'dome' around the kaolinite peak is caused by the presence of fine-grained iron oxyhydroxides occurring together with the clay minerals. These assemblages can be considered as representative of the weathered vertisols and ferrisols being deposited in the shallow Omo delta.

The mineralogy of the silts and sands also reflects source rock composition and intensity of weathering in the Ethiopian Uplands. Silt grains are dominated almost exclusively by quartz, feldspar and magnetite, with the last constituent

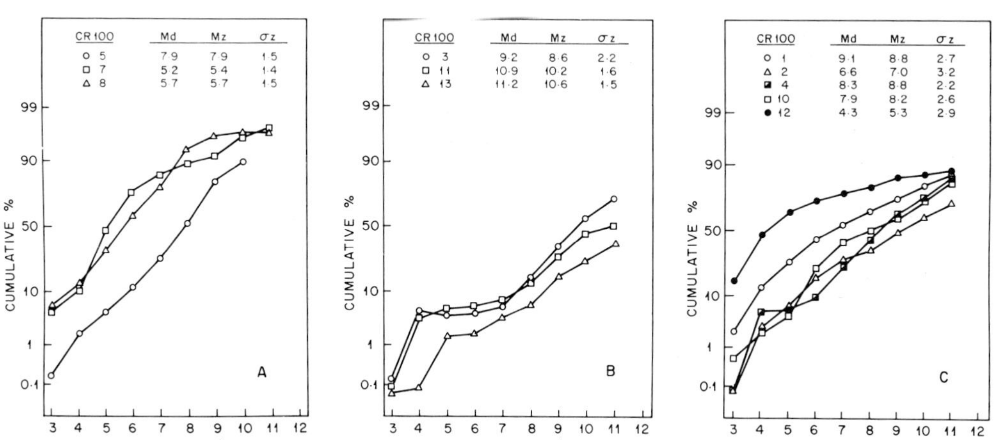

FIG. 4. Cumulative curves for sediments from nearshore areas. A. Deltaic regions. B. Central lake. C. Other nearshore areas. Sample numbers are given. Md = median grain size, Mz = graphic mean, σz = graphic standard deviation.

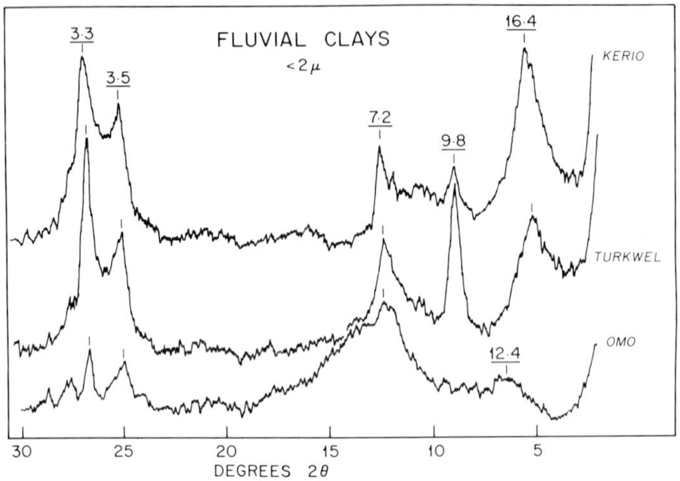

FIG. 5. Suspended clay minerals from the major rivers entering Lake Turkana. Turkwel and Kerio samples have been glycolated. Peak identification: 7.2 Å = kaolinite; 9.8 Å = illite; 12.4 Å = smectite untreated; 16.4 Å = smectite glycolated.

providing up to 20% of the silt fraction. Ferromagnesian minerals are virtually absent, presumably weathered to clays, although some biotite is found locally.

Soils and climate in the Kerio-Turkwel catchment are not as well known as those in the Omo drainage. Sandy soils predominate generally, and contain abundant quartz and feldspar (both plagioclase and K-spar) (Joubert 1966; Dodson 1971). Volcanic rocks in this region weather to rubble (Fairburn & Matheson 1970), and clay mineral development is retarded. Coupled with the source area sizes (Table 1), this causes the sediment load of the Turkwel and Kerio Rivers to differ from that of the Omo; 12Å smectite is a much more important clay in the Turkwel and Kerio Rivers (Fig. 3). Furthermore, the prevalence of illite, particularly in the Turkwel, reflects the importance of metamorphic rocks in the source area. Although the overall quartz : feldspar ratio in the silts is lower than that for Omo-derived sediments, the sand fractions are extremely quartz rich. Apparently quartz reaches the lake here as sand grains weathered from gneisses, in contrast to the smaller grains available to the Omo River from volcanic rocks.

Local source areas are very important along the eastern side of the lake. Exceptionally high smectite : kaolinite ratios (7–10) occur here, which reflect the preponderance of partially weathered volcanics. Many of the sediments high in smectite are very fine grained (more than 90% clay), and are probably altered ashes and tuffs. Sediments here also show a comparatively high calcite content (often 40–50% of the silt fraction). This indicates an isolation from the main influx of Omo River detritus. In addition, the high plant productivity associated with these protected, quiet water environments promotes bio-induced calcite precipitation and a favourable environment for large populations of ostracods, the shells of which often make a considerable contribution to the sand-sized fraction.

Sediment characteristics: profundal zone

Texture

In water depths greater than *c.* 20 m, the characteristics of Lake Turkana sediments become more uniform. Grain size ranges from 50% to >90% clay (<2μ), with the fine-grained sediments located in the deeper parts of the North Basin. Unexpectedly, the South Basin, although deeper and more distant from the major sediment inputs than the North Basin, has a noticeably coarser average grain size (*c.* 70% clay).

Mineralogy

Differences in nearshore sediments can be related to source areas, but the controls on the composition of sediments farther removed from inputs is less easily determined. One would presume that, because of the proximity to the Omo River, the North Basin sediments should possess a strong compositional resemblance to the Omo delta deposits. This is not the case. Smectite is much more

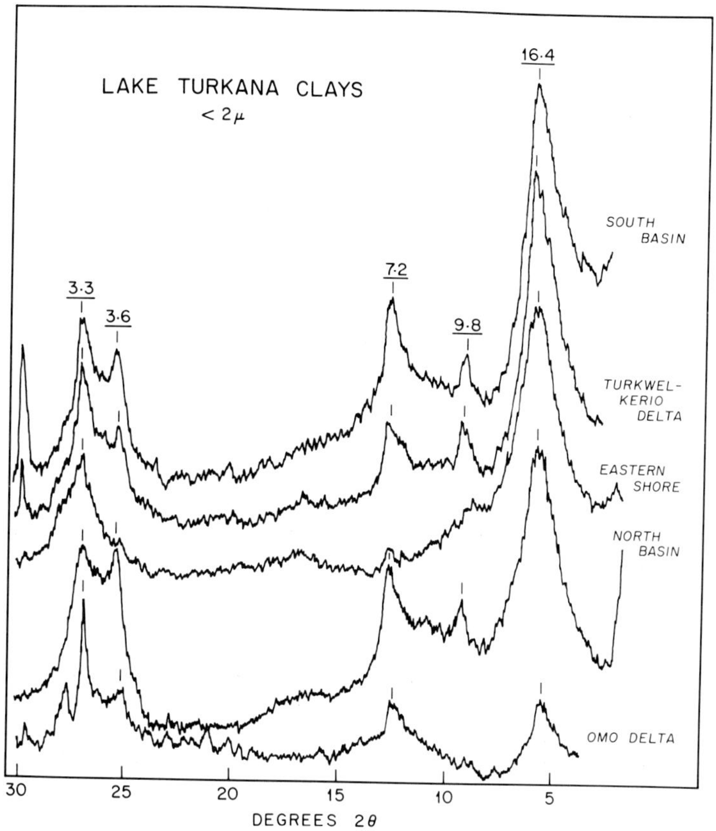

FIG. 6. Clay mineral assemblages from Lake Turkana. All samples glycolated. See Fig. 5 for peak identification.

abundant in the clay-size fraction (smectite : kaolinite ratio averages 1.5–2.0), and the profundal smectites generally exhibit higher crystallinity (Fig. 6). In the silt fraction, the quartz : feldspar ratio is 1.6–2.5, which is similar to that of the Omo delta. However, there is considerably more calcite here (as much as 30%). The amount of sand is small, and quartz is dominant, with subsidiary grains consisting of green amphiboles and plagioclase feldspar. Such an assemblage bears a greater resemblance to that found in the Kerio-Turkwel delta or along the eastern lake margin. This fact, together with the consistent size of these grains in the fine sand range (0.15 mm) and the uniform distribution of these grains throughout the profundal sediments, indicates a wind-blown origin. These minerals constitute a 'metamorphic assemblage' which is probably derived from sources to the SE and blown into the lake by the strong prevailing winds.

In the South Basin, the smectite : kaolinite ratio is also near 2 (Fig. 7), and the general mineral abundances are similar to those in the North Basin. Since soils in the local southern source areas are only poorly developed (Dodson 1963), it seems probable that the clays are spillover from more northerly sources, transported at depth by subsurface countercurrents to the northwesterly flowing surface waters (Ferguson & Harbott 1982). The silt composition of the South Basin is unique, consisting almost entirely of calcite and silica. The latter is caused largely by the abundance of diatoms (*Nitzschia* and ?*Coscinodiscus*); the calcite silt consists of acicular crystals of autochthonous origin, but whether these are inorganic precipitates or bio-induced is still uncertain. The sand-sized grains consist of numerous ostracod shells, many of which are overgrown with sparry calcite. Fish bones (otoliths) are also fairly abundant. Volcanic glass (usually with a brownish tinge) and basaltic rock fragments are important locally, particularly in the vicinity of South Island and the eastern shore. This reflects the importance of recent volcanic activity in this southernmost part of Lake Turkana.

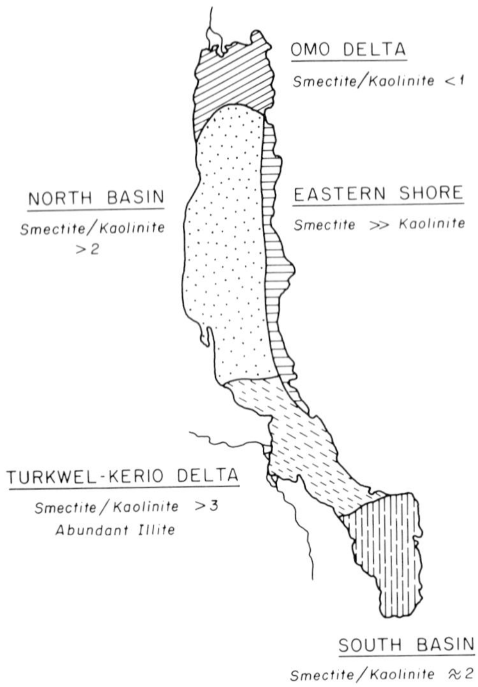

FIG. 7. Patterns of clay mineral distribution within the sediments of Lake Turkana, based on X-ray diffraction of 127 samples.

Geochemical patterns

The mineralogical associations noted previously are reinforced by trends in chemical composition of the sediment. The influence of the Omo River is illustrated by those constituents exhibiting a N–S trend. Many major and trace elements exhibit a southward decrease (Fig. 8). Fe_2O_3, TiO_2, Mn and Zn are all indicative of the weathered vertisols and ferrisols which make up the bulk of the Omo River detrital load. In contrast, CaO and MgO increase southward (Fig. 9), demonstrating the increased importance of authigenic and biogenic sedimentation in the more distal parts of the lake, away from the Omo inputs. Na_2O and K_2O distribution within Lake Turkana (Fig. 9) are controlled principally by the detritus delivered from the Turkwel and Kerio Rivers as well as local (volcanically related?) inputs on the eastern shore near Koobi Fora and Alia Bay. The trace elements Co, Ni, V and Cr show little clear relationship to either Omo or Kerio-Turkwel influences (Fig. 10). Instead, local influences from the eastern shore are evident in their distribution.

The heterogeneity in the mineralogy of Lake Turkana sediment (or any sediment, for that matter) makes it difficult to determine the specific causes for some of the geochemical changes in the

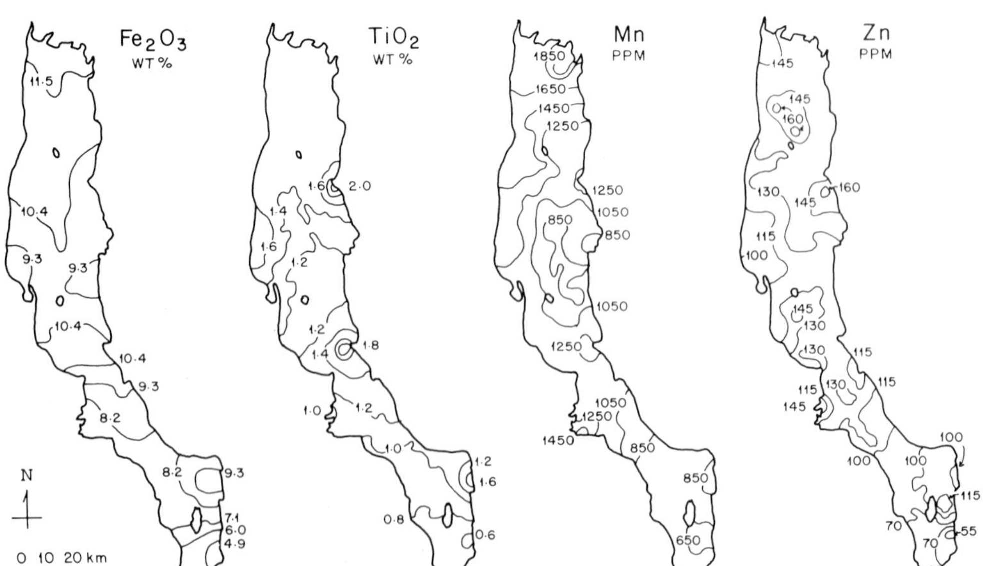

FIG. 8. Chemical components in Lake Turkana deposits which exhibit major influences from Omo River sources. The Omo enters the lake at its N end.

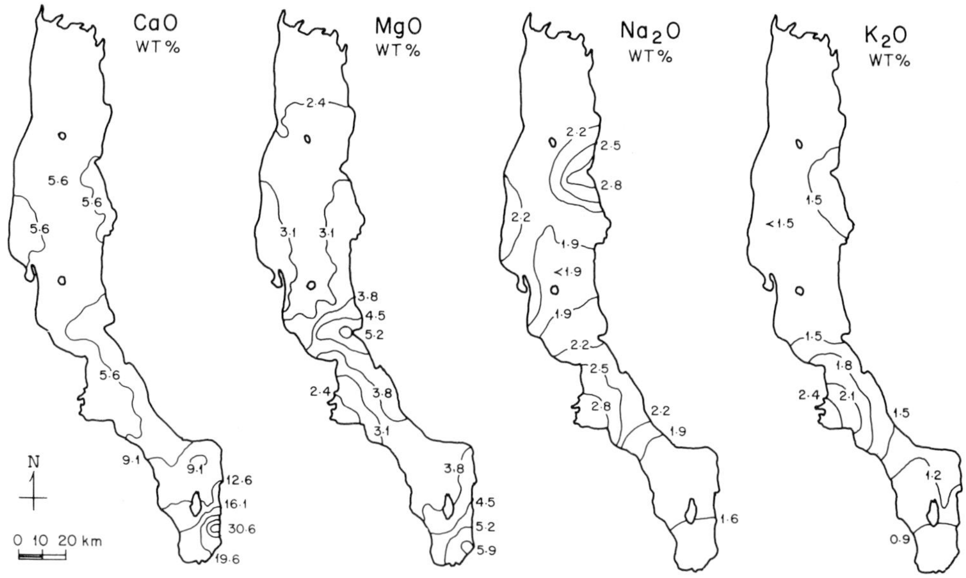

FIG. 9. Chemical components in Lake Turkana deposits which exhibit major influences from autochthonous sediments in South Basin (CaO, MgO) and Kerio-Turkwel River sources (Na_2O, K_2O).

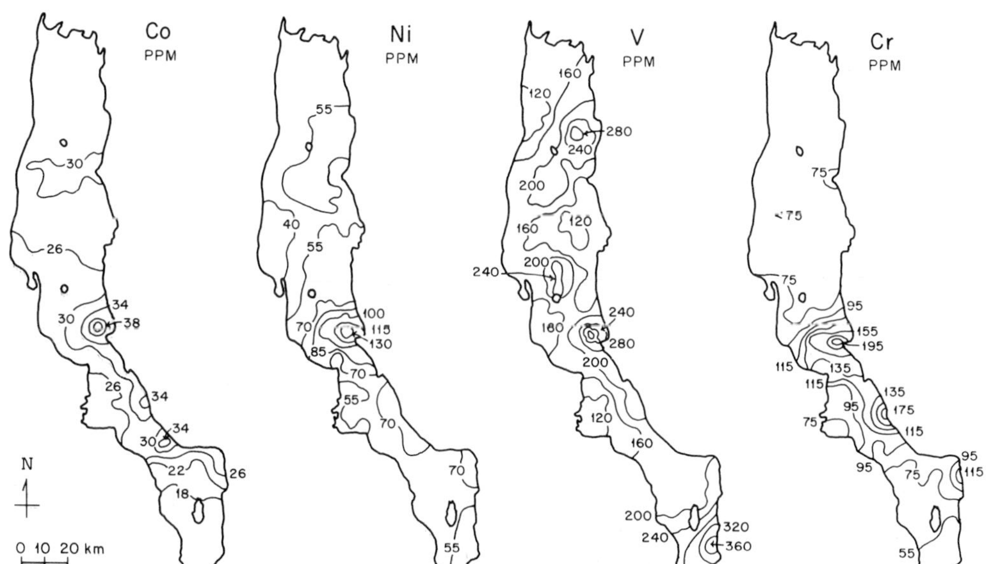

FIG. 10. Chemical components in Lake Turkana which are influenced primarily by local sources along the eastern shore.

sediments. However, multivariate statistical analysis of the sediment composition using principal components analysis (Davis 1973) can often reveal significant correlations which quantify the relationship between geochemistry and mineralogy.

Major elements

Two principal components analyses, one using only data from surface sediments, and the second using data from cores up to 3 m in length, revealed useful information about the composi-

TABLE 2. *Principal components analyses involving major elements of surface sediments*

	Principal components			
	Provenance	Silt	Clay	Texture
% clay	0.09	−0.37	0.28	0.86
K_2O	0.32	0.27	−0.61	0.23
Na_2O	0.23	0.57	−0.31	0.31
CaO	−0.45	0.30	0.12	0.09
MgO	−0.21	0.62	0.43	0.16
Fe_2O_3	0.44	0.15	0.30	−0.16
X	0.45	0.02	0.24	−0.23
Y	−0.44	−0.10	−0.32	−0.05
% variance	49.0	17.3	12.8	11.6

TABLE 3. *Principal components analysis involving major elements of sediment cores*

	Principal components			
	Clay	Smectite	Silt 1	Silt 2
% clay	0.27	0.02	−0.76	0.19
Smectite: kaolinite	0.07	0.54	−0.01	−0.81
Na_2O	0.36	0.36	0.36	0.36
K_2O	0.43	0.07	0.45	0.11
Fe_2O_3	0.50	−0.35	−0.07	−0.30
CaO	−0.58	0.22	0.06	0.10
MgO	0.15	0.64	−0.28	0.26
% variance	37.2	25.2	17.7	10.2

tion of the individual mineral phases. The four major determinants in the first analysis appear to be, in decreasing importance: (1) provenance; (2) silt composition; (3) clay composition; and (4) texture (Table 2). The provenance factor is especially important, accounting for 49% of the total variance. The positive and negative contributions to this component agree with the observed differences in mineralogy and trends in chemical composition.

The silt composition factor is indicated by large positive contributions from Na_2O and MgO, together with a negative influence from the amount of clay (Table 2). The amphiboles present in these fractions are probably the principal controls. Moreover, the ferromagnesian minerals of the Turbo-Kitale Group are sodic varieties, which satisfactorily explains the positive contributions from the alkalis. Since the amphiboles are found in about equal abundance in all parts of the lake, the geographic location of the sediment is of little consequence.

The influence of Mg and Fe-rich clays, principally smectite, is indicated by the positive contributions of Fe_2O_3, MgO and clay to the third component. Finally, the grain size factor is evidenced by the very high positive values contributed by the clay. However, the poor correlation of texture with any compositional variable, together with its low contribution to total data variance, indicates that changing grain size alone does not greatly influence the chemistry of Lake Turkana sediments.

In order to determine the relationships between geochemistry and mineralogy independent of sample location, principal components analysis of samples from 3 m long cores was done without the X and Y variables. As in the previous analysis, 90% of the data variance can be explained by four (new) principal components (Table 3). With the provenance influence removed, then clay mineral composition accounts for 37% of the variance, with significant contributions from Na_2O, K_2O, Fe_2O_3 and the amount of clay. The second component indicates that smectite controls the Mg in the clays, because of high positive covariances between the smectite: kaolinite ratio and MgO. Fe_2O_3, on the other hand, is negative and suggests that kaolinite, not smectite, is the associated clay. This would presumably be a coating on the kaolinite micelles as has been proposed by Carrol (1958).

The composition of the silt fraction, although reduced in importance with the removal of provenance, constitutes the third and fourth components. Alkali feldspar is probably the most important mineral as shown by the dominance of Na_2O and K_2O in the third component. The last significant component has a positive correlation between Na_2O and MgO, which has already been identified as a sodic amphibole or pyroxene phase in the previous analysis.

Trace elements

A principal components analysis of the trace elements, together with the major oxides, can serve to establish more precisely the distribution of minor elements in the mineral phases. Rearrangement of the 11 variables used explains 90% of the data variance in six new components (Table 4). A clay mineral component is once again the most important, and both Zn and Mn make substantial contributions here. Both these oxides have fairly high correlations with Fe_2O_3. These are probably associated with iron, as an oxide coating on the kaolinite. The second component pairs the trace elements Cr and Co with MgO and Na_2O. These major oxides composed the smectite and amphibole components of the first principal

TABLE 4. *Principal components analysis involving trace elements*

	Principal components					
	1	2	3	4	5	6
Na_2O	0.26	−0.30	−0.06	−0.30	−0.48	−0.58
K_2O	0.38	−0.22	−0.12	0.45	−0.06	−0.22
Fe_2O_3	0.43	0.24	0.17	−0.12	−0.14	0.28
CaO	−0.46	−0.08	−0.23	0.04	0.16	−0.20
MgO	0.01	−0.50	0.26	−0.31	−0.13	0.09
Mn	0.30	0.27	−0.31	0.18	0.27	−0.43
Co	0.17	−0.39	−0.39	0.12	−0.05	0.42
Cr	0.07	−0.51	0.02	0.38	0.29	0.10
Ni	−0.04	−0.05	0.75	0.36	0.09	−0.23
Zn	0.43	0.20	0.11	0.08	−0.04	0.22
V	0.28	−0.15	0.07	−0.51	0.73	−0.11
% variance	34.0	24.3	12.2	7.3	6.8	5.5

components analysis (Table 2). Although some of these trace elements may actually be in the smectites, it is more likely that these elements are contained in other Na–Mg minerals prevalent in smectite-rich localities, such as sodic amphiboles and pyroxenes. New component four has moderate contributions from Cr and Ni, together with K_2O. This may represent illite with Cr and Ni incorporated into the lattice. The strong independence of Ni and V is illustrated by components three and five, and the sixth component demonstrates a correlation between Co and Fe_2O_3.

Discussion of geochemical patterns

Examining these principal components in conjunction with the geographic variations in mineralogy and chemistry suggests several substantive patterns:

1 Provenance is the major control upon sediment composition, accounting for at least 50% of total variability. Geographic changes in silt mineralogy are generally more important than those for clays.

2 In the absence of provenance considerations, the clay component has a greater variance than the silt composition. However, in all cases there is little direct control of grain size upon chemical composition.

3 Iron is an important constituent of both clay and silt minerals. Although some Fe may be in the smectite, the majority seems to be associated with kaolinite and probably consists of an iron oxide coating on the clay mineral surface (Carrol 1958). Such an Fe oxide–kaolinite association explains the trend of Fe_2O_3 decreasing away from the Omo River in spite of the increasing abundance of smectite in the sediments. In the silts (and sands), biotite and sodic amphiboles, in addition to magnetite, make important contributions to the iron content. The abundance of these minerals is chiefly responsible for the increased iron concentrations at several places along the eastern shoreline and the high iron 'ridge' N of the Kerio-Turkwel delta (Fig. 8).

4 Magnesium is important in the composition of both amphiboles and smectite. It is the presence of the clay mineral which exerts the primary control on the percentage of MgO in the central reaches of the lake (Fig. 9). Along the eastern shore, the amphiboles and pyroxenes from nearby metamorphic outcrops are the dominant Mg-bearing phase. In the South Basin an additional source is in the calcite, which contains about 3 mol % Mg. This calcite, together with magnesium silicates in volcanic detritus and smectite, accounts for the Mg content of the southern sediments.

5 Na_2O and K_2O are primarily in non-clay minerals and are intimately associated with each other. They occur largely in feldspars, although biotite and illite account for a sizeable amount of the K_2O. The Turkwel and Kerio catchment areas are the prime source of these minerals. Some Na occurs as an exchangeable cation on the smectite (Yuretich & Cerling 1983), and there is an additional amount contained in the amphiboles. The high Na_2O province along the eastern side of the lake (Fig. 9) is a function of an increase in blue-green amphiboles in the silt and sand, plus locally high smectite concentrations.

6 Clay minerals themselves are not important carriers of trace elements, but the iron–trace element correlations suggest that oxide coatings on the clays carry large quantities of Zn and Mn

along with them. The southward-decreasing amounts of these elements (Fig. 8) is satisfied by this association.

7 Sediments with abundant illite and amphiboles are generally enriched in Co, Cr and Ni. All these are concentrated along the eastern shore of the lake (Fig. 10), suggesting derivation from local volcanic and metamorphic rocks.

Diagenesis in Lake Turkana

Yuretich & Cerling (1983) have recently summarized our collective information concerning the likely diagenetic processes in Lake Turkana. By comparing the chemistry of the inflowing river water with the present composition of the lake water, we were able to determine the major mineral–water reaction mechanisms based on mass balance and chemical equilibrium relationships. In general, we found that over 90% of the incoming cations were being removed by cation exchange, burial of interstitial water and authigenic mineral formation.

Cation exchange is of principal importance for the abundance of Na and Ca in the water. Na replaces Ca (and some Mg) as the principal exchangeable cation when fluvial clays enter the lake. This removes almost half the dissolved Na being delivered to the lake (Table 5), balanced by a corresponding increase in Ca and Mg. The high sedimentation rates in the lake also remove a significant amount of dissolved material by burial of interstitial water. Because the lake water is enriched in Na and Cl in comparison with the annual amount delivered by the rivers, the effect is greatest on these two ions, removing some 32% and 65%, respectively.

Mineral–water reactions

Of interest to the present discussion are the remainders, which must be removed by diagenetic mineral formation. The bulk of Ca is precipitated as calcite; X-ray diffraction reveals that the composition of the calcite averages 3 mol % Mg, so some of this latter ion is removed as well. Calcite is produced both as shell material (ostracods and gastropods), as well as autochthonous precipitates of micron-size acicular crystals. From the Kerio-Turkwel delta up through the northern parts of the lake, biogenic calcite is the only type recognized; the bulk of the autochthonous precipitates are found in the South Basin where detrital sedimentation rates are lower. The mass of Ca being removed corresponds well with the abundance of calcite in Lake Turkana sediments (Yuretich & Cerling 1983).

The question of authigenic aluminosilicates is still under consideration. Mass balance relationships require some sort of magnesium silicate formation to balance Mg, SiO_2 and alkalinity (T_{CO_2})(Yuretich & Cerling 1983). The abundance of smectite in the profundal sediments of the lake (Fig. 6) and its relative paucity in the sediment influx, makes this mineral a likely candidate. Our calculations suggest that the authigenic formation of stevensite, a trioctahedral smectite, provides the best balance for Lake Turkana water chemistry (Yuretich & Cerling 1983). Jones & Weir (1983) have found that diagenetic stevensite is forming in the sediments of Lake Abert in Oregon, USA. At the present time, we are characterizing the structural chemistry of the smectites in Lake Turkana. Our early results indicate that we can separate essentially pure smectites by analysing the <0.1 μm fraction. The oriented-slide specimens we have examined so far

TABLE 5. *Removal of chemical components from lake water (after Yuretich & Cerling 1983; Cerling 1986)*

	% of dissolved river input							
	Na	K	Ca	Mg	alkalinity (T_{CO_2})	Cl	SO_4	SiO_2
River input	100	100	100	100	100	100	100	100
Exchange	−45.1	−21.2	+33.8	+24.6	0	0	0	0
Burial	−30.9	−5.0	−0.3	−0.4	−6.7	−65	−3.3	−0.5
Calcite	0	0	−132	−6.3	−56.5	0	0	0
Stevensite	−7.1	−5.0	−1.5	−117.9	0	0	0	−87.3
Other*	0	−68·8	0	0	0	0	−96.7	−12.2
Accumulating in lake water	16·9	0	0	0	3.8	35	0	0

* Illite regradation, sulphate reduction and diatom sedimentation.

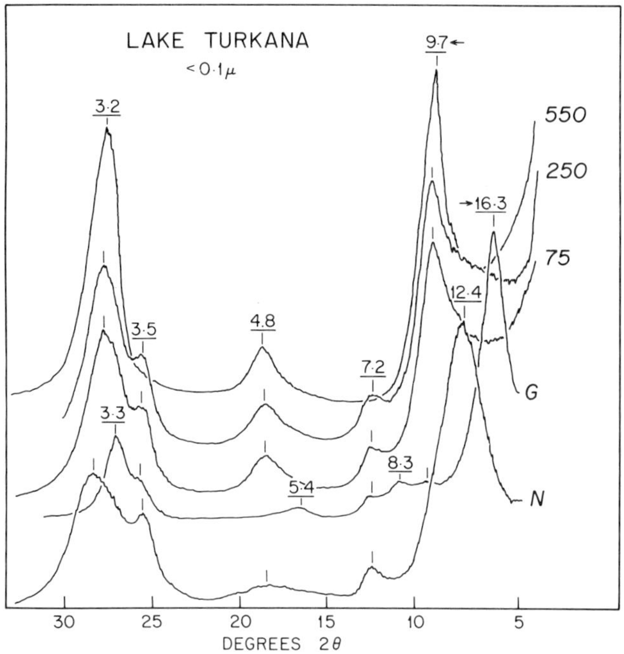

FIG. 11. X-ray diffractograms of <0.1 μm size fraction of sediments from the Kerio-Turkwel region. *N* = untreated; *G* = glycolated; 75, 250, 550 = temperature of various heat treatments.

show that the clays of the Kerio-Turkwel delta are dioctahedral, as determined by a d(060) of about 1.49 Å and the rapid collapse of the 001 reflection at low temperatures (Fig. 11). However, these clays may not be representative of those in the rest of the lake. When our X-ray and chemical analyses are completed we will be able to determine if the smectite distribution in Lake Turkana reflects authigenic processes, or if it is mainly a function of hydrodynamic segregation (Gibbs 1977).

The other reactions listed in Table 5 reflect the amount of components removed by illite regradation (K), sulphate reduction (SO_4) and diatom sedimentation (SiO_2). Such processes should occur in Lake Turkana since none of these components are accumulating in the lake water (Yuretich & Cerling 1983), but they are still in need of positive verification.

Conclusions

Lake Turkana has a water composition which permits the preservation of both detrital–clastic and authigenic signatures in the sedimentary record. In detrital sediments, it is possible to distinguish between different source areas by variations in sediment texture and composition. The majority of lake sediment is derived from weathering of volcanic rocks in the Omo River basin to the N of the lake. Detritus delivered to the lake from here is dominantly fine grained, with kaolinite as the major clay mineral. The intense weathering in this region has developed ferrisols, and the kaolinite is apparently coated with Fe-oxides which also contain significant quantities of Mn and Zn. Silts and sands have appreciable magnetite (plus ilmenite) in addition to feldspar and quartz. In general, quartz is not very abundant in the sand fraction, reflecting the dominance of alkaline rocks in the Omo drainage.

Detrital sediments from the Turkwel and Kerio River catchments reflect both the greater prevalence of metamorphic rocks as well as a more arid climate. Crystalline smectite and illite are more abundant in the clays. The alkalis Na_2O and K_2O exhibit a maximum concentration in the Kerio-Turkwel delta, contributed principally by abundant feldspars and sodic amphiboles in the silts. Quartz sand is prevalent as well. Local volcanic source areas along the eastern shore of the lake enrich the sediment in smectite, as well as in the trace elements Co, Cr, Ni and V.

Although these source area influences can also be observed in profundal sediments, some diagenetic overprinting and homogenization may occur. Crystalline smectite is the most important clay mineral in the lake. This abundant smectite may be concentrated in the distal parts of the lake by hydrodynamic segregation, yet the fluvial supply seems inadequate to account for it all. In addition, hydrochemical mass balances support the authigenic formation of some sort of magnesium silicate phase (smectite?). Research is still underway to distinguish this from detrital smectite. Calcite sedimentation is also very important, particularly in the South Basin where clastic influx is lessened.

The composition of Lake Turkana is, at present, delicately poised. Detrital influences control the major sediment patterns, yet hydrogeochemical considerations point toward incipient alkaline lake diagenesis. Except for calcite precipitation, we have not yet documented conclusively the mineral reactions which regulate water composition. However, it is clear that the possible diagenetic processes in Lake Turkana are compatible with those observed in other African rift lakes (Stoffers & Singer 1979), and that major shifts in the chemistry of lake water and climate of the surrounding watershed will leave their imprint upon the sedimentary record.

ACKNOWLEDGEMENTS: This paper has profited from discussions with Thure Cerling. Laboratory analysis of the clay minerals has progressed with the help of Janet Bailey and Joan Barry. Illustrations were prepared by Marie Litterer, Technical Illustrator, Department of Geology and Geography, University of Massachusetts.

References

BUTZER, K.W. 1971. Recent history of an Ethiopian delta. *Univ. Chicago, Res. Pap. Dept. Geogr.* **136,** 184pp.

CARROL, D. 1958. Role of clay minerals in the transportation of iron. *Geochim. Cosmochim. Acta* **14,** 1–27.

CERLING, T.E. 1986. A mass-balance approach to basin sedimentation: constraints on the recent history of the Turkana Basin. *Paleogeography, Paleclimatology, Paleoecology* (in press).

DAVIS, J.C. 1973. *Statistics and Data Analysis in Geology.* Wiley & Sons Inc., New York, 550pp.

DODSON, R.G. 1963. Geology of the South Horr area. *Rep. Geol. Surv. Kenya* **60,** 53pp.

—— 1971. Geology of the area south of Lodwar. *Rep. Geol. Surv. Kenya* **87,** 36pp.

FAIRBURN, W.A. & MATHESON, F.J. 1970. Geology of the Loiya-Loruguma area. *Rep. Geol. Surv. Kenya* **85,** 52pp.

FERGUSON, A.J.D. & HARBOTT, B.J. 1982. Geographical, physical and chemical aspects of Lake Turkana. In: HOPSON, A.J. (ed) *Lake Turkana: A Report on the Findings of the Lake Turkana Project, 1972–1975.* Overseas Development Administration, London, pp. 1–107.

FOLK, R.M. 1974. *Petrology of Sedimentary Rocks.* Hemphill Publishing Co., Austin, Texas, 170pp.

GIBBS, R.J. 1977. Clay mineral segregation in the marine environment. *J. Sed. Petrol.* **47,** 237–43.

JONES, B.F. & WEIR, A.H. 1983. Clay minerals of Lake Abert, an alkaline, saline lake. *Clays Clay Minerals* **31,** 161–72.

JOUBERT, P. 1966. Geology of the Loperot area. *Rep. Geol. Surv. Kenya* **74,** 52pp.

MCCALL, G.J.H. 1964. Geology of the Sekerr area. *Rep. Geol. Surv. Kenya* **41,** 34pp.

MOHR, P.A. 1971. *The Geology of Ethiopia.* Haile Selassie I University, Addis Ababa, 268pp.

OWEN, R.B., BARTHELEME, J.W., RENAUT, R.W. & VINCENS, A. 1982. Palaeolimnology and archaeology of Holocene deposits north-east of Lake Turkana, Kenya. *Nature* **298,** 523–9.

PALLISTER, J.W. 1971. The tectonics of East Africa. In: *Tectonique de l'Afrique (Tectonics of Africa).* UNESCO, pp. 511–42.

SEMMEL, A. 1964. Beitrag zur Kenntnis einiger Boden des Hochlandes von Godjam (Äthiopien). *Neues Jahrb. Geol. und Paläontol.* **8,** 474–87.

STOFFERS, P. & SINGER, A. 1979. Clay minerals in Lake Mobutu Sese Seko (Lake Albert)—their diagenetic changes as an indicator of the paleoclimate. *Geol. Rundsch.* **68,** 1009–24.

VINCENS, A. 1982. Recherches palynologiques dans la region du Lac Turkana (Kenya). *Paleoecol. Afr.* **14,** 85–91.

WILLIAMSON, P.G. 1981. Paleontological documentation of speciation in Cenozoic molluscs from Turkana Basin. *Nature* **293,** 437–43.

YURETICH, R.F. 1979. Modern sediments and sedimentary processes in Lake Rudolf (Lake Turkana), eastern rift valley, Kenya. *Sedimentology* **26,** 313–31.

—— 1982. Possible influences upon lake development in the East African rift valleys. *J. Geol.* **90,** 329–37.

—— & CERLING, T.E. 1983. Hydrogeochemistry of Lake Turkana, Kenya: mass balance and mineral equilibria in an alkaline lake. *Geochim. Cosmochim. Acta* **47,** 1099–109.

R.F. YURETICH, Dept of Geology & Geography, University of Massachusetts, Massachusetts 01003, USA.

Sedimentary carbonates as isotopic marker horizons at Lake Turkana, Kenya

P.I. Abell & J.P. McClory

SUMMARY: The Plio-Pleistocene sediments of the Koobi Fora region of Lake Turkana contain a variety of carbonate horizons: molluscs, stromatolites, indurated calcitic sediments and concretions. The stable oxygen isotope ratios of these carbonates are largely unchanged since deposition, and constitute a record of environmental conditions at the time of deposition. Two factors influence these isotope ratios: (1) temperature, which produces more negative $\delta^{18}O$ values with increasing temperature; and (2) evaporative concentration of the heavier oxygen isotope in the rainfall and body of the lake, which makes for more positive values of $\delta^{18}O$ under arid conditions. The major change in climate which took place at about the time of deposition of the KBS Tuff is isotopically plain in most sections, but stratigraphic control is not sufficient to correlate other significant changes. Nevertheless, the general trends in environmental change during the deposition of any particular section of sediments can be clearly perceived provided there are enough carbonate horizons. Gastropods provide the least equivocal evidence for environmental conditions, and a survey of the entire continent of Africa for stable isotope ratios in contemporary gastropods has been conducted to provide a database for comparison of environmental conditions in the fossil gastropods.

Stable isotope ratios as marker horizons

The history of the geological and palaeontological research at Koobi Fora on Lake Turkana has demonstrated very acutely the need for stratigraphic control and identifiable marker horizons. Correlation of facies from one exposure to another has been very difficult. The most obvious correlative horizons, the volcanic tuff deposits, proved deceptive as marker beds until refined chemical and isotopic techniques were employed to identify them (Cerling & Brown 1982). Thus the discovery of any other markers would be a useful exercise, particularly in the Koobi Fora situation, but also in similar sedimentary sequences elsewhere.

Cerling *et al.* (1977) were the first to use stable isotope ratios in carbonates as an adjunct to conventional stratigraphic techniques in East Africa. Their classic paper had a different aim, however, from correlation, as they were demonstrating the existence of a major climatic event along the Great Rift Valley at *c.* 1.8 Ma. It was obvious from their work, however, that stable isotopes could be used as a method of identifying potential marker horizons, and Cerling did just that in his early work with the Koobi Fora tuffs (Cerling 1977) measuring oxygen isotope ratios in pumices. At about the same time this laboratory began a stable oxygen isotope study of the various carbonate horizons of the Koobi Fora sediments. Abell *et al.* (1982) worked first with the freshwater stromatolites of the palaeo-Lake Turkana, and identified the same major climatic alteration that had been observed by Cerling *et al.* (1977). This study also demonstrated that stromatolites are effective recorders of environmental information, showing little or no alteration in their subsequent sedimentary history.

The stromatolite beds were too few in the section and too laterally confined to be of any great use as correlative units, although one of the beds does persist for a distance of *c.* 6 km. But other carbonate horizons were present and afforded the possibility of correlative use.

The molluscs in the Koobi Fora sediments are almost ubiquitous. Some molluscan species are capable of growing in a wide variety of environments (Brown 1980), and, indeed, Williamson (1981, 1982) used the Koobi Fora molluscs as a tool to demonstrate evolutionary change over the several million years of time recorded in those sediments. Using his collection of molluscs and his stratigraphic correlations, we examined the molluscs isotopically. Our study (Abell 1982) was restricted to three species of gastropods, *Cleopatra ferruginea*, *Melanoides tuberculata* and *Bellamya unicolor*, as these are the most widely distributed in the Koobi Fora sediments, are capable of growth under a wide range of temperature and salinity, and grow in good isotopic equilibrium with their aquatic environment. Again in this study, there was a demonstration of a major environmental change at about 1.8 Ma, near the KBS Tuff level, but other variations in the climate were also obvious. A period of oscillation in climatic conditions leading up to the KBS Tuff

From Frostick, L.E. *et al.* (eds) 1986, *Sedimentation in the African Rifts*, Geological Society Special Publication No. 25, pp. 153–8.

times was evident, and there were cool and/or wet periods at about the time of the deposition of the Chari Tuff (1.3 Ma) and the Lorenyang Tuff (2 Ma). But there are large gaps in the gastropod record, some where conditions (for example, high alkalinity) prohibited gastropod growth, and where nearshore sediments are not exposed.

At about this time two other laboratories were examining oxygen isotope ratios in sedimentary sequences, using both inorganic and biogenic carbonate sediments in attempts to define climatic conditions at the time of deposition of the sediments. Krishnamurthy & DeNiro (1982) examined lacustrine and pedogenic carbonates in sedimentary sequences in the Kashmir basin. They interpret the *c.* 2‰ difference between the pedogenic and lacustrine carbonates (−7‰, and −8.8‰, respectively) as a consquence of change in climate to a warmer regime. In another study Hillaire-Marcel *et al.* (1982) measured oxygen isotope ratios for a substantial series of calcitic concretions and fossil shells in the Hadar Formation of Ethiopia. Here the span of $\delta^{18}O$ ranged from −10‰ to +2‰ (PDB). The more positive values were clearly associated with evaporative conditions, the more negative with strong sources of freshwater from the Ethiopian highland plateau. Rapid changes in oxygen isotope ratios over thin sediment accumulations suggests rapid changes in the sedimentary dynamics, and they observe, as did Abell *et al.* (1982), that the molluscs from ostensibly the same bed may demonstrate a considerable range of $\delta^{18}O$. Rapid changes from freshwater to brackish water by evaporation are suggested.

Our own laboratory was at this time also looking at the many other carbonate horizons present, in addition to the stromatolites and gastropods in the Koobi Fora sediments. These additional horizons are mainly indurated sandstones, but also include indurated shell beds and calcitic concretions.

Carbonate sediments as isotopic marker horizons

It seems unlikely that all the carbonate beds at Koobi Fora are unaltered, but we have found that where there is a close stratigraphic association with unaltered stromatolites or gastropods, the indurated carbonates and concretions give nearly equivalent oxygen isotope ratios. This argues that most of the indurated horizons were formed nearly contemporaneously with the gastropods or stromatolites, or were formed of carbonate and water which had moved horizontally through these horizons and become nearly isotopically equivalent to the shells or stromatolites.

While we have some confidence that the indurated carbonate beds represent primary isotope ratios, that is not the only problem facing the use of these beds as markers. The four main problems are discussed here.

First, very few of these beds are continuous over long distances, and therefore cannot in themselves be used as marker horizons, although the environmental conditions prevailing at the time of deposition of such a bed may be manifested in other carbonates at substantially the same stratigraphic level in other exposures.

Second, the beds are often not homogeneous. In some cases they are completely void-free indurated sandstones, but in other cases the beds may contain many voids. Particularly, the indurated gastropod beds often consist largely of empty casts of the gastropods themselves, as a cavity of the indurated sandstone. What and where to sample in such a bed is a difficult question. We have not tried to place a reliability index on these various indurated beds, but only recognize that they may not always be reliable.

Third, we have encountered one major regional anomaly. In area 105 our section runs from about 35 m below the KBS Tuff to just above the KBS Tuff. All isotope ratios are very negative (*c.* −4‰ PDB), in contrast to sections further S. This probably represents an area of freshwater inflow to the lake, and is not characteristic of environmental conditions in the region as a whole. Such an anomaly may be useful in itself, but could prove confusing if that were the only section analysed.

Fourth, and by far the largest problem, is the dichotomy of temperature effects vs. evaporative effects on $\delta^{18}O$. Increasing water temperatures produce more negative values of $\delta^{18}O$ according to the empirical relationship worked out by Epstein *et al.* (1953) and Craig (1965). The change in $\delta^{18}O$ with temperature, while not quite linear, averages at *c.* −0.18‰/°C. Evaporative losses of water in precipitation and from the surface of the lake and its tributaries favours a loss of the light isotope of oxygen as $H_2{}^{16}O$. The residual lake water thus becomes isotopically enriched in the heavier oxygen isotope. In the present Lake Turkana this latter enrichment has progressed to the point where, if it were not so alkaline as to prohibit their growth, gastropods would be expected to have oxygen values in their shells of *c.* +6‰. Thus, hot and dry conditions tend to cancel each other out on measured values of $\delta^{18}O$, as will cool and wet conditions. Fortunately one or the other—temperature or evaporation—tends to dominate. This problem will be discussed later when we present our work on contemporary African gastropods and their oxygen isotope ratios.

Results

The oxygen isotope ratios of 280 samples of carbonate from 16 sections are presented in Fig. 1. The gastropods are differentiated from the other sedimentary carbonates in the figure. All isotope ratios are quoted against the PDB standard. The stratigraphic level of the KBS Tuff is indicated as the zero datum point for each section. The sections are roughly arranged in their sequence from S to N (sections A to P), and their locations shown in Fig. 2.

In ten of the sections (sections A, C, E, F, G, J, K, L, N and O) there are carbonate-containing sediments sufficiently closely spaced at the KBS Tuff level to demonstrate the shift in environmental conditions from arid to wet and back to arid, noted in previous work cited on stromatolites and gastropods (Abell 1982; Abell *et al.* 1982). The non-gastropod carbonates seem to be as effective as the gastropods in signalling that sequence of climatic changes.

This shift takes place across the erosional disconformity just above the KBS Tuff level, which marks the border between the Lower and Upper Members of the Koobi Fora Formation. Above the KBS disconformity the record becomes more difficult to read, partly from lack of precise stratigraphic control, but also because

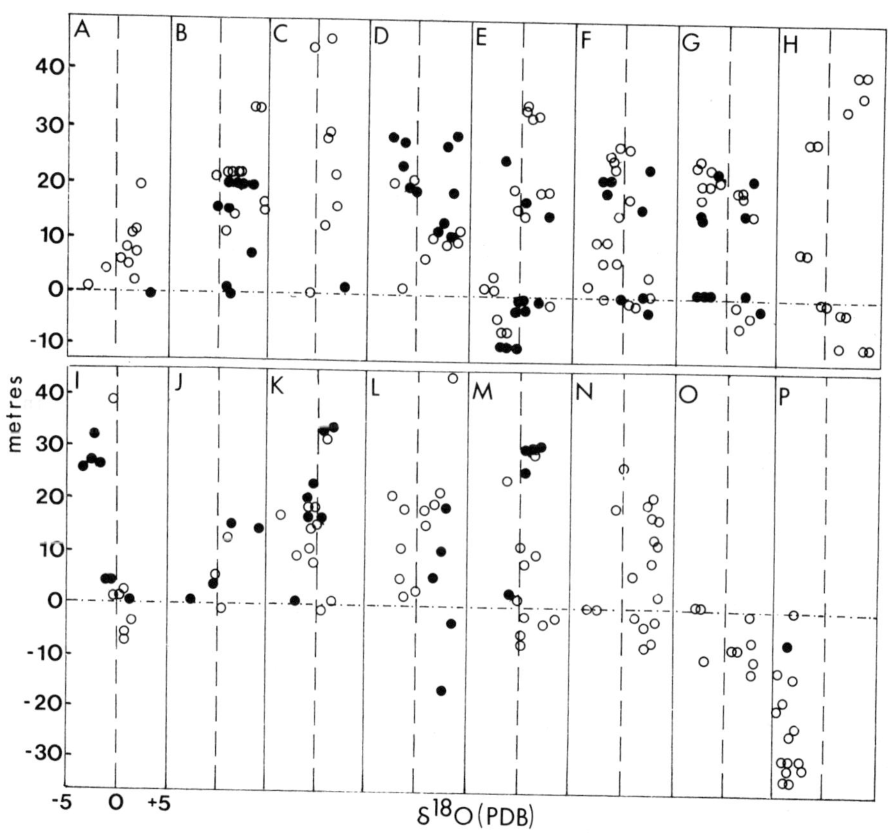

FIG. 1. 280 oxygen isotope ratios for carbonate sediments in 16 sections at Koobi Fora. Open circles = indurated sandstones and concretions; closed circles = gastropods. All sections are normalized to the real or putative level of the KBS Tuff, and are arranged approximately in their sequence from S to N. A. Area 123, Elephant site. B. Area 110, Bura Hasuma Hill. C. Area 123, E end of exposures. D. Area 106, escarpment at centre of area. E. Area 106, escarpment at Luga Bura Hasuma. F. Area 106, Rhino Canyon, N end. G. Area 106, Rhino Canyon, S end. H. Area 103, composite section, near the centre of area. I. Area 101, E side of area. J. Area 102, W of anticline. K. Area 102, E of anticline. L. Area 102, canyon on E side of area. M. Area 100, just W of fault separating areas 100 and 104. N. area 104, centre of area. O. Area 104, N end. P. Area 105, combined sections at Island Hill and opposing escarpment.

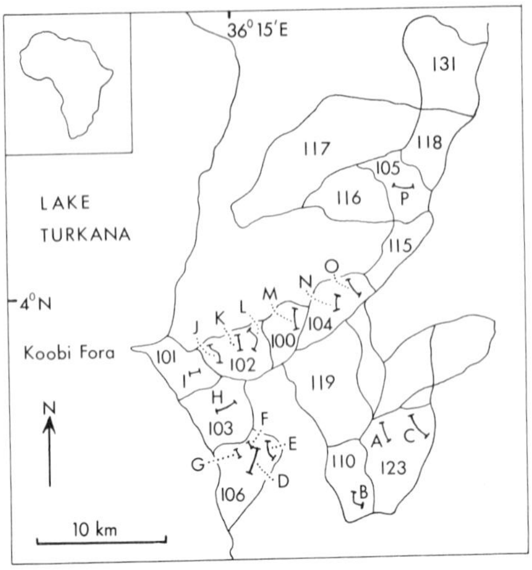

FIG. 2. Locations of sections sampled for oxygen isotope analyses. Numbered areas are defined by topography. Letters refer to sections sampled, see Fig. 1.

at this time the lake was oscillating near overflow and there were no climatic extremes. This is evident in the morphology of the molluscs (Williamson 1981), which demonstrated normal freshwater growth habits during this period.

The values of oxygen isotope ratios near the tops of these 16 sections (including a few $\delta^{18}O$ values omitted because the sections were too long to plot at this scale) are generally in the range of $\delta^{18}O = 0‰$ to $+3‰$ (sections C, E, H, I, L and M), which would indicate only mildly arid conditions. There are a few gastropods in sediments deposited between the Chari and Guomde Tuffs, collected by Williamson and analysed by Abell (1982), which show very negative oxygen isotope ratios, indicating a very cool freshwater environment. None of these sediments are included in the 16 sections analysed in this paper. Even gastropods deposited on the raised beaches *c.* 10,000 years ago do not have extremely negative values of $\delta^{18}O$ (Abell 1982).

Thus, the only common isotopic marker horizon readily recognizable in the sedimentary sequences of this paper is the one at the KBS Tuff level, signalling the same climatic changes that produced the disconformity separating the Lower and Upper Members of the Koobi Fora Formation.

Interpretations

These results, rather disappointing in terms of finding recognizable marker horizons, turn out to be very useful in terms of recognizing climatic conditions prevailing at the time a given stratum was deposited. In general, negative values of oxygen indicate freshwater conditions, and positive values indicate evaporative losses and the onset of brackish conditions. But these qualitative descriptions are inadequate unless they can be related to real situations.

Therefore, in order to gain further insight into the meaning and reliability of carbonate sediments as climatic markers and potentially as stratigraphic marker beds, we have been conducting a survey of modern African gastropods as recorders of environmental conditions in the oxygen isotope ratios of their shells. We have studied gastropod samples from 80 locations over most of the African continent, and sought correlations between modern climates and the stable oxygen isotope ratios. We give a preliminary report on this research here, but full details will be reported in another publication.

As our measures of modern climate we have used two measures: (1) Yurtsever's (1975) survey of oxygen isotope ratios in precipitation, which has been plotted as a map, Fig. 3A; and (2) NASA's (Tucker *et al.* 1985) monthly satellite imagery of vegetative cover over the African continent, summarized in Fig. 3B. Our own summary maps of isotopic data are given in Fig. 3C, showing gastropod collection sites, and Fig. 3D, showing the trends in $\delta^{18}O$. It is clear that Yurtsever's data, although of very high precision, are too widely spaced with reference to collection sites to give a very useful map. The NASA satellite data, on the other hand, give a very precise picture of changing climatic conditions, and correlate very well with our own gastropod data. The following points emerge from this work:

1 In most situations the isotope ratios of gastropods are good indicators of environmental conditions.

2 Values of $\delta^{18}O$ from zero to $+2$ or $+3\%$ may well be found in freshwater lakes such as Lakes Victoria, Edward and Albert, indicating highly fractionated rainfall rather than brackish lake conditions. This is not some manifestation of the usual 'continental effect', as observed by Förstel & Hützen (1983) in their study of isotopic ratios of oxygen in German groundwater, in which the earliest precipitation from a moving air mass is isotopically heavy, and later precipitation isotopically light. Rather, it is probably the global or latitudinal pattern (Yurtsever 1975), which involves a continuous decrease in $\delta^{18}O$ in precipitation from the equator toward the polar regions. That such a latitudinal pattern is far from

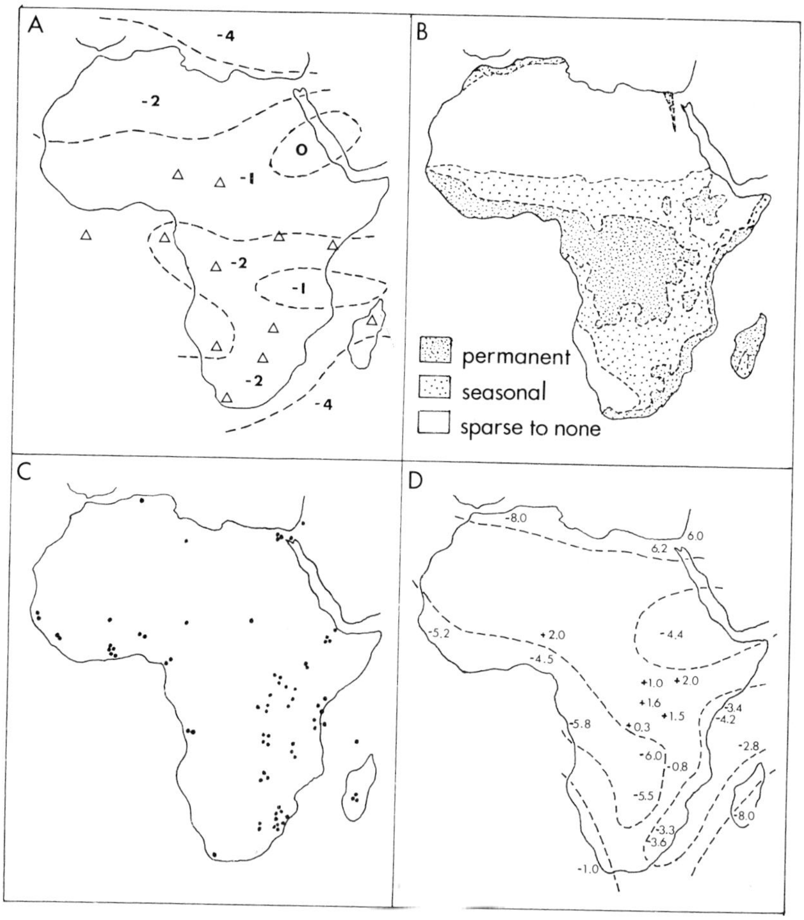

FIG. 3. A. Oxygen isotope ratios in precipitation, redrawn from Yurtsever (1975). Triangles = data collection sites. B. Vegetational map of modern Africa, redrawn from NASA satellite photographs (Tucker *et al.* 1985), showing areas of (1) little or no vegetation at any time in the year; (2) seasonal vegetation; and (3) permanent vegetation. C. Collection sites of contemporary African gastropods. D. Stable oxygen isotope ratios in contemporary gastropod shells. Isolines are approximated from the data obtained at collection sites shown in C.

universal is obvious from the oxygen isotope ratios found for West African and some South African gastropods. This failure of pattern in West Africa and South Africa is one of the reasons for this study.

3 Detailed study of samples along the spiral growth structure of the gastropod shells demonstrates that they are sensitive recorders of water temperature during their approximately one year of normal lifetime. This means that we can use the general isotopic ratios of oxygen isotopes in these freshwater gastropods to picture overall environment, and detailed studies of individual gastropods to show seasonality of climate.

Koobi Fora Palaeoclimatic interpretations

The study of contemporary gastropods of Africa was very revealing in interpreting the Lake Turkana material. By itself, a very positive value $\delta^{18}O$ in the fossil gastropods or carbonate sediments of Lake Turkana is surely indicative of highly evaporative conditions. There is no way to achieve values of $\delta^{18}O$ of around $+4‰$ except by very substantial evaporative fractionation.

On the other hand, values of $\delta^{18}O$ of 1 or 2‰ on either side of zero probably represents the East

African equatorial 'norm', and can be equated to slightly brackish to freshwater conditions, not very different from Lake Victoria or Lake Albert. To differentiate between brackish and freshwater would require evidence other than isotopic ratios.

Values of $\delta^{18}O$ from -3 to $-4‰$ normally represent true freshwater conditions, with a source of rainfall that is isotopically about the same as that for the highland areas of East Africa and Ethiopia today. That is, if the influx of rainfall runoff from the Ethiopian highlands had been substantially greater at times in the past, then the waters of the Omo River and the Sagan River issuing—through Chew Bahir (Lake Stephanie) and into Lake Turkana—would have overwhelmed the evaporative alterations in $\delta^{18}O$ of the lake. Undoubtedly this condition did exist from time to time. A look at the isotopic patterns in the growth habits of the individual shells might be useful here. Seasonal temperature changes or sudden alterations in freshwater supply should show up clearly.

The topography of the lake itself is such that isotopic oscillations are likely to be much larger than in most lakes. The inflow and outflow of Lake Turkana (when it was high enough to overflow) are both largely at the N end of the lake, and so only during periods of very substantial influx will the lake be well enough mixed from N to S to approach isotopic homogeneity. At other times substantial inhomogeneity might be expected, with more positive values of oxygen isotope ratios at the S end, and rapid oscillation between fresh and brackish water in intermediate regions near Koobi Fora and Allia Bay, as the water supply varies.

Thus, in summary, we can identify conditions of considerable aridity, conditions of major influxes of freshwater, and numerous conditions in between that may be of only local importance. The high aridity and high rainfall regimes can be expected to leave their isotopic signature as marker horizons in the carbonates. The less extreme regimes are unlikely to be recognizable in the 'noise' of local variations.

ACKNOWLEDGEMENTS: We wish to thank R.E. Leakey and the National Museums of Kenya for field support at Koobi Fora; Dr P.G. Williamson and the Museum of Comparitive Zoology of Harvard University, and Dr D.S. Brown of the British Museum (Natural History) for supplying us with contemporary gastropod specimens from their collections; and the Graduate School of Oceanography of the University of Rhode Island for use of their isotope mass spectrometer.

References

ABELL, P.I. 1982. Paleoclimates at Lake Turkana, Kenya, from oxygen isotope ratios of gastropod shells. *Nature* **297,** 321–3.

——, AWRAMIK, S.M., OSBORNE, R.H. & TOMELLINI, S. 1982. Plio-Pleistocene lacustrine stromatolites from Lake Turkana, Kenya: morphology, stratigraphy and stable isotopes. *Sed. Geol.* **32,** 1–26.

BROWN, D.S. 1980. *Freshwater snails of Africa and their medical importance.* Taylor & Francis Ltd, London.

CERLING, T.E. 1977. *Paleochemistry of Plio-Pleistocene Lake Turkana and diagenesis of its sediments.* Unpublished PhD thesis, University of California, Berkeley.

—— & BROWN, F.H. 1982. Tuffaceous marker horizons in the Koobi Fora region and the lower Omo valley. *Nature* **299,** 216–21.

——, HAY, R.L. & O'NEIL, J.R. 1977. Isotopic evidence for dramatic climatic changes in East Africa during the Pleistocene. *Nature* **267,** 137–8.

CRAIG, H. 1965. The measurement of oxygen isotope paleotemperatures. In: TONGIORGI, E. (ed) *Stable Isotopes in Oceanographic Studies and Paleotemperatures.* Consiglio Nazionale delle Recerche Laboratoria di Geologia Nuclear, Pisa, pp. 161–82.

EPSTEIN, S., BUCHSBAUM, R., LOWENSTAM, H.A. & UREY, H.C. 1953. Revised carbonate-water isotopic temperature scale. *Bull. Geol. Soc. Am.* **64,** 1315–26.

FÖRSTEL, H. & HÜTZEN, H. 1983. Oxygen isotope ratios in German groundwater. *Nature* **304,** 614–16.

HILLAIRE-MARCEL, C., TAIEB, M., TIERCELIN, J.J. & PAGE, N. 1982. A 1.2 Myr record of isotopic changes in a late Pliocene rift valley lake, Ethiopia. *Nature* **296,** 640–3.

KRISHNAMURTHY, R.V. & DENIRO, M.J. 1982. Isotope evidence for Pleistocene climatic change in Kashmir, India. *Nature* **298,** 640–1.

TUCKER, C.J., TOWNSHEND, J.R.G. & GOFF, T.E. 1985. African land-cover classification using satellite data. *Science* **227,** 369–75.

WILLIAMSON, P.G. 1981. Paleontological documentation of speciation in Cenozoic molluscs from Turkana basin. *Nature* **293,** 437–43.

—— 1982. Molluscan biostratigraphy of the Koobi Fora hominid-bearing deposits. *Nature* **295,** 140–2.

YURTSEVER, Y. 1975. *Worldwide Survey of Stable Isotopes in Precipitation.* IAEA report, Section of Isotope Hydrology.

P.I. ABELL & J.P. MCCLORY, Dept. of Chemistry University of Rhode Island, Kingston, Rhode Island 02881, USA.

Mineral precipitation and diagenesis in the sediments of the Lake Bogoria basin, Kenya Rift Valley

R.W. Renaut, J.J. Tiercelin & R.B. Owen

SUMMARY: Lake Bogoria is a perennial, saline, alkaline closed lake fed by hot springs and ephemeral runoff. Detailed studies of the sedimentology, mineralogy and chemistry of the late Quaternary sediments, both from exposures around the shoreline and from lake bottom cores, have provided abundant evidence for early diagenesis and mineral precipitation. Marginal fan-deltaic clastics exhibit carbonate (calcite, dolomite), iron oxide, opaline silica and zeolite (analcime, natrolite) cements, especially where groundwater is shallow and subject to evaporative concentration. Travertines associated with microbial activity around hydrothermal springs are predominantly calcite; high temperature, inorganically precipitated travertines are aragonitic. Most are relic deposits. In places, spring carbonates have been silicified (opal-A) or altered to fluorite in response to changing groundwater composition. Fan-delta conglomerates have been hydrothermally altered by alkaline spring waters yielding zeolites (analcime, thomsonite, clinoptilolite, mordenite), clay minerals, opaline silica and magadiite. The lacustrine sediments include prodeltaic clastics, organic oozes and sodium carbonate evaporites. Trona and nahcolite predominate in the central sub-basin. The local occurrence of thermonatrite may be associated with sub-lacustrine hydrothermal activity. In the north sub-basin, evaporites are exclusively gaylussite. Sodium silicates (magadiite, kanemite) occur in black organic oozes and formed in an anoxic environment rich in sodium carbonate and dissolved SiO_2. Other authigenic minerals in the lacustrine sediments include pyrite, siderite, fluorite, zeolites (analcime, mordenite, clinoptilolite), opaline silica (formed from a gel precursor around sub-lacustrine springs?) and vivianite.

Modern and ancient saline, alkaline lake deposits are well known for their abundant and varied mineral suites. These include minerals precipitated directly from solution, such as many salts and alkaline-earth carbonates, and authigenic minerals such as zeolites and sodium silicates, formed by the interaction of interstitial brines with sediment or fresher waters (Eugster & Hardie 1978; Eugster 1980a). Much of this knowledge stems from the detailed studies by Hay (1970), Eugster (1980b) and others working in the rift valley lake basins of East Africa.

Of the East African lakes, Magadi is often regarded as a 'type-model' for saline, alkaline rift lakes, having been studied intensively by Eugster and colleagues for nearly two decades (Surdam & Eugster 1976; Jones *et al.* 1977; Eugster 1980b, this volume). It is today essentially an ephemeral salt pan, only holding surface water briefly during the rainy seasons.

Since 1976, research has been conducted at Lake Bogoria, another saline, alkaline rift lake some 220 km to the N (Fig. 1, inset B). Unlike Magadi, which is mostly hot-spring fed and lacks significant surface inflow, Bogoria is perennial, being fed both by hot springs and seasonal streams. Furthermore, since Lake Magadi itself has little marginal relief, clastic input is minimal. Bogoria, in contrast, has high marginal relief, and clastics form a major proportion of the lacustrine sediments.

Though sharing much in common with Lake Magadi, Bogoria has several distinctive mineral occurrences and sedimentary features not previously described. The purpose of this paper is to provide a preliminary account of the occurrence and distribution of the precipitated and authigenic minerals, and some of the diagenetic processes recorded from the Bogoria sediments. Full details of the mineral genesis, and of the sedimentology and hydrochemistry of the basin will be presented elsewhere.

The results are based on a detailed study of the sedimentology, mineralogy and chemistry of the late Quaternary sediments, both from exposures around the shoreline and from lake bottom cores. The sites of three long cores drilled to depths of between 15 and 20 m (SI–SIII), are shown in Fig. 1. Details of sampling and analytical methods are given in Tiercelin (1981) and Renaut (1982).

Setting of the Bogoria basin

Lake Bogoria (formerly Lake Hannington) lies within a deep asymmetric half-graben at the E margin of the axial trough of the Kenya Rift

From FROSTICK, L.E. *et al.* (eds) 1986, *Sedimentation in the African Rifts*, Geological Society Special Publication No. 25, pp. 159–75.

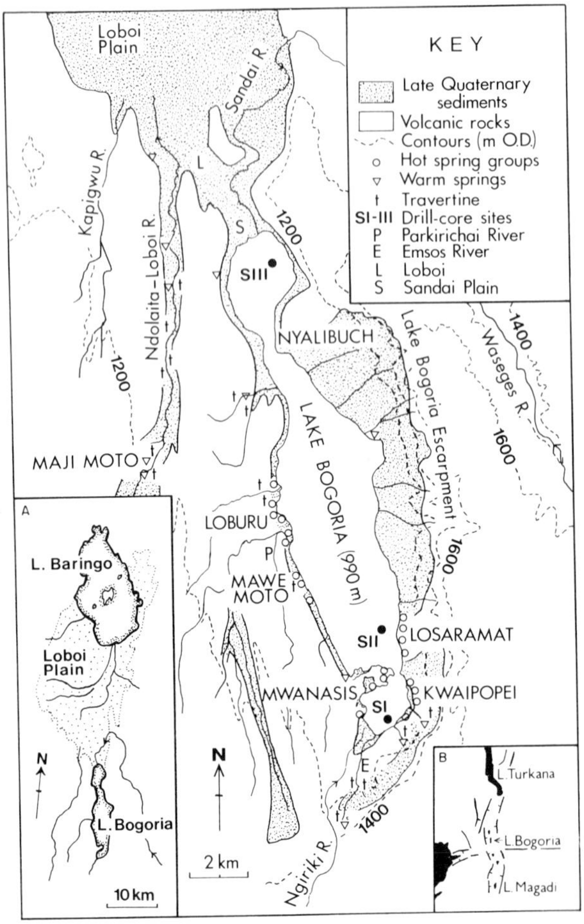

FIG. 1. Location map and setting of the Lake Bogoria basin.

FIG. 2. Lake Bogoria. Foreground shows Loburu delta-plain with several steaming hot springs. Beyond the lake lies the Lake Bogoria escarpment.

Valley, just N of the equator (Fig. 2). To the N of this half-graben, which is *c.* 60 km long, lies Lake Baringo. The two lakes are separated by the broad, locally swampy, alluvial Loboi Plain (Fig. 1, inset A).

The lake is 17 km long and varies from 0.5 to 3.5 km in width. It occupies three sub-basins within the depression—N, central and S respectively—separated by shallow sills (Fig. 1). Its surface lies at *c.* 990 m OD, and the maximum water depth of 12 m (1978) is attained in the southern central sub-basin. The saline, alkaline waters support little life today, except for algae, cyanobacteria, diatoms and rotifers which feed a thriving flamingo population.

The catchment area (705 km^2) is almost entirely volcanic. To the E, the land rises abruptly in a series of steep, NW–SE aligned fault-scarps, dominated by the Lake Bogoria escarpment. This region is composed of trachytic, phonolitic and basaltic volcanics of Miocene to Pleistocene age. Directly W of the lake, the northerly dipping rift floor comprises an intensely grid-faulted region of Pleistocene trachyphonolite and basaltic lavas. Details of regional geology can be found in McCall (1967), Griffiths (1977) and Williams & Chapman (this volume).

Origin and chemistry of the Bogoria basin waters

Table 1 gives the composition of representative samples of lake, spring, river, ground and rain waters from the Bogoria basin.

The lake waters originate from three main sources. The Sandai-Waseges River, which drains two-thirds of the catchment, is the principal fresh affluent. Though ephemeral, it approaches perennial flow in exceptionally wet years. The lake also receives sporadic inflow from *c.* 100 minor ephemeral streams. The second main source is from *c.* 200 perennial alkaline hydrothermal springs, mostly located at three sites along the lake shore (Fig. 1), and augmented by direct sub-lacustrine spring discharge. Thirdly, direct precipitation on the lake surface contributes up to *c.* 700 mm a^{-1}.

The lake waters are essentially of Na–CO_3–HCO_3–Cl composition. Their pH exceeds 10.3, and the salinity of the surface waters exceeds 40,000 mg l^{-1}. After carbonate species, Cl^- is the next most important anion, while K^+ is the subsidiary cation. Mg^{2+}, Ca^{2+} and SO_4^{2-} are relatively low. Silica concentration appears to vary with time, possibly in response to short-term fluctuations in lake level and pH, and/or its removal as a solid phase. Baker (1958) and Kilham (1971) both recorded high silica concentrations (up to 1210 mg l^{-1}), while Tiercelin (1981) and Renaut (1982) found very low values (<0.1 mg l^{-1} in 1978).

Comparing surface and lake bottom water analyses (sites 85-S and 85-B in Table 1) clearly demonstrates a salinity and chemical stratification or meromixis. The relatively stable, poorly mixed, bottom waters contain up to 90 g l^{-1} dissolved salts and may partly orginate from saline sub-lacustrine spring seepage (and/or dissolution of salts). The meromixis is enhanced during periodic transgressions when dilute surface waters come to overlie the stagnant bottom waters (Tiercelin *et al.* 1980). Lake salinity increases generally southwards with distance from the Sandai-Waseges inflow.

Local dilute river waters have similar compositions to those previously reported for volcanic catchments in East Africa (e.g. Kilham 1971). A few streams (e.g. River Emsos) that are partly fed by warm spring, have slightly higher than average Na^+, HCO_3^-, SiO_2 and F^- concentrations.

The hydrothermal waters are intermediate in concentration between the dilute inflow and lake waters. Chemically and spatially they fall into two distinctive groups, which for convenience can be termed 'warm spring' waters and 'hot spring' waters (Renaut 1982). The warm springs mostly lie on fault-line sites away from the lake shore (e.g. Ndolaita-Loboi valley, Fig. 1). Temperatures of the warm springs range from 34 to 48°C; they have a pH range of 7.1 to 7.8, and solute concentrations of 0.3–1.0 g l^{-1}. Their waters are of Na–Ca–HCO_3 composition. They are slightly more saline than local river waters, but have significantly higher SiO_2, F^-, Na^+ and HCO_3^- concentrations, and lower alkaline earths.

Most hot springs lie a few metres above the lake level at sites where fault lines either intersect or lie parallel to the lake shore (Fig. 1). There are three main groups: (1) Loburu, where *c.* 60 spring orifices pock-mark a delta plain; (2) Mawe Moto, where *c.* 45 springs emerge from fractured fissile trachyphonolites along a 350 m sector of the shoreline; and (3) Mwanasis-Kwaipopei-Losaramat (the Southeastern Group) where *c.* 75 springs issue from jointed lavas and conglomerates at the site of a complex transverse fracture system. Hot spring temperatures range from 64 to 98.5°C.

Solute concentrations range from *c.* 4 g l^{-1} at Loburu, to 5–6 g l^{-1} at Mawe Moto, and up to 15 g l^{-1} in the Southeastern Group. The increase in salinity southwards parallels the similar trend in the lake waters, suggesting that some shallow mixing may have occurred. There is, however, no clear correlation between spring temperature and salinity. The pH ranges from 8.6 to 9.9 (June 1977). The hot spring waters have a general Na–

TABLE 1. *Selected chemical analyses of Bogoria basin waters*

	Analysis number	Na	K	Ca	Mg	Alk.	Cl	SO_4	F	SiO_2	pH
Rain water, Loburu (8.77)	RF1–77	2.34	0.33	2.68	0.21	0.10	4.12	0.68	nd	nd	6.23
Sandai-Waseges R. (8.76)	SR1–76	70	3	30	9	4.03	38	5	0.5	51	7.4
Parkirichai R. (8.77)	KL1–77	101	6	26	8	5.01	46	3	2.6	45	7.5
Warm spring, Maji Moto (10.78)	E62–187	138	tr	18	1	6	35	tr	3.6	75	6.7
Hot spring, Loburu (10.77)	E64–008	1450	34	2	0	55	266	67	nd	200	nd
Hot spring, Mawe Moto (8.77)	BH10–77	1665	45	3	tr	59	365	86	83	78	8.9
Hot spring, Losaramat (8.77)	BH14–77	5474	94	0	0	190	1475	125	135	80	9.9
Lake margin groundwater, S Loburu (8.77)	GW3–77	7825	148	6.5	2.5	279	2125	285	20	27	10.1
Lake water, site 85-B (10.77)	E64–025	40900	644	34	2	1550	9040	549	nd	0	9.2
Lake water, site 85-S (10.77)	E64–026	27800	428	24	0	1050	5850	434	nd	0	9.3
Lake water, site 85-B (10.78)	E62–182	26000	390	14	1	860	9570	300	2.5	tr	10.1
Lake water, site 85-S (10.58)	E62–181	10600	195	14	4	350	3900	68	2.5	tr	10.2

All analyses are given in mg l^{-1} except for alkalinity (CO_3+HCO_3) which is expressed in mEq l^{-1}. For locations of sampling sites see Tiercelin (1981) and Renaut (1982). Site 85 lies in the northern central sub-basin (approximately at the 'E' in LAKE in Fig. 1). Date of sampling is given in column 1, (10.78) = October 1978; nd = not determined; tr = trace.

HCO_3–CO_3–Cl composition, are high in silica and fluoride, and low in alkaline earths.

The origin of the hydrothermal waters is complex and remains poorly understood. A hydrogeological survey of the Kenya Rift (Noble & Ojiambo 1976) revealed two groundwater catchment areas extending N and S of Eburru (90 km S of Bogoria). Noble & Ojiambo suggest that northward-moving groundwater, recharged from the escarpments bounding the rift valley and Lake Naivasha, may be discharged at the Bogoria springs. In contrast, McCall (1967) suggested that the hot spring waters originate primarily as lake water which seeps down 'fissures' where it encounters hot gases, including juvenile CO_2, and is returned to the surface. Further study, including stable isotope analyses, is required before conclusions can be drawn.

The relative importance of runoff and spring discharge in recharging the lake is uncertain, but the two will probably vary inversely, partly dependent on climate. During drier periods, when the lake is subject to intense evaporation (*c.* 2600 mm a^{-1} today), principal inflow will be from the perennial springs which are initially moderately saline. Thus the trend towards increased salinity is enhanced. Further details of the hydrochemistry, brine evolution and hydrogeology of the basin are given in Renaut (1982).

Modern sedimentary environments

Despite its small size, the basin displays a wide range of depositional environments, as shown in Fig. 3. The sedimentological features that define each environment have been described previously by Tiercelin *et al.* (1980, 1981, 1982), and are also summarized briefly here.

Below the Lake Bogoria escarpment, there are a series of well-vegetated, coalescent alluvial fans and fan-deltas. Incised channel sections reveal characteristic interdigitation of paraconglomerates (debris and mudflow deposits) and orthoconglomerates (stream and sheetflood deposits). Major landslips, locally producing chaotic megabreccias, and talus accumulations are widespread along the scarp front.

The 'ephemeral stream floodplain' sub-environment of Hardie *et al.* (1978) is restricted by the steep basin margins, to short ephemeral braided streams that feed fan-deltas W of the lake, and the margins of the Sandai-Waseges River. The sediments are typically of bimodal size, comprising poorly-developed channel bar conglomerates, interstratified with reddish brown, feldspathic sands and silts. Bedding structures are generally poorly developed or preserved. Frequent fluctuations in base (lake) levels have produced numerous minor unconformities and small river terraces.

The 'sandflat' and 'dry mudflat' sub-environments of Hardie *et al.* (1978) are difficult to distinguish around most of Bogoria, partly due to frequent small fluctuations in lake level. They are combined as the 'delta-plain' in Fig. 3. These areas for the most part represent emergent delta-platform surfaces (mostly silt and fine sand), cut by ephemeral, braided sand channels and veneered by poorly sorted gravel spreads, sheets of mud-cracked graded silt (sheetflood deposits?) and saline pond deposits. Much of the surface is vegetated by grasses with local sedge marshes. A narrow (few metres) 'saline mudflat' rims the

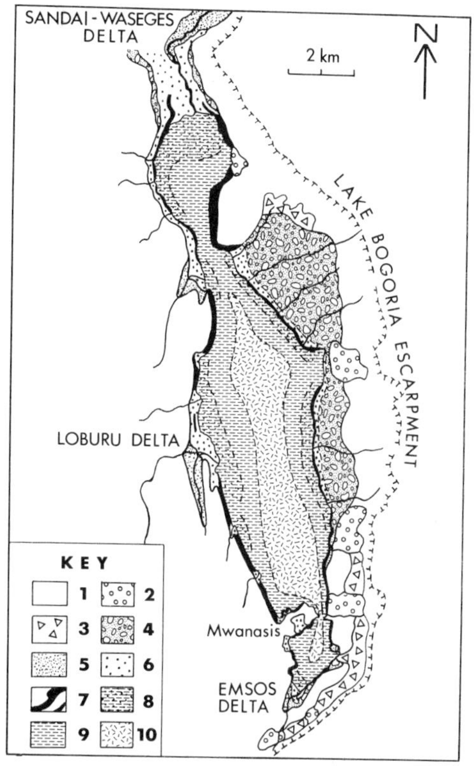

FIG. 3. Depositional environments of the Lake Bogoria basin (simplified). Key: 1 volcanic rocks, regolith, thin soils; 2 landslips; 3 talus; 4 alluvial and piedmont fans; 5 ephemeral stream floodplain; 6 delta-plain (includes 'sandflat' and 'mudflat' sub-environments of Hardie *et al.* 1978); 7 beaches and beach terraces; 8 delta-front clastics (sand–silt–clay); 9 pro-delta clastics (muds); 10 organic oozes and evaporites.

delta-plain margins, but is not shown on Fig. 3.

Approximately 80% of the lake shore is bordered by clastic beaches. On interdeltaic volcanic headlands stepped beach terraces, composed of limestone-encrusted subangular gravels, rise up to 999 m. Deltaic shorelines are bordered by narrow (0.2–2 m) beaches and spits composed of subrounded granules and fine gravels.

Three lithofacies consitute the 'perennial saline lake' sub-environment of Hardie *et al.* (1978):

1 The delta-front deposits are mostly pale brown or grey feldspathic sands and silts occurring in units a few cm to several dm thick. There are two main types. The first grades upwards from massive granular sands through medium sands to laminated silts, and these have been interpreted as low density turbidites. The second type encompasses a range of well-bedded sands,

silty sand and silts which show Gilbert-type foresets on some of the steeper fan-deltas.

2 The pro-delta sediments are massive and laminated, silty clays and clays (mostly smectite; locally illite). These are interpreted as both the distal portions of turbidity flows, and quiet clay sedimentation from hypopycnal plumes. Clays from cores contain diatoms, ostracods, fish and molluscs (locally forming coquinas)—all indicative of former fresher waters.

3 The central, deepest part of the lake is one of pelagic mud and sapropelic ooze accumulation, alternating with sodium carbonate evaporites. The oozes are massive, or laminated, green to blackish muds in units a few cm thick, containing up to 6% organic matter, most of which is sapropelic (Herbin 1979). Preserved laminae and pyrite indicate periodic anoxicity. The evaporites are described in detail below.

Tiercelin *et al.* (1980, 1982) have shown that the evaporites form during periods of relative aridity when the lake level is drawn down by reduced inflow and evaporation, whereas the organic muds form during wetter periods when the lake level rises and there is high organic productivity.

In common with the other closed lakes of the Eastern Rift (Owen *et al.* 1982), Bogoria has fluctuated in surface level and salinity throughout the late Quaternary, mostly in response to climatic change (Young & Renaut 1979; Tiercelin *et al.* 1981, 1982; Carbonel *et al.* 1983; Vincens *et al.*, this volume).

Early diagenetic processes in lake margin clastics

The fan-delta sediments marginal to the lake are mostly interbedded reddish-brown feldspathic silts, sands and silty clays, cut by lenticular units of conglomerate. The sands and silts are dominated by alkali feldspar, are very poor in quartz, and contain variable quantities of volcanic glass and mafics. The clay minerals are mostly smectites (Renaut 1982).

Five main early diagenetic processes have been recognized as occurring within the marginal delta-plain environment:

Sediment disturbance

Many deltaic and ephemeral stream floodplain sediments show evidence of post-depositional disturbance, in the form of disrupted bedding or homogeneous mixing of texturally variable sediments. Two groups of processes are held responsible—bioturbation and salt crystallization.

Biogenic disruption is principally by invertebrates. Termites and ants are locally abundant in drier parts of the plains. Homogeneous, massive sediments around termitaries can be seen to pass laterally into well-bedded silts, demonstrating the effectiveness of insects in bioturbation (see Crossley, this volume). At S Loburu, shallow channel sections reveal shallow domal zones of disruption up to 30 cm high, attributed to flamingo nest-mound construction. Many delta-plains are partially vegetated by grasses and sedges, whose root systems disrupt sediments to a depth of *c.* 40 cm.

Where shallow saline groundwater lies less than *c.* 70 cm below the surface, crystallization of efflorescent trona causes sediment disturbance (Eugster & Kelts 1983), particularly within the upper 5–10 cm. Though less effective than bioturbation in disrupting lamination today, the process may have been more significant in drier climatic phases.

Precipitation of alkaline earth carbonates

Calcite is very common as a micritic intergranular cement in shallow soils around Lake Bogoria, and locally forms vadose concretions, sparite cement, micrite veinlets or laminar crusts on gravels. Laterally extensive calcretes are uncommon although one eroded example is seen on the Sandai Plain. Most of the shallow carbonate is attributed to capillary precipitation from shallow groundwaters or surface evaporation (cf. Eugster 1980b).

'Soil carbonates' are most common on the proximal parts of delta-plains and their feeder channels where waters are relatively dilute. Towards the lake shore, shallow groundwaters become increasingly more saline (pH 8.5–9.5) and poor in alkaline earths. Carbonate cements are consequently less abundant, but local lenses of micrite form where downward percolating, Ca-bearing, dilute runoff encounters alkaline groundwaters with high pH (Renaut 1982).

Carbonate rootcasts are abundant around much of the lake shore, particularly at Loburu, Mwanasis and the Sandai Plain. Though many are clearly relict, forming winnowed surface lags, some are forming today. Calcite-encrusted rhizomes of living *Cyperus laevigatus* were found at depths of 2–10 cm on the Loburu delta-plain.

Minor fine-grained dolomite cement occurs in the toe zone of the eastern alluvial fans, but only where the host sediment is basaltic and not trachytic detritus. The basalts may have supplied a local enrichment of Mg^{2+}, allowing dolomite to form after the Mg:Ca ratio had been raised by calcite precipitation. Extensive dolomite fringes, recorded from other saline lakes, are absent.

Opaline silica cementation

On the lakeward margin of some delta-plains (notably Loburu), white authigenic opaline silica (opal-A) occurs within the upper metre of sediment. The opal occurs as intergranular cement, veinlets and fine knobby concretions 1–3 mm in diameter. Eugster (1980b) described a similar occurrence at Magadi. It probably originates through evaporative concentration and upward capillary movement of shallow silica-rich groundwaters with high pH. Silicified wood (twigs) occurs locally.

Cementation by iron oxides

Many lake margin sediments are reddened by iron oxides (hematite, goethite). Petrographic study shows that in some cases, this is an early diagenetic cement.

Authigenic zeolites

On the Sandai Plain (Fig. 1), soft root-marked fluvio-deltaic claystones and siltstones of probable late Pleistocene age have been extensively reddened and altered to indurated zeolitic rocks (up to *c.* 40% zeolites, by volume) to a depth of up to 1 m below the present land surface (Hay 1970). The zeolites, dominantly analcime ($NaAlSi_2O_6 \cdot H_2O$), but with minor natrolite ($Na_2Al_2Si_3O_{10} \cdot 2H_2O$), formed in a delta-plain soil profile where saline, alkaline, lake-marginal groundwater lay at shallow depth ($< c.$ 1.5 m) below the land surface.

The zeolites formed by the reaction of poorly crystalline clays with interstitial sodium carbonate-rich pore waters concentrated close to the surface by capillary evaporation (Renaut 1982). The zeolitic rocks locally contain up to *c.* 15% illite, which is a minor component of the original claystones (<2 to *c.* 8%), and is also believed to be authigenic. Calcium released during the reaction has been reprecipitated within the profile as sparite and micrite cement (locally dolomicrite), and may have contributed to the formation of rhizoliths and calcrete that locally overlie the zeolitic profile.

Spring deposits

Travertine

Travertine is present at six Loburu springs and occurs extensively at Mawe Moto. At Loburu, it forms orifice mounds, pool-rim dams, small terraced-mounds, pisoids, and basal crusts in outflow channels and spring pools. All the travertines rest on a substratum of fan-delta clastics. At Mawe Moto, travertine has been precipitated within near-surface joints and fractures in trachyphonolites, and locally forms sloping mounds along the shore zone. Travertine and porous plant and moss tufas also occur at several of the warm spring sites distant from the lake, where in most cases they are being eroded (e.g. Maji Moto, Fig. 1).

Most precipitation of $CaCO_3$ as travertine can be attributed to the removal of CO_2 from saturated and supersaturated solutions (Slack 1967; Julia 1983). As is commonly the case, it is difficult to demonstrate whether precipitation resulted from inorganic or biochemical processes, or both. Field, petrographic, SEM and geochemical evidence suggests that inorganic processes, notably CO_2 degassing, but also agitation and evaporation, are dominant proximal to the orifice of springs whose water temperatures exceed *c.* 92°C; whereas bacteria and cyanobacteria may play a significant role at lower temperatures (Renaut 1982; Casanova, this volume).

Mineralogically, the hot spring travertines comprise both low-Mg calcite and aragonite. Warm spring travertines and tufas are almost exclusively low-Mg calcite. Loburu travertines are predominantly low-Mg calcite (*c.* 85–90% of all the carbonate by volume). Aragonite is normally only present where precipitation appears to have been the most recently active, or where stagnant, saline, alkaline groundwater saturates the porosity. At Mawe Moto, aragonite comprises *c.* 60–70% of the travertine at the northern, most active, end of the site, but is less abundant at the southern end where the deposits are eroding and are mostly low-Mg calcite. Some of the most recent aragonite is strontian aragonite, containing up to 4.1 mol% $SrCO_3$.

The factors that control which polymorph is precipitated remain imperfectly understood. Several studies have demonstrated that aragonite is favoured over calcite with increasing temperature up to 100°C (e.g. Kitano 1962). Others have shown that Mg^{2+}, among other ions, can inhibit calcite crystallization favouring aragonite (Berner 1975).

Aragonite may have formed at spring orifices principally because of the high temperature of precipitation (presently 70–98°C). This may have occurred even though the spring waters today, and probably in the past, have a low Mg:Ca ratio (Table 1). Reams (1974) showed experimentally that waters poor in Mg^{2+} can precipitate aragonite if stirred to induce loss of CO_2, suggesting that rapid CO_2 degassing could also favour aragonite precipitation. Calcite nucleation is slower than its rate of crystallization, and kinetic factors tend to favour aragonite crystallization over calcite. Thus where inorganic precipitation is rapid, as inferred for many fibrous meshwork travertines at Mawe Moto, aragonite may again be the preferred polymorph. Where a biochemical role in precipitation can be demonstrated or inferred, low-Mg calcite is commonly the preferred carbonate. However, petrographic and geochemical evidence suggests that some calcite travertines showing bacterial structures may originally have been aragonite, having undergone diagenesis.

Travertine diagenesis

Lying within only a few metres of lake level, the hot spring travertines have not only been subject to vadose processes, but also to phreatic diagenesis during periodic submergence by alkaline lake waters of fluctuating salinity and variable chemical composition. Furthermore, there is evidence to show that the spring water composition has itself varied through time.

Four groups of early diagenetic processes have been identified:

1 *Alteration of aragonite to calcite*. Several lines of evidence indicate that more of the hot spring travertine was originally aragonitic, and has since transformed to more stable low-Mg calcite: (1) Aragonite only occurs in orifice mounds at Loburu where waters are actively discharging at temperatures close to boiling point. Older and abandoned mounds are mostly calcite. (2) There is petrographic evidence to support aragonite to calcite neomorphism; e.g. fibrous calcites containing linear crystal inclusions similar to those described by Folk & Assereto (1976). Microprobe analyses have revealed higher strontium concentrations within the inclusions than in the host crystal. (3) The Sr content of some calcites is much higher than predicted by partitioning coefficients for the inferred temperatures of precipitation (Renaut 1982).

This is not meant to imply that all Bogoria hot spring travertines were originally aragonitic. With sustained high temperature, favourable pore-water composition and limited circulation, aragonite may persist (Lippman 1973; Longman 1980). If, however, environmental factors change, for example by the leaching of the travertine by dilute runoff, aragonite may transform by solution/reprecipitation or neomorphism to calcite.

2 *Calcite to calcite*. Evidence for probable dissolution and reprecipitation of calcite, mostly as pore-filling sparite or micrite, has been seen in many thin sections.

3 *Silicification*. Pervasive or patchy replacement of calcite and aragonite by fibrous opaline silica (opal-A), which in places has recrystallized to microquartz, has been recorded at Mawe Moto and at one Loburu spring. Opaline silica forms a rare cement in some travertines.

4 *Alteration to fluorite*. At Mawe Moto, fluorite occurs as veins, cement and concordant lenses within travertines, and locally pervasively replaces or pseudomorphs calcite or aragonite.

The significance of the silicification and alteration to fluorite are discussed below.

Opaline silica crusts

Forming a rim at the shallow margins of many of the Loburu spring pools and their outflow channels, are thin (1–6 mm), white crinkly crusts composed of opaline silica (opal-A). Locally underlying the crusts are white, milky suspensions of colloidal silica that grade into denser silica gels. The dissolved silica content of the Loburu springs ranges from about 60 to 200 mg l^{-1} SiO_2, averaging about 80 mg l^{-1} (McCall 1967; Tiercelin 1981; Renaut 1982). When analysed in 1977, springs rimmed by crusts were undersaturated with respect to amorphous silica.

The crusts commonly lie up to 15 mm above water level in the adjacent spring or outflow channel, but are absent below water level. They form by evaporative concentration and the cooling of spring water-saturated marginal sediments. By these processes, precipitation may occur even when spring waters are undersaturated with respect to amorphous SiO_2.

Interpretation of the spring deposits

The modern hot spring waters are poor in alkaline earths, averaging <4 mg l^{-1} Ca+Mg (Table 1). Although opaline silica crusts are clearly forming today, most travertines are relic features, recording precipitation from waters formerly much richer in calcium. However, strontium fractionation data, using the coefficients of Kinsman (1969), indicate that some aragonite may be forming inorganically today at northern Mawe Moto (Renaut 1982).

Most travertine mounds at Loburu are encrusted by stromatolitic limestones, produced during lacustrine transgression dated at *c*. 4000 BP (Tiercelin *et al*. 1981; Vincens *et al*., this volume), demonstrating that the main period of travertine formation preceded this transgressive phase. However, most of the structures at Loburu are clearly subaerially formed (cf. Chafetz & Folk 1984). This is supported by a paucity of clastic contaminants, which would be expected if the structures were deposited subaqueously in a delta-front setting.

There is then an apparent problem in explaining more dilute, calcium-rich spring waters during a period when the lake level was as low or lower than that of today. A second problem is that many orifice travertines at modern boiling springs show evidence of biogenic structures, suggesting that cooler waters existed during travertine precipitation. This may be explained by a lack of synchronization between the groundwater and lake water compositions, as summarized in Fig. 4.

During more humid conditions the groundwater may become charged with alkaline earths, reflecting increased weathering by CO_2-charged waters and only limited precipitation of soil carbonates. The lake level will then rise (Fig. 4B). At such times the proportion of dilute waters within the geothermal system will increase, either by direct recharge of the reservoirs, or more probably, by greater mixing of non-thermal waters with deeper thermal waters. With the onset of drier conditions, dilute recharge may be attenuated and the lake level will fall. The spring waters may still retain alkaline earths inherited from the wetter period. These may be precipitated rapidly as travertine, until the waters are depleted of most of their calcium (Fig. 4C). Cooling of the thermal waters by non-thermal waters also increases the potential for biogenic travertine precipitation.

During drier periods, dilute recharge will diminish. More of the spring water will thus be derived from greater depths, where higher salinities may exist, reflecting higher temperatures and greater residence time, and consequent increase in reaction rates between minerals and solution. Emergent spring waters will be hotter, and have a higher pH. High concentrations of silica and fluoride may evolve (though silica concentration will probably be controlled in part by equilibrium relationships with subsurface volcanic wall rocks).

Today opaline silica precipitates at the surface. During wetter phases, shallow mixing may cause subsurface silica precipitation by lowering the temperature and pH. With little calcium to remove fluoride in the thermal waters today, high concentrations can evolve, and fluorite may form today by reaction with earlier travertine. The reaction is not ubiquitous and some evaporative concentration (within the porosity) may be required.

Much of the hot spring travertine may be associated with the wetter phase that led to formation of the stromatolites at *c.* 4000 BP. Two periods of crust formation have been recognized, separated by a regressive phase (Vincens *et al.*, this volume). The subaerial travertines may have formed during the period of low lake levels while spring waters were still charged with alkaline earths, prior to the second phase of stromatolitic encrustation. Those travertines lacking the encrustation would then post-date the second phase.

Hydrothermal alteration

Extensive alteration of conglomerates and *in situ* lavas by hot alkaline waters and fumerole gases, has occurred along the southern and eastern shores of Mwanasis (Fig. 1), at Kwaipopei, and at N Loburu.

Detailed analysis of the rocks has yet to be undertaken. Preliminary X-ray diffraction and petrographic analyses have revealed the following probable reaction products: iron oxides (hematite), pyrolusite (which lines fumerole vents), opaline silica, quartz (pervasively replacing lava), anatase, smectite, illite, amorphous clays, analcime, thomsonite, mordenite and clinoptilolite. The zeolites may result from reactions between alkaline hydrothermal fluids and trachytic detritus. Minor magadiite is present within the matrix of conglomerates at Kwaipopei (cf. Icole & Périnet 1984). Whether it is a reaction product or a precipitate has not yet been determined.

Minerals formed in the lacustrine sediments

Evaporite minerals

Evaporite minerals are widespread in the Bogoria basin. They occur in the long cores (Fig. 5) and also form efflorescent crusts on the lake-marginal deltaic plains.

Mineralogy and morphology

The evaporites are predominantly sodium carbonates—trona ($NaHCO_3 \cdot Na_2CO_3 \cdot 2H_2O$), nahcolite ($NaHCO_3$), thermonatrite ($Na_2CO_3 \cdot H_2O$) and also gaylussite ($Na_2Ca(CO_3)_2 \cdot 5H_2O$). Halite ($NaCl$) is locally present.

Five types of deposit have been distinguished in the drill-cores:

1 Sub-compacted black (2.5YR 2/0) ooze with small or large (up to several cm) acicular crystals of trona, characterized by the (100) cleavage plane parallel to crystal elongation along the (010) direction. Thermonatrite and halite may also occur in this sediment.

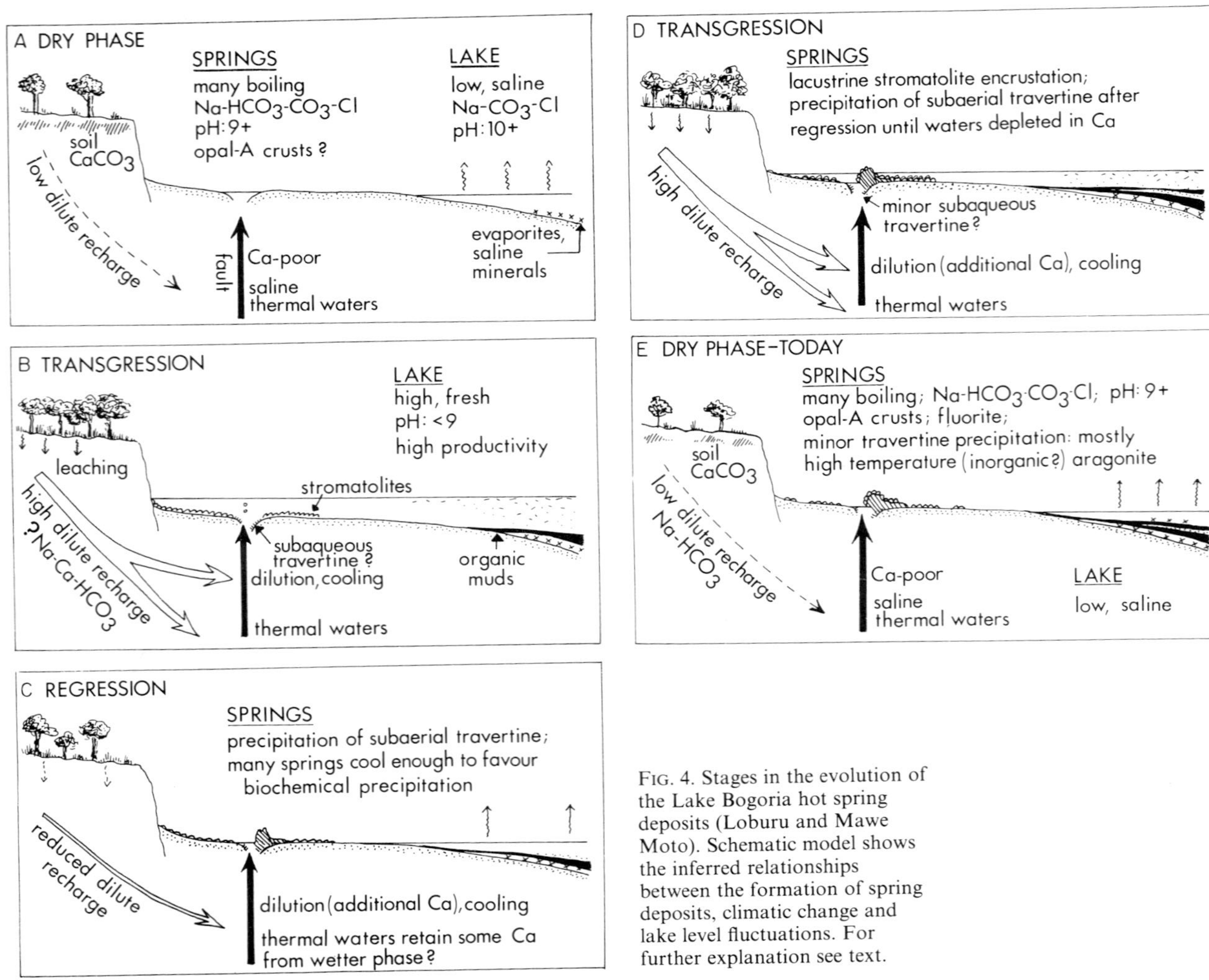

FIG. 4. Stages in the evolution of the Lake Bogoria hot spring deposits (Loburu and Mawe Moto). Schematic model shows the inferred relationships between the formation of spring deposits, climatic change and lake level fluctuations. For further explanation see text.

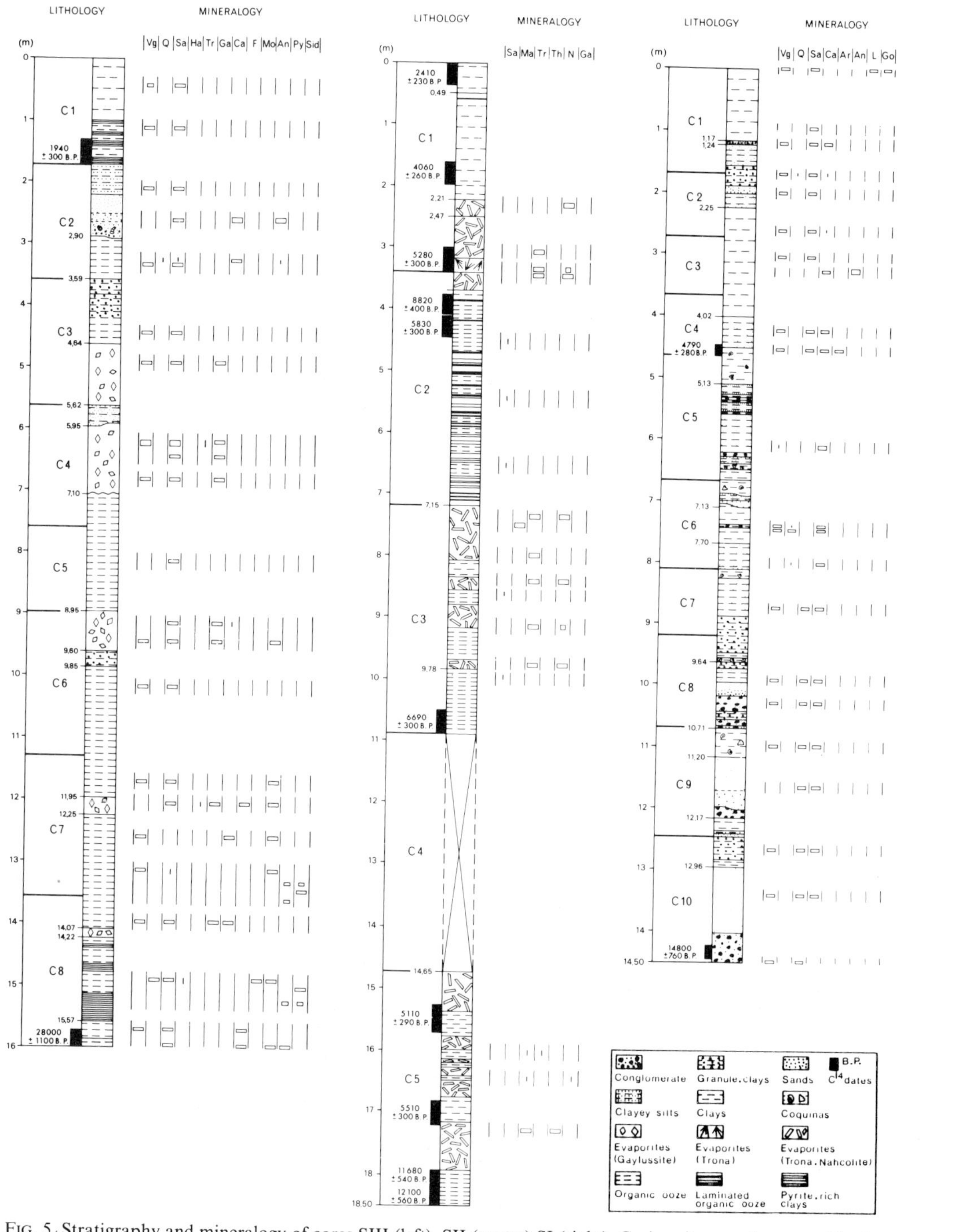

FIG. 5. Stratigraphy and mineralogy of cores SIII (left), SII (centre) SI (right). Coring sites are shown on Fig. 1. Vg = volcanic glasses; Q = quartz; Sa = sanidine; Ha = halite; Tr = trona; Ga = gaylussite; Ca = calcite; F = fluorite; Mo = mordenite; An = analcime; Py = pyrite; Sid = siderite; Ar = aragonite; L = lepidocrocite; Go = goethite; Ma = magadiite; Th = thermonatrite; N = nahcolite.

2 Sub-compacted to compacted black ooze with trona nodules or geodes about a centimetre in diameter.
3 Sub-compacted black ooze with rare to abundant, large, prismatic nahcolite crystals, elongated parallel to the (001) direction, with well-developed (110) and (010) faces.
4 Masses of large acicular crystals of trona, and less commonly nahcolite, forming competent beds a decimetre or more thick.
5 Grey (2.5YR 5/2) to brown (10YR 2/1) clays with prismatic crystals of gaylussite about a centimetre long, elongated parallel to (001), with well-developed (110) and (011) faces.

Black oozes with trona, nahcolite and thermonatrite crystals, and massive beds of trona and nahcolite, occur in the deeper parts of the lake, including the S and central sub-basins. In core SIII, these evaporite facies alternate with black (2.5Y 2/0), grey (5Y 4/1) or blackish green (5Y 4/3) organic oozes that are commonly laminated (Tiercelin *et al.* 1982).

Gaylussite crystals have only been recorded from drill-core SIII, recovered offshore from the Sandai-Waseges delta in water 4.7 m deep. They occur at several levels in grey to brown clays in beds a few decimetres to a metre thick (Fig. 5). Small gaylussite euhedra were observed on the margins of the lake near freshwater springs (Cerling 1979); small (1–5 mm) crystals also occur in surficial efflorescent crusts on the Sandai delta-plain.

Processes and reactions

Gaylussite is the first sodium carbonate mineral to be precipitated when mildly saline, alkaline water is concentrated by evaporation; it succeeds the precipitation of calcite (Rankama & Sahama 1964; Cheverry 1974). The mineral can also form secondarily by the reaction of alkaline brines with primary calcite/aragonite, waters containing Ca^{2+} and HCO_3^-, or calcite-rich sediments (Surdam & Eugster 1976; Eugster & Hardie 1978). In the presence of calcite, gaylussite can form readily by the reaction:

$$CaCO_3 + 2Na^+ + CO_3^{2-} + 5H_2O = Na_2CO_3 \cdot CaCO_3 \cdot 5H_2O$$

Both types of reaction have probably occurred at Lake Bogoria. Gaylussite in the clays of core SIII may have formed by the reaction of alkaline brines with calcite during periods of low lake level. The gaylussite crystals found on the surface of the delta-plains may originate by evaporative concentration during regressive phases. Alternatively, the gaylussite may have been washed or blown onshore, the crystals having formed offshore at a chemical delta where calcium-bearing river waters mixed with alkaline, sodic lake brine.

Precipitation of sodium carbonate phases is strongly dependent on P_{CO_2} (Eugster 1966; Bradley & Eugster 1969). Trona and nahcolite precipitate when there is no calcium source to allow gaylussite to form. The stability field of trona straddles the P_{CO_2} of the atmosphere (Eugster 1980b), and trona is the most common salt in cores from the S and central sub-basins, and in efflorescent crusts.

Nahcolite forms when CO_2 partial pressures are elevated substantially above those in the atmosphere. High P_{CO_2} may develop in the lacustrine environment where organic matter is abundant and intensive bacterial activity

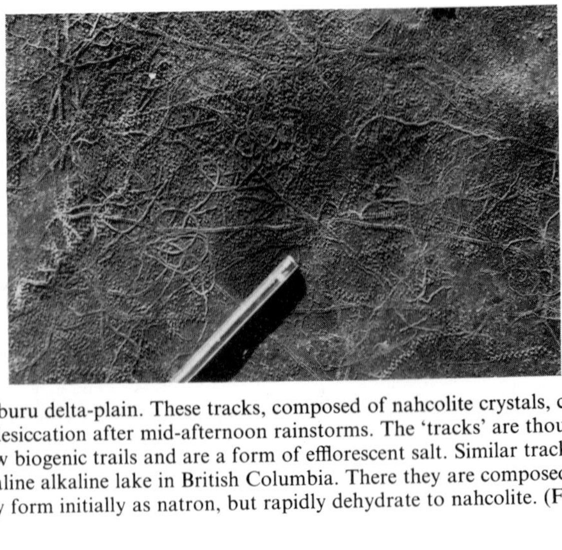

FIG. 6. 'Salt-tracks', Loburu delta-plain. These tracks, composed of nahcolite crystals, cover the surface rapidly (< 30 minutes) during desiccation after mid-afternoon rainstorms. The 'tracks' are thought to be a surface manifestation of shallow biogenic trails and are a form of efflorescent salt. Similar tracks form today around Goodenough Lake, a saline alkaline lake in British Columbia. There they are composed of natron and trona. The Bogoria tracks may form initially as natron, but rapidly dehydrate to nahcolite. (For further details see Renaut 1982.)

degrades organic matter, yielding CO_2. Such processes probably account for nahcolite in the Bogoria organic oozes. Nahcolite also occurs in soils where P_{CO_2} is high (Al-Droubi 1976). At the Loburu delta-plain, nahcolite is present at the land surface in the form of 'salt-tracks' (biogenic traces composed of salt; Fig. 6), and as small intra-sediment crusts just below the surface. Its local occurrence there, instead of the more typical trona, may be due to upward diffusion of CO_2 associated with local hot springs, through the soil profile (Renaut 1982).

Thermonatrite precipitation requires low P_{CO_2}, high Na^+ concentration and a temperature >35°C. Though uncommon in the Bogoria basin, it occurs in a few cores. Local sub-lacustrine hydrothermal spring seepages (Naylor 1972) may be a factor in raising the temperature. The mineral also occurs in efflorescent crusts on the sun-baked delta-plains.

Halite is commonly associated with trona (Maglione 1974). At Bogoria this association is found locally in cores and efflorescent crusts.

Silica mineral precipitation

Sodium silicates

Of the four hydrous sodium silicates so far discovered—makatite, kenyaite, magadiite and kanemite—the latter pair have been found in recent sediments of Lake Bogoria (Périnet *et al.* 1982).

The occurrence of kanemite ($NaSi_2O_5 \cdot 3H_2O$) at Bogoria is only the second record for the mineral, first discovered at Kanem in Chad (Johan & Maglione 1972). Mineralogical characteristics of Bogoria and Chad kanemites are identical. The mineral occurs as white masses associated with small trona crystals, or as small (1 cm) nodules associated with magadiite and trona. Fluorite, analcime and mordenite are also present. The two occurrences of kanemite are from black organic oozes in cores from the central sub-basin, dated at *c.* 4000–5000 BP (Tiercelin *et al.* 1981).

Magadiite ($NaSi_7O_3(OH)_3 \cdot 3H_2O$), first discovered in Pleistocene lacustrine sediments at Lake Magadi (Eugster 1967), is common at Lake Bogoria and shows the same mineralogical characteristics as that from Magadi. The mineral appears in organic oozes from several cores, dated between 10,000 and 4000 BP. It generally forms nodules (*c.*1 cm diameter) that are black with a grey cortex of trona, magadiite and a few feldspar laths.

The conditions required for magadiite and kanemite to form seem to be very similar. Diminishing P_{CO_2}, probably related to decreasing organic matter in the sediment, may explain precipitation of firstly magadiite, then kanemite. Increasing temperature, perhaps related to geothermal phenomena, may also explain the stability of kanemite with respect to magadiite (Al-Droubi 1976).

Cherts, similar to those found associated with magadiite at Lake Magadi, have been found along the shoreline at Kwaipopei (Fig. 1), but not *in situ*.

Opaline cherts

Along the southern shoreline of the Mwanasis peninsula, adjacent to a linear group of shoreline hot springs, a white porcellanous chert cements trachytic beach gravels and locally forms a laterally continuous crust up to 15 cm thick. Composed mostly of amorphous silica (opal-A), the chert is crowded with planktonic diatoms that have 'fused' with the opaline matrix (Fig. 7).

The chert is believed to have formed from a gel-like or colloidal precursor of silica or sodium-silica composition, that was precipitated within gravels around a linear group of sub-lacustrine hot spring orifices, when hot silica-rich alkaline spring waters encountered cooler dilute waters during one of the high-level phases of the lake during the Holocene (Renaut 1982). Planktonic diatoms rained into the soft gel and were preserved. Subsequent regression allowed the gel to lithify and slightly contract. The denser interior of two samples contains minor quantities of opal-CT, precipitated as poorly structured blades within the interior of diatom frustules and other pores. The origin of the opal-CT is uncertain, but it appears to post-date exposure and leaching of the rock by runoff. Local elevation of tempera-

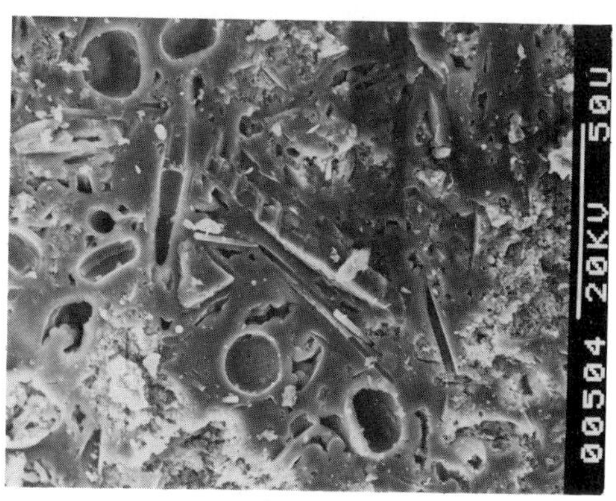

FIG. 7. Opaline cherts from southern Mwanasis. Diatoms lie in a matrix of opaline silica (opal-A). (SEM photograph, scale 50 μm).

ture by hydrothermal activity may have been a factor in its formation.

It is possible that a sodium silicate (e.g. magadiite) may have been the precursor, but magadiite normally transforms to quartz, not opaline silica. The chert lacks many of the attributes of normal Magadi-type chert.

Alkaline earth carbonates

Much of the shoreline, up to the level of the modern Bogoria–Baringo watershed (999 m OD) is encrusted by grey, botryoidal stromatolitic limestones (Young & Renaut 1979). They are products of biochemical (algal) precipitation during a late Holocene period of higher precipitation, when the lake waters carried significantly more dissolved calcium than today (Renaut 1982; Vincens *et al.*, this volume).

Local elevation of the pH has caused calcification of diatoms, sponge spicules and phytoliths within lacustrine clays (Tiercelin *et al.* 1982). Micritic cements also occur locally in marginal clastics from core profiles.

Other minerals

Iron minerals

Pyrite (FeS_2) occurs in massive or finely laminated grey and green clays at the base (13–16 m below sediment surface) of core SIII from the N sub-basin. Cubes, framboids, aggregates or tubules are 1–2 mm in length and locally abundant.

Siderite ($FeCO_3$) occurs as grains or grain aggregates, commonly associated with pyrite, in the same grey and green clays. The pyrite forms in an anaerobic reducing environment, where organic matter is broken down by heterotrophic bacteria with the release of H_2S (Jones & Bowser 1978). Precipitation of siderite needs an environment with low calcium and sulphide (Kelts & Hsu 1978). Good evidence for the gradual transformation of pyrite to siderite has been observed, and is discussed in detail by Tiercelin *et al.* (1982).

Iron oxides (goethite, lepidocrocite) are also present in recent Lake Bogoria sediments. Vivianite ($Fe_3(PO_4)_2 \cdot 8H_2O$) occurs as small brown crusts (1–2 mm thick) on corroded catfish vertebrae in pro-deltaic sediments at Loburu. It formed by dissolution and reprecipitation of bone phosphate under reducing, alkaline conditions (cf. Muller & Forstner 1973).

Zeolites

Analcime and mordenite occur in the finely laminated grey and green clays at the base of core SIII. These clays are rich in volcanic glasses and feldspars. The zeolites formed by reaction of the latter with interstitial alkaline brines (Hay 1970; Surdam & Sheppard 1978). Clinoptilolite, and analcime recorded in delta-front clays at S Loburu, probably had a similar origin.

Fluorite

In addition to its occurrence in travertines, fluorite is present in core samples (e.g. SIII). It also occurs in lake-marginal sites where it forms translucent, pale brown, crudely laminated crusts (1 cm thick) on lavas or cherts, infills mud cracks, and locally forms cements in some deltaic sediments. Fluorite rhizoliths occur at Loburu.

Fluorite can form in lake sediments authigenically by reaction of fluoride-rich brines with calcite (Surdam & Eugster 1976), or as a primary precipitate (Sheppard & Mumpton 1984). High F^- concentrations develop in alkaline brines because when Ca^{2+} is absent, there is no mineral to remove it (Eugster 1980b). Measured fluoride concentrations in Bogoria waters have been highly variable through time, ranging from <3 mg l^{-1} (Tiercelin 1981) to >1100 mg l^{-1} (Baker 1958). The low values may reflect brine dilution during wetter periods (e.g. 1978) or may indicate quantitative removal of fluoride as fluorite, with increased runoff supplying calcium ions.

The origin of the Bogoria fluorites remains uncertain, but their greatest development in lake-marginal deposits occurs at Mawe Moto and Mwanasis–Kwaipopei, both hot spring sites. The level of fluorides in hot spring waters today is high (Table 1), and they are probably a major factor in the large fluoride concentrations. Although in travertines and rhizoliths replacement of calcite/aragonite is demonstrable petrographically, in lake-marginal sites fluorite encrusts, and is encrusted by, carbonates with sharp contacts suggesting that it may be a primary precipitate, possibly by evaporative concentration.

Conclusions

Fig. 8 shows, diagrammatically, the distribution of the precipitated and authigenic minerals within the Bogoria basin, according to depositional environment. This survey demonstrates the importance of hydrothermal activity in mineral genesis at Bogoria. The hot spring waters not only contributed directly to mineral formation, either by orifice precipitation or hydrothermal alteration, but to a great extent control the chemistry of the lake waters and the chemical processes

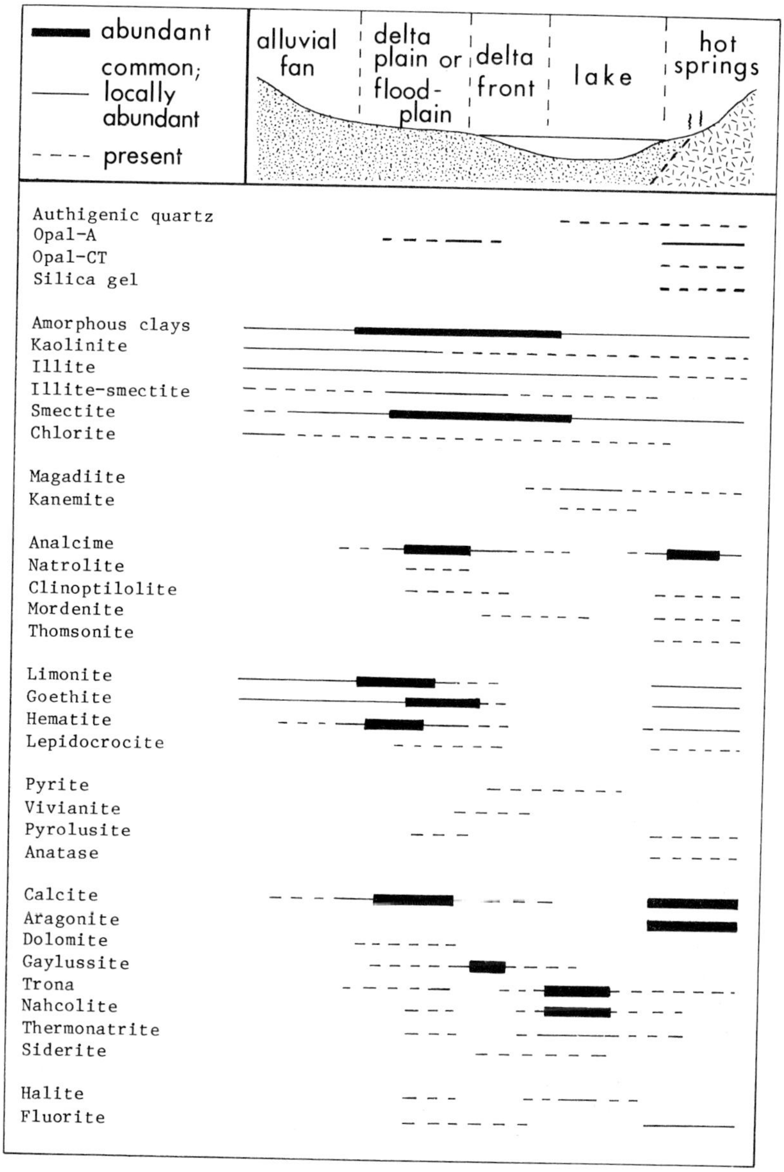

FIG. 8. Summary chart showing the distribution of authigenic and precipitated minerals within the Bogoria basin, according to depositional environment. Except for clay minerals, many of which are allochthonous, most minerals listed have formed within the environment(s) indicated.

occurring within them. The possible influences of hydrothermal activity should not therefore be underestimated when examining ancient analogous rift sediments.

Travertine and hydrothermal alteration might provide indicators of former hydrothermal activity in a fossil situation. However, at many modern springs in the Kenya Rift System neither is present (e.g. much of Loburu). Small, saline, alkaline springs, poor in alkaline earths, commonly issue directly from fractures in volcanic rocks or from shallow pools excavated within sediments. If spring activity ceased, the pools might cave in or infill with sediment of similar grade, and their former existence might not be easily detected. However, their control on lacustrine sedimentation, hydrochemistry and mineralogy may have been paramount.

Although most of the authigenic minerals found to date at Bogoria have been discussed, little attention has yet been given to clay minerals. Reconnaissance examination of clays in recent soils in the catchment and the suspended sediment of the rivers, showed considerable variety, including kaolinite, smectite, amorphous clays, illite, chlorite and mixed layer clays. In contrast, lacustrine and deltaic clays are heavily dominated by smectites, many of which are poorly ordered. Processes of clay regradation as described from Lake Turkana (Yuretich & Cerling 1983; Yuretich, this volume) are also probable at Bogoria.

ACKNOWLEDGEMENTS: This work has been variously supported by grants from NERC and the Royal Society (UK), CNRS and Elf-Aquitaine (Production) (France), and NSERC (Canada).

References

Al-Droubi, A. 1976. Géochimie des sels et solutions concentrées par évaporation. Modèle thermodynamique de simulation. Application aux sols salés du Tchad. *Mém. Univ. Louis Pasteur, Strasbourg* **46,** 177pp.

Baker, B.H. 1958. Geology of the Magadi area. *Rep. Geol. Surv. Kenya* **42,** 81pp.

Berner, R.A. 1975. The role of magnesium in the crystal growth of calcite and aragonite from sea water. *Geochim. Cosmochim. Acta* **39,** 489–504.

Bradley, W.H. & Eugster, H.P. 1969. Geochemistry and paleolimnology of the trona deposits and associated authigenic minerals of the Green River Formation of Wyoming. *Prof. Pap. US Geol. Surv.* **496B,** 71pp.

Carbonel, P., Grosdidier, E., Peypouquet, J.P. & Tiercelin, J.J. 1983. Les ostracodes, témoins de l'evolution hydrologique d'un lac de rift: exemple du lac Bogoria, rift Gregory, Kenya. In: Popoff, M. & Tiercelin, J.J. (eds) *Rifts et Fossés Anciens—Tectonique, Volcanisme, Sédimentation. Apports de l'actuelisme.* Bull. Centres Rech. Explor. Prod. Elf-Aquitaine **7,** 301–13.

Cerling, T.E. 1979. Paleochemistry of Plio-Pleistocene Lake Turkana, Kenya. *Palaeogeogr. Palaeoclimatol. Palaeoecol.* **27,** 247–85.

Chafetz, H.S. & Folk, R.L. 1984. Travertines: depositional morphology and bacterially constructed constituents. *J. Sed. Petrol.* **54,** 289–316.

Cheverry, C. 1974. *Contribution à l'étude pédologique des polders du lac Tchad. Dynamique des sels en milieu continental subaride dans des sediments argileux et organiques.* Unpublished PhD thesis, ORSTOM, Strasbourg, 275pp.

Eugster, H.P. 1966. Sodium carbonate, bicarbonate minerals as indicators of P_{CO_2}. *J. Geophys. Res.* **71,** 3369–77.

—— 1967. Hydrous sodium silicates from Lake Magadi, Kenya. Precursors of bedded cherts. *Science* **157,** 1177–80.

—— 1980a. Geochemistry of evaporitic lacustrine deposits. *Ann. Rev. Earth Planet. Sci.* **8,** 35–63.

—— 1980b. Lake Magadi, Kenya and its precursors. In: Nissenbaum, A. (ed) *Hypersaline Brines and Evaporitic Environments.* Elsevier, Amsterdam, pp.195–232.

—— & Hardie, L.A. 1978. Saline lakes. In: Lerman, A. (ed) *Lakes: Chemistry, Geology, Physics.* Springer, New York, pp.237–93.

—— & Kelts, K. 1983. Lacustrine chemical sediments. In: Goudie, A.S. & Pye, K. (eds) *Chemical Sediments and Geomorphology.* Academic Press, London, pp.321–68.

Folk, R.L. & Assereto, R. 1976. Comparative fabrics of length-slow and length-fast calcite and calcitized aragonite in a Holocene speleothem, Carlsbad Caverns, New Mexico. *J. Sed. Petrol.* **46,** 486–96.

Griffiths, P.S. 1977. *The geology of the Lake Hannington—Perkerra River area, Kenya Rift Valley.* Unpublished PhD thesis, University of London, 187pp.

Hardie, L.A., Smoot, J.P. & Eugster, H.P. 1978. Saline lakes and their deposits: a sedimentological approach. In: Matter, A. & Tucker, M.E. (eds) *Modern and Ancient Lake Sediments.* Spec. Publ. Int. Assoc. Sedimentol. No. 2, Blackwell Scientific Publications, Oxford, pp.7–41.

Hay, R.L. 1970. Silicate reactions in three lithofacies of a semi-arid basin, Olduvai Gorge, Tanzania. *Spec. Pap. Miner. Soc. Am.* **3,** 237–55.

Herbin, J.P. 1979. Sédimentation de rift: géochimie organique des sédiments récents du lac Bogoria. *Rapp. Inst. Fr. Pétrol.* **26,** 881pp.

Icole, M. & Périnet, G. 1984. Les silicates sodiques et les milieux évaporitiques carbonatés bicarbonatés sodiques: une revue. *Rev. Geol. Dyn. Geogr. Phys.* **25,** 167–76.

Johan, Z. & Maglione, G.F. 1972. La kanemite, nouveau silicate de sodium hydraté de néoformation. *Bull. Soc. Fr. Minéral. Crystallogr.* **95,** 371–82.

Jones, B.F. & Bowser, C.J. 1978. The mineralogy and related chemistry of lake sediments. In: Lerman, A. (ed) *Lakes: Chemistry, Geology, Physics.* Springer, New York, pp.179–235.

——, Eugster, H.P. & Rettig, S.L. 1977. Hydrochemistry of the Lake Magadi Basin, Kenya. *Geochim. Cosmochim. Acta* **41,** 53–72.

Julia, R. 1983. Travertines. *Mem. Am. Assoc. Petrol. Geol.* **33,** 64–72.

Kelts, K. & Hsu, K.J. 1978. Freshwater carbonate sedimentation. In: Lerman, A. (ed) *Lakes: Chemistry, Geology, Physics.* Springer, New York, pp.295–323.

Kilham, P. 1971. *Biogeochemistry of African lakes and rivers.* Unpublished PhD thesis, Duke University, Durham, NC 199pp.

Kinsman, D.J. 1969. Interpretation of Sr^{2+} concentrations in carbonate minerals and rocks. *J. Sed. Petrol.* **39,** 486–508.

Kitano, Y. 1962. A study of the polymorphic formation of calcium carbonate in thermal springs with an emphasis of the effect of temperature. *Bull. Chem. Soc. Jap.* **35,** 1980–5.

LIPPMANN, F. 1973. *Sedimentary Carbonate Minerals.* Springer, Berlin, 228pp.

LONGMAN, M.W. 1980. Carbonate diagenetic textures from near-surface diagenetic environments. *Bull. Am. Assoc. Petrol. Geol.* **64,** 461–87.

MAGLIONE, G. 1974. *Géochimie des évaporites et silicates néoformes en milieu continental confiné.* Unpublished PhD thesis, University of Paris, 331pp.

MCCALL, G.J.H. 1967. Geology of the Nakuru-Thomson's Falls, Lake Hannington area. *Rep. Geol. Surv. Kenya* **78,** 122pp.

MULLER, G. & FORSTNER, U. 1973. Recent iron ore formation in Lake Malawi. *Mineral. Deposita* **8,** 278–90.

NAYLOR, I. 1972. *Geology of the Lake Hannington geothermal prospect.* Unpublished report, United Nations/Kenya Power Company, 16pp.

NOBLE, J.W. & OJIAMBO, S.B. 1976. Geothermal exploration in Kenya. *2nd UN Symp. Develop. Use Geotherm. Energy* **1,** 189–204.

OWEN, R.B., BARTHELME, J.W., RENAUT, R.W. & VINCENS, A. 1982. Palaeolimnology and archaeology of Holocene deposits north-east of Lake Turkana, Kenya. *Nature* **298,** 523–9.

PÉRINET, G., TIERCELIN, J.J. & BARTON, C. 1982. Présence de kanemite dans les sédiments récents du lac Bogoria, Rift Gregory, Kenya. *Bull. Minéral.* **105,** 633–9.

RANKAMA, R. & SAHAMA, T.G. 1964. *Geochemistry.* University of Chicago Press, Chicago, 911pp.

REAMS, M.W. 1974. Rate of carbon dioxide loss and calcite-aragonite deposition in caves. *Proc. 4th Conf. Karst Geol. Hydrol.* (West Virginia Geol. Surv.), 179–82.

RENAUT, R.W. 1982. *Late Quaternary geology of the Lake Bogoria fault-trough, Kenya Rift Valley.* Unpublished PhD thesis, University of London, 498pp.

SHEPPARD, R.A. & MUMPTON, F.A. 1984. Sedimentary fluorite in a lacustrine zeolitic tuff on a Gila Conglomerate near Buckhorn, Grant County, New Mexico. *J. Sed. Petrol.* **54,** 853–60.

SLACK, K.V. 1967. Physical and chemical description of Birch Creek, a travertine depositing stream, Inyo County, California. *Prof. Pap. US Geol. Surv.* **549A,** 1–19.

SURDAM, R.C. & EUGSTER, H.P. 1976. Mineral reactions in the sedimentary deposits of the Lake Magadi region, Kenya. *Bull. Geol. Soc. Am.* **87,** 1739–52.

—— & SHEPPARD, R.A. 1978. Zeolites in saline, alkaline lake deposits. In: SAND, L.B. & MUMPTON, F.A. (eds) *Natural Zeolites: Occurrence, Properties, Use.* Pergamon Press, Oxford, pp.145–74.

TIERCELIN, J.J. 1981. *Rifts continentaux. Tectonique, climats, sédiments. Exemples: la sédimentation dans le nord du rift Gregory (Kenya) et dans le rift de l'Afar (Ethiopie) depuis le Miocène.* Unpublished PhD thesis, Université Aix-Marseille II, 260pp.

——, LE FOURNIER, J., HERBIN, J.P. & RICHERT, J.P. 1980. Continental rifts: modern sedimentation, tectonic and volcanic control. Example from the Bogoria-Baringo graben, Gregory Rift, Kenya. In: *Geodynamic Evolution of the Afro-Arabic Rift System.* Acad. Nazionale dei Lincei, Rome, **47,** pp.143–64.

——, PÉRINET, G., LE FOURNIER, J., BIEDA, S. & ROBERT, P. 1982. Lacs du rift est-africain, exemples de transition eaux douces-eaux salées: le lac Bogoria, rift Gregory, Kenya. *Mém. Soc. Geol. Fr.* **144,** 217–30.

——, RENAUT, R.W., DELIBRIAS, G., LE FOURNIER, J. & BIEDA, S. 1981. Late Pleistocene and Holocene lake levels in the Lake Bogoria basin, northern Kenya Rift Valley. *Palaeoecol. Afr.* **13,** 105–20.

YOUNG, J.A.T. & RENAUT, R.W. 1979. A radiocarbon date from Lake Bogoria, Kenya Rift Valley. *Nature* **278,** 243–5.

YURETICH, R.F. & CERLING, T.E. 1983. Hydrogeochemistry of Lake Turkana, Kenya: mass balance and mineral reactions in an alkaline lake. *Geochim. Cosmochim. Acta* **47,** 1099–109.

R.W. RENAUT, Dept of Geological Sciences, University of Saskatchewan, Saskatoon, Saskatchewan S7N OWO, Canada.

J.J. TIERCELIN, GIS Océanologie et Géodynamique, Université de Bretagne Occidentale, 29287 Brest, Cedex, France.

R.B. OWEN, Dept of Geography & Earth Science, Chancellor College, University of Malawi, PO Box 280, Zomba, Malawi.

Lake Magadi, Kenya: a model for rift valley hydrochemistry and sedimentation?

H.P. Eugster

SUMMARY: Previous studies on Lake Magadi, Kenya, and the Green River Formation of Wyoming, are reviewed in order to delineate the differences between these two settings, to determine whether Lake Magadi can serve as a reference for rift valley environments and to define a set of diagnostic criteria for recognizing saline alkaline lake deposits in the geologic column. Magadi inflow waters are influenced strongly by weathering of volcanic rocks and are characterized by $Na \approx Ca$ and $(Ca + Mg)/HCO_3 \ll 1$. Compared to the Green River Formation, calcite production is low to moderate and occurs largely in the soil zone. Sediments are dominated by volcaniclastic material transported during flood events. In contrast, Eocene Lake Gosiute and its mudflats had more gentle gradients, and inflow chemistry was such that vast amounts of calcite and dolomite were produced before brines became alkaline. Sedimentation in both cases is represented by the ephemeral floodplain–saline lake complex; a preliminary subfacies map for Magadi has been constructed. Magadi possesses the typical attributes of a rift valley basin in terms of hydrology, hydrochemistry and sedimentation, as well as the indices characterizing saline alkaline lakes: sodium carbonate evaporites or their pseudomorphs, Magadi-type chert, and tuffaceous rocks altered to zeolites and K-feldspar. Many similar environments have been recognized in the Pleistocene lake beds of North America. Magadi-type cherts were found from the Pleistocene to the Jurassic, and are also present in the Cambrian Officer Basin of Australia. Proterozoic Magadi–Green River-type environments have been suggested for the Duruchaus Formation of Namibia and HYC shale of the McArthur River area of North Australia. Many more Proterozoic examples remain to be discovered.

Geologic field studies at Lake Magadi, Kenya, were carried out during 1966–73 and the results have been summarized in Eugster (1980). Initially chosen to shed light on the mechanisms of trona deposition, Lake Magadi was to be a model for the depositional environments of the Eocene Green River Formation of Wyoming, Utah and Colorado. While the analogy remains valid with respect to hydrochemistry and brine evolution in a lacustrine high-pH environment, there are profound differences with respect to sedimentary processes and products, which are related to differences in inflow, hydrology and tectonics. Work at Lake Magadi has illuminated these differences and has demonstrated a mechanism for the inorganic formation of bedded chert derived from the minerals magadiite and kenyaite as first described from Magadi.

Also observable at Magadi is the large-scale precipitation of Na-Al-Si gels (Eugster & Jones 1968), precursors of zeolites (Surdam & Eugster 1976). The hydrochemistry of Lake Magadi inflow and brines has provided us with the best-documented example of an alkaline closed basin system, representing an important branch of our brine evolution scheme (Hardie & Eugster 1970; Eugster & Hardie 1978). Finally, however, the greatest significance of this work will be derived from the fact that saline alkaline lake deposits are found with increasing frequency in the geologic record, from the Pleistocene to the Proterozoic, when continental crust and rifting episodes were widespread. In this brief summary two main points will be addressed: (1) similarities and differences between Lakes Magadi and Gosiute (Eocene, Wyoming); and (2) how can one recognize Magadi–Green River-type settings in the geologic past?

Lake Magadi and its Pleistocene precursors

In the Magadi trough (Fig. 1) deposits from three separate lakes can be recognized, from oldest to youngest: Lake Oloronga (>780,000 BP), High Magadi Lake (10,000 BP) and Lake Magadi, the active lake (Eugster 1980). Lake Oloronga is dated from the glassy selvage of a trachyte lava flow which was extruded over the Oloronga Beds, recognizable by their green cherts and caliche cap. Oloronga Beds were also encountered in the Natron basin, suggesting that Lake Oloronga at one time may have occupied both basins as a single body of water. More work is necessary to verify this possibility, as well as the suggestion that the Moinik Formation of the Peninj Group

From FROSTICK, L.E. *et al.* (eds) 1986, *Sedimentation in the African Rifts*, Geological Society Special Publication No. 25, pp. 177–89.

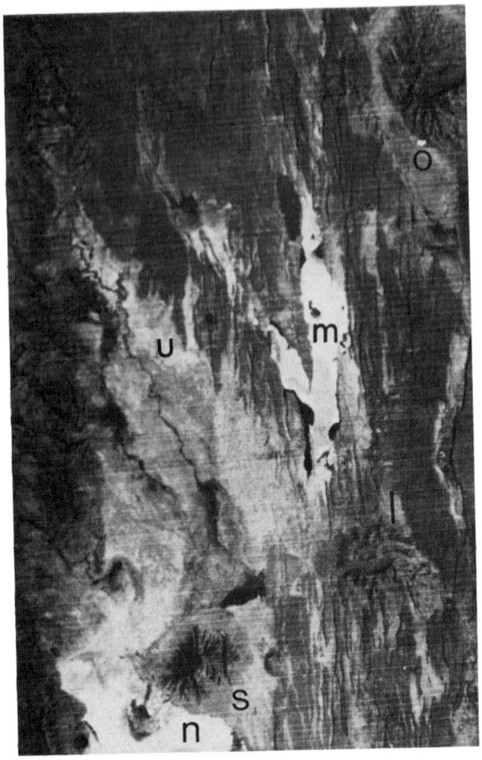

FIG. 1. Landsat photograph of Magadi area. Lake Magadi (m), Lake Natron (n), Uaso Ngiro (u) and the volcanoes Shompole (s), Lenderut (l) and Olorgesaile (o) are clearly visible, as is the large floodplain between Lakes Natron and Magadi. Dark areas surrounding Magadi are the perennial lagoons, whereas the dark spot N of the letter m is the trona dredge. Little Magadi is visible NW of Magadi.

correlates with Oloronga time (Eugster 1980, p.224). It is not known whether Lake Oloronga dried up prior to the trachyte flow or whether it continued to exist well into the Upper Pleistocene. In any case, the thick caliche cap present on every Oloronga section, including that below the lava flow, indicates a long depositional hiatus between Oloronga and High Magadi time.

Oloronga sediments consist of volcaniclastic sands, silts and clays, along with interbedded cherts derived from magadiite beds precipitated from Lake Oloronga. Short stratigraphic sections of nearshore outcrops are given in Eugster (1980, Fig. 15.20); none are thicker than 10 m. However, in core F6, 45 m of Oloronga Beds were penetrated without reaching the bottom of the section (Surdam & Eugster 1976). It is reasonable to assume that the maximum thickness is at least 50 m. Although the lake had to be alkaline prior to magadiite precipitation, it may have been dilute for much of its life. Dilute episodes are indicated by the presence of phillipsite in the outcrops NE of Little Magadi.

Judging from the existing old shorelines and the distribution of the High Magadi Beds, High Magadi Lake was only 4–5 m above the present trona surface and did not extend far beyond Lake Magadi. Typical outcrop sections are given in Eugster (1969), and a maximum thickness of 68 m has been measured in core F6 (Surdam & Eugster 1976, Table 7). The nature of the sediments is very similar to that of the Oloronga Beds, with volcaniclastic sands, silts and muds making up the largest portion. Bedded cherts are present throughout the section as indicated by the drill core logs. Some of the magadiite horizons from which these cherts formed are still preserved in outcrops. In fact, in many places around the lake it is possible to locate a magadiite bed by digging a trench in the valley floor, and to observe its conversion to chert as it emerges at the surface (see Eugster 1969, Fig. 5).

High Magadi and Oloronga Beds can be distinguished by differences in induration and colour, but the safest guide is the caliche cap of the Oloronga outcrops mentioned earlier (Fig. 2). During High Magadi time a large amount of Oloronga sediment was eroded and redeposited in the High Magadi Lake. This cannibalism has produced a complicated overlap of lake beds, which depends largely on local topography and can be mapped only on a very large scale. Areas where one type of lake bed predominates over another have been indicated in Eugster (1980, Fig. 15.3). The age of High Magadi Lake has been reviewed by Butzer *et al.* (1972) who concluded that 10,000 years was the most probable age for its largest expansion and greatest dilution. This estimate is based on dates for carbonized fish remains from surface outcrops, and on a corrected date from drill core trona 47.5 m below the present surface. High Magadi Lake could of course have become established much earlier.

Except for differences in size and age, there is great similarity between Lakes Oloronga and High Magadi Lake with respect to sediment content and water chemistry. Judging from the many chert intervals, both lakes were probably alkaline throughout their lives, without ever reaching a stage of extensive trona precipitation. On the other hand, the chert beds require that the pH had risen above 9.0 in order to store enough silica for lake-wide magadiite precipitation (see Eugster 1967; Jones *et al.* 1967; Bricker 1969). Considering that Lake Oloronga at one time probably occupied both the Magadi and Natron basins, a significant hydrologic difference must have existed between Lake Oloronga and High

FIG. 2. High Magadi-Oloronga unconformity in the Dry Lagoon, SE of Lake Magadi. High Magadi Beds on the right (light colour). My wife is sitting on the caliche cap of the Oloronga Beds. In the background is a dry mudflat–sandflat area connected to Lake Magadi through a passage cut into trachyte lava flows.

Magadi Lake, a difference illustrated today by Lakes Natron and Magadi. A precursor of the Uaso Ngiro which descended from the Mau escarpment must have been a major perennial source of water and sediment for Lake Oloronga, as the Uaso Ngiro is today for Lake Natron. High Magadi Lake, on the other hand, must have received its inflow largely from ephemeral storm runoff and perennial springs similar to the current situation at Lake Magadi.

Finally we must briefly touch on the Holocene evaporite series and the sediments accumulating today. The chief mineral is trona, $NaHCO_3 \cdot Na_2CO_3 \cdot 2H_2O$, and thicknesses of up to 40 m have been encountered in drill cores (Eugster 1980). The trona is bedded on the centimetre scale and consists of a very porous, sparry framework of large trona blades with the interstices filled by saturated brine. Bedding is marked by thin horizons of volcaniclastic sand and silt, mostly of windblown derivation. Precipitation of trona through evaporative concentration leads to bicarbonate depletion and a rise in pH to values above 11.0. During the dry season the level of the interstitial brine falls several metres below the trona surface, whereas during the rainy seasons flooding occurs with dilute storm runoff covering the trona surface and dissolving the surface crust. This flood layer evaporates after a few months, precipitating and recrystallizing the surface layer. Yearly net accumulation has been estimated at 2–3 mm (Eugster 1980).

Hydrology, hydrochemistry and brine evolution at Lake Magadi

The hydrology and hydrochemistry of the active Lake Magadi have been discussed by Eugster (1970), Jones *et al.* (1977), Eugster & Hardie (1978), Eugster & Jones (1979), and Eugster (1980). Inflow is derived primarily from two sources: (1) ephemeral storm runoff; and (2) perennial saline springs which issue through the High Magadi Beds and trachyte lava flows surrounding the lake. Spring temperatures range from ambient to 86°C, and total dissolved solids from 10,000 to 35,000 ppm. The Na:Cl ratio, on the other hand, is remarkably uniform. A deep, hot groundwater reservoir has been identified as one of the three major spring inputs which is mixed in various proportions with dilute, cold, shallow groundwater and with concentrated, alkaline (bicarbonate-depleted), cold groundwater representing recirculated lake brines. Spring runoff feeds lagoons, year-round bodies of open brine rimming the lake, which are best characterized as a salt pan.

Solutes are also contributed by runoff dissolving the trona-thermonatrite-halite efflorescent crusts coating the Pleistocene lake beds, crusts which form and reform during the dry seasons through capillary evaporation of shallow groundwater. These wetting and drying cycles (see Drever & Smith 1978) are probably the major mechanism of evaporative concentration and solute acquisition in arid terranes, and Magadi is no exception. Efflorescent crusts form wherever groundwater in porous sediments is close enough to the surface to permit capillary draw. However, as indicated by the Ndopa and Ngejepa cattle wells N of Magadi (see Jones *et al.* 1977, Fig. 2), groundwater aquifers with substantial flow need not become very saline (1000–2000 ppm TDS).

Recharge of the groundwater system occurs not only through storm runoff, but also through perennial streams descending the western scarp of the rift valley. Whereas the Uaso Ngiro flows as a perched stream for more than 50 km along the western floor of the valley and eventually drains into Lake Natron, a number of perennial freshwater streams disappear into their alluvial fans before they reach the valley floor. These streams, principally Oloibortoto and Endosapia, are listed by Jones *et al.* (1977) as rim streams, and together with the Uaso Ngiro are taken as representative of the dilute-water recharge of the Magadi groundwater system.

There is no need to repeat the detailed documentation of the evolution of water composition in the Magadi system given in Eugster (1970) and Jones *et al.* (1977), and only the major conclusions will be presented here. Dilute inflow consists of Na-Ca-HCO_3 waters with initially roughly equimolar amounts of Na and Ca. This is a higher Na:Ca ratio than normally results from weathering of granitic rocks (Drever 1982), but is compatible with dissolution studies of partially crystallized basalts (Gislason & Eugster 1985). In such rocks, through crystallization of Ca-phases, the residual glass is enriched in Na, and on contact with water Na is released more rapidly than Ca. In trachytes, this effect would be still greater. As the solute load is enhanced through evaporative concentration, the Na:Cl ratio of the dilute inflow increases from 0.025 to 1, indicating input of Na^+ and HCO_3^- through silicate weathering reactions (Eugster & Jones 1979, Fig. 5). At this point chloride has increased only from 0.1 to 0.25 mmol kg^{-1} representing a two and a half-fold concentration factor.

From here on, waters feeding the Magadi system disappear underground until they emerge as hot springs, and this part of their chemical evolution must be inferred from a limited number of groundwater samples. The dilute inflow streams (Jones *et al.* 1977, Table 1a) have an average molar (Ca+Mg):HCO_3 ratio of 0.28. This is 'the critical evolutionary parameter' (Hardie & Eugster 1970, p. 279). This four-fold dominance of bicarbonate over alkaline earths ensures that the latter are depleted early during evaporative concentration, with calcite precipitation acting as the chemical divide (see Drever 1982, ch.9). Bicarbonate, as a consequence, is enriched, as are Na and Cl. Using Cl as a conservative solute, Eugster & Jones (1979) documented a gradual depletion in total carbon species with respect to chloride from dilute inflow to hot springs and then to lake brines. During this depletion, carbonate alkalinity (m HCO_3^- + 2m CO_3^{2-}) changes little (Eugster & Jones 1979, Fig. 7), which points to a mechanism involving the release of CO_2, such as: $2HCO_3^- \rightarrow CO_3^{2-} + CO_2$. During this process, charge balance (alkalinity) is preserved, while P_{CO_2} is regulated by the atmosphere in the long run. Degassing of CO_2 is the cause of the large pH increase observed during evaporative concentration of HCO_3^- water, with photosynthesis acting as an important intermediate mechanism.

In saturated lake brines, precipitation of trona, $NaHCO_3 \cdot Na_2CO_3 \cdot 2H_2O$, removes HCO_3^- and CO_3^{2-} on an equimolar basis, a further mechanism for enrichment of CO_3^{2-} over HCO_3^-, and hence a rise in pH. In consequence, many of the residual brines are so depleted in HCO_3^- that further trona precipitation is halted and can proceed only by addition of CO_2. For such brines, Monnin & Schott (1984) calculated P_{CO_2} values as low as 10^{-6}, undersaturated with respect to the atmosphere by three and a half orders of magnitude! Obviously, the addition of CO_2 from the atmosphere is a very slow process, and CO_2 production by decay of organic matter has not kept pace with trona precipitation. The residual brines are also enriched preferentially in Cl and depleted in Na through trona precipitation.

Brine evolution at Magadi is particularly easy to follow because the watershed lithologies are exclusively igneous and metamorphic rocks, with rapidly weathering volcanics dominant; consequently inflow is compositionally homogeneous. Chloride input is moderate (4–8 ppm) and chloride is conservative throughout evaporative concentration, including the saturated lake brines which normally do not reach halite saturation. Br, F, B and PO_4 are also largely conserved and become strongly enriched in the residual brines, even though their inflow concentrations are not unusually high. For instance, typical inflow values for fluorine cluster around 1 ppm, whereas brines can contain over 2000 ppm, and precipitation of villiaumite, NaF, has been reported. Such enrichment causes authigenic reactions to form fluorite. Sulphate, on the other hand, is depleted in the brines largely through bacterial reduction. Even if sulphate depletion had not occurred, Magadi could not have produced sulphate-rich brines, because its dilute inflow is low in sulphate (2–6 ppm). Sulphate-rich systems, such as Great Salt Lake, Utah, have high sulphate inflow due to oxidation of sulphides or dissolution of $CaSO_4$ evaporites. With initial supply of both chloride and sulphate low to moderate and Na more abundant than Ca, the evolution of an alkaline system at Magadi is inevitable. The importance of rapidly weathering volcanics has been championed by Jones (1966). Because at Magadi the volcanics are predominantly alkaline (trachytes) and basic (basalts), boron does not play as important a role as it would if the rift valley had extensive acid volcanism.

Finally, we must consider the claim that the source of the substantial trona accumulation at Lake Magadi is the natrocarbonatite ash observed from Ol Doinyo Lengai: 'The trona evaporites of Lakes Magadi and Natron are derived from natrocarbonatite ash from the still-active Ol Doinyo Lengai volcano.' (Baker, this volume). Ash contributed to the Natron–Magadi basin will indeed yield Na-HCO_3 waters upon leaching, just as the trachytes and basalts do upon weathering. Hay (this volume) has suggested that such ashes were more common during the Pleistocene, but I do not believe that they were, or are, major solute sources for Lake Magadi. If they were, Lake Natron should be the major trona deposit rather than Magadi. Accumulation of the trona evaporites at Magadi is the inevitable consequence of the weathering of basic and alkaline lavas of which natrocarbonatites form a minute fraction. This conclusion is supported by the fact that none of the other active or near-active trona lakes (e.g. Owens Lake, Searles Lake, Deep Springs Lake, California) are connected with natrocarbonatite volcanoes, and it would be misleading to stipulate such a connection for the Eocene Green River Formation. In contrast, I have suggested (Eugster 1970, p.229) that the natrocarbonatite lava and ash of Lengai are due to the interaction of the alkaline silicate magma with a saline alkaline groundwater reservoir.

Diagenetic reactions

The alkaline carbonate brines at Magadi interact, and have interacted in the past, with the volcanogenic sediment with which they come into contact, producing a suite of authigenic minerals characteristic of such a high pH environment. Many of these minerals, especially the zeolites and cherts, are much less soluble than trona and hence provide a more permanent record of Magadi-type environments in the geologic record.

Assemblages of authigenic minerals from Magadi were discussed by Surdam & Eugster (1976), based on samples from unconsolidated sediment, outcrops and drill cores. The most frequently encountered minerals are erionite, analcime, gaylussite, fluorite and calcite, together with the Na-Al-Si gels first described by Eugster & Jones (1968) (Fig. 3). Erionite is a Na-Si-rich zeolite typical of alkaline lake deposits (Surdam & Sheppard 1978) which forms in environments rich in silica and sodium and poor in alkaline earths. At Magadi it can form simply by the hydration of trachyte glass, and it does so rapidly and ubiquitously. No Na^+ need be added from the brine, but their high pH and ability to dissolve silica and alumina enhance the conversion kinetics. In cases where silica activity is lowered by the transformation of magadiite to chert, analcime forms at the expense of erionite by the release of silica and the addition of Na^+ from solution. Analcime can also crystallize directly from Na-Al-Si gels of similar composition. Such gels have been encountered frequently and in substantial quantities on the present sediment surface, in outcrops and in core samples. They form from trachytes and trachyte glass by hydration, addition of Na^+ and release of silica, assuming that Al is conserved during the reaction. They represent an important intermediate reaction product and may well be precursors for the other zeolites as well.

As mentioned previously, in a Magadi-type setting calcium is removed from solution early

FIG. 3. Na-Al-Si gels (panga for scale) coated with efflorescent crusts and located in a stream channel N of Little Magadi.

during brine evolution. Consequently Ca-bearing authigenic minerals such as gaylussite, $Na_2CO_3 \cdot CaCO_3 \cdot 5H_2O$, and fluorite, CaF_2, form by reaction of alkaline brines with sedimentary Ca-bearing particles. The trachyte debris and tuffaceous material which constitute the bulk of the sediment contain roughly 1% CaO, and represent a large calcium reservoir. Evaporation of dilute runoff or groundwater which has extracted calcium from this source leads to calcite precipitation. Coatings of inflow channel pebbles and the calcite pisolite pavement observed near the Magadi shoreline (see Eugster 1980, Fig. 15.22) are formed in this manner. In the High Magadi tuffs capillary evaporation of similar waters is responsible for the calcite cement reported by Surdam & Eugster (1976). In the core samples, calcite is common only in the deeper sections which have been assigned to the Oloronga Beds.

Insight into possible reaction mechanisms can be gained by comparing the mineralogy of modern muds with that of the High Magadi Beds. While analcime is common in the High Magadi and Oloronga Beds, the modern muds, silts and sands 'contain abundant unaltered glass shards, gaylussite, and erionite but no analcime . . . they also contain gelatinous material' (Surdam & Eugster 1976, p.1741). Magadi erionite contains about half of the CaO found in the trachyte (Surdam & Eugster 1976, Table 4). The common association of erionite with gaylussite seems to indicate that the Ca liberated during erionite growth is consumed by the formation of gaylussite or fluorite. Authigenic calcite mentioned above is another potential source for Ca. Fluorite was encountered frequently in the core samples, particularly those of Oloronga age, but is scarce in outcrops; whereas gaylussite seems to be restricted to surface and near-surface samples, where it grows in brine-soaked unconsolidated sediment.

Sedimentation at Lake Magadi

Sedimentation in hydrologically closed basins was discussed by Hardie *et al.* (1978) in terms of depositional complexes consisting of individual subfacies, with each subfacies characterized by specific depositional processes and a set of diagnostic sedimentary structures. Alluvial fans, sandflats, dry mudflats, saline mudflats, salt pans, perennial saline lakes, perennial stream floodplains, ephemeral stream floodplains, spring deposits and shoreline features were mentioned as examples of such subfacies, and they provide the building blocks for various depositional complexes. The alluvial fan–ephemeral saline lake complex fits the Magadi situation best. In discussing the Wilkins Peak member of the Green River Formation of Wyoming, Smoot (1983) refined this approach by introducing an additional subfacies labelled EL which represents sporadically exposed lake margin conditions. Fieldwork at Magadi was completed before these sedimentological studies were initiated and a detailed sedimentological analysis of Lake Magadi and its precursors is lacking. Some suggestions, however, can be made as to what might be expected.

The following subfacies can be recognized in the Magadi basin: (1) salt pan (see Fig. 4); (2) spring marsh and perennial lagoon; (3) saline mudflat (see Fig. 5); (4) dry sandflat and mudflat; (5) delta; (6) spring deposit; (7) ephemeral stream floodplain; and (8) alluvial fan. Because of the episodic occurrence of flowing surface waters, lack of resistant sand grains and steep topographic gradients, there seems to be no clear separation of sandflats from mudflats and they have been lumped together in a single subfacies. Most of these flats are in fact old lake beds consisting of clays, zeolites, tuffaceous material, trachyte debris, magadiite and chert fragments. The saline mudflats are present around the perimeter of the salt pan and are associated with perennial springs. Gaylussite is a characteristic mineral of these brine-soaked mudflats, as are Na-Al-Si gels. Deltas are being constructed out onto the saline mudflats and salt pan where small channels are incised in the lava flows by ephemeral streams close to the lake. No permanent buildup occurs around spring openings, largely because the waters are saline, devoid of alkaline earths and undersaturated with respect to amorphous silica. Locally, gel accumulations are found around spring orifices. An exceptionally well-developed ephemeral stream floodplain of 4 km length (Fig. 6) stretches from the N end of Lake Magadi to the Ndopa Wells (site 1020 in Jones *et al.* 1977). In the Magadi trough itself, alluvial fans are not common, probably because of the lack of ephemeral surface channels providing sediment point sources (Fig. 7). Magnificent fans are present, however, on the W side of the rift valley where the perennial rim streams descend the 1500 m scarp formed by metamorphic rocks and basalt flows. Fig. 8 shows a preliminary sketch of the depositional subfacies of the Magadi trough.

Sedimentary processes in closed basins are determined by climate, topography and nature of the bedrock, which in turn define the local hydrology. At Magadi, sediment transport and deposition occurs by ephemeral runoff and wind. Storm runoff is by sheetwash and ephemeral streams. Such streams are usually very short

FIG. 4. Salt pan (hammer for scale) showing crystallization of trona. The ridge on which the Magadi houses are located consists of trachyte lava.

FIG. 5. Saline mudflat grading into perennial lagoon, SE of Lake Magadi, looking to the W scarp of the rift valley (flamingoes for scale).

because of the extensive fracturing and high porosity of the trachyte lava flows. Hence solutes and water are supplied to the salt pan largely by perennial springs. A similar situation exists in the rift valley of NE Iceland, even though precipitation in the form of rain or snow is very common. Nevertheless, because of the porosity of the largely unweathered lava flows, and the presence of many hyaloclastite horizons, no surface streams exist in the area of the post-glacial

FIG. 6. Ephemeral stream floodplain N of Magadi. To the right and separated by a lava horst, is Little Magadi; in the background, Shompole.

FIG. 7. Alluvial fan built by an ephemeral stream into a lagoon NE of Magadi. Note the extensive fringe of efflorescent crusts.

basalts (Gislason & Eugster 1985). The pre-glacial Pleistocene basalts, on the other hand, have sufficient clay and zeolite seals to permit surface streams to become established.

Green River Formation geochemistry and sedimentation

The Eocene Green River Formation of Wyoming, Utah and Colorado contains bedded trona and halite deposits as well as extensive accumulations of oil shale (Bradley 1964; Bradley & Eugster 1969). The Wilkins Peak Member of Wyoming contains 42 major trona beds, some over 3 m thick and of large areal extent. Mining these beds now provides the major commercial supply of sodium carbonate in the US, and there is little doubt that the geochemical environment in which these beds were deposited *c.* 45 Ma must have been similar to Lake Magadi today. This inference is supported by the presence of Magadi-type chert, although the chert beds are surprisingly thin and scarce (Surdam *et al.* 1972). Zeolites, analcime and K-feldspar, on the other hand, are abundant in the many tuffaceous beds,

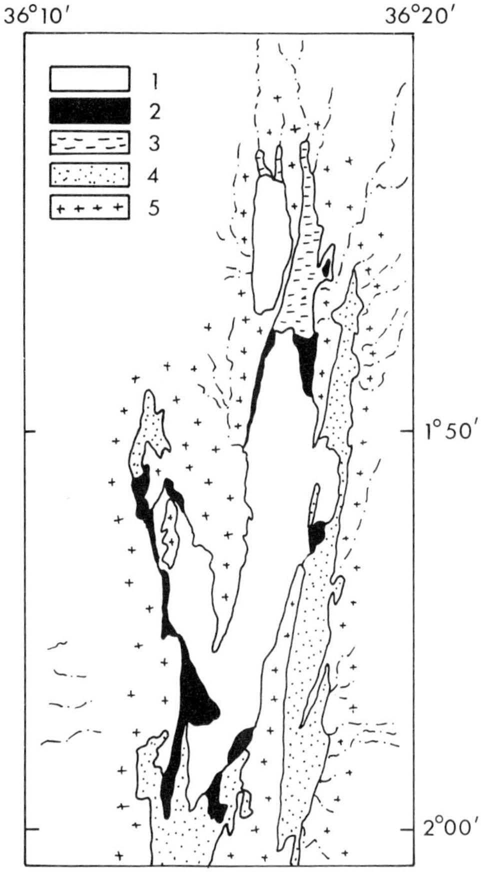

FIG. 8. Preliminary map of depositional subfacies of Lake Magadi. 1 = salt pan; 2 = saline mudflats and lagoons; 3 = ephemeral stream floodplains; 4 = dry mudflats and old lake beds; 5 = trachyte.

and reveal the characteristic zonal pattern interpreted as due to lateral salinity differences (Surdam & Parker 1972). Even if the trona beds had not been preserved, the high silica, alkaline zeolites along with the cherts would indicate a high pH geochemistry. Tuffs of the most saline central regions are largely replaced by K-feldspar, followed outwardly by the analcime zone which, in turn, gives way to the least saline zone characterized by a variety of zeolites, such as clinoptilolite, mordenite and others. Authigenic minerals which grew by interaction of the interstitial brines with the sediments also point to alkaline brines, particularly the abundant shortite ($Na_2CO_3 \cdot 2CaCO_3$), pirssonite ($Na_2CO_3 \cdot CaCO_3 \cdot 2H_2O$) (implied to have been present from crystal moulds), dawsonite ($NaAl(CO_3)(OH)_2$), fluorite (CaF_2) and a number of rarer carbonates, such as eitelite ($Na_2Mg(CO_3)_2$).

Lake Gosiute, in which the Wilkins Peak Member was deposited, was very much larger and longer-lived than Magadi ever will be, and yet Magadi-like hydrochemical conditions established and re-established themselves repeatedly. Major inflow was from the nearby Uinta Mountains to the S, the emerging Wind River Range to the NE and the thrust Belt to the NW. This inflow must have been dominated by Na-Ca-HCO_3 waters derived from the granitic and metamorphic terranes of the Wind River and Uinta Mountains and perhaps the Absaroka volcanics. Lake Gosiute hydrology was quite different from that of Lake Magadi, primarily because the basin was huge and had very gentle gradients. Its closure and depression was probably caused by foreland loading in response to overthrusting from the W rather than by rifting.

Based on sedimentological studies, Eugster & Hardie (1975) advocated the presence of a perennial central lake which episodically shrank to a salt pan during periods of trona accumulation. This lake was surrounded by wide mudflats which were seasonally wet and subaerially exposed much of the time. During wetter periods, the lake would transgress over the mudflats and initiate deposition of oil shale, organic-rich laminated carbonate rocks. Large bodies of siliciclastic sands extended far out towards the centre of the basin, and they have been attributed to ephemeral braided streams which probably supplied the bulk of the inflow to Lake Gosiute. Indications of spring activity are present, but such springs were abundant mainly at the toes of alluvial fans, and they do not seem to have played as crucial a role (compared to solute and inflow sources) as they do at Magadi. Nevertheless, except for flood times, circulation must have been largely subsurface. Efflorescent crusts were important solute sources where the groundwater was close enough to the surface for capillary draw. At the edges of the basin, particularly to the N and S, huge alluvial fans were present, as indicated by thick conglomerates (see Smoot 1983).

The most significant hydrochemical difference between Lakes Gosiute and Magadi are the $Ca:HCO_3$ and $Ca:Na$ ratios of the inflow waters. Although tuffaceous rocks are abundant in the Green River Formation, volcanic glass and alkaline lavas were not dominant rock types in the watershed as they are at Magadi. Consequently, Lake Gosiute inflow waters probably were calcium bicarbonate rich with $Ca > Na$ on a molar basis and $1 > Ca:HCO_3 \gg 0.25$, that is the latter ratio must have been considerably larger than the Magadi average. Consequently, upon evaporative concentration, a large amount of calcite was precipitated, much of it near the spring outflows and on the exposed mudflats. During high stands of the lake, algal blooms would also precipitate annual increments of perhaps 0.2–0.5 mm $CaCO_3$, the thickness of Fossil Syncline laminae. Preferential removal of Ca increased the Mg:Ca ratio in the water and led to the formation of low-Mg calcite, then high-Mg calcite and finally protodolomite. Details of the carbonate production in Wilkins Peak was discussed by Smoot (1978), and the Mg:Ca ratio of the carbonate to be expected from a given solution has been determined by Füchtbauer & Hardie (1980).

Seasonal changes in lake chemistry in a calcite-producing lake were reported by Weber (1981) and reviewed by Eugster & Kelts (1983), whereas conditions leading to oil shale accumulation have been outlined in Eugster (1985). Lake Magadi is fed by waters with $HCO_3 \gg Ca + Mg$ (path I in Eugster & Hardie 1978), and relatively minor calcite precipitation leads to Ca depletion and a rise in pH. Lake Gosiute inflow, on the other hand, must have been dominated by inflow in which HCO_3 only slightly exceeded the alkaline earths (path IIIA in Eugster & Hardie 1978), and a very large amount of (Ca+Mg)-carbonate minerals had to be precipitated upon evaporative concentration to finally reach bicarbonate dominance and a rise in pH. Considering the differences in bedrock lithologies, this inference makes sense. Once high-pH brines were established in the Gosiute system, the subsequent developments, including trona precipitation and authigenic mineral growth, can be modelled by reference to Lake Magadi. One final difference should be mentioned: Lake Gosiute eventually reached halite saturation and extensive beds of halite accumulated (see Deardorff & Mannion 1971), whereas most Lake Magadi brines are undersaturated with respect to halite (Gueddari 1984).

With such differences in hydrochemistry and hydrology between Lakes Gosiute and Magadi one should not be surprised to find also substantial differences in sedimentary processes and products. Whereas volcanic sands, silts and muds, along with cherts and trona, predominate at Magadi, the Wilkins Peak Member consists largely of siliciclastic sands, carbonate mudstones, carbonate sandstones, oil shales and trona, reflecting differences not only in inflow water compositions and hydrology, but also in basin morphology and tectonic settings. Nevertheless, the detailed analysis carried out at Lake Magadi helped immensely in understanding closed-basin deposits in general, and the Wilkins Peak Member in particular. The scarcity of cherts in the Green River Formation is understandable partly in terms of differences in basin morphology. At Magadi, flooding of saturated brines leads to stratification and magadiite supersaturation, whereas at Lake Gosiute, gentle gradients produced mixing of saturated brines with dilute inflow, thus preventing magadiite supersaturation and the eventual formation of chert. Precipitation of borosilicates, such as searlesite, can also inhibit magadiite precipitation by decreasing silica activity.

Talbot (pers. comm.) argues that Lake Victoria provides a plausible basin analogue for Green River oil shale accumulation. Clastic sedimentation is reduced by dense vegetation and uniform climate. Stable thermal stratification in this deep and dilute lake favours preservation of organic matter, which is a mixture of planktonic and terrestrial material. Lake Victoria has the correct size and water chemistry for a Green River analogue, but its hydrology, depositional environments and climate do not fit either Lakes Gosiute or Uinta, both of which were hydrologically closed, had steep basin margin gradients and experienced repeated evaporitic episodes. Some of the abundant and thick oil shales deposited in Lake Uinta may have been the product of a dilute to brackish lake, but the particle flux was always alternating between inorganic carbonates and organic matter. From our experience of productivity and organic matter preservation at Great Salt Lake, Utah (Spencer *et al.* 1984), I have argued (Eugster 1985) that in Lake Gosiute the richest oil shales are formed during a regressive cycle just prior to an evaporitic stage. It is possible that Lake Victoria did accumulate rich oil shales during the early Holocene, but I would caution that the analogy with Green River oil shales not be carried too far or it will obscure some of the very real differences.

One final point: Hertz (1984) has shown that the organic matter of a variety of Green River oil shales is a complex mixture of planktonic and terrestrial material, as suggested by Talbot (pers. comm.) for basins with dense vegetation cover. However, Baedeker (1985) has documented, for Great Salt Lake, Utah sediments that the organic matter eventually preserved is strongly influenced by its response to bacterial degradation. Short-chained lipids of planktonic derivation are the most easily destroyed; terrestrial plant debris is the most refractory fraction and is preserved preferentially.

Other ancient Magadi-type environments

Searles Lake, California, contains trona of Pleistocene age, and its stratigraphy and geochemistry have been discussed by Smith (1979). The deposit has not yet become lithified and contains large amounts of interstitial brine which is being exploited commercially. These brines are high-pH solutions similar to Magadi brines, but with some important differences. Chloride, sulphate, borate and potassium are higher, whereas carbonate is distinctly lower. This is expressed in the fact that at Searles Lake halite is abundant and minerals such as burkeite ($2Na_2SO_4 \cdot Na_2CO_3$), hanksite ($9Na_2SO_4 \cdot 2Na_2CO_3 \cdot KCl$), aphthitalite ($K_3Na(SO_4)_2$) and borax ($Na_2B_4O_7 \cdot 10H_2O$) are present, pointing to differences in inflow compositions. Phillipsite, analcime, K-feldspar and fluorite are common authigenic minerals (see also Hay & Moiola 1963), as is searlesite, ($NaBSi_2O_6 \cdot H_2O$). The latter mineral keeps silica activity in the brines low and is responsible for the absence of magadiite or chert, an effect noted by Surdam & Sheppard (1978) at Teels Marsh, California. If the evaporites were removed, it is not certain that an ancient Searles Lake-like deposit would be recognized as a high-pH environment, though the presence of boron-rich minerals should be sufficient to indicate that cherts need not be present.

In a series of studies on Pleistocene lake deposits in western North America, Sheppard (1969, 1973), Sheppard & Gude (1968, 1969, 1974) and Sheppard & Mumpton (1984) have identified many fossil alkaline lakes, including Lake Tecopa, Barstow Formation, Big Sandy Formation, Rome Beds (Oregon) and Gila Beds (New Mexico). Identifications are based on the presence of tuff beds which have been converted to zeolites or K-feldspar and the presence of Magadi-type cherts. The latter were encountered in the Rome and Gila Beds, both of which are also rich in fluorite. Many of these deposits show the characteristic zeolite–K-feldspar zonation, and the alkaline environment is easily recognizable even though no evaporite minerals are preserved. Magadi-type cherts were encountered in many older sediments, including the Jurassic Morrison formation (Surdam *et al.* 1972).

There is no doubt that many more such examples will come to light as criteria for recognition of fossil high-pH environments become refined. Precambrian shield areas are likely candidates, because of the prevalence of igneous and metamorphic terranes and the presence of stable continental basins. Pitt *et al.* (1980) and White & Youngs (1980) have identified the Officer basin, South Australia, and specifically the Cambrian Observatory Hill Beds, as such a setting. On the basis of pseudomorphs after evaporite minerals, Magadi-type cherts and the presence of kerogen-rich carbonates, White & Youngs (1980, p. 1284) concluded that these beds 'were deposited in a Cambrian playa-lake environment very similar to the younger playa-lake environments of the Green River Basin and Lake Magadi'.

The Proterozoic Duruchaus Formation of the Damara Orogen in Namibia has been called a meta-playa sequence by Behr *et al.* (1983) and is thought to have been deposited in an alkaline rift valley basin. This conclusion is based largely on albite pseudomorphs after carbonate evaporite minerals and the presence of saline fluid inclusions. Magadi-type cherts appear to be missing, but abundant tourmaline may indicate that here also boron-rich waters inhibited magadiite formation. The assumption is that the predominance of albite is due to the metamorphic conversion of analcime and other zeolites.

Recently I have suggested (Eugster 1985) that the pyritic, dolomite HYC shale of the Proterozoic McArthur Group of the Gulf of Carpentaria of North Australia formed some 1500 Ma in a Magadi-like rift valley setting. Contemporaneous metal-rich alkaline brines deposited vast Zn-Pb-Ag ore bodies as conformable lenses within the shale. They have been identified as belonging to the Green River type (GRT) by their association with K-feldspar tuffs and Magadi-type cherts which show the characteristic soft-sediment deformation features.

More speculative is the suggestion that the cherts of the 2200 Ma old Outokumpu Zone of the Karelian schist area of Finland (Eugster 1985) were derived from magadiite beds associated with serpentine weathering. I have also argued (Eugster 1969; Eugster & Chou 1973) that cherts of some Precambrian banded iron formations were initially precipitated as magadiite beds, still a plausible and likely mechanism, but without definite proof.

Recognition of Magadi-type environments in the geologic record

Work on saline alkaline lakes and on deposits formed from such lakes has made it possible to assemble a set of criteria for the recognition of such environments in the past, even after soluble salts have been lost or replaced by pseudomorphs. Mineralogical as well as sedimentological evidence must be assembled in order to document the existence of a closed-basin deposit. The morphology of the basin and its tectonic setting may be deduced from the associated lithologies, such as bedrock fault blocks, alluvial fans, river deposits, dunes, springs and others, whereas the presence of alkaline brines is reflected in salt casts,

Magadi-type chert and a suite of authigenic minerals, such as K-feldspar, zeolites and fluorite. Careful study of the subfacies and their spatial distribution should reveal a zonal arrangement; asymmetry is common and points to tilting of the basin floor. Vertical successions are often cyclic, because closed-basin lake levels respond rapidly to changes in climate and tectonics, resulting in many sequences of transgression and regression. Thickness and lateral extent of closed-basin deposits vary greatly but can reach hundreds of metres vertically and hundreds of kilometres horizontally. Igneous and metamorphic rocks provide ideal bedrocks for the production of the required inflow waters with $(Ca+Mg)/HCO_3<1$. Volcanic rocks are particularly effective because of the rapid release of sodium and silica from residual glasses. Weathering of acid volcanics, on the other hand, may lead to boron domination, whereas sulphate enrichment is often attributable to sulphide weathering. The presence of gypsum or anhydrite is not compatible with an alkaline environment, because with HCO_3^- predominance calcium never builds up sufficiently to reach gypsum or anhydrite saturation.

We now know enough about closed-basin hydrology, hydrochemistry and sedimentation to permit recognition of such environments throughout the geologic column, provided that structural deformation and metamorphic overprint are not too severe. Examples will become known with increasing frequency, particularly from areas of stable continental crust.

ACKNOWLEDGEMENTS: This review does not contain new data, but the detailed comparison between Lake Magadi and the Green River Formation has not been presented before. I am grateful to R.W. Renaut for encouraging me to think once again about Magadi, although I have not visited that area for a dozen years. Work on the Green River Formation, on the other hand, is continuing in the Fossil Syncline. Nobody has contributed more to my understanding of closed basins than have my colleagues B.F. Jones and L.A. Hardie. My wife Elaine has participated actively in the field-work for over 16 years with untiring enthusiasm, patience and good cheer. Financial support was provided by the National Science Foundation and the Petroleum Research Fund administered by The American Chemical Society with a series of grants starting in 1965.

References

BAEDEKER, M.J. 1985. *Organic material in sediments of Great Salt Lake, Utah: influence of changing environments*. Unpublished PhD thesis, George Washington University, Washington, DC.

BEHR, H.J., AHRENDT, H., MARTIN, H., PORADA, H., ROHRS, J. & WEBER, K. 1983. Sabkha-playa environments of Upper Proterozoic age in the basal Damara sediments and their significance for the evolution of the Damara Orogen. In: MARTIN, H. & EDER, F.W. (eds) *Intracontinental Fold Belts—Case Studies in the Variscan Belt and Damara Belt*. Springer, Berlin and Heidelberg, pp. 577–610.

BRADLEY, W.H. 1964. Geology of the Green River Formation and associated Eocene rocks in southwestern Wyoming and adjacent parts of Colorado and Utah. *Prof. Pap US Geol. Surv.* **496A,** 87–110.

—— & EUGSTER, H.P. 1969. Geochemistry and paleolimnology of the trona deposits and associated authigenic minerals of the Green River Formation of Wyoming. *Prof. Pap. US Geol. Surv.* **496B,** 71pp.

BRICKER, O.P. 1969. Stability constants and free energies of formation of magadiite and kenyaite. *Am. Mineral.* **54,** 1026–33.

BUTZER, K.W., ISAAC, G.L., RICHARDSON, J.L. & WASHBOURN-KAMAU, C. 1972. Radiocarbon dating of East African lake levels. *Science* **175,** 1069–76.

DEARDORFF, D.L. & MANNION, L.E. 1971. Wyoming trona deposits. *Wyom. Univ. Contrib. Geol.* **10,** 15–23.

DREVER, J.I. 1982. *Geochemistry of Natural Waters*. Prentice-Hall, Englewood Cliffs, NJ.

—— & SMITH, C.L. 1978. Cyclic wetting and drying of the soil zone as an influence on the chemistry of ground water in arid terrains. *Am. J. Sci.* **278,** 1448–54.

EUGSTER, H.P. 1967. Hydrous sodium silicates from Lake Magadi, Kenya; precursors of bedded chert. *Science* **157,** 1177–80.

—— 1969. Inorganic bedded cherts from the Magadi area, Kenya. *Contrib. Mineral. Petrol.* **22,** 1–31.

—— 1970. Chemistry and origin of the brines of Lake Magadi, Kenya. *Spec. Publ. Mineral. Soc. Am.* **3,** 215–35.

—— 1980. Lake Magadi, Kenya and its precursors. In: NISSENBAUM A. (ed) *Hypersaline Brines and Evaporites*. Developments in Sedimentology No. 28, Elsevier, Amsterdam, pp. 195–232.

—— 1985. Oil shales, evaporites and ore deposits. *Geochim. Cosmochim. Acta* **49,** 619–35.

—— & CHOU, I. 1973. The depositional environments of Precambrian banded iron formations. *Econ. Geol.* **68,** 1144–68.

—— & HARDIE, L.A. 1975. Sedimentation in an ancient playa-lake complex: the Wilkins Peak Member of the Green River Formation of Wyoming. *Bull. Geol. Soc. Am.* **86,** 319–34.

—— & —— 1978. Saline lakes. In: LERMAN A. (ed) *Chemistry, Geology and Physics of Lakes*. Springer-Verlag, New York, pp. 237–93.

—— & JONES, B.F. 1968. Gels composed of sodium aluminum silicate, Lake Magadi, Kenya. *Science* **161,** 160–4.

—— &—— 1979. Behaviour of major solutes during closed-basin brine evolution. *Am. J. Sci.* **279,** 609–31.

—— & KELTS, K. 1983. Lacustrine chemical sediments. In: PYE, K. & GOUDIE, A. (eds) *Chemical Sedimentation and Geomorphology*. Academic Press, London, pp. 321–68.

FÜCHTBAUER, H. (ed) & HARDIE, L.A. 1980. Comparison of experimental and natural magnesian calcites (abstract). *Int. Ass. Sediment. Europ. Meeting Bochum, 1980*, 167–9.

GISLASON, S.R. & EUGSTER, H.P. 1985. Interaction of basalt with meteoric water: a field and laboratory study. *Eos* **66,** 401–2.

GUEDDARI, M. 1984. *Géochimie et thermodynamique des évaporites continentales. Etude du Lac Natron en Tanzanie et du Chott El Jerid en Tunisie.* Memoire 76, Univ. Louis Pasteur, Strasbourg, 143pp.

HARDIE, L.A. & EUGSTER, H.P. 1970. The evolution of closed-basin brines. *Spec. Publ. Mineral. Soc. Am.* **3,** 273–90.

——, SMOOT, J.P. & EUGSTER, H.P. 1978. Saline lakes and their deposits: a sedimentological approach. In: MATTER, A. & TUCKER, M. (eds) *Modern and Ancient Lake Sediments*. Int. Ass. Sedimentol. Spec. Publ. No. 2, pp.7–41.

HAY, R.L. & MOIOLA, R.J. 1963. Authigenic silicate minerals in Searles Lake, California. *Sedimentology* **2,** 312–32.

HERTZ, T.L. 1984. *Organic geochemistry of the Wilkins Peak, Laney, and Fossil Butte members of the Green River formation, Wyoming*. Unpublished PhD thesis, Johns Hopkins University, Baltimore.

JONES, B.F. 1966. Geochemical evolution of closed basin waters in the western Great Basin. *2nd N. Ohio Geol. Soc. Symp. Salt*, **1,** 181–200.

——, EUGSTER, H.P. & RETTIG, S.L. 1977. Hydrochemistry of the Lake Magadi Basin, Kenya. *Geochim. Cosmochim. Acta.* **41,** 53–72.

——, RETTIG, S.L. & EUGSTER, H.P. 1967. Silica in alkaline brines. *Science* **158,** 1310–14.

MONNIN, C. & SCHOTT, J. 1984. Determination of the solubility products of sodium carbonate minerals and an application to trona deposition in Lake Magadi (Kenya). *Geochim. Cosmochim. Acta* **48,** 571–81.

PITT, G.M., BENBOW, M.C. & YOUNGS, B.C. 1980. A review of recent geological work in the Officer Basin, South Australia. *J. Aust. Petrol. Explor. Assoc.* **20,** 209–20.

SHEPPARD, R.A. 1969. Diagenesis of tuffs in the Barstow Formation, Mud Hills, San Bernardino County, California. *Prof. Pap. US Geol. Surv.* **634,** 35pp.

—— 1973. Zeolites and associated authigenic silicate minerals in tuffaceous rocks of the Big Sandy Formation, Mohave County, Arizona. *Prof. Pap. US Geol. Surv.* **830,** 36pp.

—— & GUDE, A.J., 3rd. 1968. Distribution and genesis of authigenic silicate minerals in tuffs of Pleistocene Lake Tecopa, Inyo County, California. *Prof. Pap. US Geol. Surv.* **597,** 38pp.

—— &—— 1969. Authigenic fluorite in Pliocene lacustrine rocks near Rome, Malheur County, Oregon. *Prof. Pap. US Geol. Surv.* **650D,** 69–74.

—— &—— 1974. Chert derived from magadiite in a lacustrine deposit near Rome, Malheur County, Oregon. *J. Res. US Geol. Surv.* **5,** 625–30.

—— & MUMPTON, F.A. 1984. Sedimentary fluorite in a lacustrine zeolitic tuff of the Gila conglomerate near Buckhorn, Grant County, New Mexico. *J. Sed. Petrol.* **54,** 0853–60.

SMITH, G.I. 1979. Subsurface stratigraphy and geochemistry of late Quaternary evaporites, Searles Lake, California. *Prof. Pap. US Geol. Surv.* **1043,** 130pp.

SMOOT, J.P. 1978. Origin of the carbonate sediments in the Wilkins Peak Member of the lacustrine Green River Formation (Eocene), Wyoming, USA. In: MATTER, A. & TUCKER, M.E. (eds) *Modern and Ancient Lake Sediments*. Int. Assoc. Sedimentol. Spec. Pap. No. 2, pp.109–27.

—— 1983. Depositional subenvironments in an arid closed basin; the Wilkins Peak Member of the Green River Formation (Eocene), Wyoming, USA. *Sedimentology* **30,** 801–27.

SPENCER, R.J., BAEDEKER, M.J., EUGSTER, H.P., FORESTER, R.M., GOLDHABER, M.B., JONES, B.F., KELTS, K., MCKENZIE, J., MADSEN, D.B., RETTIG, S.L., RUBIN, M. & BOWSER, C.J. 1984. Great Salt Lake, and precursors, Utah: the last 30,000 years. *Contrib. Mineral. Petrol.* **86,** 321–34.

SURDAM, R.C. & EUGSTER, H.P. 1976. Mineral reactions in the sedimentary deposits of the Lake Magadi region, Kenya. *Bull. Geol. Soc. Am.* **87,** 1739–52.

——, —— & MARINER, R.H. 1972. Magadi-type chert in Jurassic and Eocene to Pleistocene rocks, Wyoming. *Bull. Geol. Soc. Am.* **83,** 2261–5.

—— & PARKER, R.D. 1972. Authigenic aluminosilicate minerals in the tuffaceous rocks of the Green River Formation, Wyoming. *Bull. Geol. Soc. Am.* **83,** 689–700.

—— & SHEPPARD, R.A. 1978. Zeolites in saline, alkaline-lake deposits. In: SAND, L.B. & MUMPTON, F.A. (eds) *Natural Zeolites Occurrence, Properties, Use*. Pergamon Press, Oxford, pp. 145–74.

WEBER, H.P. 1981. *Sedimentologische und geochemische Untersuchungen im Greifensee* (Kanton ZH/Schweiz). Unpublished PhD thesis, ETH-Zurich, Nr. 6811.

WHITE, A.H. & YOUNGS, B.C. 1980. Cambrian alkali playa lacustrine sequence in the northeastern Officer Basin, South Australia. *J. Sed. Petrol.* **50,** 1279–86.

H.P. EUGSTER, Dept of Earth & Planetary Sciences, Johns Hopkins University, Baltimore, Maryland 21218, USA.

Sedimentation by termites in the Malawi Rift Valley

R. Crossley

SUMMARY: Homogeneous red clayey sand sheets, averaging 5 m thick, cover a discontinuous area totalling 8800 km^2 in the Karonga section of the Malawi Rift Valley floor. The sheets accumulated in a terrestrial environment, sedimentation was slow and gradual, and fossil termite mounds occur at the base of the sheet in one area. Topographic considerations rule out deposition by overland flow, and texture is incompatible with aeolian sedimentation. Modern termites (*Macrotermes falciger*) are transporting sediment from beneath the sand sheets onto the ground surface. The textures of fossil and modern *Macrotermes* mound deposits are very similar to the textures of the sand sheets. It is concluded that the sand sheets have accumulated through vertical transfer of sediment by *Macrotermes*. Scrutiny of the literature suggests that similar deposits may be widespread in African savanna and rainforest areas, both within the African Rift System and in epeirogenic basins. Equivalent sediments must occur in ancient tropical sequences since fossil termites are first recorded in the mid-Cretaceous.

Termite numbers typically average 1000–4000 m^{-2} in tropical savanna and tropical rainforest systems (Wood & Sands 1978). Most species of termites build nest structures, predominantly of mineral matter, on or in the soil, and in extreme cases termite mounds may occupy as much as 30% of the land area (e.g. mounds of *Macrotermes* in the central Congo; Meyer 1960). Mineral matter is also used to build foraging runways and sheeting on the ground, and for infilling galleries within eaten out branches and twigs (Lee & Wood 1971). Lepage (1974) showed that *Macrotermes subhalinus* in Senegal annually used 675–950 kg ha^{-1} of mineral matter for building foraging runways and sheeting, and *c*. 1000–1400 kg ha^{-1} annually for nest mound construction.

The mineral matter used by termites is obtained from either the soil or subsoil (Lee & Wood 1971). Termites are known in some cases to penetrate to depths of 10–30 m in search of water (e.g. West 1970; Komanda 1975), and the associated transfer of subsoil to the surface contributes to the accumulation of the soil unit. Estimates of the rate of subsoil to soil surface transfer are in the order of 25–200 mm ka^{-1} (e.g. Nye 1955, West Africa; Watson 1960, Zimbabwe; De Ploey 1964, Lower Congo).

If these rates apply generally in African savanna and rainforest regions then the mineral matter in mature soil mantles in these areas has largely accreted through termite activity (Watson 1974). It also follows that sedimentary units which accumulated through termite activity might be found in areas of terrestrial deposition, such as epeirogenic basins or rift valley troughs.

This paper arises mainly out of fieldwork undertaken between 1979 and and 1984 on late Cenozoic sediments in the Karonga area of the Malawi Rift Valley (Fig. 1). The stratigraphy of the area was largely completed by Dixey (1927), though refinements were added by Stephens (1966), Clark *et al.* (1970) Kaufulu *et al.* (1981) and Crossley (1982). The succession comprises a Plio-Pleistocene sequence of shallow lacustrine and lake margin sediments, the Chiwondo Formation, overlain unconformably by gravels and sands of the Chitimwe Formation.

Laterally discontinuous sheets of homogeneous red clayey sand, totalling *c*. 8800 km^2 (Fig. 1) and typically about 5 m thick, cap several flat or very gently sloping hilltops in the area. Clark *et al.* (1970) described this sand as part of the upper unit of the Chitimwe Formation. Texturally, the sand sheets fall into the clayey sand and sand classes of Shephard (1954) since they contain 73–82% sand and 12–20% clay. Equivalent pedological classifications such as loamy sands and sandy loams are not used here since they fail to highlight the low silt contents (*c*. 7%) in the sand sheets. The process by which the sand was deposited has not been specified, but Haynes (1970) suggested that the lack of structure was possibly due to bioturbation by soil insects. This paper is primarily concerned with the origin of these red clayey sands.

Red clayey sand sheets of the Karonga area

Relations with underlying units

Usually the underlying deposits are gravels of the Chitimwe Formation, which rest in turn on the Chiwondo Formation. However, in the Sungwa area the sand sheet rests on gravels which overlie Mesozoic sediments, whilst in the Mwenirondo

From Frostick, L.E. *et al.* (eds) 1986, *Sedimentation in the African Rifts,* Geological Society Special Publication No. 25, pp. 191–9.

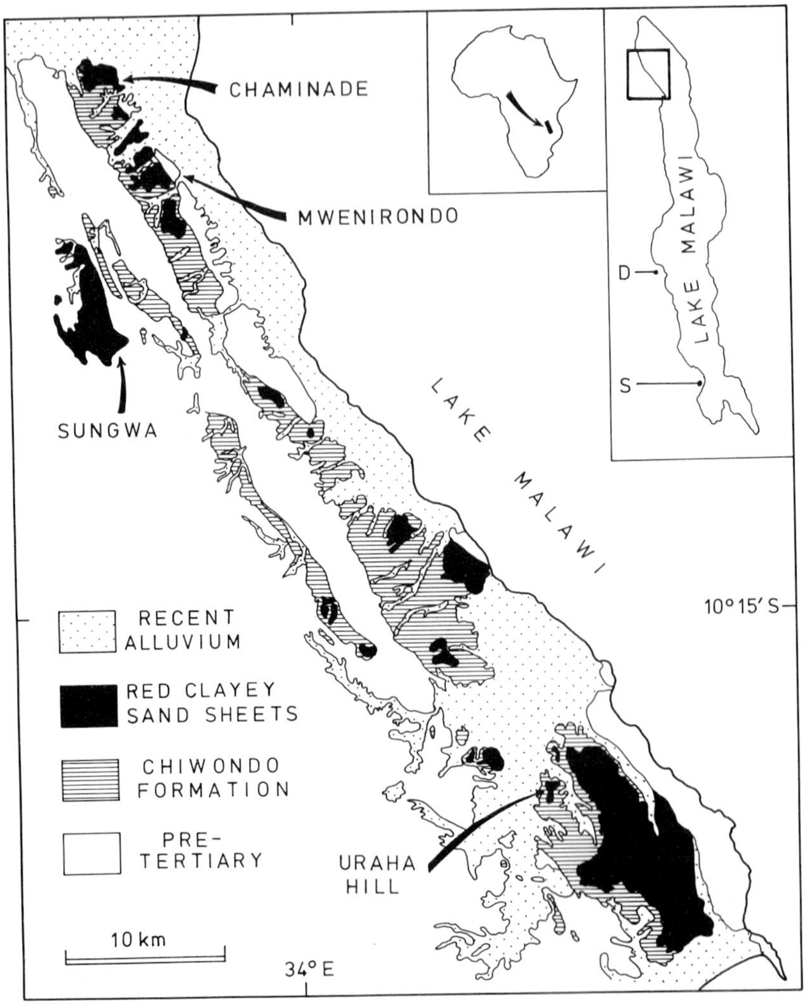

FIG. 1. Maps of locations and of the geology of the Karonga lake shore. Inset map of lake Malawi shows the location of Dwangwa Delta (D), and Senga Bay (S).

area the eastern margin of the sand sheet overlies gravels on metamorphic basement. Contacts with underlying units slope at angles $<3^\circ$.

In some places, such as at Sungwa and at Chaminade locality 2 (Fig. 2), the underlying gravels are thin, structureless and unsorted with a clayey sand matrix. Unabraded stone artefacts occur on, and in, these gravels at Sungwa and Chaminade. These gravels are interpreted as lags left during erosion by overland flow and rainsplash.

Elsewhere, for example at Chaminade localities 1 and 3 (Fig. 2) and at Uraha Hill (Fig. 3), the gravels are both clast-supported and clayey sand matrix-supported. They show discontinuous horizontal stratification and tabular and trough cross-bedding, and are between a few decimetres and several metres thick. Abraded stone artefacts occur within the bedded gravels at Chaminade. Unabraded stone artefacts rest on the bedded gravels at Chaminade and Uraha Hill (Figs. 2 and 3). These gravels are interpreted as the results of fluvial deposition.

Further indications that the tops of the gravels represent terrestrial environments are to be found in the Chaminade area. Here dome-shaped piles up to 3 m high and 6 m across, formed of irregular carbonate concretions in a brown clayey sand matrix, rest on the gravels (Fig. 2). Carbonate nodules are common in *Macrotermes* mounds even when surrounding soils contain little or no pedogenic carbonate (Lee & Wood 1971). These

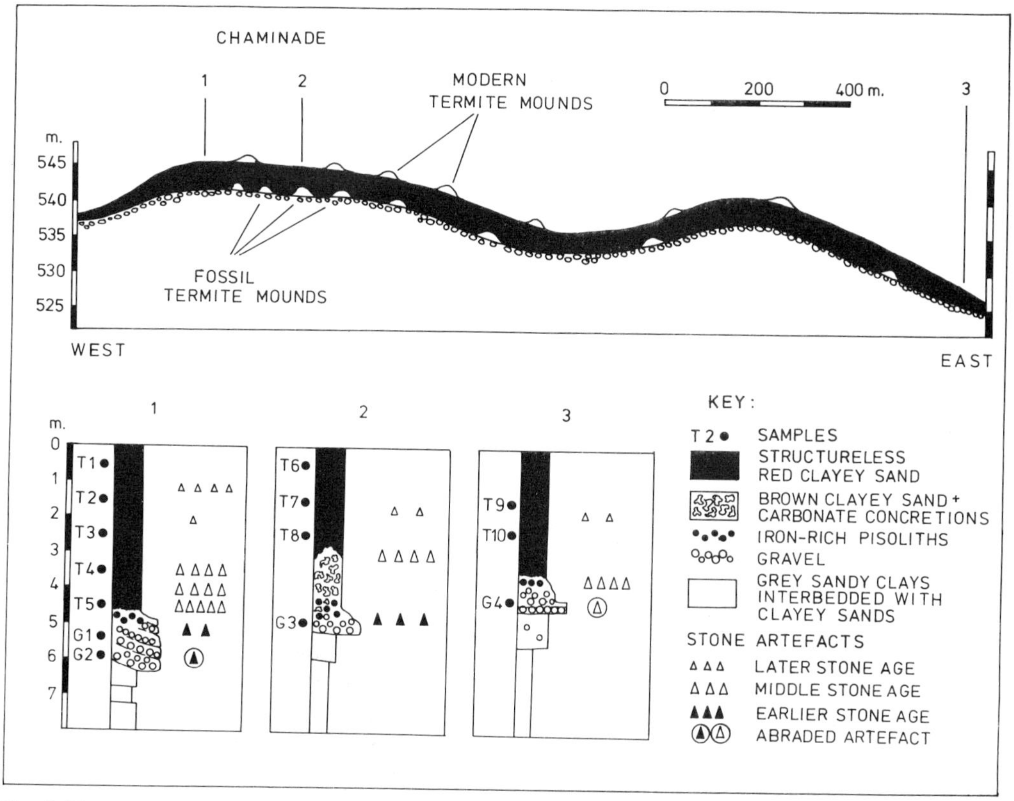

FIG. 2. Transverse section of the Chaminade sand sheet and the microstratigraphy at sites 1, 2 and 3. Location of samples taken for textural analysis are shown.

piles have therefore been interpreted as fossil *Macrotermes* mounds (Crossley 1984) with carbonate deposition occurring (by analogy with mounds in Kenya (Hesse 1955) and in Malawi) whilst groundwater levels were high.

In addition, an extensive but laterally impersistent zone of iron pisoliths occurs within the upper part of the gravels around Chaminade (Fig. 2). Some of the iron pisoliths beneath the fossil termite mounds are enclosed by the basal carbonate nodules of those mounds, so the iron pisoliths cannot have formed by a much later groundwater mobilization process, such as that documented by Goldbery & Beyth (1984). The pisoliths are therefore taken to represent a fossil soil which was affected by seasonal fluctuations in the water table (MacFarlane 1976).

The Chaminade sand sheet occurs at lower altitudes than the others. Evidence of shallow groundwater levels, either in the form of fossil termite mounds or pisolithic iron, has not been found in the units underlying the Uraha, Mwenirondo or Sungwa sand sheets.

The only consistent features about the units underlying the sand sheets therefore appear to be that their tops represent terrestrial environments with palaeoslopes of less than 3°.

Character of the sand sheets

The uniformity of the different sand sheets means that the following general description applies widely.

The basal contact of the sand is gradational over a narrow zone—the sand coarsens downwards into gravel typically over a depth of a few decimetres. The sand sheets are usually between 3 and 6 m thick and are often consistent in thickness over considerable areas, even when draping subdued topography (e.g. Fig. 2).

The sands are virtually homogeneous in appearance in the field. The colour is typically red (2.5YR 4/8), though the top of the unit is generally paler (2.5YR 6/8) and the lower part a little darker (2.5YR 3/6). The sediment is a clayey sand and there is no discernible bedding. The degree of lithification by clay/iron cementation increases downwards, so that collection of coher-

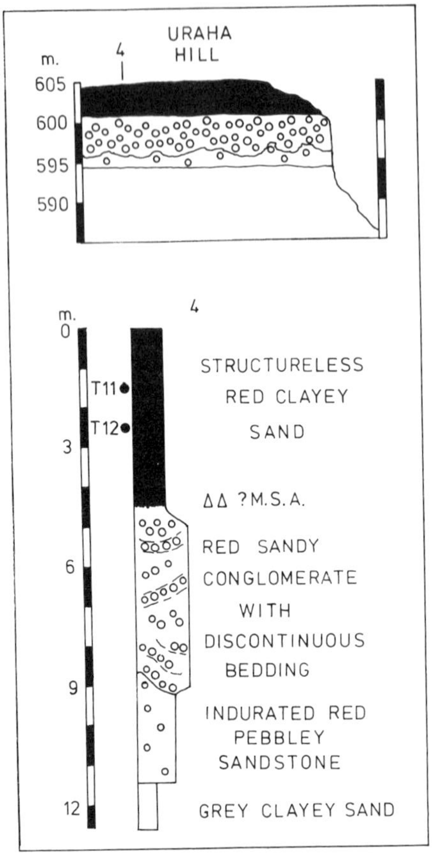

FIG. 3. Transverse section of Uraha Hill and the microstratigraphy of site 4. Location of samples taken for textural analysis are shown.

ent handspecimens is possible in the mid to lower part of the unit. The mid to lower part of the unit shows indistinct, widely-spaced, rectilinear jointing. Small (*c.* 4 mm diameter) burrows occur sparsely throughout the deposit.

Near-horizontal lines of scattered stone artefacts occur at various depths within the sands. The artefacts are unabraded, and stone flakes collected from some horizons at Chaminade have been successfully re-fitted (Clark 1970a; and present work). It therefore follows that sedimentation of the clayey sand was gentle, and the successive horizons of artefacts of differing industries are evidence of deposition in a nearly flat-lying terrestrial environment.

Age and rate of accumulation

Clark (1970a) reports that the assemblage of stone artefacts found directly on the gravels at the base of the red sand sheet in the Chaminade–Mwenirondo area resemble material described as 'Sangoan' at Kalambo Falls in northern Zambia. The artefact horizons within the basal part of the sand sheet at Chaminade are ascribed to the 'Middle Stone Age', but unfortunately the ages of these industries are problematical. The 'Sangoan' materials at Kalambo Falls have been dated at between 45,000 and 35,000 BP (Clark 1969), and Clark (1970b) puts the onset of the African 'Middle Stone Age' at *c.* 35,000 BP. In contrast, the onset of the 'Middle Stone Age' in South Africa is placed at 180,000 BP by P. Beaumont (pers. comm.), and is thought to have ended around either 60,000 BP (Beaumont *et al.* 1978) or 30,000 BP (Deacon and Thackeray 1984).

The base of the sand sheet in the Chaminade area might therefore be anything between *c.* 200,000 and 35,000 BP. Average rates of accumulation for a typical 5 m thickness would be 25 mm and 143 mm ka^{-1} respectively. Charcoal provided a date of $10{,}400 \pm 300$ BP (Haynes 1970). The precise depth was not reported, but if the charcoal is assumed to have come from a depth of 2.5 m then an average accumulation rate of 240 mm ka^{-1} is indicated for the upper part of the deposit. Although the dating is not precise, even the most conservative age estimates indicate slow accumulation of the clayey sand unit.

Origin of the red clayey sand sheets

Possible non-biological sedimentation

It is possible to envisage two non-biological processes that might produce the sedimentary deposits described above. These are: (1) the transport of material from neighbouring hills by overland flow; and (2) aeolian sedimentation.

Detailed mapping of the base of the sand sheet at a scale of 1:2500 and using a 2 m contour interval, shows that the area around Chaminade school has been a low hill top from early in the history of accumulation of the sand sheet. Overland flow could not, therefore, have deposited the sand sheet in that area.

In order to test the possibility of aeolian sedimentation the textures of recent dune sands from the Lake Malawi shore were analysed (Fig. 4, A) but were found to be very different from those of the sand sheets (Fig. 4, T).

Modern biological sedimentation

The most prominent termite mounds in the Chaminade area are those of *Macrotermes falciger* (Gertsacker) which occur at a density of 2–6 ha^{-1}. Twelve mounds in the vicinity of Chaminade school showed accumulations of

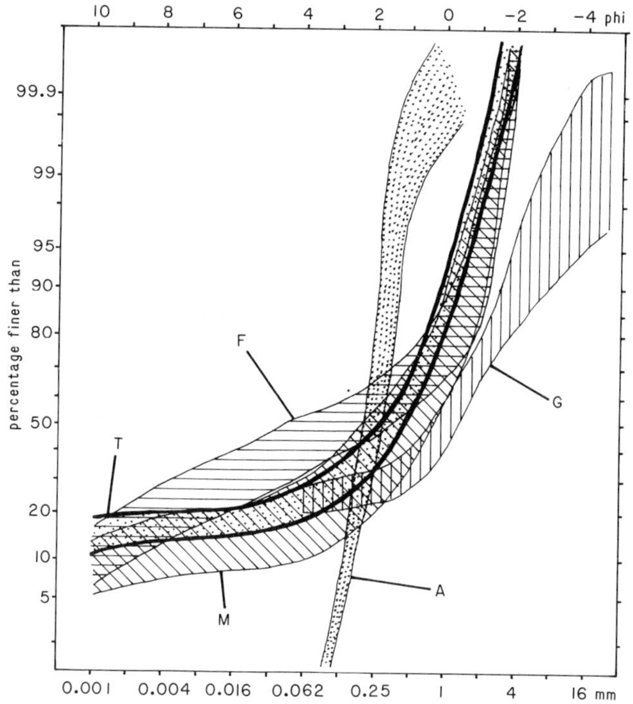

FIG. 4. Particle size distributions. Samples from Karonga: T = red clayey sand sheets (stippled field enclosed by heavy lines), n = 15; G = sandy gravels of the Chitimwe Formation, n = 4; M = modern *M. falciger* mounds on top of the red clayey sand sheet, n = 6; F = fossil termite mounds on Chitimwe gravels, n = 2, and modern *M. falciger* mounds on re-exposed Chitimwe gravels, n = 6; A = modern aeolian sands from Senga Bay, n = 4.

fresh deposits at the time of mapping. The deposits on five of the mounds contained a mixture of red and grey sediment, with the grey component varying between *c*. 20 and 70%. These grey sediments must have come from the Chiwondo Formation, which occurs at depths of at least 5 m below the surface of the red sand sheet. The termites at these five mounds were, therefore, in the process of contributing to the accretion of the Chaminade sand sheet. The remaining seven mounds showed only red fresh deposits, and it must be concluded that those mounds were merely bioturbating the sand sheet at that time.

Other organisms, including ants and different species of termites, are also active in reworking the top of the Chaminade sand sheet. However, there was no evidence that any animals apart from *M. falciger* were adding material to the top of the sand sheet from the underlying Chiwondo formation. These other organisms are, therefore, only bioturbating the sand sheet and so are unlikely to change its overall granulometry. If the surficial accumulation of sediment transferred vertically by *M. falciger* is significant in the long term, then the texture of the sand sheets should reflect selection by this organism alone.

Granulometric analyses

Procedures described by Folk (1974) were followed during wet seiving of the sand fraction at 0.5 ϕ unit intervals, and the silt and clay determinations were made using the hydrometer method at *c*. 0.25 ϕ intervals. Samples from the fossil termite mounds were treated with dilute hydrochloric acid to remove carbonates before analysis.

Samples from modern *M. falciger* mounds in the Chaminade area were taken either from fresh deposits that had been recently added to the external wall or from mound interiors. Other workers have noted that the walls of *Macrotermes* mounds sometimes contain less clay than the interiors (Nye 1955; Maldague 1959). It is possible that such differences are due to preferential loss of silt and clay by weathering processes, so old mound wall textures may not accurately reflect the size fractions originally deposited. Old mound walls were, therefore, not analysed.

Results and interpretation

The sand sheet in the Chaminade area was sampled at three exposures in order to check for vertical changes, and samples were collected from Mwenirondo and Uraha Hill to look for any lateral variations. The texture of the sand sheets showed little variation, either vertically or laterally (Fig. 4, T). At the three vertical exposures, the amount of clay increased by between 5 and 10% from the top to the base of the sheet. This corresponds to a downward increase in both lithification and rubification, all of which might be the result of the downward translocation of fines in these permeable soils.

The results from the fossil termite mounds which rest on the gravels of the Chitimwe Formation at Chaminade (Fig. 4, F) are indistinguishable from modern mounds which sit on the re-exposed gravels. The well-defined size cut-off of 4 mm ($-2\ \phi$) in these deposits must represent the maximum size that the termites use in mound building, since larger grains are present in the immediately underlying gravels (Fig. 4, G).

The *M. falciger* mounds on top of the sand sheet at Chaminade contain more sand in the 1.41–0.35 mm size range (Fig. 4, M) than the mounds on the gravels. This contrast is inferred to be due to selection by the termites, and not to differences in the character of the substrate immediately beneath the two groups of mounds. There are three reasons for this inference. Firstly, two of the samples that lie in envelope M of Fig. 4 contain more than 50% grey sediment which must have come from the Chiwondo Formation. Secondly, there is a greater range of textures in the termite mound analyses than in the sand sheet analyses and the mean value of the termite mound data places it in a sandier category than the mean of the sand sheets, thereby implying that the termites manipulate sand/silt/clay by their activity. Thirdly, there are many accounts that show that the texture of termite deposits is not governed by the texture of the substrate (e.g. Lee & Wood 1971). It is not clear, however, exactly why the termites have selected different grain sizes in the two locations.

The identical maximum grain size of 4 mm, in both the sand sheets and in the termite mounds, suggests that the deposits have a common origin. Furthermore, the sand sheet texture is intermediate between the two different types of termite deposit. The restricted textural variation in the sand sheets is interpreted as the result of the homogenizing influence of bioturbation. Comparison of the termite and sand sheet data does reveal some differences (Fig. 4), but these are small, and the sand sheets can be interpreted as being very largely the product of termite activity.

Overland flow and rainsplash rework these deposits, as does modern bioturbation, but their roles should be seen as largely helping to even out the micro-topography. On slopes greater than 3°, overland flow and rainsplash presumably become sufficiently effective agents of erosion to prevent long-term accumulation of termite deposits. This is inferred from the fact that sand sheets are only found where the pre-existing ground slope was of low gradient.

Implications for subsurface subsidence

It has been inferred that the sediment in the sand sheets has largely been derived from below the Chitime gravel horizon, but the sediments below that horizon are not riddled with cavities. It can be speculated that the termite burrows collapse when water gets into the sediment during each wet season, and that re-excavation to reach groundwater occurs during each dry season. This would imply that the gravel horizon subsides as sediment is removed, and that collapse occurs. The amount of subsidence may be less than the 5 m average thickness of the sand sheets, because the bulk density of the sediment/soil material that comprises the sheets was judged to be less than that of the sediment beneath the gravels, though no precise measurements were made.

Stone artefacts are too large to be moved upwards by most soil animals. As bioturbation of the accumulating sand sheet progresses, so the artefacts will tend to settle downwards through the profile as finer material is removed from beneath them. Cahen & Moeyersons (1977) inferred that artefacts descended several metres through a clayey sand sheet in Zaire by this process. However, these displacements could be explained in a different way if it could be assumed that the Zaire clayey sand sheet had *accreted* through termite activity. One can only note at present that downwards settling of artefacts may be plausible, but that the rate of movement will depend upon the rate and depth of bioturbation. This has so far not been quantified.

Sedimentation by termites in other areas

An area 3 km W of the Dwangwa delta in Malawi (Fig. 1) shows sediment sheets similar to those around Karonga. The sheets of red clayey sand are *c.* 1 m thick. Unabraded stone artefacts occur along the base of the unit, and it overlies gravels

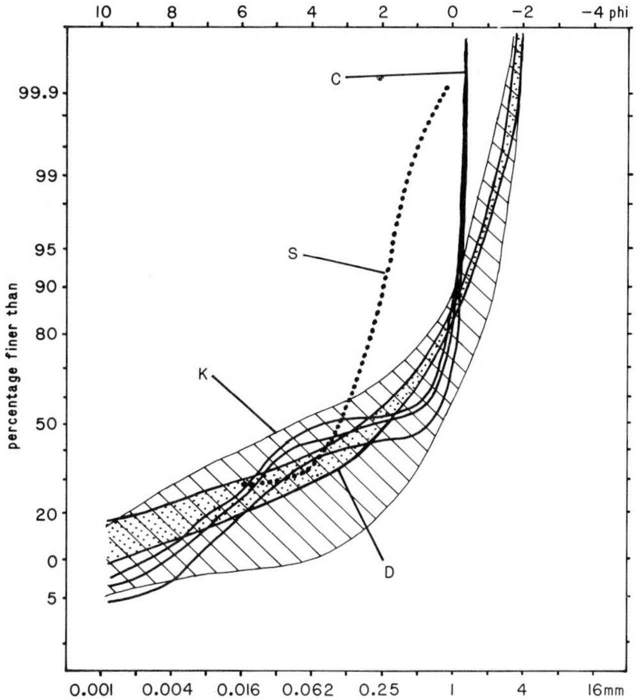

FIG. 5. Particle size distributions. Samples from Dwangwa, Senga Bay, and Karonga (Lake Malawi), and from Zaire: D = modern *Macrotermes* mounds, n = 2, and red clayey sand sheet, n = 2; C = modern *Cubitermes sankurensis* mounds, three samples plotted; K = all types of modern *M. falciger* mounds from the Karonga area; S = modern *M. natalensis* mound from Zaire (after Stoops 1964).

which rest on weathered metamorphic basement. The results of textural analyses of a *Macrotermes* mound which protrudes through the sand sheet, and of two samples from the sand sheet (Fig. 5, D), are essentially similar to those of the Karonga area (Fig. 5, K).

Red clayey sand sheets containing artefact horizons and (as far as field observations alone can be used to judge) having textures similar to the Karonga sheets, are common between the rift escarpment and the modern lake shore in Malawi. Similar sediment sheets, *c.* 3–5 m thick, also occur on the eastern shore of Lake Tanganyika near Kogoma (R. Raynolds, pers. comm.). Comparable clayey sand sheets deposited by termites have also been reported by De Ploey (1964) in the lower Congo. Termite-deposited sediment sheets may therefore be a widespread feature of terrestrial sedimentation in the more humid parts of Africa.

Most published analyses of African termite deposits are expressed only as percentages of sand, silt and clay. It is, therefore, not possible to compare the textural characteristics of *Macrotermes* deposits in detail. However, the maximum grain size deposited by *M. bellicosus* in Nigeria (Nye 1955) and by *M. falciger* in southern Zaire (Komanda 1975) is 4 mm, which is the same size limit found in this study. In contrast, Stoops (1964) found very few grains greater than 1 mm in diameter in mounds of *M. natalensis* in the lower Congo (Fig. 5, S) but it was not stated whether this reflected a selection by the termites or a limitation imposed by the grain size of the substrate.

In an area of the Lake Malawi shore around Senga Bay (Fig. 1), mound building is dominated by *Cubitermes sankurensis* (Wasm), whose worker caste is about half the size of *Macrotermes*. The maximum grain size in these *Cubitermes* mounds (Fig. 5, C), is only 1.5 mm. This represents selection by *Cubitermes*, because particles up to gravel size occur in the deposits beneath the mounds. An additional textural contrast with the analysed *Macrotermes* deposits is that the *Cubitermes* mounds contain only small amounts of sand in the 0.5–0.09 mm size range (Fig. 5, C). Even on the basis of this limited evidence, these textural variations suggest that it may be possible to make comments about the size, if not the genus, of termites which deposited ancient sheets suspected to be of biogenic origin.

Conclusions

1 Transport of mineral matter from the subsoil to the soil surface by termites, during deep excavation for groundwater, is significant in African tropical savanna and tropical rainforest areas.
2 In Malawi examples analysed here, the genus *Macrotermes* transports material with the following textural characteristics: a maximum grain size 4 mm, and a size distribution that gives 48–91% sand, 3–22% silt, and 7–28% clay.
3 These termite-transported sediments accumulate in areas with slopes of less than 3° to give deposits up to 6 m thick.
4 In the long term, bioturbation of the accumulating sediment results in a deposit of relatively consistent texture. Analysis of the Karonga deposits gave a maximum grain size 4 mm, and narrow ranges of each size class—73–82% sand, 5–9% silt, and 12–20% clay.
5 Comparison with data from elsewhere in Africa suggests that the 4 mm maximum grain size is a general characteristic of deposits transported by *Macrotermes*.
6 The maximum grain size of deposits transported by the smaller termite genus *Cubitermes* is 1.5 mm.
7 The preservation potential of termite-deposited sediments in the geological record would be similar to that of soils, with the best prospects lying in the interfluve areas of low-relief landscapes.
8 We do not know when termites evolved the capacity for deep excavation, but termite fossils of relatively advanced character occur in the mid-Cretaceous sediments of Labrador (Emerson 1968), and termites were widely distributed by the early to mid-Tertiary (Wilson 1971). Termite-transported sediments might, therefore, occur in appropriate palaeolatitudes through a substantial portion of the geological column.

ACKNOWLEDGEMENTS: I wish to thank the following: the University of Malawi for funding, S. Bacchus of the British Museum for termite identification, the Geological Survey of Malawi for assistance with granulometry, and W.A. Sands, J.P. Watson and R. Raynolds for literature and encouragement.

References

BEAUMONT, P.B., DE VILLIERS, H. & VOGEL, J.C. 1978. Modern man in sub-Saharan Africa prior to 49,000 years BP. *South Afr. J. Sci.* **74,** 409–19.

CAHEN, D. & MOEYERSONS, J. 1977. Subsurface movements of stone artefacts and their implications for the prehistory of Central Africa. *Nature* **266,** 812–15.

CLARK, J.D. 1969. *Kalambo Falls Prehistoric Site, 1.* Cambridge University Press, London.

——1970a. Interim report on the archaeology of the Malawi, Rungwe and southern Rukwa regions. *Quaternaria* **13,** 331–53.

——1970b. *The Prehistory of Africa.* Thames & Hudson, London.

——HAYNES, C.V., MAWBY, J.E. & GAUTIER, A. 1970. Interim report on palaeo-anthropological investigations in the Lake Malawi rift. *Quaternaria* **13,** 305–6.

CROSSLEY, R. 1982. Late Cenozoic stratigraphy of the Karonga area in the Malawi rift. *Palaeoecol. Afr.* **15,** 139–44.

——1984. Fossil termite mounds associated with stone artefacts in Malawi, Central Africa. *Palaeoecol. Afr.* **16,** 397–401.

DE PLOEY, J. 1964. Stone lines and clayey-sandy mantles in lower Congo: their formation and the effect of termites. In: BOUILLON, A. (ed) *Etudes sur les Termites Africains.* Leopoldville University Press, Leopoldville, pp. 399–414.

DIXEY, F. 1927. The Tertiary and post-Tertiary lacustrine sediments of the Nyasan rift valley. *J. Geol. Soc.* **83,** 432–47.

DEACON, H.J. & THACKERAY, J.F. 1984. Late Pleistocene environmental changes and implications for the archaeological record in southern Africa. In: VOGEL, J.C. (ed) *Late Cainozoic Palaeoclimates of the Southern Hemisphere.* Balkema, Rotterdam, pp. 375–90.

EMERSON, A.E. 1968. Cretaceous insects from Labrador 3, A new genus and species of termite (*Isoptera hodotermitidae*). *Psyche* **74,** 276–9.

FOLK, R.L. 1974. *Petrology of Sedimentary Rocks.* Hemphill Publishing Co., Austin, Texas.

GOLDBERY, R. & BEYTH, M. 1984. Lateritization and groundwater alteration phenomena in the Triassic Budra Formation, south western Sinai. *Sedimentology* **31,** 575–94.

HAYNES, C.V. 1970. Interim report on the Quaternary geology of northern Malawi and southern Tanzania. *Quaternaria* **13,** 307–18.

HESSE, P.R. 1955. A chemical and physical study of the soils of termite mounds in East Africa. *J. Ecol.* **43,** 449–61.

KAUFULU, Z., VRBA, E.S. & WHITE, T.D. 1981. Age of the Chiwondo Beds, northern Malawi. *Ann. Transvaal Mus.* **33,** 1–8.

KOMANDA, A. 1975. Le role des termites dans la mise en

place des sols de plateau dans le Shaba méridional. In: ALEXANDRE, J. (ed) *Géomorphologie Dynamic Dans Les Régions Intertropicales*. Presses Universitaires du Zaire, pp. 81–93.

LEE, K.E. & WOOD, T.G. 1971. *Termites and Soils*. Academic Press, London.

LEPAGE, M. 1974. *Les termites d'une savane sahelienne (Ferlo Septentrional, Senegal): peuplement, populations, consommation, rôle dans l'écosysteme*. Unpublished PhD thesis, University of Dijon.

MACFARLANE, M.J. 1976. *Laterite and landscape*. Academic Press, London.

MALDAGUE, M.E. 1959. Analyses de sols et matériaux de termitieres du Congo Belge. *Insectes sociaux* **6,** 343–59.

MEYER, J.A. 1960. Résultats agronomiques d'un essai de nivellement des termitieres réalisé dans la Cuvette centrale Congolaise. *Bull. Agr. Congo Belge* **51,** 1047–59.

NYE, P.H. 1955. Some soil-forming processes in the humid tropics. 4. The action of the soil fauna. *J. Soil Sci.* **6,** 73–83.

SHEPARD, F.P. 1954. Nomenclature based on sand-silt-clay ratios. *J. Sed. Petrol.* **24,** 151–8.

STEPHENS, E.A. 1966. Geological account of the north-west coast of Lake Malawi between Karonga and Lion Point, Malawi. *Spec. Publ. Anthropol. Recent Studies in Palaeoanthropology* **68,** 49–58.

STOOPS, G. 1964. Application of some pedological methods to the analysis of termite mounds. In: BOUILLON, A. (ed) *Etudes sur les Termites Africains*. Leopoldville University Press, Leopoldville, pp. 379–98.

WATSON, J.P. 1960. Some observations on soil horizons and insect activity in granite soils. *Proceedings of the 1st Federation Science Congress, Salisbury, Rhodesia*, pp. 271–6.

——1974. Termites in relation to soil formation, groundwater, and geochemical prospecting. *Soils Fert.* **37,** 111–14.

WEST, W.F. 1970. The Bulawayo symposium papers 2. Termite prospecting. *Chamber of Mines J. (Rhodesia)* **12,** 32–5.

WILSON, E.O. 1971. *The Insect Societies*. Harvard University Press, Cambridge, Mass.

WOOD, T.G. & SANDS, W.A. 1978. The role of termites in ecosystems. In: BRIAN, M.V. (ed) *Production Ecology of Ants and Termites*. Cambridge University Press, London, pp. 245–92.

R. CROSSLEY*, Dept of Geography & Earth Science, Chancellor College, University of Malawi, PO Box 280, Zomba, Malawi.

* Present address: Roberton Research International Ltd, Llandudno, Gwynnedd LL30 1SA, UK.

East African Rift stromatolites

J. Casanova

SUMMARY: The East African Rift System contains many lacustrine basins in which biologically precipitated stromatolite carbonates are found. The development of these carbonates results from low-Mg calcite precipitation by cyanophytes and bacteria; the binding and trapping of sediments are not involved. A general outline for the distribution of stromatolites has been established by analysing data from different sites in East Africa. In the fluviatile environment, travertines are associated with a biotype of flowing waters with high calcium content. During exceptional rainy seasons, inundation of floodplains promotes seasonal growth of oncolites. Chimneys and pool-rim dams associated with hydrothermal springs may be sub-lacustrine, or may occur along lake shores. In the lacustrine environment, stromatolites form shorelines that indicate former high stands. The study of Quaternary stromatolites and their modern equivalents reveals their exceptional potential for reconstructing sedimentary palaeoenvironments in rifts. Stromatolites represent a particular continental carbonate facies, somewhat analogous to marine reefs. Furthermore they constitute an important source of other geological data: material for isotopic dating; determining altitudes of former lake levels; and providing evidence for hydrological and physico-chemical conditions in palaeolakes. All these may be combined to assist palaeoclimatic reconstruction.

Recent stromatolites in continental environments have received considerable attention. Comparison of their occurrence with their marine equivalents has led to new palaeoecological interpretations of fossil stromatolites (Monty 1976, 1977; Semikhatov *et al.* 1979).

The zones of subsidence in the East African Rift include numerous lakes in which biogenic carbonates that are considered to be stromatolitic have formed. Although several authors have briefly described 'calcareous crusts' or 'algal carbonates' (Taieb 1974; Gasse 1975; Holdship 1976; Street 1979), there have been few systematic studies of the Quaternary stromatolites of this region. The majority of the stromatolite occurrences that have been described have been from lacustrine environments with emphasis placed on the stromatolites' relationship to former lake levels (Grove & Goudie 1971; Gasse & Rognon 1973; Gasse *et al.* 1974; Grove *et al.* 1975; Grove & Dekker 1976; Tiercelin 1981). Additionally, they have been used in stable isotope analysis (Fontes & Pouchan 1975; Hillaire-Marcel *et al.* 1982; Camara *et al.* 1984). The best known stromatolite occurrences are those from the Plio-Pleistocene deposits of northeastern Lake Turkana (Johnson 1974; Johnson & Raynolds 1976), where they have recently been used to interpret the palaeoclimate of the environment of fossil hominids (Abell *et al.* 1982).

The purpose of this paper is to review the different types of stromatolite and their occurrence within a continental rift, and to show the close relationship that exists between the morphology of the encrustation and the environment colonized by Cyanobacteria. This study is based on samples and observations from the eastern branch of the African Rift System (Fig. 1), notably at Lake Abhe (Republic of Djibouti), the Suguta valley, Lakes Bogoria and Magadi (Kenya) and Lakes Natron and Manyara (Tanzania).

Fluvial channel and spring environment

Travertines occurring in fluvial environments between springs and lakes, are similar in all respects to those described from Europe (Casanova 1981). At Lake Bogoria, there are several travertine terraces associated with former springs. At Lake Natron, travertines are currently forming a waterfall dam at a break of slope in the Engarescero River (Fig. 2A).

Cyanophytes and bacteria, occurring as tufted mats or as isolated colonies, are the organisms that have built these travertines by extracellular precipitation of low-Mg calcite (Fig. 2B). The gross morphology of the different types of travertine depends on the nature of the environment in which they form, whereas the morphological features of the stromatolite itself are controlled by the nature and shape of the substrate (Casanova 1982). Spring environments generally give rise to cylindrical or planar encrustations, whereas waterfall dams characteristically show massive encrustations that originate from bushes of bryophytes. Downstream, the carbonate that is precipitated along the river channel displays characteristics of calm (encrustations of reed beds and

From FROSTICK L.E. *et al.* (eds) 1986, *Sedimentation in the African Rifts*, Geological Society Special Publication No. 25, pp. 201–10.

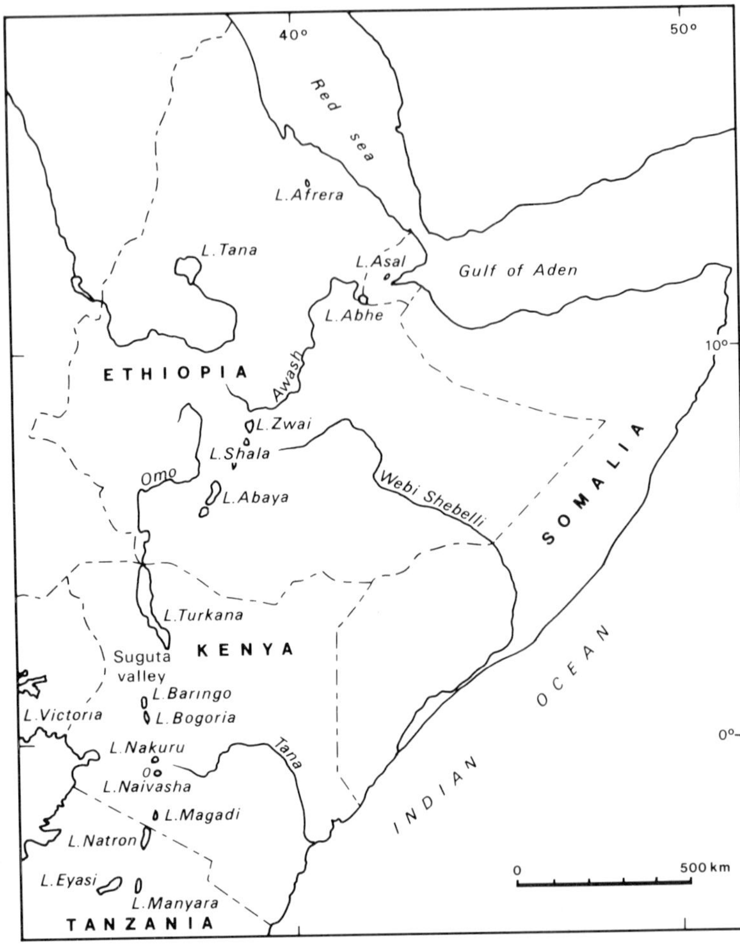

FIG. 1. Location of the main lakes in the eastern branch of the East African Rift System.

gastropods) or agitated (pebble encrustations, spheroidal oncolites) hydrodynamic conditions.

The algal microflora of these travertines comprises several species of *Phormidium*, *Homeothrix*, *Schizothrix* and *Rivularia*, and includes numerous bacteria associated with the algal mucus. The nature and the shape of the precipitated crystals are specific to the encrusting Cyanobacteria. Despite frequent recrystallization of these carbonates, *Phormidium incrustatum* has been identified as the dominant species of the association in the exceptionally well-preserved samples from Lake Bogoria (Vincens *et al.*, this volume). This is similar to the situation commonly encountered in fluvial environments in Europe. The similarity of African and European travertines testifies to the ecological meaning of these deposits. The travertines have a specific importance. Their formation corresponds to particular biotopes, fed by groundwater that has dissolved abundant calcium during its subsurface flow (e.g. Chafetz & Folk 1984).

Floodplains

Lakes occupy much of the rift valley floor, so that floodplain environments are relatively rare. Where they do occur, such as at Lake Magadi, this type of environment may be represented by a wide area that is covered characteristically by a thin sheet of water during exceptionally heavy rainy seasons. Inundation of the plains bordering Lake Magadi generally takes place during the longest rainy seasons (March–April) (East African Meteorological Dept 1975). The basin floor receives direct precipitation as well as surface runoff. This thin layer of water is only a few

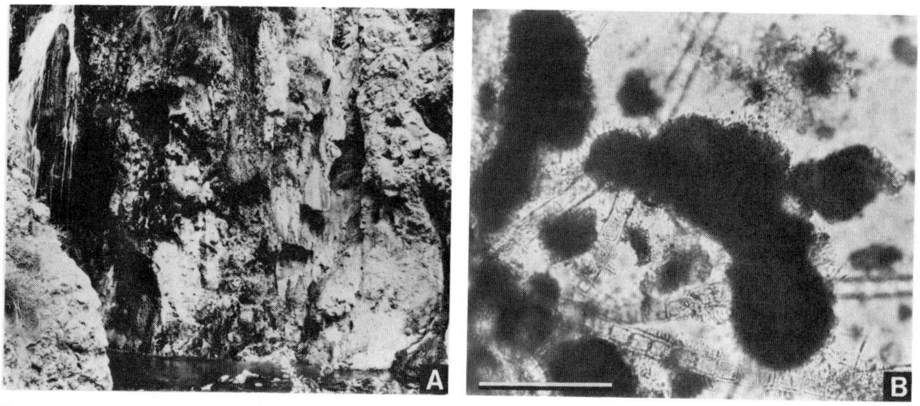

FIG. 2. Modern travertines from Engarescero River, Lake Natron, Tanzania. A. Travertine dam forming a 40 m high waterfall. B. Algal micrite generated by *Homeothrix janthina*, epiphytic of *Chantransia* sp. (scale bar = 100 μm).

centimetres deep and has no hydrodynamic potential. It is then subject to evaporation, but remains long enough for oncolites to form around a nucleus (Casanova & Tiercelin 1982). The nuclei could be detrital grains derived from local trachytes or biologically consolidated aggregates of carbonate mud.

Oncolites from Lake Magadi (Fig. 3A), previously described as pisolites by Eugster (1980), commonly take the form of discs > 1–3 cm in diameter, with one or two concave faces (Fig. 3B). No unusual morphologies were observed. Pill-shaped or irregular forms are rare.

In thin section, the oncolites show micritic or microsparitic laminae 100–500 μm thick (Fig. 3C). They all contain abundant cyanophyte filaments arranged in clusters normal to the substrate (Fig. 3D). Some finely laminated fan-shaped structures derived from bacteria may be intercalated among the filaments. The laminae are defined by dark micritic layers, 10–20 μm thick, that do not show clearly recognizable biological structures. The presence of filaments that cut across several laminae, the discontinuity of the micritic layers, and major recrystallization, all combine to obscure the details of the rhythmic lamination of the stromatolite.

These oncolites characteristically display concave upper and lower surfaces, and therefore show thickening towards their lateral margins. In some examples, the concavity of the lower face reflects erosion at the interface between the oncolite and sediment. However, where the lamination is well preserved, it appears to continue around the nucleus and shows lateral hypertrophy (Fig. 3B). This abnormal thickening of the laminae towards the margins of the oncolite may be a response of the micro-organisms to the environment. In a very thin layer of water subjected to intense evaporation, the zone of maximum wetness will occur along the margins of the oncolite.

Hydrothermal environments

Intense hydrothermal activity is favoured by the presence of numerous vertical faults, associated with thinning of the lithosphere. Hot springs and geysers are commonly located at the margins of lakes where faults intersect the shoreline. These thermal waters are alkaline and relatively enriched in dissolved silica and fluoride. Their pH may vary from 7.8 to 9.9, and the temperature at the spring orifice can vary from 40 to 100°C (the range for modern springs at Lakes Bogoria, Magadi and Natron). Within this environment, populations of cyanophytes and bacteria are commonly present, in the form of globular, flat or fibrous mats. Around Lake Bogoria the maximum observed temperatures are 72°C for blue-green algae and above 92°C for bacterial colonies. No carbonates have been found directly associated with these micro-organisms in modern settings, except for aragonitic deposits that have a physico-chemical origin.

However, there are many fossil carbonate deposits, in the form of pool-rim dams (Fig. 4A) and 'chimneys' that have been related to hot springs, such as those at Lake Abhe (Fontes *et al.* 1975), Lake Bogoria (Renaut 1982) and Lake Magadi (Casanova, in press). These hydrothermal constructions can reach several tens of metres in height and can be built both in a sub-lacustrine environment (Lake Abhe) and in shoreline settings (Lakes Bogoria and Magadi). These crusts consist of low-Mg calcite and show stromatolitic

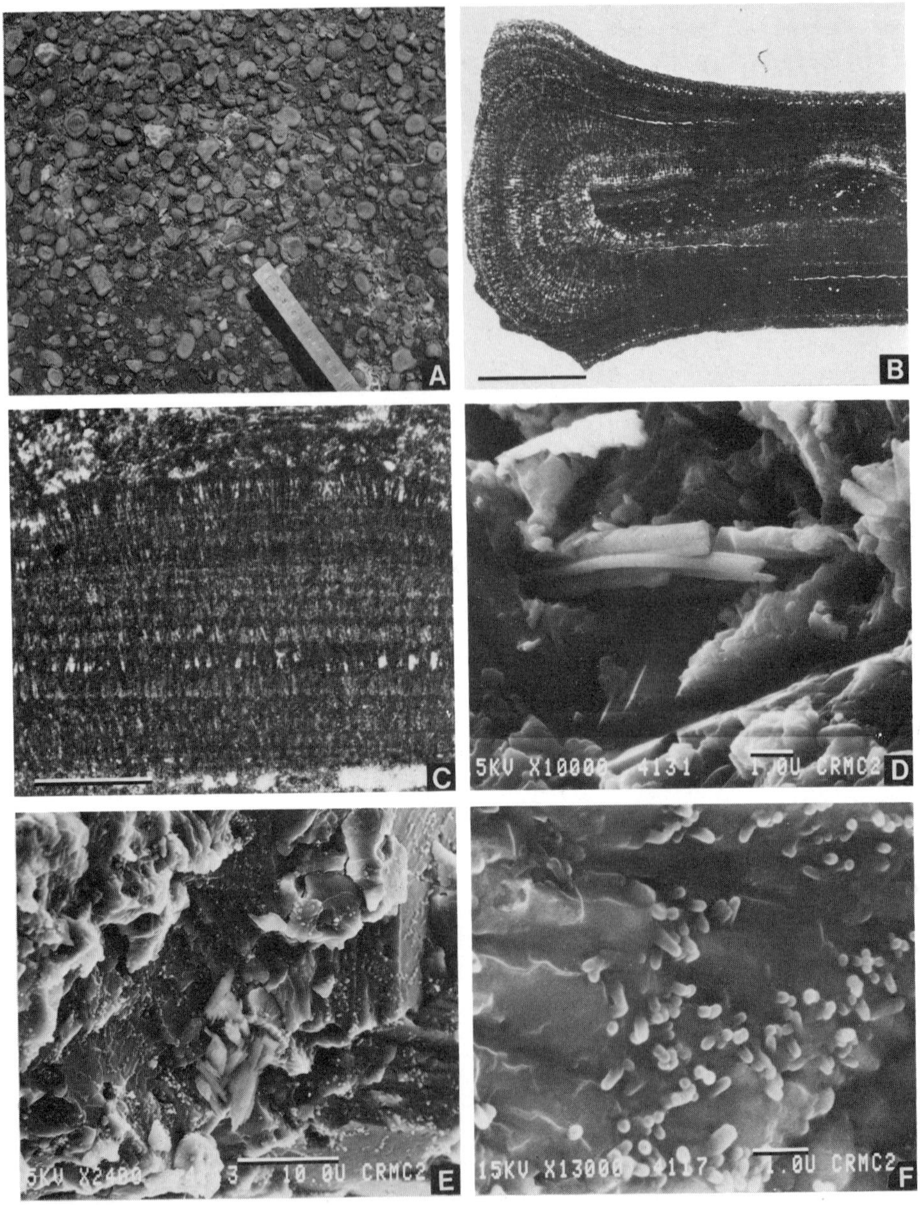

FIG. 3. Oncolite facies from Lake Magadi floodplains, Kenya. Floodplain oncolites (A) are characterized by a hypertrophic lateral growth (B, scale bar = 3 mm) related to maximum moisture. Lamination (C, scale bar = 1 mm) is composed of algal and bacterial laminae; plane-polarized light. Bundles of *Schizothrix*-like filaments (D, SEM, scale bar = 1 μm). Micro-laminated sparitic fabric (E, SEM, scale bar = 10 μm) generated by bacterial colonies (F, SEM, scale bar = 1 μm).

features (Fig. 4B). Laminae are coarse (200 μm to several mm) and reflect a stable hydrothermal environment during the biological activity. At Lake Bogoria, X-ray diffraction analyses have revealed the presence of aragonite in a few samples; it is possible that some original aragonite has been replaced by calcite. The laminar structure consists of interlocked (Fig. 4D) or bushy (Fig. 4C) needles with a sparitic texture. The dam-building organisms are either isolated or colonial bacteria (Fig. 4E) commonly arranged into a polygonal network (Fig. 4F).

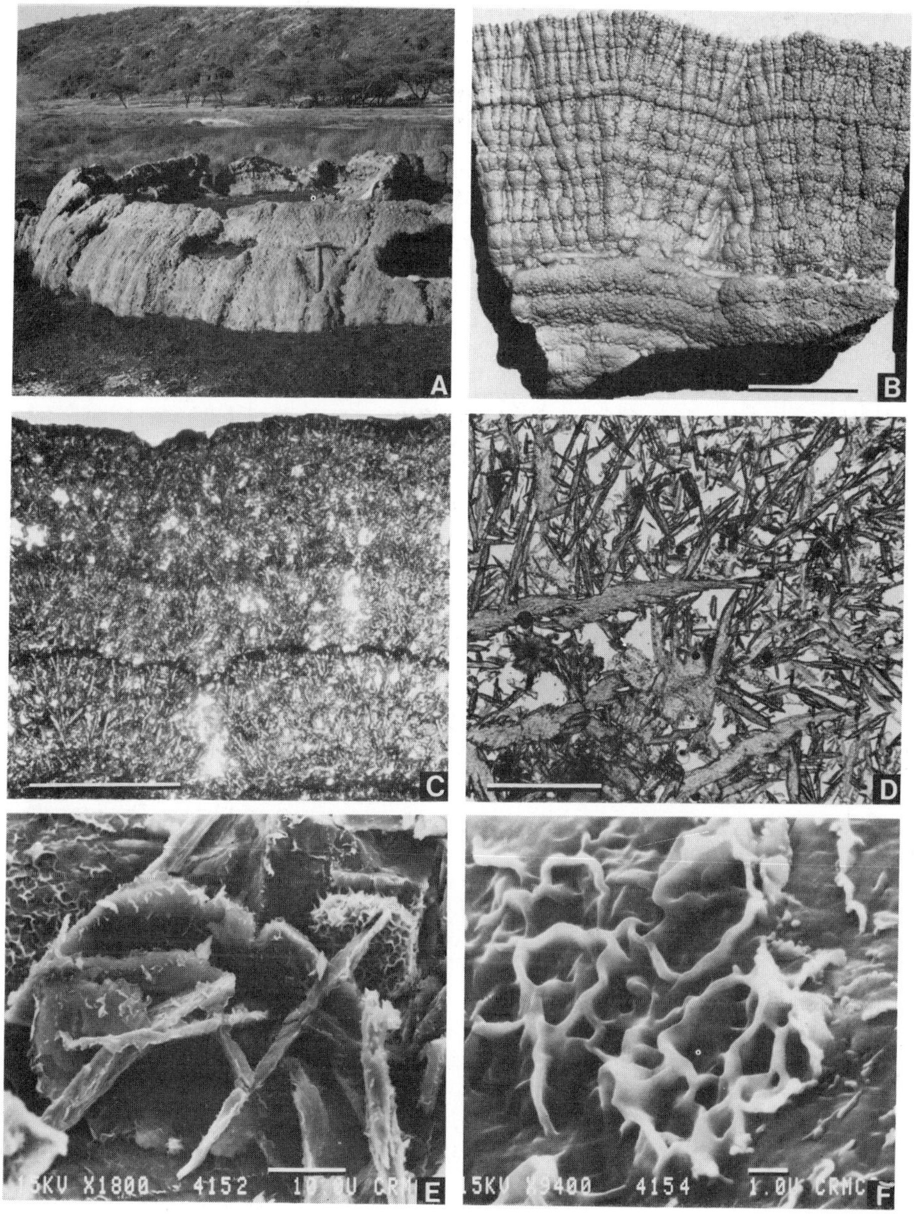

FIG. 4. Hydrothermal stromatolites from Lake Bogoria, Kenya. Pool-rim dam and orificial mound from fossil hot spring (A, hammer = 35 cm) are composed of roughly laminated stromatolitic encrustations (B, scale bar = 4 cm). Laminae have a fibrous fabric (C, scale bar = 2 mm), while acicular crystals in massive units show no clear pattern of orientation (D, scale bar = 400 μm); plane-polarized light. All crystals are of a bacterial origin, independent of their structure, most of them showing a polygonal organic framework (E, SEM, scale bar = 10 μm). Close up of bacterial colony (F, SEM, scale bar = 1 μm).

Lacustrine environments

There are no recorded examples of stromatolites currently forming in the sodium carbonate-bicarbonate lakes of the East African Rift System. Where studied, fossil stromatolites have been used to indicate former higher lake levels associated with climatic variations.

These encrustations consist of low-Mg calcite precipitated *in situ* by blue-green algal mats and/

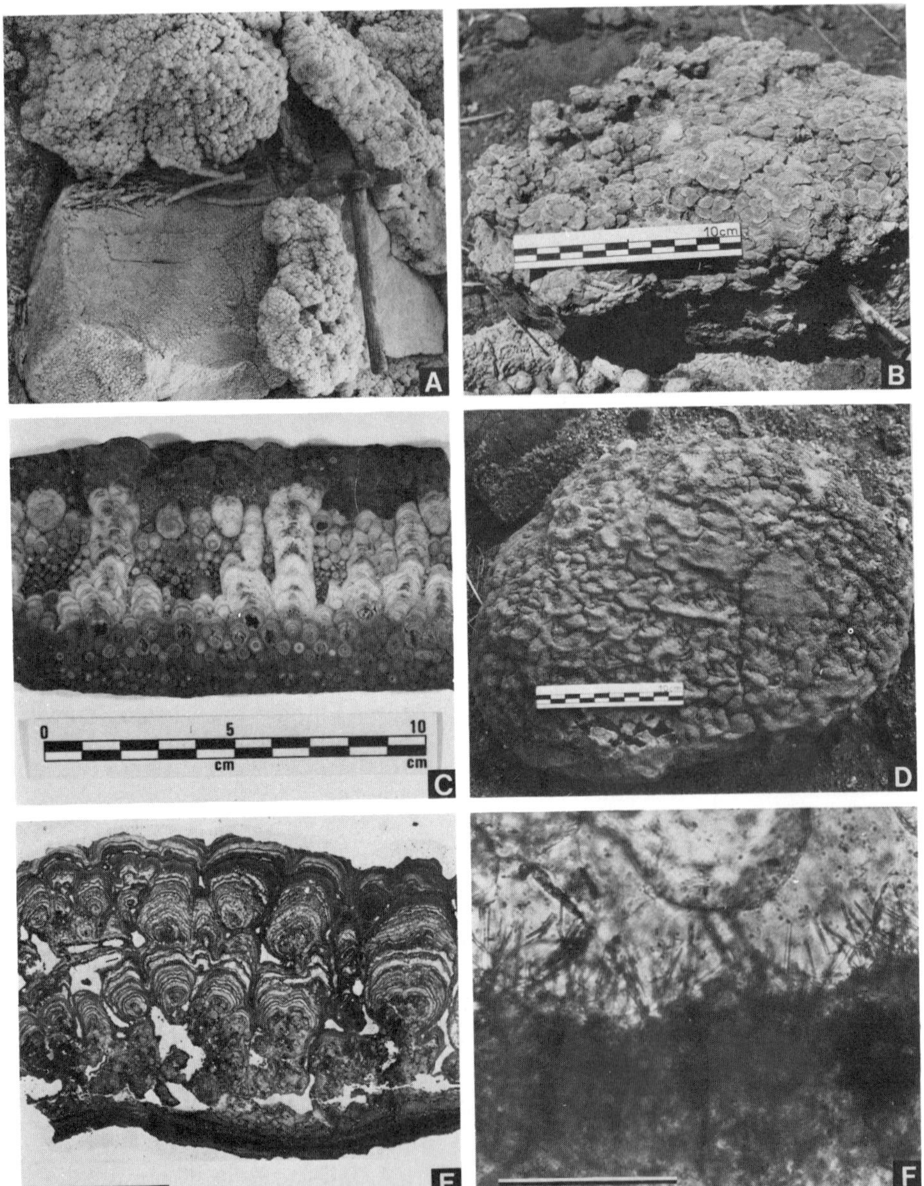

FIG. 5. Lacustrine stromatolites. A. Encrusted surface (blocks and twigs) from the Holocene shoreline of Lake Bogoria, Kenya (hammer = 35 cm). B. Isolated bioherm with rich ornamentation; bathymetric zone 0–2 m from the Middle Pleistocene shoreline of Lake Natron, Tanzania. C. Oncolite facies related to palaeochannels; bathymetric zone 7–10 m from the Middle Pleistocene shoreline of Lake Natron. D. Thin and smooth encrustation of block; bathymetric zone 7–10 m from the terminal Pleistocene–Holocene shoreline of Lake Natron. Most of the lacustrine stromatolites have a columnar structure (E, scale bar = 1 cm) generated by competitive growth of erect calcified filaments (F, scale bar = 100 μm); plane-polarized light.

or bacterial colonies. The shoreline stromatolites provide evidence for freshwater palaeolakes that contained significant dissolved calcium carbonate and whose waters remained stable throughout the period when the stromatolites were forming. If the duration of the lacustrine episode is relatively brief (of the order of 10–100 years), the stromatolites envelope the shape of the hard substrate uniformly (boulders, pebbles or vegetal remains) and thus form a surface crust that differs only slightly in morphology from the underlying substrate. This is the case for the Holocene high levels of Lake Bogoria (Tiercelin 1981; Renaut 1982; Vincens *et al.*, this volume) (Fig. 5A) and Lake Abhe (Gasse & Rognon 1973), and the late Pleistocene high level of the Natron–Magadi basin (Casanova, in press; Hillaire-Marcel *et al.*, in press). In contrast much longer lacustrine phases (of the order of 10^3–10^4 years) can result in the development of stromatolitic 'reefs' whose size and form varies as a function of water depth. This is the case for the Middle Pleistocene high lake levels of the Suguta valley, the Natron–Magadi basin (Hillaire-Marcel *et al.*, in press) and for Lake Manyara.

Regardless of the duration of the lacustrine episodes, stromatolites can provide detailed information on the nature of the former shorelines, since different ecological zones can be recognized from the lake surface down to a depth of 12 m. Therefore, even though the individual stromatolites for each lacustrine episode may differ, it is possible to recognize the different depth zones of a palaeolake from the morphological form of the carbonate.

The bathymetric zone between depths of 0 and −2 m favours biological competition and is characterized by stromatolites with strong ornamentation (Fig. 5B) that results from columnar growth of algal mats. This layer of water contains the highest dissolved oxygen and highest light intensity within the lake. The forms most commonly encountered are isolated bioherms with complex ornamentation, mammilated oncolites and similar encrustations of cobbles and vegetal remains.

The zone from −2 to −7 m is a zone of stability. Throughout the duration of the lacustrine episode, Cyanobacteria construct large flattened bioherms with low ornamentation, arranged in lines parallel to the shoreline. They may also form thickened encrustations which colonize the entire substrate. The morphology as well as the micro-facies that develop, are identical to those of the lake margins and are relatively little influenced by the local ecological conditions.

The zone from −7 to −11 m is characterized by relatively thin stromatolites lacking any characteristic ornamentation. This deeper zone is less favourable in terms of dissolved oxygen and light intensity. Locally, sub-lacustrine currents may favour oncolite development (Fig. 5C) in channels perpendicular to the shoreline. This facies could extend outwards to form large oncolitic, polyphased surfaces. These oncolites are subject to constant agitation and have a spheroidal form due to isotropic algal growth. They are a few mm to a few cm thick and lack ornamentation. Periods of immobilization favour vertical algal growth corresponding to incident sunlight. In such cases, the oncolites develop an asymmetric form that resembles a columnar structure (Fig. 5C). Without sufficient current activity, the stromatolites only consist of thin and smooth encrustations showing flexuous grooves (Fig. 5D).

Conclusions

The study of Plio-Pleistocene stromatolites from the East African Rift, combined with comparative studies of recent algal structures, demonstrates their exceptional potential for palaeoenvironmental reconstruction within the rift. In fact, stromatolites may occur in almost all of the continental environments, each of which has its own characteristic morphological type. The various studies from different sites in East Africa have been combined to develop a model for the occurrence of stromatolites within a continental rift setting (Fig. 6).

Fluvial environments are mainly colonized by travertines. The immense morphological diversity of these stromatolites, as well as their well-known ecology (Golubic 1969; Casanova 1981; Chafetz & Folk 1984), allows the recognition of numerous biotopes. It may be possible to correlate the travertine terraces with a major recharge of phreatic groundwater and augmentation of surface runoff (Vincens *et al.*, this volume). However the main significance of travertines is palaeoecological and the possibilities for palaeoclimatological interpretation are limited. Furthermore, the transportation of dissolved carbon by groundwater, and kinetic factors during precipitation of these carbonates, make radiocarbon dating unreliable and stable isotope analysis (^{13}C and ^{18}O) difficult in travertine-forming environments.

The oncolites that develop on floodplains have a distinctive morphology, and so cannot be confused with those that originate in fluvial (channel) and lacustrine environments. The modern example of Lake Magadi shows that these oncolites are not associated with an extensive phase of lacustrine expansion. On the contrary, they could be used to indicate a phase of relative aridity. The prevailing hydrological conditions during the development of each lamina are sufficiently restrictive to make an analysis of seasonal variability possible, and also to allow

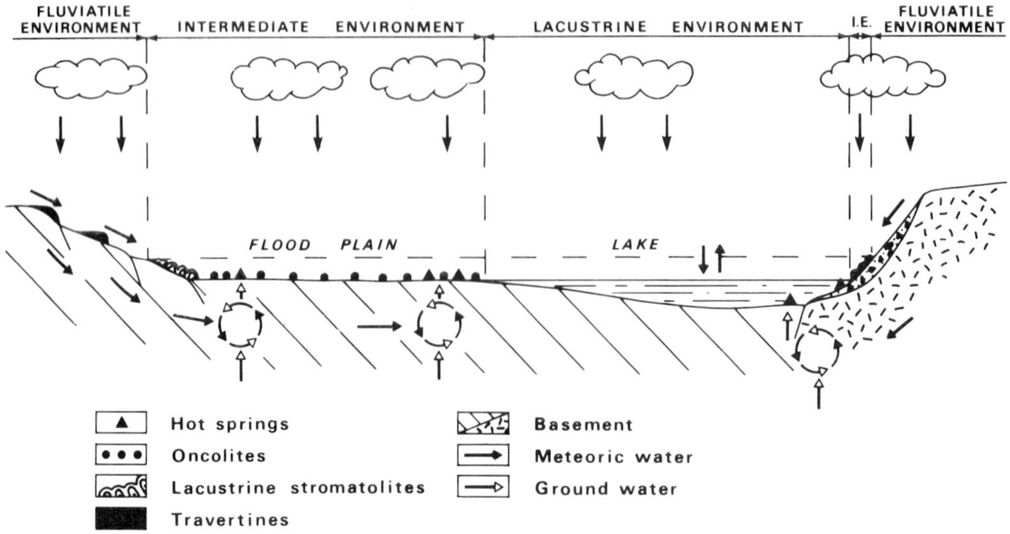

FIG. 6. Schematic distribution of stromatolitic environments in the East African Rift System.

recognition of the periodicity of exceptionally heavy rainfall events in ancient environments.

The formation of stromatolites associated with hydrothermal activity provides the best sedimentological and geochemical evidence for recognizing hydrothermal regimes in ancient environments. Fontes & Pouchan (1975), for example, have proposed a limnological and climatic interpretation for the development of sub-lacustrine chimneys in Lake Abhé, using isotopic analysis.

However, the most significant use of stromatolites in the rift is associated with the lacustrine environment, for two main reasons. Firstly, these bioherms represent a continental carbonate facies that in many respects resembles reef development, and therefore they can provide considerable geological and limnological information. Secondly, by indicating former shorelines, they can provide evidence for the age, altitude and several hydrological and physico-chemical characteristics of palaeolakes (Casanova *et al.*, in press). Together, these pieces of evidence can be used indirectly to help reconstruct palaeoclimates (Abell *et al.* 1982; Hillaire-Marcel *et al.*, in press).

When defining shorelines, however, great care must be exercised in interpreting the bathymetric significance of stromatolites. Although many examples of marine stromatolites from shallow, intertidal zones are well known to geologists, it is incorrect to extend the inference of water depth to all stromatolite environments. The existence of ecological zones extending locally to depths of 15 m in Lake Natron or −30 m in former Lake Suguta (Casanova, unpub. data), and similarly the discovery of stromatolites growing at a depth of −60 m in Lake Wondergat in South Africa (Gow 1981), clearly demonstrate the danger of assigning all lacustrine stromatolites to the eulittoral zone.

Laminae are the basic units which determine the microstructure of stromatolites. They are directly related to the composition of the microbial population and define the limits of the ecological framework in which they develop. We have seen that these microstructures differ considerably according to the site in which they form, but also that the general morphological characteristics are similar within the major environmental groups. In the fluvial and floodplain environments, growth always occurs by successive planar laminae, and columnar forms are absent. In the hydrothermal environment, bacteria are the principal stromatolite builders and the microstructure is characterized by bushy growth of bacterial colonies. In the lacustrine environment, microstructures generated by columnar growth are most widespread (Fig. 5E and F) while planar or digitate fabrics are of lesser importance.

All the analysed samples from the rift provide evidence for a biological origin of the stromatolitic carbonate, regardless of the environment in which it formed. This is contrary to the hypothesis of detrital particle trapping suggested by Johnson (1974) at Lake Turkana. This is significant for stable isotope analyses of these stromatolites because the results would make little sense if the analysed carbonate was not contemporary with the structure. Furthermore, bearing in mind the limitations given for the fluvial environments, isotopic analysis can be applied to all rift environments where stromatolites have formed. Thus they may be considered a valuable source of palaeoecological and palaeohydrological information.

This study has demonstrated how important and useful stromatolites can be in the East African Rift lacustrine environment. Most occurrences have still received little attention to date. It sets the stage for more detailed studies that will increase our knowledge of the prevailing limnological conditions in palaeolakes, and help refine the details of Quaternary climatic fluctuations in East Africa.

ACKNOWLEDGEMENTS: This research was made possible thanks to the assistance of the Compagnie Française des Pétroles and the CNRS (Laboratoire de Géologie du Quaternaire). The author would also like to thank Drs K. Kelts and R. Renaut for translating the text into English, management of CFP Total for authorizing this publication, and the Kenya and Tanzania Governments for granting research permission.

References

ABELL, P.I., AWRAMIK, S.M., OSBORNE, R.H. & TOMELLINI, S. 1982. Plio-Pleistocene lacustrine stromatolites from Lake Turkana, Kenya: morphology, stratigraphy and stable isotopes. *Sed. Geol.* **32**, 1–26.

CAMARA, E., FONTES, J.C., GASSE, F., SALIEGE, J.F. & STEINBERG, M. 1984. Le lac Abhé (Ethiopie, Djibouti) à l'Holocène supérieur. In: SOC. GÉOL. FR. PARIS (eds) *10th R.A.S.T., Bordeaux 1984*. I.G.B.A. Talence, p.111.

CASANOVA, J. 1981. *Etude d'un milieu stromatolitique continental: les travertins Plio-Pléistocènes du Var (France)*. Unpublished PhD thesis, University of Aix-Marseille II, Marseille, 192pp.

—— 1982. Morphologie et biolithogénèse des barrages de travertins. In: NICOD, J. (ed) *Formations Carbonatées Externes, Tufs et Travertins*. Mem. Ass. Fr. Karst. No. 3, pp.45–54.

—— (in press). Les stromatolites et hauts niveaux lacustres Pléistocènes du Bassin Natron-Magadi (Tanzanie-Kenya). *Sci. Géol.*

—— & HILLAIRE-MARCEL, C. (in press). Chronologie et paléohydrologie des hauts niveaux quaternaires du Bassin Natron-Magadi (Tanzanie-Kenya) d'après la composition isotopique (^{18}O, ^{13}C, ^{14}C, U/Th) des stromatolites littoraux. *Sci. Géol.*

—— & TIERCELIN, J.J. 1982. Construction stromatolitiques en milieu carbonaté sodique: les oncolites des plaines inondables du lac Magadi (Kenya). *CR Acad. Sci. Paris* **295**, 1139–44.

CHAFETZ, H.S. & FOLK, R.L. 1984. Travertines: depositional morphology and the bacterially constructed constituents. *J. Sed. Petrol.* **54**, 289–316.

EAST AFRICAN METEOROLOGICAL DEPARTMENT 1975. *Climatology Statistics for East Africa. 1, Kenya.* East Africa Community, Nairobi.

EUGSTER, H.P. 1980. Lake Magadi, Kenya, and its precursors. In: NISSENBAUM, A. (ed) *Hypersaline Brines and Evaporitic Environments*. Developments in Sedimentology No. 28, Elsevier, Amsterdam, pp.195–232.

FONTES, J.C. & POUCHAN, D. 1975. Les cheminées du lac Abhé (TFAI): stations hydroclimatiques de l'Holocène. *C.R. Acad. Sci. Paris* **280**, 383–6.

——, ——, SALIEGE, J.F. & ZUPPI, G.M. 1980. Environmental isotope study of ground water systems in the Republic of Djibouti. In: *Arid-zone hydrology: investigations with isotope techniques*. IAEA-AG. No. 158, pp.237–62.

GASSE, F. 1975. *L'évolution des lacs de l'Afar Central (Ethiopie et T.F.A.I.) du Plio-Pléistocène à l'Actuel.* Unpublished PhD thesis, University of Paris VI, Paris, 390pp.

——, FONTES, J.C. & ROGNON, P. 1974. Variations hydrologiques et extension des lacs holocènes du désert Danakil. *Paleogeogr. Paleoclimatol. Paleoecol.* **15**, 109–48.

—— & ROGNON, P. 1973. Le Quaternaire des bassins lacustres de l'Afar. *Rev. Géogr. Phys. Géol. Dynam.* **4**, 405–14.

GOLUBIC, S. 1969. Cyclic and non-cyclic mechanisms in the formation of travertine. *Verh. Intern. Verein. Limnol.* **17**, 956–61.

GOW, C.E. 1981. Unusual occurrence of biogenic (cyanobacterial) carbonate sediments from a sinkhole lake in the Western Transvaal, South Africa. *South Afr. J. Sci.* **77**, 564–5.

GROVE, A.T. & DEKKER, G. 1976. Late Quaternary lake levels in the rift valley of southern Ethiopia. In: *VIIth Panafr. Congr. Preh. Quatern. Stud. Addis Ababa, 1971.* Ministry of Culture Press.

—— & GOUDIE, A.S. 1971. Secrets of Lake Stephanie's past. *Geogr. Mag.* **43**, 542–7.

——, STREET, F.A. & GOUDIE, A.S. 1975. Former lake levels and climatic change in the rift valley of Southern Ethiopia. *Geogr. Mag.* **141**, 177–202.

HILLAIRE-MARCEL, C., CARRO, O. & CASANOVA, J. (in press). ^{14}C and Th/U dating of Pleistocene and Holocene stromatolites from East African paleolakes. *Quaternary Research.*

——, TAIEB, M., TIERCELIN, J.J. & PAGE, N. 1982. A 1.2 Myr record of isotopic changes in a late Pliocene rift lake, Ethiopia. *Nature* **296**, 640–2.

HOLDSHIP, S.A. 1976. *The paleolimnology of Lake Manyara, Tanzania: a diatom analysis of a 56 meter sediment core.* Unpublished PhD thesis, Duke University, Durham, NC, 120pp.

JOHNSON, G.D. 1974. Cainozoic lacustrine stromatolites from hominid-bearing sediments east of Lake Rudolf, Kenya. *Nature* **247**, 520–3.

—— & RAYNOLDS, R.G.H. 1976. Late Cenozoic environments of the Koobi Fora Formation: the Upper Member along the western Koobi Fora

Ridge. In: Coppens, Y., Howell, F.C., Isaac, G.L. & Leakey, R.E.F. (eds) *Earliest Man and Environments in the Lake Rudolf Basin*. Chicago University Press, Chicago, pp.115–23.

Monty, C.L.V. 1976. The origin and development of cryptalgal facies. In: Walter, M.R. (ed) *Stromatolites*. Developments in Sedimentology No. 20, Elsevier, Amsterdam, pp.193–250.

—— 1977. Evolving concepts on the nature and the ecological significance of stromatolites: a review. In: Flügel, E. (ed) *Fossil Algae. Recent Results and Developments*. Springer-Verlag, Berlin, pp.15–36.

Renaut, R.W. 1982. *Late Quaternary Geology of the Lake Bogoria Fault-Trough, Kenya Rift Valley*. PhD Thesis, University of London, 498pp.

Semikhatov, M.A., Gebelin, C.D., Cloud, P., Awramik, S.M. & Benmore, W.C. 1979. Stromatolite morphogenesis—progress and problems. *Can. J. Earth Sci.* **16,** 992–1015.

Street, F.A. 1979. *Late Quaternary lakes in the Ziway-Shala basin, Southern Ethiopia*. Unpublished PhD thesis, Cambridge University, 457pp.

Taieb, M. 1974. *Evolution quaternaire du bassin de l'Awash (Rift éthiopien et Afar)*. Unpublished PhD thesis, University of Paris VI, Paris, 390pp.

Tiercelin, J.J. 1981. *Rifts continentaux, tectonique, climats, sédiments. Exemples: la sédimentation dans le Nord du Rift Gregory (Kenya) et dans le Rift de l'Afar (Ethiopie) depuis le Miocène*. Unpublished PhD thesis, University of Aix-Marseille II, Marseille, 260pp.

J. Casanova, Laboratoire de Géologie du Quaternaire, CNRS—Case 907 Luminy, 13288 Marseille, Cedex 9, France.

Mire sediments in East Africa

A. Hamilton & D. Taylor

SUMMARY: Mires (peat-forming systems) are very localized within East Africa, being concentrated in areas of higher rainfall and lower temperature. The main centre is the high ground of the wetter part of the Western Rift, in southwestern Uganda, Rwanda, Burundi and probably Zaire. Peat-forming species include trees, such as *Erica* and *Syzygium*, sedges and *Sphagnum*. Mire stratigraphy is very variable between sites, reflecting a diversity of environmental contexts. Many changes in the characteristics of mire sediments in particular profiles appear to be related to climatic events. Some recent changes in mire vegetation are the result of human influence.

General distribution of mires

The climate of East Africa is generally unsuitable for the accumulation of dead organic matter, and organic sediments of Holocene or Pleistocene age are of very local occurrence. Here attention is focused on the sediments found beneath modern mires (defined as peat-accumulating ecosystems); such sediments can include various other types of sediments in addition to peat, for example, lacustrine deposits.

Mires are favoured by a cold, moist climate and are more common along the Western than the Eastern Rift. The type of mire most dependent on climate and most independent of topography is blanket bog, well known in cool, oceanic NW Europe. Well-developed mires of this type seem to be restricted in East Africa to the wettest of the high mountains, Ruwenzori (Thompson & Hamilton 1983). The dominant peat-forming species is *Carex runssoroensis*.

Normally, mires in East Africa only occur in sedimentary basins, where the impediment of waterflow can create conditions favourable to peat formation. Suitable sedimentary basins have been formed by a variety of processes, including earth movements, volcanism and glaciation.

Earth movements, sometimes aided by volcanic activity, have created sedimentary basins within and along the margins of many parts of the African rift, and some of these contain wetland plant communities. However, not all wetland plant communities form peat. The major mire area in Africa is the high ground (reaching to *c.* 2400 m) of the rift edge in the wetter parts of the Western Rift, especially in southwestern Uganda, Rwanda, Burundi and probably Zaire (Thompson & Hamilton 1983). The lower altitude of extensive peat occurrence is given by Deuse (1966) for Rwanda as 1600–1700 m, which may be taken as applicable to the area as a whole. The actual floor of the rift valley, even in this relatively high rainfall area, is usually too dry for peat formation, though there are some mires at an altitude of *c.* 2000 m near the Virunga volcanoes. The Eastern Rift is climatically drier, and peat deposits seem to be absent both from its floor and from sedimentary basins along its margins at altitudes similar to those containing peat along the Western Rift. Thus rift-margin valleys in the Kikuyu and Eldoret areas of Kenya appear to lack peat. The Aberdares and Cherangani Hills, reaching to higher altitudes, are exceptional; there is extensive peat formation in the plateau area of the Aberdares at an altitude of *c.* 3000 m (Hamilton 1982; Perrott & Street-Perrott 1982) and peat deposits are found at a height of 2900 m on the Cherangani Hills (Coetzee 1967). Further S, peat is present in dambos on the Nyika Plateau (2100–2500 m) of Malawi (Meadows 1984).

Most basins containing mires along the Western Rift owe their origin solely to earth movements but a few are known from southwestern Uganda which have been formed, at least in part, as a result of damming by lava or pyroclastics (Morrison 1968; Morrison & Hamilton 1974). At least one volcanic crater, on the summit of Mt Mgahinga, Virunga volcanoes, is occupied by a mire.

Past glacial activity had created ice-scooped or moraine-dammed basins on several of the high mountains of East Africa. Mires are present on the wetter mountains, but reportedly absent from dry Kilimanjaro (Thompson & Hamilton 1983). An investigation of sediment stratigraphy beneath Lake Bujuku, at a height of 3920 m on Ruwenzori, the origin of which had previously been ascribed to glacial processes, showed peat present below lacustrine deposits and it has been suggested that a landslide may have been responsible for raising the water table (Livingstone 1967).

Mires studied

The dating of mire sediments in East Africa is

From Frostick, L.E. *et al.* (eds) 1986, *Sedimentation in the African Rifts*, Geological Society Special Publication No. 25, pp. 211–17.

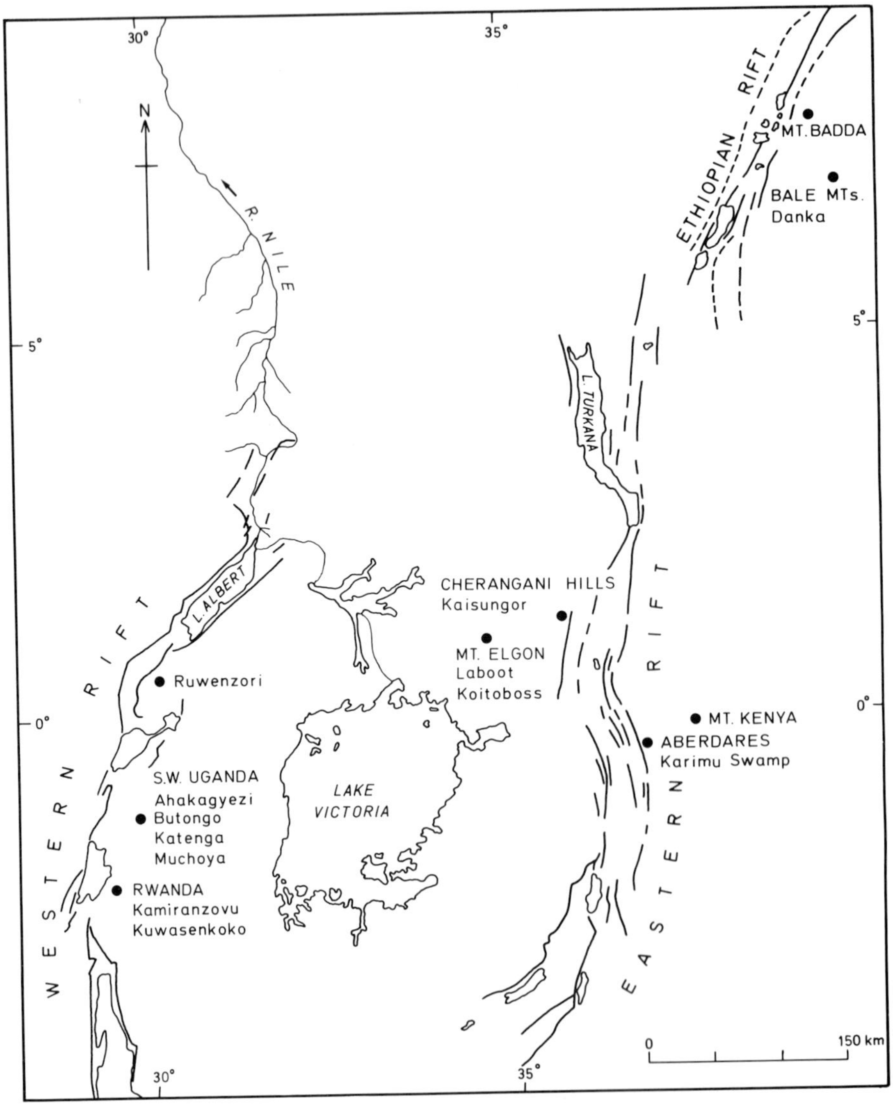

FIG. 1. East Africa, showing the distribution of mires whose stratigraphies are shown on Fig. 2.

generally very poor. Attention is restricted here to sediment sequences which have been dated, however incompletely, by radiocarbon or pollen analysis. The geographical distribution of sites is shown in Fig. 1, and stratigraphies are sketched in Fig. 2. It may be noted that in some cases the dating of deposits has been revised from original publications on the basis of advances in knowledge.

Both sediment stratigraphy and the rate of sediment accumulation vary greatly from site to site. Clearly much depends on the local environment which can be very variable both in space and time. Sediment stratigraphy is much more diverse than is, for example, the case with mire deposits in Ireland with its more uniform modern climate and climatic history. It has been found that predictions of stratigraphy made before examining sediments by coring have often proved inaccurate, as also have guesses of the age of sediments made during fieldwork. All this underlines our inadequate knowledge of the controls over sedimentation in these mire basins. The interpretation of sedimentary changes at any particular site in terms of environmental events remains to some extent a matter of speculation.

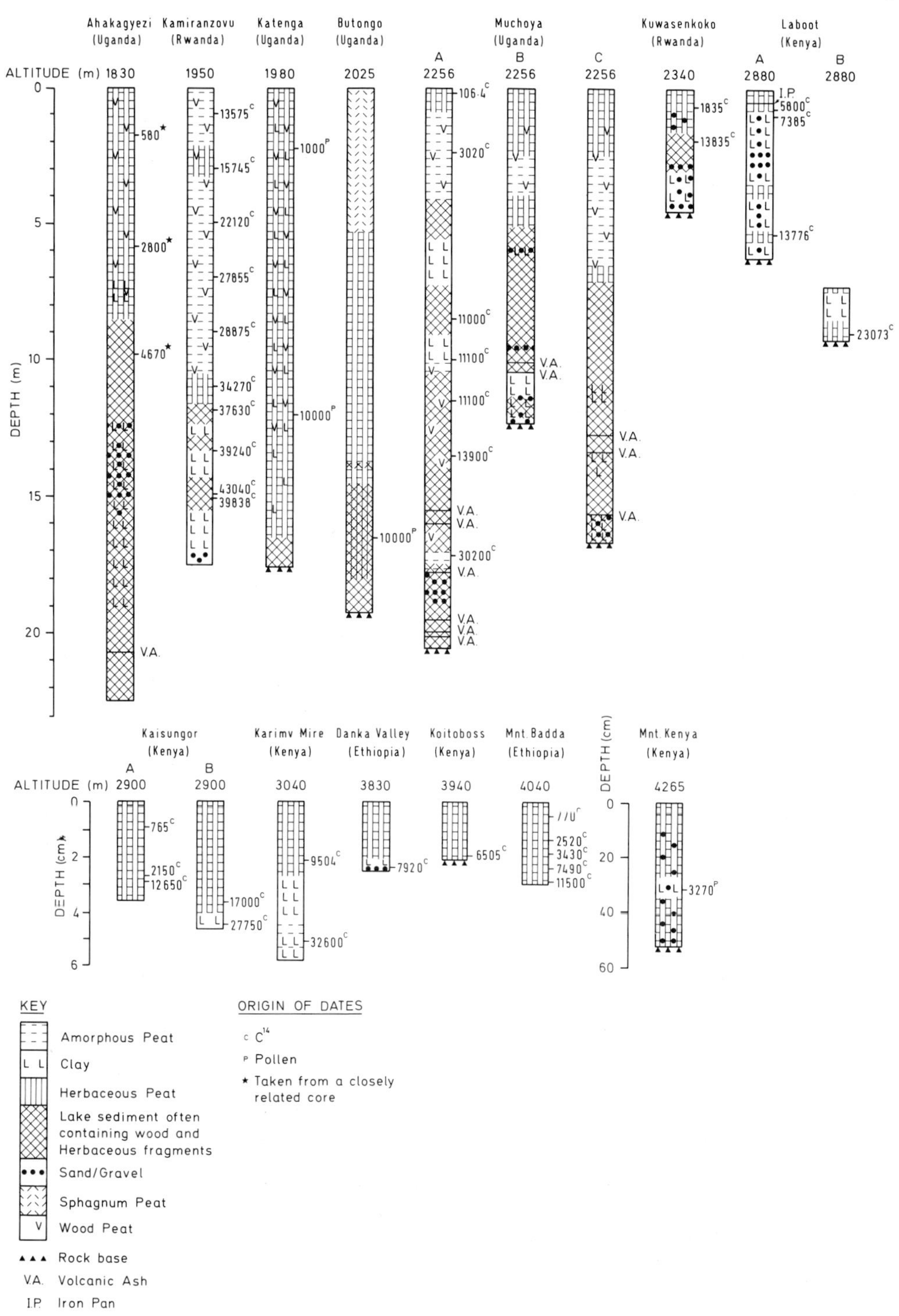

FIG. 2. Mire stratigraphies in East Africa. Dates in years BP. All the Ugandan sites are in southwestern Uganda (Kabale District), Laboot and Koitoboss are on Mt Elgon, Kaisungor is on the Aberdares and Danka valley is on the Bale Mountains. References: Mt Badda, Danka, Kamiranzovu, Koitoboss, Kuwasenkoko and Laboot (Hamilton 1982); Butongo and Katenga (Morrison & Hamilton 1974); Karimu (Perrott & Street-Perrott 1982); Ahakagyezi and Muchoya (Taylor & Hamilton, unpub. data).

Sediment types of mires

A wide range of organic and inorganic sediments is found beneath mires in East Africa. Recognition of the sedimentary environment is not always easy from field characters alone, and can be aided by laboratory analyses such as determination of organic matter content, examination of macrofossils and microfossils (diatoms, pollen), and chemical and magnetic studies.

The principal types of mire vegetation known or believed to form peat are listed below. There are problems in compiling this list. For instance, the dominance of a species on a mire today does not necessarily imply that the species itself has formed the peat in which it is rooted. Another point is that the identification of peat-formers by pollen is complicated by difficulties in knowing whether pollen in peat is local or more long distance in origin and, in any case, pollen is usually identifiable only to the genus or family rather than the species. Also, very little macrofossil work has been undertaken on East African peats. From all of this it follows that the list below may require substantial modification as knowledge grows. In addition, more work needs to be undertaken on the relationships between vegetation type and environment to augment the meagre existing information (Lind 1956; Deuse 1966; Thompson & Hamilton 1983).

1 *Syzygium swamp forest*

Syzygium cordatum, and the probably conspecific *S. staudtii*, are small trees (*c.* 8 m tall) recorded as peat-formers at Ahakagyezi swamp in southwestern Uganda (1830 m) and Kamiranzovu swamp in Rwanda (1950 m). The sediments are rich in wood and Myrtaceae pollen. *Syzygium* swamp forest may be seen today at Ahakagyezi and Kamiranzovu (Deuse 1966; Bouxin 1974). It is a rather lush, species-rich type of swamp vegetation and this, as well as sedimentary evidence (Hamilton 1982), suggests that it requires a relatively mesotrophic environment.

2 *Myrica swamp forest*

Myrica kandtiana has a wide distribution in wet places in East Africa (Lind & Morrison 1974), with its highest known altitudinal occurrence being Muchoya swamp (2256 m) in southwestern Uganda. It is abundant as an understorey tree beneath *Syzygium* at Ahakagyezi swamp. There is abundant *Myrica* pollen associated with *Erica* pollen at Muchoya (Morrison 1968), suggesting that the species was at one time a dominant or co-dominant.

3 *Erica swamp forest*

Peat with wood and frequent Ericaceae pollen is known from the Kamiranzovu and Muchoya swamps. The species was probably *Erica kingaensis*, which is known today up to an altitude of 3200 m on Ruwenzori. *Erica* peat suggests an oligotrophic environment.

4 *Carex bog*

Tussock species of the sedge *Carex*, principally *C. runssoroensis* but more locally *C. monostachya*, are believed to be the main peat-formers at high altitude. The altitudinal range of modern *C. runssoroensis* mires is given as 3200–4500 m by Thompson & Hamilton (1983); there are extensive *C. monostachya* mires at *c.* 3000 m on the Aberdares. The peat contains abundant fibrous herbaceous fossils and much Cyperaceae pollen. An oligotrophic environment is indicated.

5 *Pycreus nigricans bog*

Pycreus nigricans is another tussock-forming sedge which may be regarded as being to some extent a lower altitude replacement of *Carex*. Peat rich in Cyperaceae pollen and often conspicuously fibrous, is known from Laboot bog (2880 m) on Mt Elgon and Muchoya swamp, both of which are dominated by *Pycreus* today; *Pycreus* is assumed to have been the peat-forming plant.

6 *Gramineae and Cyperaceae other than Carex and Pycreus*

Wetland grasses are generally dominant in more mesotrophic environments than sedges and are not known definitely to be peat-formers in East Africa. Amongst the grasses, *Miscanthus violaceus* is ecologically exceptional in that it can be abundant in wetter acidic swamps. It is reported sometimes to be a peat former (Thompson & Hamilton 1983), for example, possibly at Butongo swamp (2025 m) in southwestern Uganda (Morrison & Hamilton 1974). The large tussock-forming grass *Pennisetum mildbraedii* is abundant on the margins of Muchoya swamp (Morrison 1968), where it has probably contributed to peat accumulation.

As for sedges other than *Carex* and *Pycreus*, Cyperaceae pollen is abundant at some horizons in Kamiranzovu sediments, where it could be derived largely from tall-growing, non-tussock species such as *Cyperus latifolius*. *C. latifolius* is common in wet marginal parts of Kamiranzovu swamp today; it is also said to be a species which is favoured by anthropogenic activities (Deuse 1966; Morrison 1968). Papyrus (*Cyperus papyrus*) is common in swamps in central Africa up to altitudes of *c.* 1900 m, with its highest altitudinal occurrences recorded as 2050 m in Rwanda (Deuse 1966) and 2073 m in Uganda (Morrison 1968). Lower altitude papyrus swamps, such as those which are so common around Lake Victoria, do not accumulate peat (Morrison 1968), but at higher altitudes papyrus is sometimes found on peat and could be a peat-former. On the Grand Marais of the Rwanda–Burundi border it is associated with *Cladium* (Thompson & Hamilton 1983). According to Deuse (1966), *Cladium*, *Phragmites* and *Typha* are present on sites of relatively high pH (see also Lind 1956), *Cyperus* is associated with moderate pH, and *Miscanthus*, *Sphagnum* and *Xyris* are characteristic of the most acidic sites.

7 *Typha swamp*

Typha peat, recognized by the presence of coarse herbaceous fossils and abundant *Typha* pollen, is known from Katenga swamp (1980 m) in southwestern Uganda. *Typha* is favoured by a high trophic status (Thompson & Hamilton 1983).

8 *Sphagnum bog*

Sphagnum is known today as a minor element of *Carex*, *Erica* and *Pycreus* mires, but only very locally as a more extensive dominant. *Sphagnum* lawns are found at high altitudes on Ruwenzori and on Butongo swamp in southwestern Uganda. *Sphagnum* is not restricted to high altitudes, being present in nutrient-deficient wetlands at 1350 m around Lake Victoria (Lind & Morrison 1974), but peat formation by *Sphagnum* at this altitude has not yet been demonstrated. One site where *Sphagnum* is known to have formed peat is Butongo swamp (Morrison & Hamilton 1974). Although not noted in the present vegetation from which it has probably been excluded by eutrophication, there are 10.5 m of peat with *Sphagnum* spores and *Erica* and *Xyris* pollen starting just below the surface at Kamiranzovu swamp. The association of the *Sphagnum* spores with the acid-loving *Erica* and *Xyris* strongly suggests that *Sphagnum* contributed to peat formation. The presence of *Sphagnum* in East Africa indicates a highly oligotrophic environment.

Some mires contain underlying lake sediments. At Muchoya swamp there is a thick layer of organic-rich sediment containing variable quantities of herbaceous macrofossils below the upper layers of *Pycreus* and *Erica* peats. This was interpreted during fieldwork as peat, but the rather uniform size of the macrofossil fragments and the topographic context suggest, on reflection, that it is more likely to be detrital or lacustrine. Other types of deposit are more obviously lacustrine. For example, further down in the Muchoya sequence there is a diatom-rich and relatively inorganic stratum containing pollen of the aquatic plant *Potamogeton*: this was clearly deposited beneath a lake.

At Muchoya and on the Aberdares very hard, almost lignitic, organic deposits (which include both peat and organic lake sediments) have been found under thick layers of clay. The sediments fracture conchoidally and require a drill for sampling, being too firm for a Hiller borer. It is noticeable at Muchoya that sediments of the same age and depositional type, but not under clay, are not nearly so hard. A further indication that age in itself is not a factor causing hardening is the fact that sediments dated at 37,000 BP from Kamiranzovu, where there is no clay capping, are soft and fresh in appearance. It is assumed that the weight of the clay has played a major role in the hardening process, but how this occurs is unclear. The upper part of the 10 m column of hardened sediment at Muchoya does not have a reduced water content compared with softer sediments of similar depositional type, so hardening is not necessarily associated with dehydration.

Layers of clay or, more rarely, coarser-grained inorganic material are recorded from several mire sequences. Clays can either show an absence of obvious internal structure, as at Muchoya, where radiocarbon determinations indicate that deposition may have been very rapid, or be laminated. At Karimu and Kimilili the thick clay horizons are banded, displaying a wide range of colours. These colours are generally lost rapidly on exposure to the atmosphere.

At Muchoya and Ahakagyezi thin volcanic ash bands are present within organic sediments and are useful for correlation. The sources of the eruptions could probably be traced chemically.

The climate of East Africa during the late Pleistocene and Holocene

The outline of the climatic history of East Africa during the last 30,000 years is reasonably well established from a variety of evidence, and a brief review is useful before considering the evolution of sites now occupied by mires. This summary is extracted largely from Hamilton (1982).

The period from *c.* 30,000 to 21,000 BP was a time of relatively cool, moist climate, with high water levels known in some lake basins and a depression of montane vegetation zones. Minimum temperatures (reduced by *c.* 6°C compared with the present) were experienced between *c.* 21,000 and 14,500 BP, a time also notable for widespread aridification and reduction in the area of mesic climate. Glaciers were probably near maximum extension.

The period 14,500–10,000 BP was a time of climatic transition, with rising temperatures and an increasingly moist climate. Details remain unclear and there were probably large climatic fluctuations, with, for example, a brief return to arid conditions just before 10,000 BP.

Temperatures seem to have remained rather constant between *c.* 10,000 BP and the present, with the possibility of a slight depression during the initial part of the period. On the whole the climate was relatively wet although it did vary, as is evident from the major fluctuations in water levels which are known to have occurred in many closed lake basins. Doubt remains as to the importance to terrestrial ecosystems of some of the climatic variations responsible for lake-level changes. However, a transition to lower water levels at roughly 4000 BP was accompanied by major alterations in many plant communities, and ecologically this was probably the most significant climatic event of the last 10,000 years.

Influences of climatic change on mire development

Four mires in southwestern Uganda (Butongo, Ahakagyezi, Katenga and Muchoya) are known to contain aquatic sediments at depth, and the transition from lake to mire can be attributed, at least to some extent, to shallowing with some sediment accumulation. At Muchoya and Ahakagyezi the transition to mire occurred at roughly 4000 BP, a time when it is known that the climate of tropical Africa became drier, and thus climatic change could have contributed to mire spread.

There is further evidence that the more or less flat mires along the Western Rift hills are responsive to climatic events. For example, the last glacial maximum was marked at both Muchoya and Kamiranzovu by particularly wet conditions, with deposition of diatom-rich sediments at Muchoya and extensive *Sphagnum* bog at Kamiranzovu. In both cases these wetter-type sediments are sandwiched between drier sediment types. Incidentally, these sites are unusual in that the glacial maximum in tropical Africa was usually a time of dryness, with many lakes at their lowest levels. This can probably be attributed to a combination of relatively high altitudes and the disposition of these sites in a region of Africa which continued to receive a relatively high rainfall during glacial times.

High-altitude mires in East Africa have accumulated peat relatively slowly and, at least at times, apparently rather regularly. The peat is generally of a uniform type, suggesting a lack of responsiveness to more minor climatic events and perhaps also a rather wide ecological amplitude for dominant species such as *Carex runssoroensis*. However, this apparent uniformity can break down under closer examination. At Badda bog in Ethiopia there are great fluctuations in the diatom flora of a superficially homogeneous peat, with alternations between a diatom association indicating humid, acidic and oligotrophic conditions, and another assemblage indicating greater alkalinity (Gasse 1978).

The deposition of clay bands in mire sediments is also thought often to have been a response to climate. Karimu swamp on the Aberdares shows clay deposition intervening between two periods of peat formation, one extending from before 32,600 BP to after 24,000 BP, and the other beginning soon before 9500 BP (Perrott & Street-Perrott 1982). The clay deposition could have been a consequence of a low vegetation cover in the catchment during ice age aridity. A similar explanation could account for the thick clay deposits at Laboot bog at 2880 m on Mt Elgon (Hamilton 1982). Here low vegetation cover during the glacial maximum is indicated by very high values of grass pollen and low values of tree pollen in the clays. There is evidence from several sites of a high rate of inorganic sedimentation just before the beginning of the Holocene (Hamilton 1982). The clay bands at coring site A on Muchoya swamp (Fig. 2) date to this time. These are of restricted latitudinal extent in the sediments and probably represent channel deposits formed when vegetation cover in the catchment was reduced.

Succession is one of the fundamental concepts in ecology. One of the frequently quoted examples of succession is the series of aquatic and mire communities which are said to replace one another in freshwater basins, for example, as a consequence of sediment infill and shallowing water. Deuse (1966) has advanced hypotheses concerning successional relationships of various wetland communities in Rwanda, one of his major contentions being that *Syzygium* swamp forest is the major end-point of succession in mires at altitudes of *c.* 1700–2000 m. This view was based mainly on a study of the modern vegetation, with little stratigraphic work, and he was strongly influenced by the extensive *Syzygium* forest on Kamiranzovu, a very large swamp with a considerable thickness of peat. However, examination of sediments at Kamiranzovu shows that *Syzygium* is growing on a surface with an age of *c.* 12,600–11,000 BP, and that the underlying peat, dating back to 32,000 BP, was formed by a *Sphagnum–Erica–Xyris* community and not by *Syzygium* forest (Hamilton 1982). *Syzygium* did formerly occur on two previous occasions, both pre-37,000 BP and both associated with relatively inorganic sediments. It seems more appropriate to regard *Syzygium* as a species which responds to more nutrient-rich conditions, rather than a necessary end-point of succession. Kamiranzovu and other relatively flat mires along the Western Rift appear to be responsive to climatic events. In general, the concept that there are normal pathways of hydroseral development seems to be of limited utility. This conclusion is broadly similar to that reached by Walker (1970) for the British Isles.

It is striking that the peat surface at Kamiranzovu dates to a time near the Pleistocene–Holocene boundary when lake levels in many basins in eastern Africa were rising. It is possible that the surface was once higher than at present and it has become lowered through decomposition or erosion, but the simplest explanation is that cessation of peat growth was a result of a rise in temperature. The effect of the temperature rise on the peat-forming equation more than compensated for that of the increase in rainfall.

Man has certainly influenced some mire ecosystems through, for example, increasing the rate of sedimentary input from surrounding slopes, burning and (recently) drainage for agriculture. The replacement of Cyperaceae by *Typha* at Katenga swamp is related to increased siltation following forest clearance (Morrison & Hamilton 1974). New data on the distribution of charcoal in peat at Muchoya clearly shows that the change from *Erica* swamp forest to *Pycreus* bog was a result of burning, rather than a consequence of any of the other possible processes suggested by Morrison (1968).

In summary, succession, human influences, and especially climatic change are all known to have contributed to the course of ecosystem development on sites now occupied by mires in East Africa. The sedimentary expression of environmental events can vary greatly from place to place, depending on variations in degree of infilling, the baseline climate and the local geomorphology.

ACKNOWLEDGEMENTS: This work, was funded in part by the Natural Environment Research Council, the Royal Society and the University of Ulster. We are grateful to the governments of Kenya, Rwanda and Uganda for permission to carry out research. We are also indebted to the Departments of Botany at Makerere and Nairobi Universities for use of their facilities. The studies in Kenya and Rwanda were carried out by A.H. together with Alan Perrott, whose assistance is gratefully acknowledged. Many thanks are due to Tony Katende, Adrian Mwesigye, Mr Dutschi, Africano and others for help with fieldwork in Uganda.

References

BOUXIN, G. 1974. Étude phytogéographique des plantes vasculaires du marais Kamiranzovu (forêt de Rugege, Rwanda). *Bull. Jard. Botan. Belg.* **44,** 141–59.

COETZEE, J.A. 1967. Pollen analytical studies in East and Southern Africa. *Palaeoecol. Afr.* **3,** 1–146.

DEUSE, P. 1966. Contribution à l'étude des tourbières du Rwanda et du Burundi. *Inst. Nat. Réch. Scient., Butare, Rwanda* **4,** 115pp.

GASSE, F. 1978. Les diatomées holocènes d'une tourbière (4040 m) d'une montagne Ethiopienne: le mont Badda. *Rev. Algol.* **13,** 105–49.

HAMILTON, A.C. 1982. *Environmental History of East Africa.* Academic Press, London and New York, 328 pp.

LIND, E.M. 1956. Studies in Uganda swamps. *Uganda J.* **20,** 166–76.

——& MORRISON, M.E.S. 1974. *East African Vegetation.* Longman, London, 257pp.

LIVINGSTONE, D.A. 1967. Postglacial vegetation of the Ruwenzori Mountains in Equatorial Africa. *Ecol. Mon.* **37,** 25–52.

MEADOWS, M.E. 1984. Past and present environments of the Nyika Plateau, Malawi. *Palaeoecol. Afr.* **16,** 353–90.

MORRISON, M.E.S. 1968. Vegetation and climate in the uplands of south-western Uganda during the later Pleistocene Period, 1. Muchoya Swamp, Kigezi District. *J. Ecol.* **56,** 363–84.

—— & HAMILTON, A.C. 1974. Vegetation and climate in the uplands of south-western Uganda during the later Pleistocene Period, 2. Forest clearance and other vegetation changes in the Rukiga Highlands during the past 8000 years. *J. Ecol.* **62,** 1–31.

PERROTT, R.A. & STREET-PERROTT, F.A. 1982. New evidence for a late Pleistocene wet phase in northern intertropical Africa. *Palaeoecol. Afr.* **14,** 57–75.

THOMPSON, K. & HAMILTON, A.C. 1983. Peatlands and swamps of the African continent. In: GORE, A.J.P. (ed) *Mires: Swamp, Bog, Fen and Moor, B. Regional studies.* Elsevier, Amsterdam, pp. 331–73.

WALKER, A. 1970. Direction and rate in some British Post-glacial hydroseres. In: WALKER, D. & WEST, R.G. (eds) *Studies in the Vegetational History of the British Isles.* Cambridge University Press, Cambridge, pp. 117–39.

A. HAMILTON & D. TAYLOR, Dept of Environmental Sciences, University of Ulster, Coleraine, Co. Londonderry, Northern Ireland.

Sedimentary History of African Rift Basins

The Pliocene Hadar Formation, Afar depression of Ethiopia

J.J. Tiercelin

SUMMARY: Along the western margin of the Afar depression there is a fluvio-lacustrine sequence, well known for its hominid fossils, which consists of a 150–300 m thick layer of coarse to fine detrital sediments with interbedded pyroclastics and basalt. This series, deposited between at least 4.2 and 2.4 Ma, is called the Hadar Formation. Sedimentary and microfossil evidence allows the reconstruction of sedimentary, volcano-tectonic and climatic events in the evolution of the Hadar basin. During 1.8 Ma, the palaeogeography of the Hadar region appears to have been an intricate succession of fluvial, marshy and lacustrine environments, subject to rapid lake-level fluctuations and a wide range of hydrological conditions. Tectonic activity is confined to subsidence. However, important vertical movements of the western escarpment occur towards the end of the sedimentary succession in the Hadar basin.

The Hadar Formation, in the Afar depression of Ethiopia, is renowned for its superb mammalian fossils, its hominid fossil *Australopithecus afarensis* and the Kada Gona artefact sites. Intensive investigation was carried out by geologists and palaeontologists of the International Afar Research Expedition (IARE) from 1973 to 1977, and completed during 1980. The 150–300 m thick Hadar Formation belongs to a 1 km thick sedimentary sequence situated along Ethiopia's Middle Awash valley (Fig. 1). This is the Awash Group and dates from Miocene to late Pleistocene (Kalb *et al.* 1982b). It has yielded rich vertebrate faunas, fossil hominids and Acheulian or Oldowan artefacts at several places (Kalb *et al.* 1980; Kalb *et al.* 1982a; Clark *et al.* 1984). The lower levels of this Awash Group include the Ch'orora Formation. This consists of a 30 m thickness of volcanic and biogenic deposits, dated at 10.7–10.5 Ma and containing an interesting mammalian fauna. It indicates the outlines of a 2400 km^2 SW–NE lacustrine basin, between the southeastern margin of the Afar depression (Hararghe escarpment) and the volcanic ridge of Assabot, Mieso and Afdem (40°18′–40°20′E, 8°53′–8°55′N) (Sickenberg & Schönfeld 1975; Tiercelin *et al.* 1979, 1980).

North of 10°N along the Ethiopian escarpment, the > 70 m thick Lower Bodo Beds (Kalb *et al.* 1980), contain a widely traceable grey lapilli tuff (Cindery Tuff), dated at 3.8–4.0 Ma, that indicates the existence of a very shallow freshwater Pliocene lake (Clark *et al.* 1984) (Fig. 1). Further N, between 13° and 15°N, there are several hundred metres of conglomerates, sands, sandstones, and red, green or multi-coloured clays, with interbedded basaltic lavas and acidic intrusions of the Red Series, which are the oldest sedimentary rocks of the Afar with ages ranging from 24 to 5.4 Ma (Bannert *et al.* 1970). They were laid down in the depression that was formed by the spreading and rotating of the Danakil Alps crank-arm (Hutchinson & Engels 1970; Sichler 1980).

Between 11° and 11°20′N, the 150–300 m thick Hadar Formation comprises fine to coarse detrital sediments with some interbedded volcanic layers. It is exposed along the Awash River, along its left-bank tributaries, the Kada Gona, Sidi Hakoma, Denen Dora, Kada Hadar, Ounda Hadar and Ourda, and along its right-bank tributaries, the Oudaleita, Andedo and Kabara (Fig. 2). Previously described as a 120–140 m thick sequence by Taieb *et al.* (1976), it was found to be nearly 300 m thick during the 1975–76 field season (Tiercelin & Taieb 1976).

The Hadar Formation has been conveniently subdivided into four members, separated by tuff horizons. These members are, in order of succession: (1) the Basal Member (BM), 40 m exposed; (2) the Sidi Hakoma Member (SH), 45–130 m thick, defined at its base by a series of white tuff and volcaniclastic sands (Sidi Hakoma Tuff—SHT); (3) the Denen Dora Member (DD), 20–35 m thick, with a series of 3–5 thin tuffs at its base (Triple Tuff—TT); and (4) the Kada Hadar Member (KH), 30–80 m thick, with a single tuff at its base (Kada Hadar Tuff—KHT).

Geological mapping in 1976 by IARE geologists (Fig. 3), together with structural observations along the supposed margins of the sedimentary basin and sedimentological studies, have increased our knowledge of the Hadar Formation. The renowned 1973–76 palaeoanthropological discoveries (A.L. 129, an articulated knee joint; A.L. 200–1, a complete maxilla; A.L. 288–1, the partial skeleton of *Lucy*; the 13 individuals of A.L. 333/333 W) were all assigned to the new species *Australopithecus afarensis* (Johanson *et al.* 1978b; Johanson & White 1980). They have been

From Frostick, L.E. *et al.* (eds) 1986, *Sedimentation in the African Rifts*, Geological Society Special Publication No. 25, pp. 221–40.

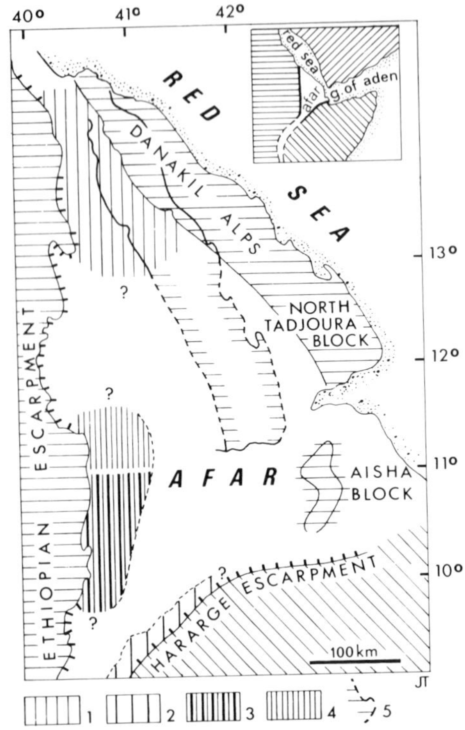

FIG. 1. Major Mio-Pliocene sedimentary formations of the Afar depression. 1: Red Series; 2, 3, 4: Awash Group; 2: Ch'orora Formation; 3: Bodo Formation; 4: Hadar Formation; 5: location of the Danakil crank-arm *c.* 4 Ma.

vividly described by Johanson & Edey (1981), and help to build a picture of the palaeoenvironment. Likewise, the 1976 IARE field season was remarkable for the discovery of Oldowan-type artefacts within the upper Kada Hadar Member conglomerates (Fig. 2). Stratigraphic correlation suggests a 2.6 Ma age for these artefacts (Roche & Tiercelin 1977, 1980).

All this sedimentological and micropalaeontogical evidence now makes it possible to refine the model of depositional environments originally proposed by Taieb & Tiercelin (1979, 1980).

Chronology

Attempts at dating the Hadar Formation have involved two volcanic horizons: (1) the Bouroukie Tuff 2 (BKT-2) in the upper part of the Kada Hadar Member; and (2) the Kada Damoum basalt (KDM), a moderately low potassium (*c.* 0.6% K_2O) tholeiite in the upper part of the Sidi Hakoma Member (Taieb *et al.* 1976; Aronson *et al.* 1977) (Fig. 4). Recent K-Ar measurements on the Kada Damoum basalt have produced ages of 3.65 ± 0.15 Ma (Walter & Aronson 1982), the previously published age being 3.0 Ma.

The age of the base of the Hadar Formation is as yet unclear. A 4.0 ± 0.4 Ma age has been postulated for the Sidi Hakoma Tuff (SHT), based on zircon fission tracks (Walter 1980). However, correlation of the SHT with the Tulu

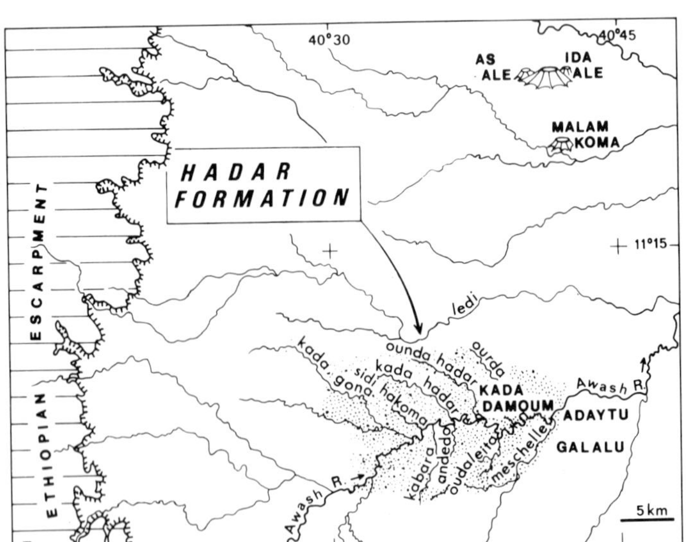

FIG. 2. The Hadar region, Afar depression.

Bor Tuff of East Turkana which lies 40 m below the 3.18 Ma Ninikaa Tuff (Brown & Cerling 1982), calls into question the accuracy of both the 4.0 Ma SHT age and the 3.6 Ma age proposed for the Kada Damoum basalt (Brown 1982, 1983; Aronson *et al.* 1983). Nevertheless, there is no doubt about the fact that the KDM basalt lies in the Hadar Formation. This was firmly established during the 1976 field season (Fig. 5), and therefore Brown's hypothesis (1983) about the KDM plateau as a horst is not correct.

The age of the primary air-fall tuff BKT-2 has been revised from 2.65 to 2.9 Ma (Walter & Aronson 1982) (Fig. 4). Tracing of beds by Taieb, Tiercelin and Walter during January 1980 confirmed the correlation of the BKT-2 and the Kada Gona Artefact Site Tuff ('cinerite I' or 'AST-1') as first proposed by Roche & Tiercelin (1977) and later by Walter (1980) and Johanson & Edey (1981).

Sedimentary series

Basal Member—facies description

The Basal Member is best developed in Oudaleita Wadi, a right-bank tributary to the Awash River (Fig. 6). It comprises at least 40 m of massive or stratified coarse to fine sands or sandstones, clayey silts, and massive or laminated clays. West of Oudaleita Wadi, and on the left bank of Awash River, the Basal Member is present but poorly exposed (Fig. 6). In all, four facies are recognized.

Coarse-grained sand facies

This facies comprises largely unstratified, or occasionally small scale, planar cross-bedded, or horizontally stratified grey to white coarse sand beds in layers 0.2–1.6 m thick. These sands are mainly quartzose and feldspathic (K-feldspar). Bedded gritty spheroids (0.2–1.0 m in diameter) occur locally, as do beds of very rounded but oblate brown clay clasts (usually 0.5–1.0 cm in diameter), finely comminuted plant debris, rootcasts, and scattered volcanic rock pebbles.

Medium to fine-grained sand facies

These are laterally extensive, medium to fine, grey to yellow sand or sandstone beds (0.05–3.0 m thick). These sands are massive, horizontally or obliquely stratified, and sheeted or locally rippled (E 988, Fig. 7). Gravels, clay clasts and plant debris are sometimes present. Freshwater gastropods (*Melania tuberculata*), ostracods (*Cyprideis* sp. SH 1 [spotted], *Stenocypris*, *Darwinula*, *Iliocypris*; Peypouquet *et al.* 1983) and Characea occur in flaggy sandstones (0.05–0.2 m thick) which are laterally extensive at the top of the Basal Member (E 536, Andedo; E 433, Sidi Hakoma; see Fig. 6).

Clayey silt facies

This facies comprises 0.02–0.2 m thick massive or sheet-like, grey or pale brown, clayey silts interbedded in clay successions. Calcareous spherical or oblate nodules, septarias 0.02–0.1 m in diameter, clay clasts, and plant debris are often associated with this facies.

Massive-laminated clay facies

This facies is common in the Basal Member. It comprises massive or finely-laminated grey, pale brown, green, brown, or red-brown clays with numerous slickensides or hydroplastic faults. Various biogenic remains are present, including plant debris, fish scales and fragments, and gastropod debris. A monospecific ostracod fauna (*Cyprideis* sp. SH 1 [spotted]; Peypouquet *et al.* 1983) associated with the gastropod *Melania tuberculata* occurs in a grey-brown clay at Ourda, 8 m below the top of Basal Member (E 1173, Fig. 6). The clay minerals are dominated by smectite.

Facies organization in the Basal Member

Owing to the dip of the series in this area, the greatest exposure of the Basal Member (40 m) occurs in Oudaleita Wadi. Two tuffs (the Oudaleita Tuffs, OT) are interbedded in the lower part of the member, which is formed of clayey silts, massive or laminated clays and some sand layers rich in plant debris.

Above the unit, the Basal Member consists of alternating massive or laminated clays and clayey silts capped by massive medium to coarse sands. North or W of Oudaleita, only the upper part of the Basal Member is exposed: 4 m in Sidi Hakoma, 8 m in Kenia Koma, 8 m in Ourda, and 13 m in Andedo. At each of these sites it is formed of alternating clayey silts with calcareous nodules, massive or laminated clay and cross-bedded fine to coarse sands or sandstones (Fig. 6).

Sidi Hakoma Member—facies description

The Sidi Hakoma Member varies in thickness between 45 and 130 m from W to E, and is defined at its base by the Sidi Hakoma Tuff. The member crops out largely in the Hadar region. In all, 11 facies are recognized (Fig. 8).

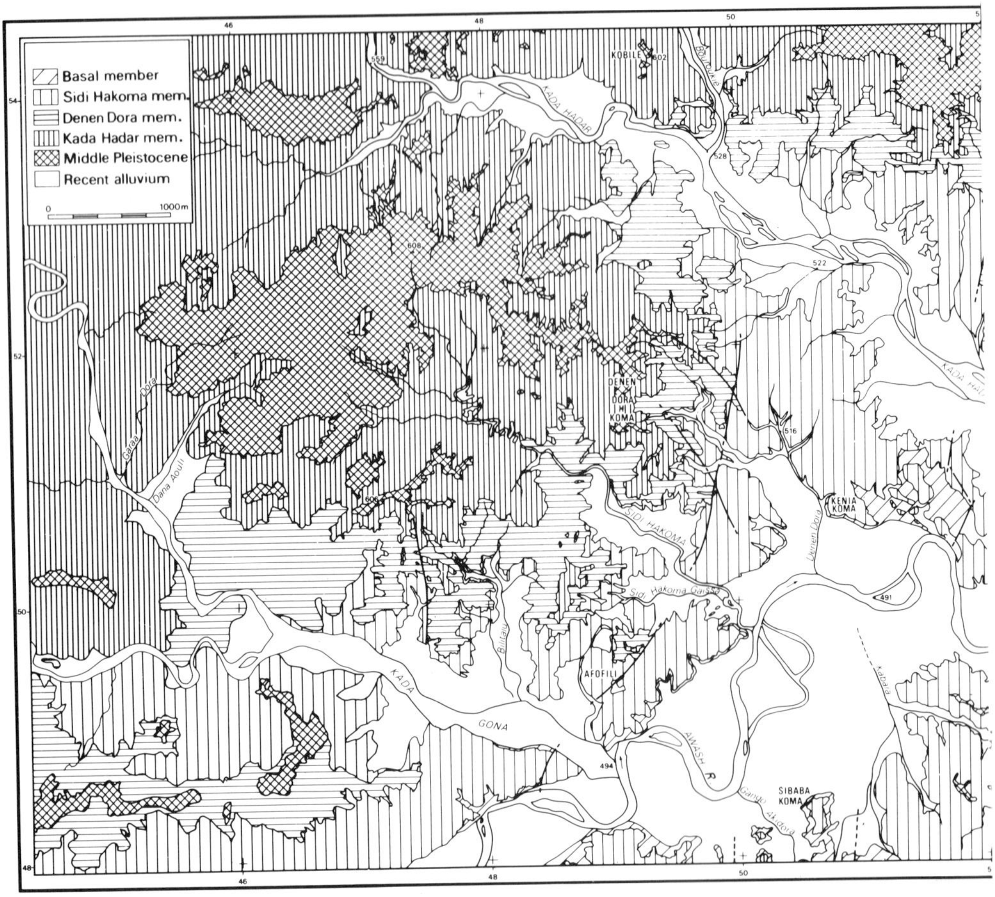

FIG. 3. Simplified geological map of the Hadar Formation. Surveys between the Kada Gona and Ourda Wadis by N. Page, M. Taieb, J.J. Tiercelin, E. Tola and R.C. Walter. Contours W of Kada Gona and E of Ourda are derived from aerial photographs by N. Page. Topographic survey by P. Planques and M. Decobert.

Sidi Hakoma Tuff facies

The laterally extensive Sidi Hakoma Tuff generally occurs as a chalky white material, 0.20–1.25 m thick. A laminated subfacies W of Sibaba Koma often clothes an erosional surface and comprises a chalky white tuff with pink or brown laminations. This is associated with a blue-grey or white horizontally-bedded or cross-bedded tuff. A sandy, sanidine-rich, massive or cross-bedded subfacies forms large channel fills (1.25–1.4 m thick) at Kenia Koma and Sibaba Koma (Fig. 8).

Conglomerate facies

This comprises unstratified, polymict, matrix-supported conglomerate beds, 0.2–8.0 m thick. The matrix is fine-grained, being sandy or gritty. Pebbles are of volcanic origin (black patina basalt near Andedo, section 4–12), and range in size from 4 to 20 cm.

Coarse-grained sand facies

This facies comprises 0.2–3.0 m thick, yellow-grey sands, showing cross or parallel and low-angle bedding. Gravel or clay clast layers and large grit spheroids (0.1–1.0 m) are abundant.

Medium to fine-grained sand facies

This includes yellow-grey to green massive, flaggy, cross or horizontally-bedded medium to fine sands. Gravels, clay clasts, mollusc shells (*Melania*, articulated *Unio*), remains of fish (*Siluridae*) and reptiles (*Crocodilia*, *Chelonia*), bone fragments and root casts are sometimes

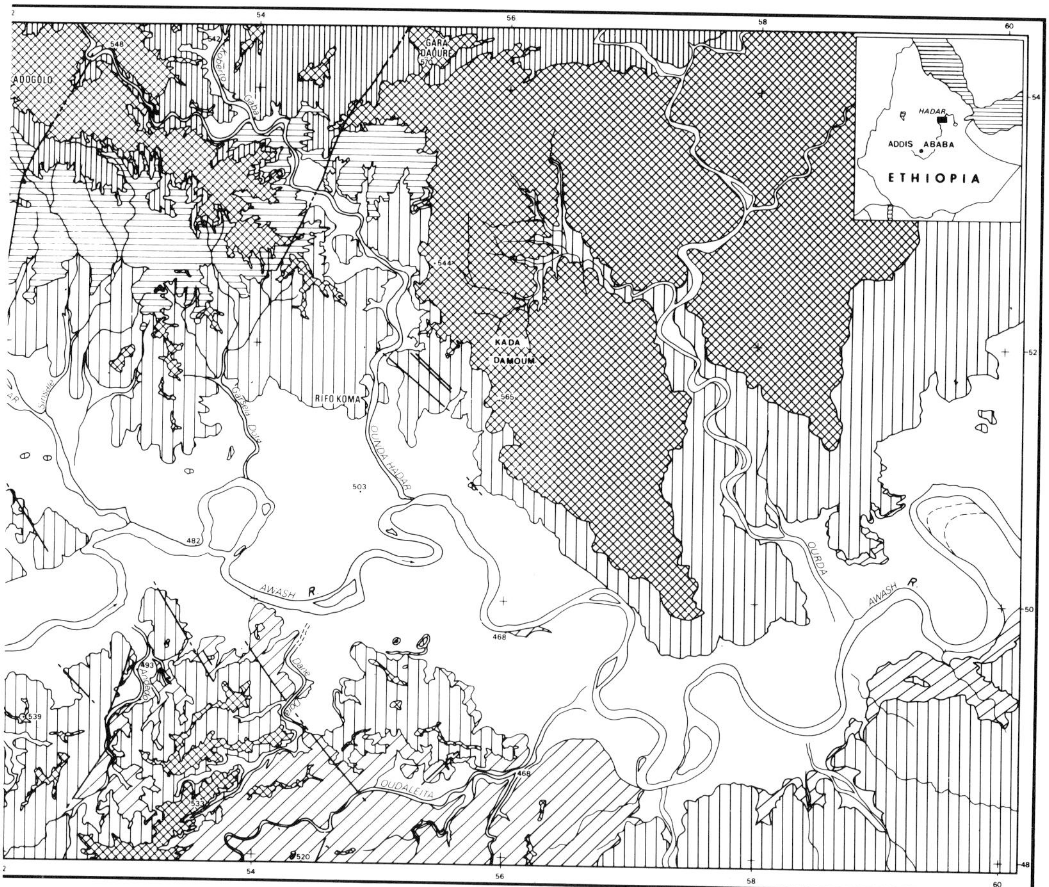

FIG. 3. Continued.

abundant, chiefly in massive or rippled sand or sandstone beds 0.05–0.15 m thick (Sabatier 1979a).

Clayey silt facies

Grey to whitish-grey, massive or sheet-like clayey silts occur as 0.7–2.0 m thick beds. Parallel-laminated silts contain finely comminuted plant debris, small gastropod shells, and clay clasts. Rootmarks and carbonate nodules are common in massive silts. Abundant gastropods (*Bellamya* sp., *Cleopatra* sp.), bivalves (*Unio* sp.), reptile eggs, and rodent skulls have been found in grey to brown flaggy silt in Ounda Hadar Wadi (E 524, Fig. 8).

Massive-laminated clay facies

This facies comprises brown, black, grey to green laminated or massive clays. Numerous hydroplastic faults, sometimes with haematite coating, occur in massive subfacies. Biogenic remains, including plant debris, leaf prints, root marks, gastropods, and fish and insect debris occur frequently. The clay minerals are chiefly smectite, with occasional traces of illite and kaolinite.

Shale facies

Grey, whitish grey or brown shales form 0.2–4.0 m thick beds. This facies is characterized by abundant biogenic remains (plant debris, wood fragments, fish scales and gastropod debris), sometimes forming true coquinas. A well preserved freshwater fauna contains gastropods (*Melania* sp., *Bellamya* sp., *Cleopatra* sp.) and a rich ostracod fauna (*Cyprideis* sp. SH1 [reticulate species], *Cyprideis* sp. KH1 [with phenotypic tubercules]; Peypouquet *et al.* 1983) (E 1084–1093, Ourda; Fig. 8).

Diatom shale facies

Half a metre above the Sidi Hakoma Tuff, a laterally extensive whitish-grey shale 30 cm thick

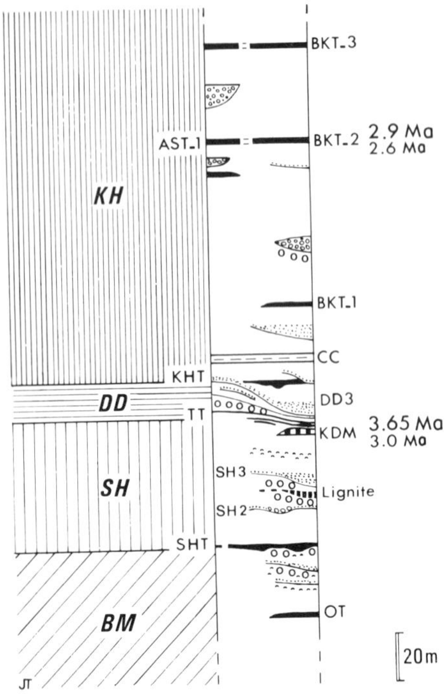

FIG. 4. General log of the Hadar Formation. BM: Basal Member; SH: Sidi Hakoma; DD: Denen Dora Member; KH: Kada Hadar Member; other abbreviations are explained in the text.

contains numerous diatoms (*Melosira granulata*) (E 438, Sidi Hakoma Wadi; Figs. 8 and 9). West of Andedo, the diatom flora is replaced by diagenetic cristobalite spheres.

Coal facies

This facies comprises a brown-purple clay overlain by a coal layer, 20–40 cm thick. The mineral fraction of this coal consists of smectite, halite, quartz, feldspars and microcrystalline natrojarosite ($NaFe_3[SO_4]_2[OH]_6$), which also constitutes the fillings of large root casts (Périnet *et al.* 1980) (E 1086). The organic fraction (total organic carbon is 1.45%) contains humic coal materials, dominated by vitrinite, reflecting and evolving parenchymes, cutinites, rare fluorinites and an important fraction of inertinite. The vitrinite reflectance value is 0.5% (the lignite-coal limit of the French classification), which denotes a strong thermal transformation (0.3 km burial) (Robert 1979, 1980). North of Kenia Koma, this facies consists of green shale interbedded with thin coal layers (total thickness 0.2 m).

Oncolite facies

This facies is found N of Kenia Koma. Here, a yellow sand containing abundant small bivalve casts, and fish, crocodile, and rodent remains, overlies a coal layer. Oncolites 8–15 cm in diameter occur laterally at the top of this sand and are covered by finely laminated beige and brown clays (E 874, Figs. 8 and 10).

Calcareous slab facies

Five to 10 cm thick calcareous slabs are frequently found capping clayey silts. A pulmonate gastropod was found in this facies at Ourda (E 1135).

Facies organization in the Sidi Hakoma Member

In the western part of Hadar, the 45 m thick Sidi Hakoma Member consists of a 30 cm thick diatom shale above the Sidi Hakoma Tuff, followed by a massive clay succession which is interrupted by lenticular cross-bedded medium to coarse sands (SH2 and SH3 units) (Johanson *et al.* 1978a). Further up in the succession the member consists of clayey silts, carbonate slabs interbedded with shales, one coal layer, and a medium to fine sand with bivalves and oncolites. To the E, the Sidi Hakoma Member is 130 m thick between the Sidi Hakoma tuff and the Kada Damoum tholeiite which is itself interbedded in the upper unit of the member. Here, the succession consists of clayey silts, massive or laminated claystones and shales. Two coal layers occur 30 m above the tuff. Cross-bedded sands with reworked tholeiite cobbles unconformably overlie the Kada Damoum lava flow (Fig. 5). Above these, the member is formed by clays and clayey silts (Fig. 8).

Denen Dora Member—facies description

The Denen Dora Member is defined at its base by the Triple Tuff Marker Bed (TT), and outcrops on the left bank of the Awash River between the Kada Gona and Ourda Wadis (Fig. 11). Its maximum thickness is 35 m in the Kada Hadar Wadi, and it decreases to 20–25 m both to the W and the E.

Triple Tuff facies

The Triple Tuff Marker Bed (TT) consists of 3–5 white or pink tuff layers, each 5–30 cm thick. Beds TT 4 and 5 are often thick (0.2–0.3 m) and lenticular. Abundant acicular baryte concretions, 1–3 cm in diameter, were found at the top of bed

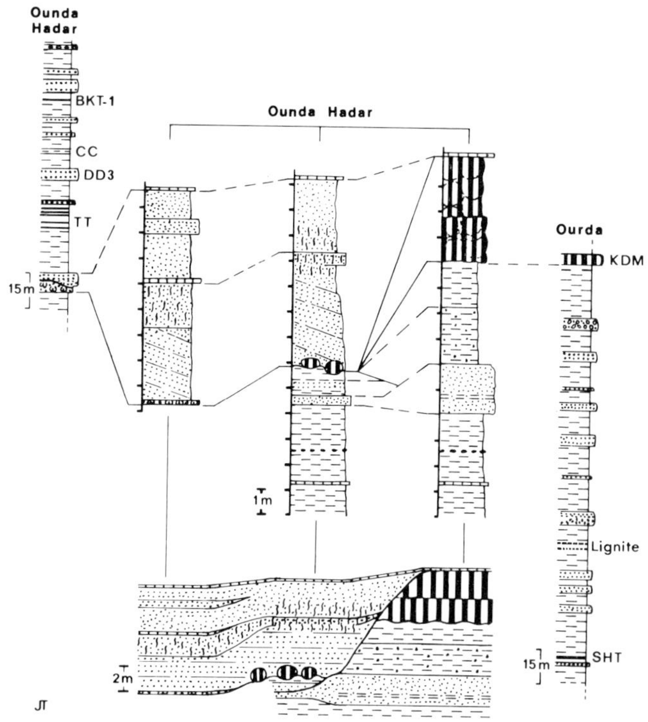

FIG. 5. Stratigraphic relation and place of the Kada Damoum basalt (KDM) in the Hadar Series.

TT 2 in the Kada Hadar Wadi. Geochemical analysis of this baryte shows strong strontium enrichment (>2000 ppm), which may be of volcanic or hydrothermal origin. Cu, Co and Zn contents are not unusual (15 ppm Cu, 39 ppm Co, 24 ppm Zn) (James 1967; Reimer 1978).

DD3 conglomerate-sand facies

This facies consists largely of an unstratified conglomerate with an orange-coloured sandy matrix, that lies on an erosional surface, or a red-brown medium to coarse sand layer. The sand layer is often cross-bedded, and is 4–9 m thick. A calcareous micro-conglomerate, 0.5 m thick, occurs occasionally at the base of the sands. Layers of clay clasts of various sizes and bone fragments are frequent. Pebbles are of volcanic origin. Fossiliferous black fine sands, 1–2 m thick, with numerous large root casts, cap the red-brown sands.

Fine-grained sand facies

This consists of 0.4–1.0 m of clayey or gravelly fine sands, that are grey or beige in colour. They are often obliquely bedded, and sometimes lie on erosional bases. Clay clasts and mollusc shells (*Melania* sp., *Unio* sp.) are frequent.

Clayey silt facies

These beige or yellow, horizontally or obliquely bedded, clayey silts occur in beds 0.2–5.0 m thick. Gastropod shells are frequent. A sponge spicule-rich silt was found in the Ounda Hadar Wadi (E 493, Figs. 11 and 12).

Massive or laminated clay facies

This facies comprises either grey, beige, green or brown, massive or laminated clay beds 0.5–3.0 m thick. Gastropods (*Melania* sp., *Bellamya* sp.) and fish remains are often abundant. Root marks and leaf prints sometimes occur. The clay fraction is mainly smectite.

Ostracod shale facies

This facies consists of beige, green or dark green shale, that varies between 0.5 and 0.8 m in thickness. It is locally massive at its base and is marked by slickensides. Plant debris, fish remains, gastropods (*Melania* sp.) and ostracods

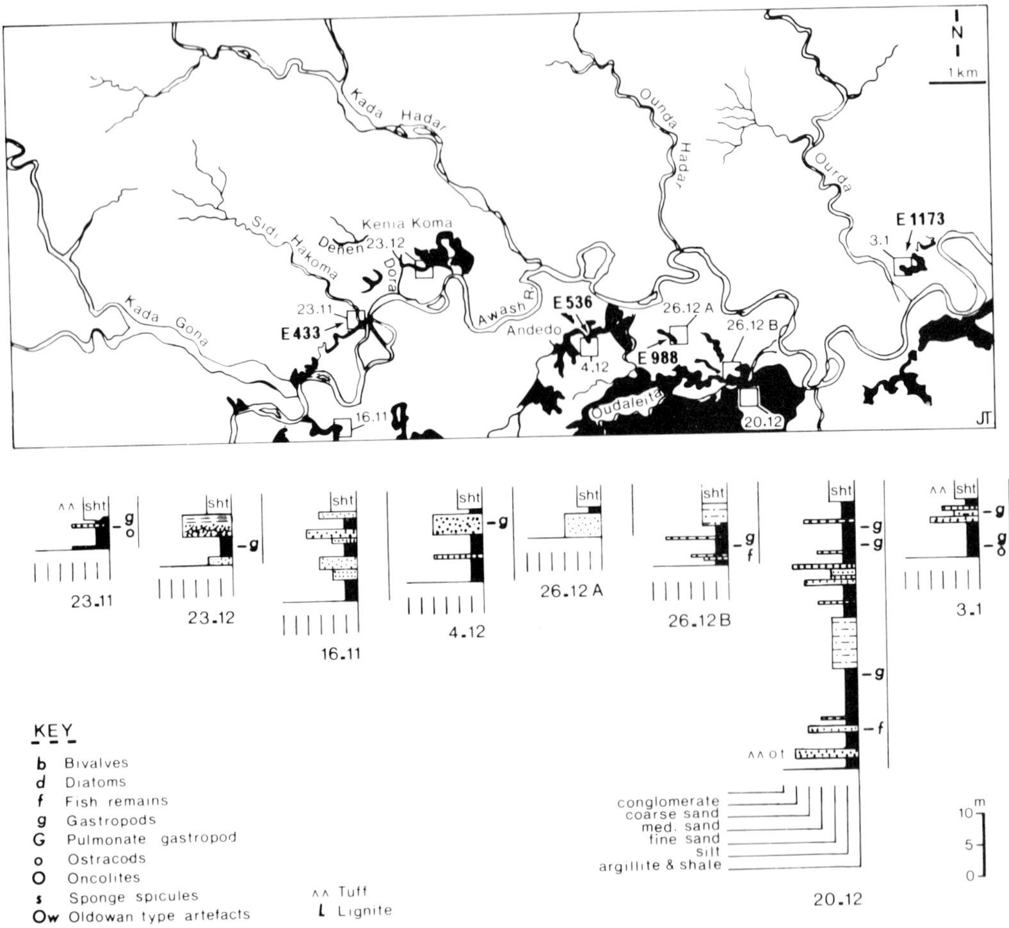

FIG. 6. Basal Member of the Hadar Formation. Outcrops and simplified lithology.

FIG. 7. Rippled sandstone slab, 10 cm thick (E of Denen Dora). Direction of ripple crests is 50–55°N. Possible crocodile or fish casts on ripple crests. E988, Basal Member.

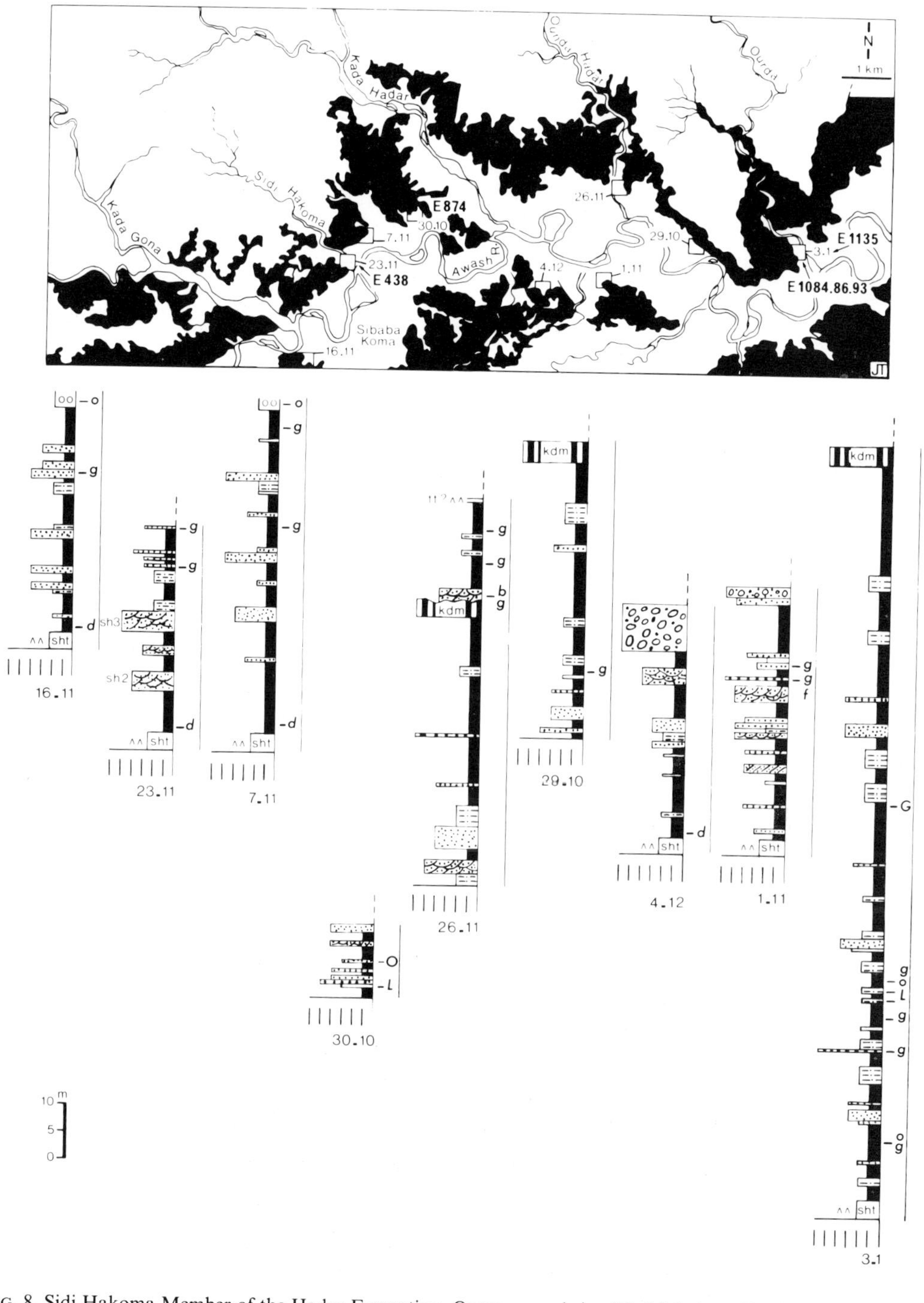

FIG. 8. Sidi Hakoma Member of the Hadar Formation. Outcrops and simplified lithology. For key see Fig. 6.

FIG. 9. Diatom shale facies (*Melosira granulata*). SEM photograph (×780). E 438, Sidi Hakoma Wadi, SH Member.

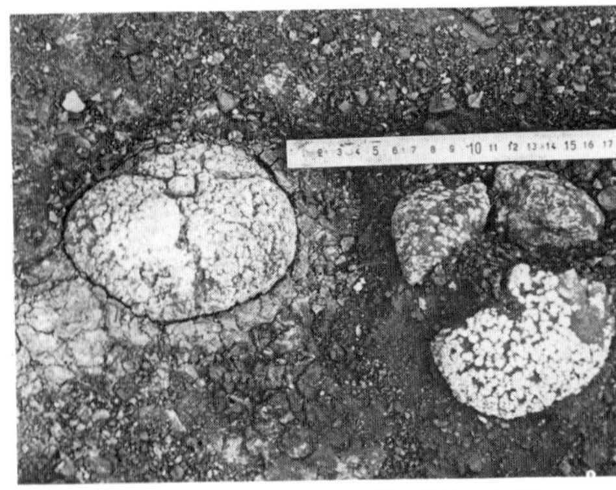

FIG. 10. Oncolite facies. E 874, N of Kenia Koma, SH Member. Scale in cm.

are numerous and sometimes form real coquinas. The ostracod fauna consists of *Cyprideis* sp. KH1, which is very abundant and has no phenotypic tubercules on the shells (Peypouquet *et al.* 1983). Spherical concretions, 2 cm in diameter and brown with a white grey cortex, occur frequently on top of this facies. These concretions consist entirely of apatite with a nucleus of fish debris. A badly preserved diatom flora (? *Thalassiosira* sp.) occurs locally (E 299; Fig. 11).

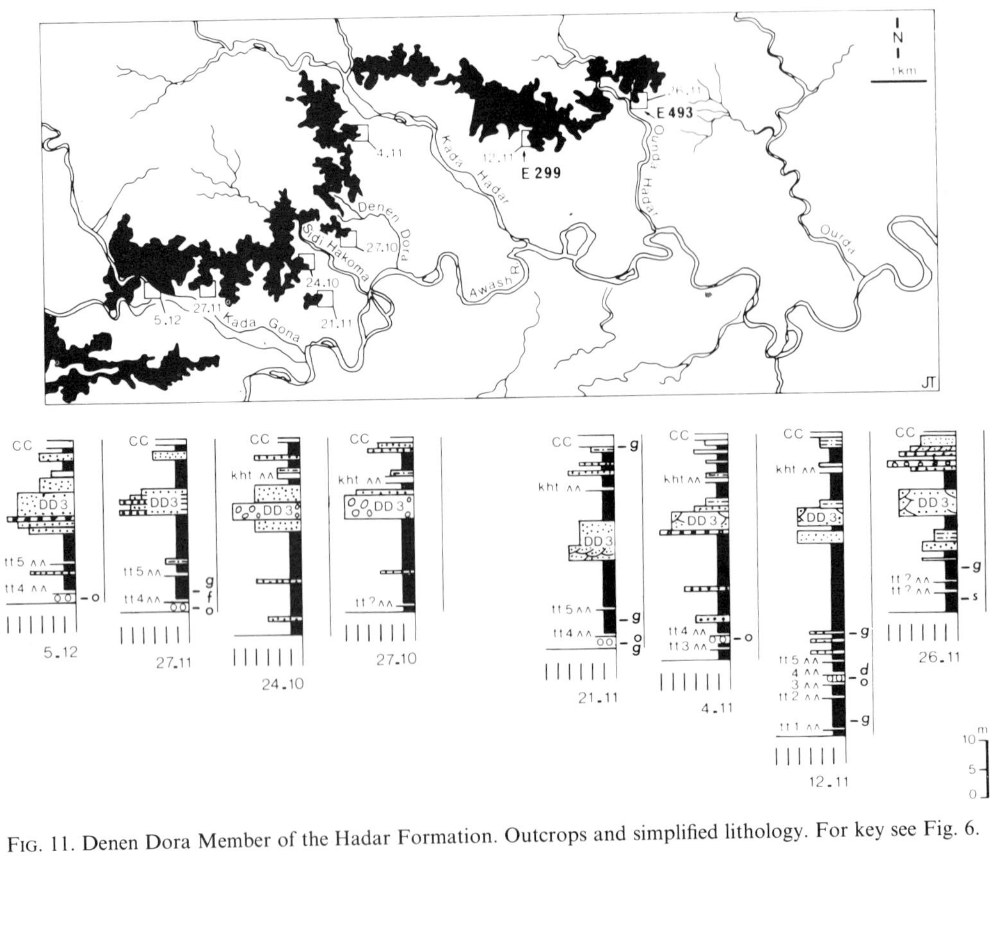

FIG. 11. Denen Dora Member of the Hadar Formation. Outcrops and simplified lithology. For key see Fig. 6.

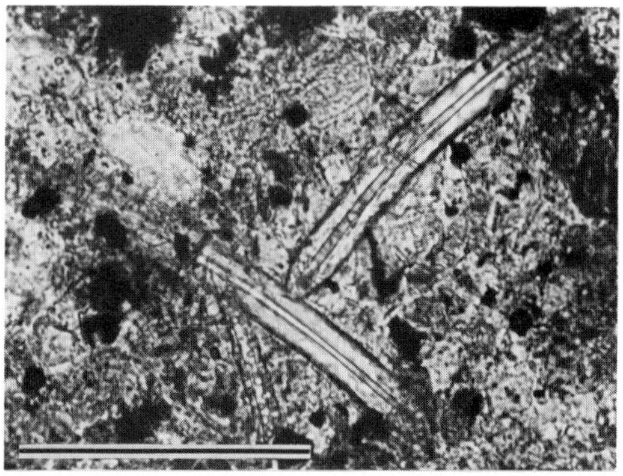

FIG. 12. Thin section of a sponge spicule-rich silt. E 493, Ounda Hadar Wadi, Denen Dora Member. Scale bar = 0.1 mm.

Shale facies

This comprises grey-green, beige, brown or purple shale beds, 0.6–1.0 m thick. Fish and gastropod remains are uncommon. Apatite nodules are frequent.

Facies organization in the Denen Dora Member

To the W the 20 m thick Denen Dora Member, above the Triple Tuff (beds TT4–5) and ostracod shale marker beds, consists of laminated or massive beige to brown clays. Further up the succession, the sandy and conglomeratic DD3 unit is 9.6 m thick and is capped by clays, silts and fine sands. In the Denen Dora Wadi and to the E, 20 m of thick grey, green or brown clays overlie

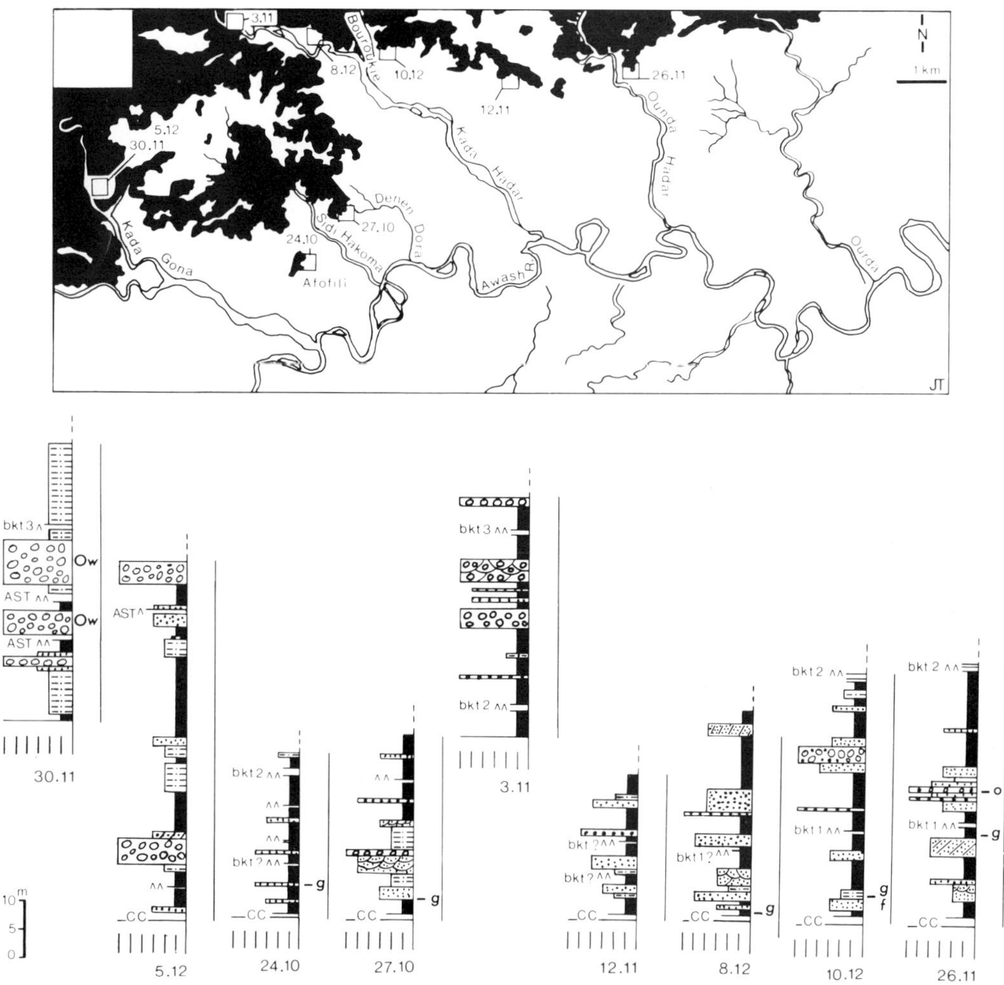

FIG. 13. Kada Hadar Member of the Hadar Formation. Outcrops and simplified lithology. For key see Fig. 6.

the TT1–5 and ostracod shale marker beds. The sandy DD3 unit is 5–6 m thick. The upper part of the member consists of clayey silts and brown massive clays of 5–8 m thickness (Fig. 11).

Kada Hadar Member—facies description

The base of the Kada Hadar Member is defined by the lenticular Kada Hadar Tuff 'Confetti Clay' (CC)–green shale association. Its maximum thickness is 80 m in the Kada Gona Wadi, and it decreases to 30–40 m in the Sidi Hakoma and Ounda Hadar Wadis (Fig. 13).

Kada Hadar Tuff facies

This white-grey sandy tuff occurs as a channel fill, 2 m thick, in the upper Kada Hadar. Elsewhere the Kada Hadar Tuff (KHT) consists of a white chalky tuff 0.8–1.0 m thick (at Afofili and Sidi Hakoma Wadi), or a laminated beige tuff with a grey sandy layer 0.75 m thick (E of Kada Hadar).

Bouroukie Tuffs facies

The three Bouroukie Tuffs have been defined in the Bouroukie Wadi, a tributary of the Kada Hadar (Taieb & Tiercelin 1980; Walter 1980). BKT-1 occurs as two beige and yellow clayey tuff layers, 3–10 cm thick, about 17 m above the 'Confetti Clay', and is seen in Ounda Hadar Wadi. In the Kada Hadar Wadi, BKT-2 consists of three successive and distinct layers: (1) a pistachio-green and brown laminated clay or fine sand, with disturbed bedding. Leaf impressions have been found in this subfacies in the Kada Gona Wadi; (2) a chalky white tuff, 6 cm thick; and (3) a yellow tuff, 3 cm thick.

In the Kada Gona Wadi, the lenticular white, beige or pink 'Artefact Site Tuffs' (AST-1 and AST-2), with a maximum thickness of 2 m, are equivalent to the two upper BKT-2 tuffs. BKT-3 tuff occurs in the upper Kada Hadar Wadi as a beige, horizontally-bedded, sandy tuff 15–60 cm thick. Root casts are sometimes abundant. In the Kada Gona Wadi, BKT-3 consists of two blue-grey tuff layers 15 and 20 cm thick, 10 m above AST-1 and AST-2.

Conglomerate facies

This facies is widely represented in the Kada Hadar Member. Four subfacies have been identified:

1 *Massive or graded, matrix-supported microconglomerate.* This is between 5 and 50 cm thick, occasionally horizontally or obliquely bedded, and lies on an erosional base. This sediment contains small pebbles, granules, rounded claystone clasts, and root casts in an abundant sandy matrix. An ostracod fauna has been found in one bed in the Ounda Hadar Wadi.

2 *Cross-bedded conglomerate.* This is polymict, often cross-bedded, and pebble-rich. Lenses 1–3 m thick of beige clayey silt, black laminated fine sands, gravels, and purple brown clays with calcareous concretions are frequent, as are clay boulders. The maximum thickness of this conglomerate is 40 m.

3 *Artefact site intermediate conglomerate.* This is pebble-rich and lenticular, and lies on an erosional base. It is overlain by layers of smaller size pebbles with large lenses of yellow sands. The maximum thickness is 4.0 m. Oldowan artefacts have been found in this subfacies.

4 *Artefact site upper conglomerate.* This is sand matrix-supported and 4.0 m thick. It consists of medium to small pebbles and lies on an erosional base. It is overlain by a coarse pebble-rich conglomerate, 3–4 m thick. Oldowan artefacts have also been found in this facies.

Coarse sand facies

This facies comprises massive, cross-bedded coarse sand beds 0.4–3.0 m thick, which are sometimes lenticular, and have scattered lenses of pebbles and granules.

Very fine to fine sand facies

Ochre, grey to beige, fine to very fine, sands or sandstones, massive or cross-bedded, form beds 0.1–4.0 m thick. Bivalve and gastropod shells, fish remains, fossil wood, and clay clasts are frequent and sometimes abundant.

Silty clay facies

Massive or horizontally-bedded brown to beige silty clays, 0.4–1.0 m thick, are generally rich in plant debris and roots or burrow casts. Gastropods and fish scales are sometimes present.

Massive laminated clay facies

Beige, white-grey, brown or green clay beds are 0.4–5.0 m thick. Numerous hydroplastic faults affect these beds. Micro-fauna is uncommon. Carbonate nodules, probably of pedogenetic origin, are abundant.

'Confetti Clay' shale facies

The 'Confetti Clay' marker bed consists of a green or grey shale, 0.8–2.0 m thick, which in weathered exposures appears in the form of fine flakes or confetti. Brown nodules (probably apatite) are frequent. The mineral fraction of the shale is smectite, with significant amounts of illite (5%) which is unusual for the series.

Facies organization in the Kada Hadar member

In the Kada Hadar and Kada Gona Wadis, the Kada Hadar Member is 80 m thick. It lies above the Kada Hadar Tuff; the basal units comprise silty clays, massive clays with calcareous concretions, and 0.1–1.0 m thick layers of fine sand. Further up, the series consists of massive clays, silty clays and interbedded conglomerates 5–40 m thick. Six volcanic tuffs (the Bouroukie Tuffs) are interstratified in this series. In the Ounda Hadar, only the lower 40 m of the Kada Hadar Member are exposed and comprise massive or laminated clays and shales with interstratified silty clays, fine to coarse sands, which are often cross-bedded, and micro-conglomerates (Fig. 13).

Interpretation of the sedimentary environments of the Hadar Formation

Fluvio-lacustrine environments became dominant in the northern Afar Rift after the Oligo-Miocene and once the spreading and rotation of the Danakil Block (Sichler 1980) had produced the Danakil depression (Fig. 1). These environments, represented by sediments of the Red Series, persisted in the Danakil depression until *c*. 5 Ma.

In the southern Afar Rift, fluviatile and lacustrine sediments accumulated up to 10 Ma (Ch'orora Formation), following the spreading of the western and southeastern Ethiopian escarpments and the volcanism associated with these crustal movements. After 9 Ma, the locus of sedimentation seems to have migrated towards the N, along the western escarpment, until 2.5 Ma (Tiercelin 1981; Kalb *et al.* 1982b). This migration coincided with the appearance of N-NNW trending volcano-tectonic ridges, formed by fissure-eruption of tholeiitic basalts. These lavas belong to the Afar Stratoid Series (Tiercelin *et al.* 1980; Walter 1980) and poured out along an axis of expansion which was probably connected either with the sinistral rotation of the Danakil crank-arm or with an approximate W–E drift of the North Tadjoura continental block away from the western escarpment (Jacout 1984).

From *c*. 4.2 Ma onwards a lacustrine environment stretched along the western escarpment, probably from Bodo to the Hadar (Fig. 1). Tholeiitic volcanism preceded these conditions. Afar Series basalts dated at 4.5 Ma appear at the base of the Hadar Formation (the Meschelle basalt, Kalb *et al.* 1982b) and along the western escarpment (Aronson *et al.* 1980; Tiercelin 1981).

4.2–4.0 Ma BP

At this time (Basal Member) Lake Hadar was apparently shallow, of small extent and probably distinct from neighbouring Lake Bodo (Fig. 15; see Fig. 14 for key to Figs. 15–22). Laminated or shaley smectitic clays represent low energy deposition. Temporary regression is suggested by the impingement of swamp environments (plant remains and gastropod shell-bearing clays and silts) and by low to moderate energy shoreline or fluvial deposits (coarse to fine sands, which are massive, bedded or rippled). The monospecific ostracod fauna (*Cyprideis* sp. SH1; E 1173, Fig. 6) indicates a lake environment characterized by variable salinity (ranging as high as mesohaline) with probable seasonal variations (Peypouquet *et al.* 1983). This is confirmed by $\delta^{18}O_{PDB}$ values corresponding to highly variable palaeohydrological conditions that included phases of very high evaporation (Hillaire-Marcel *et al.* 1982).

The influence of a fluvial system is evident at the top of the Basal Member in the form of (commonly) cross-bedded coarse to fine sands with abundant plant debris and clay clasts (Fig. 16). A diversified ostracod fauna associated

FIG. 14. Key to Figs. 15–22.

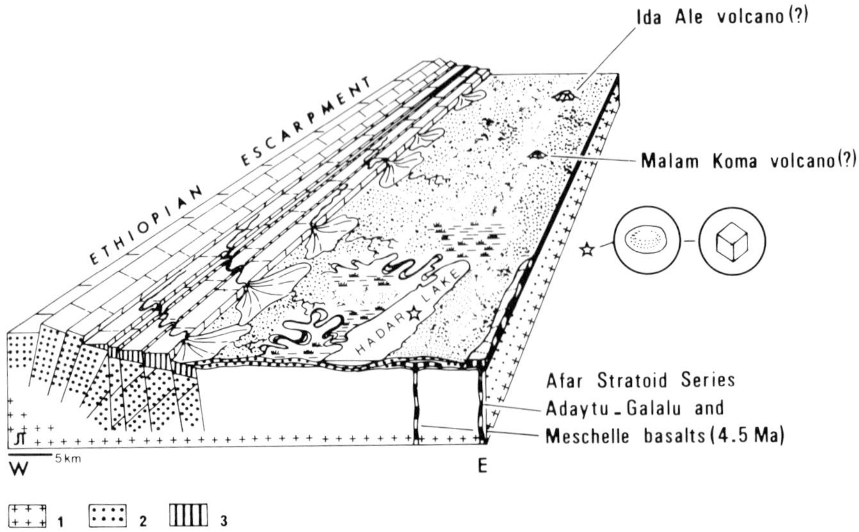

FIG. 15. Schematic representation of the Hadar region at 4.2 Ma. 1: Afar Basement; 2: Oligo-Miocene volcanics; 3: Miocene volcanics (after Zanettin & Justin-Visentin 1975). For key see Fig. 14.

with gastropods confirms the proximity of oligohaline waters (E 433, Fig. 6). However, the abundance of *Cyprideis* sp. SH1 (E 536, Fig. 6) suggests that there was no appreciable physico-chemical change in the lake environment. Isotopic studies indicate that the lake was mainly fed by large floods coming from the western escarpment and plateau, carrying clay minerals such as illite and kaolinite. The fluviatile facies of the Sidi Hakoma Tuff support this suggestion, and also partially support Brown's hypothesis (1982) of a distant volcanic eruption and long-distance water transport by an ancestral Awash. The input of such waters may explain the impressive development in the lake of the diatom *Melosira granulata* (an indicator of turbid but fresh waters) and the presence of the algae *Pediastrum* and *Botryococcus* (Bonnefille *et al.* 1983). Palynological inferences based on analyses of modern pollen rain (Bonnefille *et al.* 1983; Bonnefille & Vincens 1985) imply the proximity of mountain forest and a relief of 1300–1500 m within 10–20 km of the

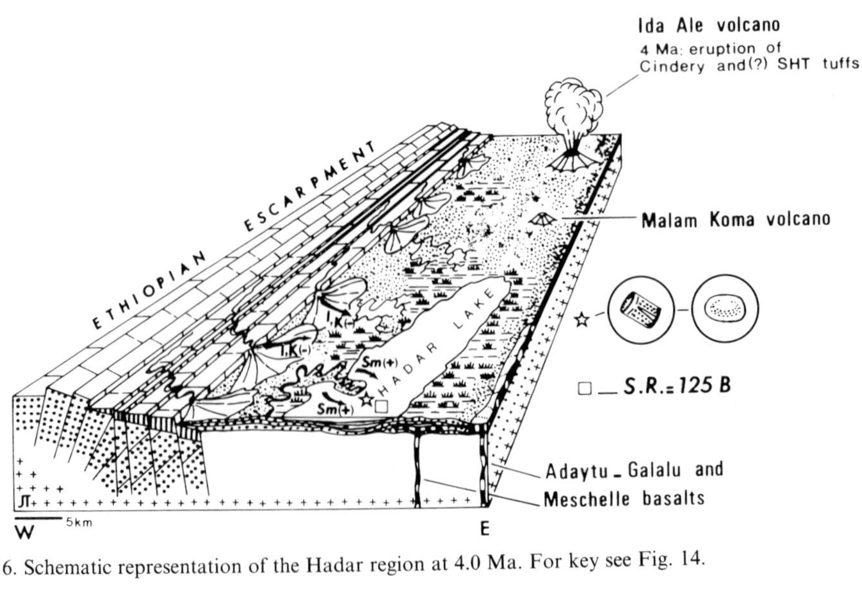

FIG. 16. Schematic representation of the Hadar region at 4.0 Ma. For key see Fig. 14.

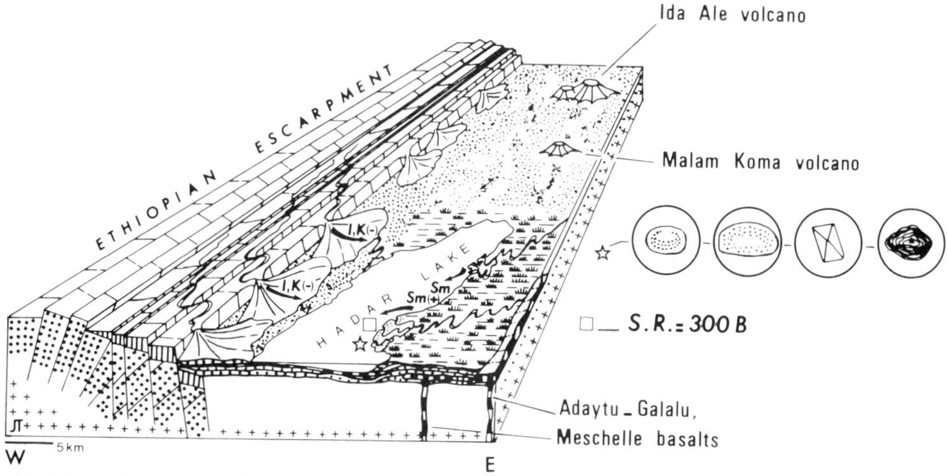

FIG. 17. Schematic representation of the Hadar region at *c.* 3.9 Ma. For key see Fig. 14.

lake. This is also indicated by the presence of the transported remains of the rodents *Praomys* and *Tachyoryctes* (Sabatier 1978, 1979a, b).

4.0–3.65 Ma BP

The influences of rivers persisted on the western shore of the lake during the period 4.0–3.65 Ma, with deposition of the Sidi Hakoma conglomeratic and sandy units (SH2, SH3). To the E of the lake the sediments reveal the development of a delta characterized by fine sands, silty clays and clays that have a mollusc fauna (Fig. 17). Subsequent deposition of silty clays, grey clays, shales with an abundant gastropod fauna, root marks, and coals suggest an abandoned lobe or a distal part of an interdistributary bay (Elliott 1974a,b). The ostracod fauna (*Cyprideis* SH1–KH1 association, E 1084–1088) suggests an oligohaline environment (salinity < 5‰) with a high input of river water and widely distributed humic-humolimnic acids (Peypouquet *et al.* 1983). This is confirmed by the abundance of aquatic plants such as *Typha* and *Cyperaceae* (Bonnefille *et al.* 1984). Fluctuations between fresh and highly saline waters allowed oncolites to develop (Hillaire-Marcel *et al.* 1982).

The presence of natrojarosite in a coal bed (E 1086) is consistent with an anoxic, reducing, acidic environment without clastic influx (Périnet *et al.* 1980), and with strong activity of Na^+ ions in interstitial waters, probably due to the proxi-

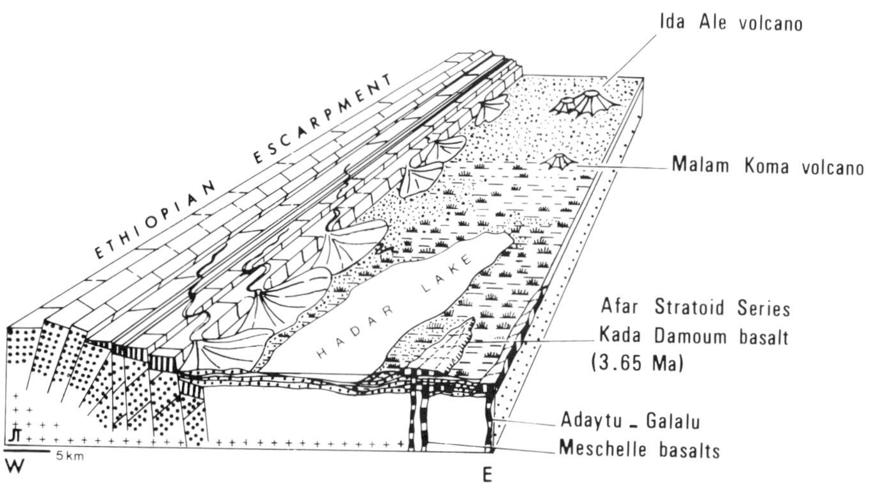

FIG. 18. Schematic representation of the Hadar region at 3.65 Ma. For key see Fig. 14.

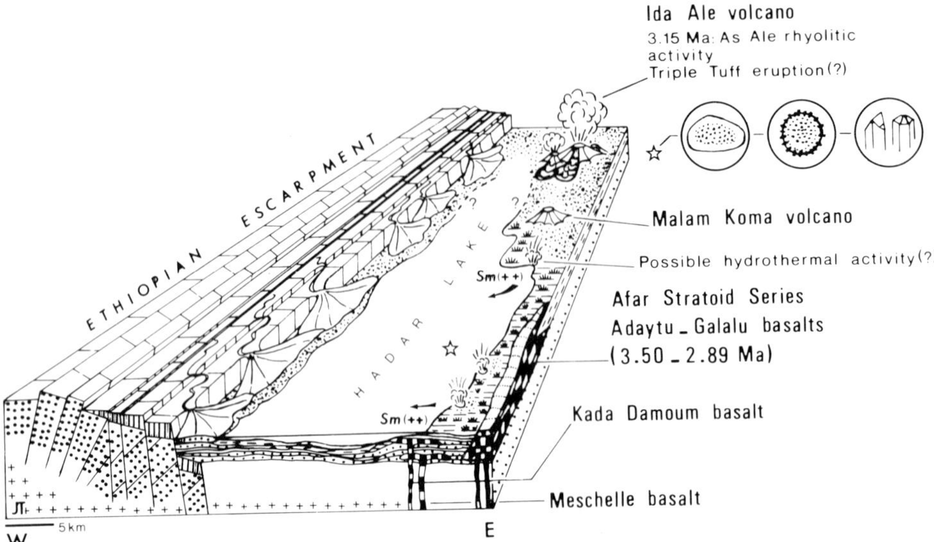

FIG. 19. Schematic representation of the Hadar region at *c.* 3.5–3.15 Ma. For key see Fig. 14.

mity of NaCl-rich lake waters (Gruas Cavagnetto *et al.* 1980). The ostracod fauna (E 1093) suggests that the environment is gradually becoming more lacustrine, with salinity possibly reaching the mesohaline level. The humid tropical climate postulated as necessary for natrojarosite formation is also indicated by the contemporary rodent fauna (Sabatier 1978, 1979b).

Swamps subsequently expanded mainly along the eastern shore, with fluvial influence and emergence being suggested by the pulmonate gastropod fauna. Towards 3.65 Ma, the Kada Damoum basalt erupted in this wet environment, baking sediments to a depth of 2–3 m (Walter & Aronson 1982) (Fig. 18). Contact metamorphism of the underlying coal layer (E 1086) suggests a proximity of the eruption fissure.

3.65–2.9 Ma BP

At this time, shorelines and marshes became submerged by the lake which expanded to occupy the whole Hadar region (Fig. 19). Volcanic activity continued along the margins of the basin, with tholeiitic fissure eruptions at Adaytu-Galalu *c.* 3.50 and 2.89 Ma, and at Mile *c.* 2.99 Ma, and rhyolitic activity of the volcanoes Ida Ale and As Ale occurring at *c.* 3.15 Ma (Aronson *et al.* 1980; Taieb & Tiercelin 1980; Walter 1980). Multiple pyroclastic eruptions occurred around 3.6 Ma in

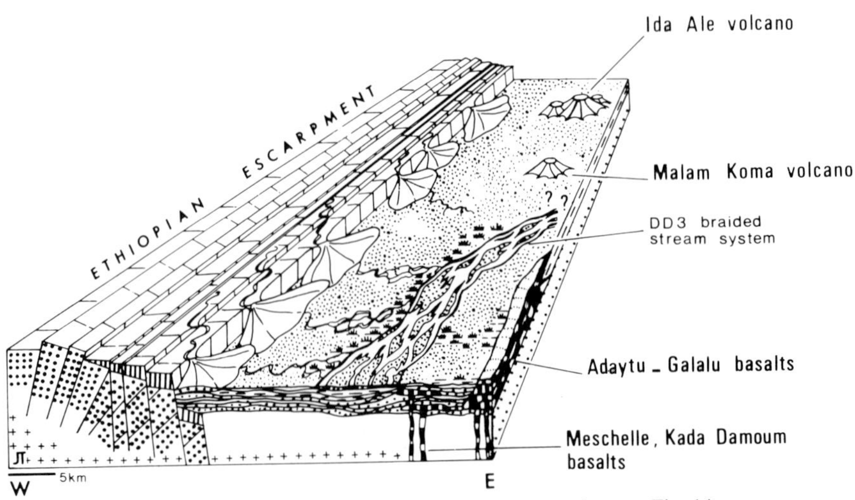

FIG. 20. Schematic representation of the Hadar region at *c.* 3.0 Ma. For key see Fig. 14.

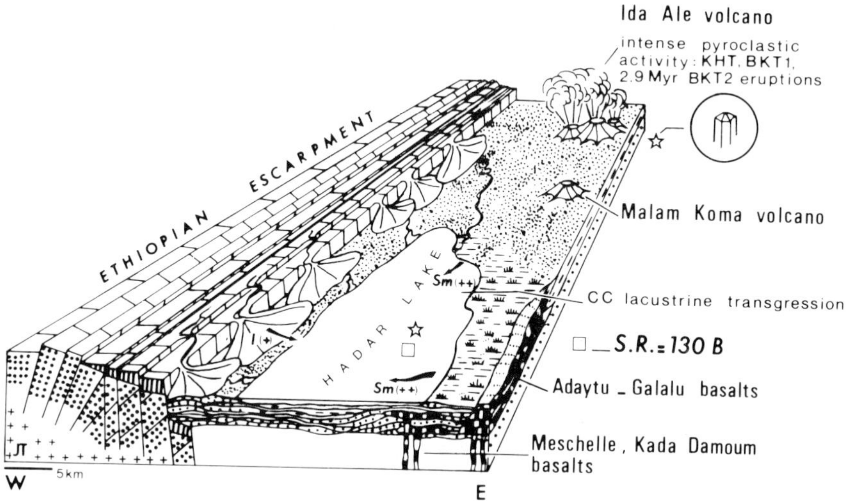

FIG. 21. Schematic representation of the Hadar region at 2.9 Ma. For key see Fig. 14.

the Hadar region (probably Ida Ale) and are represented in Lake Hadar by the TT1-5 tuff. The transgressive lacustrine environment was first characterized by an important change in hydrochemistry. High levels of salinity and dissolved nutrients, and silica and phosphatic-rich acidic waters are indicated by the ostracod fauna (large size *Cyprideis* sp. KH1), the diatom flora (?*Thalassiosira* sp., suggesting high salinity; Hecky & Kilham 1973) and by the presence of such minerals as apatite and baryte which could be taken to indicate local hydrothermal activity.

A temporary lacustrine regression and increasing fluvial activity is marked by the massive clays and silty clays, but most especially by the Denen Dora conglomeratic sand (DD3, Fig. 20). This unit probably represents a climatically-induced and widespread braided river system, with bushy and wooded banks but grassy valley gravel train (Johanson *et al.* 1982). Marsh and minor periods of fluvial activity then persisted, interrupted by a lake transgression which left the laterally extensive 'Confetti Clay' (Fig. 21). A tectonic origin may be suggested for this event, since there is an unusual quantity of illite that is possibly derived from the uplifted basement of the Afar. Rhyolitic

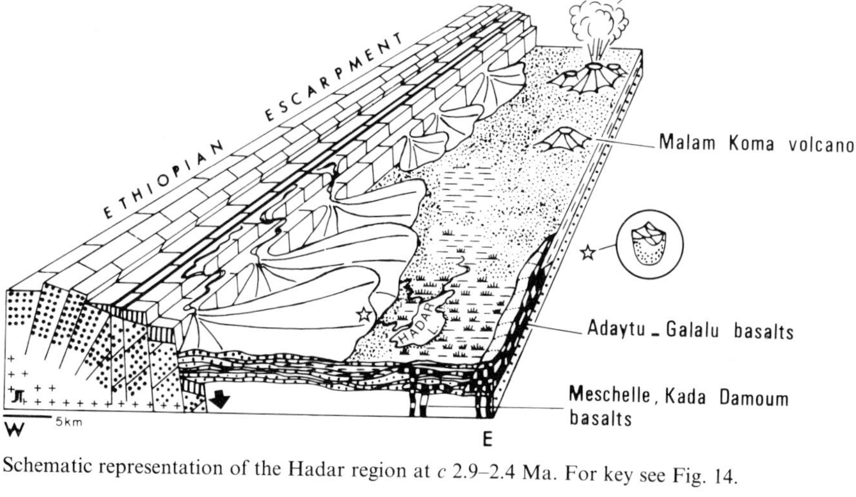

FIG. 22. Schematic representation of the Hadar region at *c* 2.9–2.4 Ma. For key see Fig. 14.

activity of the Ida Ale volcanic centre produced numerous tuffs (KHT, BKT-1, BKT-2) which accumulated in the lacustrine, marshy and fluvial environments of the Hadar.

2.9–2.4 Ma BP

The appearance of thick conglomerate–sand–silt sequences in the upper part of the Kada Hadar Member may indicate widespread development of alluvial fan and fan-delta environments along the western escarpment (Bluck 1964) (Fig. 22). Lacustrine or swamp areas may have formed a significant landscape element, since local gastropod and ostracod faunas and an abundance of aquatic plants (Bonnefille & Vincens 1985) persisted until the complete infilling of the Hadar basin had occurred around 2.4 Ma.

The lack of major isotopic changes in the coarse deposits, point to a rapidly drained system but no unduly strong influence of evaporation (Hillaire-Marcel *et al.* 1982). The sudden input of coarse clastic deposits and the development of wide alluvial fans suggests major tectonic movements along the western Ethiopian escarpment from 2.9 Ma. Basaltic activity on the northeastern margin about 2.6 Ma (Aronson *et al.* 1980) and pyroclastic eruptions *c.* 2.4 Ma from Ida Ale (represented by BKT-3) may indicate recurrent faulting.

Conclusions

From 4.2 to 2.4 Ma, the palaeogeography of the Hadar region appears to be characterized by an intricate succession of lacustrine, marshy and fluvial environments in a N–S elongate depression formed by the W–E spreading of the Danakil and North Radjoura blocks.

During this period, there were rapid lake level fluctuations and changes in the water balance, probably related indirectly or directly to volcanic activity and hydrothermal input along the margins of the depression. However, the lake and associated marshes were mainly fed by large floods emanating from the western Ethiopian escarpment or from uplands located to the E and belonging to the Danakil and North Tadjoura blocks.

During a period of 1.3 Ma, tectonic activity in the Hadar basin appears to have been limited to quiet subsidence, particularly along the eastern margin, where the thickest sediments are encountered. Only minor deformations in the sediments, such as 'hydroplastic faults', reflect this subsidence, which was probably encouraged by the continued spreading of the Danakil crank-arm.

Major tectonic activity took place between 2.9 and 2.4 Ma along the western escarpment, and is marked by a large input of detrital sediments. The present elevation of the Hadar basin (500 m) may have been attained at the end of this tectonic phase, which signifies the completion of basin infilling.

ACKNOWLEDGEMENTS: This work has been supported by the CNRS and NSF. Thanks are due to Johannes, the mechanic, and Ababa Ali, the Afar warden. Fig. 3 was drawn by M. Decobert.

References

ARONSON, J.L., SCHMITT, T.J., WALTER, R.C., TAIEB, M., TIERCELIN, J.J., JOHANSON, D.C., NAESER, C.W. & NAIRN, A.E.M. 1977. New geochronologic and paleomagnetic data for the hominid-bearing Hadar Formation of Ethiopia. *Nature* **267,** 323–7.

——, WALTER, R.C. & TAIEB, M. 1983. Correlation of Tulu Bor Tuff at Koobi Fora with the Sidi Hakoma Tuff at Hadar. *Nature* **306,** 209–10.

——, ——, —— & NAESER, C.W. 1980. New geochronological information for the Hadar Formation and the adjacent central Afar, Ethiopia. In: LEAKEY, R.E. & OGOT, B.A. (eds) *Panaf. Congr. Prehist. Quat. Studies Nairobi, 1977.* TILLMIAP, Nairobi, pp.47–52.

BANNERT, P., BRINCKMANN, J., KADING, K.C., KNETSCH, G., KURSTEN, M. & MAYRHOFER, H. 1970. Zur Geologie der Danakil-Senke (Nördliches Afar-Gebiet, NE-äthiopien). *Geol. Rundsch.* **59/2,** 409–43.

BLUCK, B.J. 1964. Sedimentation of an alluvial fan in Southern Nevada. *J. Sed. Petrol.* **34,** 395–400.

BONNEFILLE, R. & VINCENS, A. 1985. Apport de la palynologie à l'environnement des Hominidés d'Afrique orientale. In: Masson (ed) *L'environnement des Hominideś au Plio-Pléistocène.* Fondation Singer-Polignac, pp.237–78.

——, —— & BUCHET, G. 1983. Problème tectonique et paléogéographique posé par la palynologie à Hadar, Rift Ethiopien. In: *Bassins sédimentaires en Afrique.* Rés. Commun. Séance spécial. Soc. Géol. Fr., Marseille, 1983. Trav. Lab. Sci. Terre, St-Jérôme, Marseille, No. 15.

——, —— & —— 1984. Vegetation history of the Hadar Hominid site (Abstract). *8th Int. Conf. Ethiop. Stud.* Addis Ababa, 1984.

BROWN, F.H. 1982. Tulu Bor Tuff at Koobi Fora correlated with the Sidi Hakoma Tuff at Hadar. *Nature* **300,** 631–3.

—— 1983. Replies to Aronson, J.L. *et al. Nature* **306,** 210.

—— & Cerling, T.E. 1982. Stratigraphical significance of the Tulu Bor Tuff of the Koobi Fora Formation. *Nature* **229,** 212–15.

Clark, J.D., Berhane Asfaw, Getaneh Assefa, Harris, J.W.K., Kurashina, H., Walter, R.C., White, T.D. & Williams, M.A.J. 1984. Palaeoanthropological discoveries in the Middle Awash valley, Ethiopia. *Nature* **307,** 423–8.

Elliott, T. 1974a. Abandonment facies of high-constructive lobate deltas, with an example from the Yoredale Series. *Proc. Geol. Assoc.* **85,** 359–65.

Elliott, T. 1974b. Interdistributary bay sequences and their genesis. *Sedimentology* **21,** 611–22.

Gruas Cavagnetto, C., Laurain, M. & Meyer, R. 1980. Un sol de mangrove fossilisé dans les lignites du Soissonnais (Yprésien) à Verzenay (Marne). *Géobios* **13,** 795–801.

Hecky, R.E. & Kilham, P. 1973. Diatoms in alkaline, saline lakes: ecology and geochemical implications. *Limnol. Oceanogr.* **18,** 53–71.

Hillaire-Marcel, C., Taieb, M., Tiercelin, J.J. & Page, N. 1982. A 1.2 Myr record of isotopic changes in a late Pliocene rift lake, Ethiopia. *Nature* **296,** 640–2.

Hutchinson, R.W. & Engels, G.G. 1970. Tectonic significance of regional geology and evaporite lithofacies in northeastern Ethiopia. *Phil. Trans. R. Soc. Lond.* **267A,** 313–29.

Jacout, P. 1984. *Etude paléomagnétique et géochronologique en Afar (Contribution à l'étude géodynamique du sud de la Mer Rouge).* Unpublished Diplôme Et. App., Université de Bretagne occidentale, Brest, 32pp.

James, T.C. 1967. Thermal springs in Tanzania. *Trans. Inst. Min. Metal.* **76B,** 1–18.

Johanson, D.C. & Edey, M.A. 1981. *Lucy: The Beginnings of Humankind.* Simon & Schuster, New York.

——, Taieb, M. & Coppens, Y. 1982. Pliocene hominids from the Hadar Formation, Ethiopia (1973–1977): stratigraphic, chronologic, and paleoenvironmental contexts, with notes on Hominid morphology and systematics. *Am. J. Phys. Anthropol.* **57,** 373–402.

——, ——, Gray, B.T. & Coppens, Y. 1978a. Geological framework of the Pliocene Hadar Formation (Afar, Ethiopia) with notes on paleontology including hominids. In: Bishop, W.W. (ed) *Geological Background to Fossil Man.* Scottish Academic Press, Edinburgh, pp.549–64.

—— & White, T.D. 1980. On the status of *Australopithecus afarensis. Science* **207,** 1104–5.

——, —— & Coppens, Y. 1978b. A new species of the genus *Australopithecus* (Primate: Hominidae) from the Pliocene of eastern Africa. *Kirtlandia* **28.**

Kalb, J.E., Jolly, C.J., Assefa Mebrate, Sleshi Tebedge, Smart, C., Oswald, E.B., Cramer, D., Whitehead, P., Wood, C.B., Conroy, G.C., Tsrha Adefris, Sperling, L. & Berhane Kana. 1982a. Fossil mammals and artefacts from the Middle Awash valley, Ethiopia. *Nature* **298,** 25–9.

——, Oswald, E.B., Sleshi Tebedge, Assefa Mebrate, Tola, E. & Peak, D. 1982b. Geology and stratigraphy of Neogene deposits, Middle Awash valley, Ethiopia. *Nature* **298,** 17–25.

——, Wood, C.B., Smart, C., Oswald, E.B., Assefa Mabrete, Sleshi Tebedge & Whitehead, P. 1980. Preliminary geology and palaeontology of the Bodo D'Ar hominid site, Afar, Ethiopia. *Palaeogeogr. Palaeoclimatol. Palaeoecol.* **30,** 107–20.

Périnet, G., Taieb, M. & Tiercelin, J.J. 1980. Présence de Natrojarosite dans les sédiments lacustres d'Hadar (Ethiopie). *8è Réun. Ann. Sci. Terre, Marseille,* 276. Soc. Geol. Fr. éd.

Peypouquet, J.P., Carbonel, P., Taieb, M., Tiercelin, J.J. & Perinet, G. 1983. Ostracoda and evolution process of paleohydrologic environments in the Hadar Formation (the Afar depression, Ethiopia). In: Maddocks, R.F. (ed). *Applications of Ostracoda.* Univ. Houston Geosc., pp.277–85.

Reimer, T.O. 1978. Detrital barytes in the Karoo supergroup of Southern Africa. *Mineral. Deposita* **13,** 235–44.

Robert, P. 1979. Classification des matières organiques en fluorescence. Application aux roches-mères pétrolières. *Bull. Cent. Rech. Explor. Prod. Elf-Aquitaine* **3,** 223–63.

—— 1980. *Etude d'un échantillon de charbon, Hadar (Ethiopie).* Internal report Soc. Natn. Elf-Aquitaine (Product.), Lab. Géol. Boussens, France.

Roche, H. & Tiercelin, J.J. 1977. Découverte d'une industrie lithique ancienne *in situ* dans la formation d'Hadar, Afar Central, Ethiopie. *C. R. Acad. Sci. Paris* **284D,** 1871–4.

—— & —— 1980. Industries lithiques de la formation plio-pléistocène d'Hadar, Ethiopie (campagne 1976). In: Leakey, R.E. & Ogot, B.A. (eds) *Proc. 8th Panaf. Congr. Prehist. Quat. Studies, Nairobi, 1977.* TILLMIAP, Nairobi, pp.194–9.

Sabatier, M. 1978. Un nouveau *Tachyoryctes* (Mammalia, Rodentia) du bassin pliocène de Hadar (Ethiopie). *Géobios* **11,** 95–9.

—— 1979a. *Les rongeurs des sites à hominidés de Hadar et Melka-Kunture (Ethiopie).* Unpublished PhD thesis. 3è Cycle Evolution et biosystématique. Université des Sciences et Techniques du Languedoc.

—— 1979b. Les rongeurs fossiles de la formation de Hadar et leur intérêt paléoécologique. *Bull. Soc. Géol. Fr.* **21,** 309–11.

Sichler, B. 1980. La biellette danakile: un modèle pour l'évolution géodynamique de l'Afar. *Bull. Soc. Géol. Fr.* **22,** 925–33.

Sickenberg, O. & Schonfeld, M. 1975. The Ch'orora Formation. Lower Pliocene limnical sediments in the southern Afar (Ethiopia). In: Pilger, A. & Rösler, A. (eds) *Afar Depression of Ethiopia.* Schweizerbart, Stuttgart, pp.277–83.

Taieb, M., Johanson, D.C., Coppens, Y. & Aronson, J.L. 1976. Geological and palaeontological background of Hadar hominid site, Afar, Ethiopia. *Nature* **260,** 289–97.

—— & Tiercelin, J.J. 1979. Sédimentation pliocène et paléoenvironnements de rift: exemple de la formation à Hominidés d'Hadar (Afar, Éthiopie). *Bull. Soc. Géol. Fr.* **21,** 243–53.

—— & —— 1980. La stratigraphie et paléoenvironnements sédimentaires de la formation d'Hadar, dépression de l'Afar, Ethiopie. In: LEAKEY, R.E. & OGOT, B.A. (eds) *Proc. 8th Panaf. Congr. Prehist. Quat. Studies Nairobi, 1977.* TILLMIAP, Nairobi, pp.109–14.

TIERCELIN, J.J. 1981. *Rifts continentaux, tectonique, climats, sédiments. Exemples: la sédimentation dans le Nord du Rift Gregory (Kenya) et dans le Rift de l'Afar (Ethiopie) depuis le Miocène.* Unpublished D.Sc thesis, Université d'Aix-Marseille II, Marseille, 260pp.

——, MICHAUX, J. & BANDET, Y. 1979. Le Miocène supérieur du Sud de la Dépression de l'Afar, Ethiopie: sédiments, faunes, âges isotopiques. *Bull. Soc. Géol. Fr.* **21,** 255–8.

—— & TAIEB, M. 1976. La série lithostratigraphique d'un bassin sédimentaire d'âge plio-pléistocène du rift de l'Afar à Hadar, vallée de l'Awash, Ethiopie. *4è Réun. Ann. Sci. Terre, Paris*, 380. Soc. Géol. Fr. éd.

——, —— Faure, H. 1980. Continental sedimentary basins and volcano-tectonic evolution of the Afar Rift. In: *Geodynamic Evolution of the Afro-Arabic Rift System.* Accad. Nazionale dei Lincei, Rome, pp.491–504.

WALTER, R.C. 1980. *The volcanic history of the Hadar Early Man site and the surrounding Afar region of Ethiopia.* Unpublished PhD thesis, Case Western Reserve University, Cleveland, 426pp.

—— & ARONSON, J.L. 1982. Revisions of K/Ar ages for the Hadar hominid site, Ethiopia. *Nature* **296,** 122–7.

ZANETTIN, B. & JUSTIN-VISENTIN, E. 1975. Tectonic and volcanological evolution of the western Afar margin. In: PILGER, A. & RÖSLER, A. (eds) *Afar Depression of Ethiopia.* Schweizerbart, Stuttgart, pp.300–9.

J.J. TIERCELIN, GIS Océanologie et Géodynamique, Université de Bretagne Occidentale, 29287 Brest, Cedex, France.

Depositional context of Plio-Pleistocene hominid-bearing formations in the Middle Awash valley, southern Afar Rift, Ethiopia

M.A.J. Williams, Getaneh Assefa & D.A. Adamson

SUMMARY: The Awash flows from the Ethiopian Highlands via the Ethiopian Rift into Lake Abhe in the Afar Rift. Over 8000 km^2 of Neogene fluvio-lacustrine deposits crop out in the Middle Awash valley, of which we have surveyed 300 km^2 E of the river between 10°25′N and 10°45′N. Pleistocene sediments adjacent to the Awash floodplain consist mostly of unfossiliferous brown clays with sand and gravel interbeds which contain vertebrate (including hominid) fossils and prehistoric stone artefacts (Oldowan and Acheulian). Palaeocurrent directions in these fossiliferous sands invariably parallel the ephemeral sand-bed channels that flow W from the basalt escarpment to the Awash. We interpret the brown clays as palaeo-Awash floodplain deposits, in contrast to the fossiliferous gravel-sands, which are mainly of local provenance. Away from the river, alternating Pliocene clays and diatomites, with minor fossiliferous gravel-sands, are capped by a 3.8–4.0 Ma primary air-fall tuff. Facies changes are more frequent above this 'Cindery Tuff', indicating tectonic disruption of the early Pliocene lakes, reflected also in a progressive increase in fluvio-volcanic gravels from Pliocene to early Pleistocene.

Pliocene hominids have only been found in Africa so far. Similarities in plasma protein structures and in immunological response point to a divergence between pongid and hominid lineages as recently as 5 Ma (Sarich & Wilson 1967). The earliest dated *Australopithecus afarensis* fossils, recovered in 1981 from the Middle Awash valley in the western Afar Rift of Ethiopia (Figs. 1 and 2), come from 11 m beneath a primary air-fall tuff that we have helped to trace laterally for over 25 km (Clark *et al.* 1984). This tuff is dated by $^{40}Ar/^{39}Ar$ and by zircon fission track analyses to 3.8–4.0 Ma (Hall *et al.* 1984). In this paper we consider the depositional environments in which the Plio-Pleistocene hominid-bearing formations of the Middle Awash valley may have accumulated (Figs. 3–10). Our conclusions are necessarily provisional, being based on a reconnaissance survey of *c.* 300 km^2 of the Middle Awash valley during October–November 1981, and upon a semi-detailed stratigraphic study of 16 Pliocene and 16 Pleistocene sections, supplemented by field examination of 8 archaeological sections.

The oldest stone tool-making known at present is dated later than 2.6–2.8 Ma, and is in the Middle Awash valley of Ethiopia (Roche & Tiercelin 1977, 1980; Harris 1983). The tools come from Kada Gona, a small valley near Hadar, an important locality for fossil hominids. At Hadar, fluvio-lacustrine clays and sands contain Australopithecine fossils dated between 2.8 Ma and either 3.6 Ma (Walter & Aronson 1982) or 3.1 Ma (Schmitt & Nairn 1984), with apparently correlative vertebrate fossil and tephrochronological evidence favouring the younger age.

In the Ethiopian Uplands, pollen collected from diatomites laid down in Pliocene Lake Gadeb (Williams *et al.* 1979) reveals that the climate became cooler and drier towards 2.5–2.35 Ma (Bonnefille 1983), a conclusion consistent with detailed palaeoecological analysis of contemporary diatom associations (Gasse 1980).

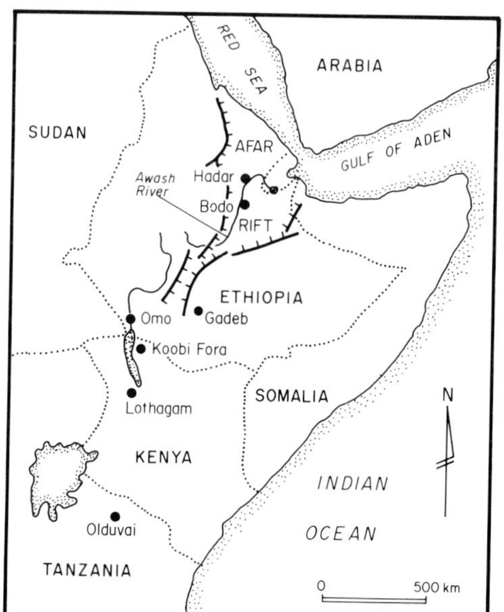

FIG. 1. Some key Plio-Pleistocene prehistoric sites in East Africa. Gadeb is in the southeastern uplands of Ethiopia; Hadar and Bodo are in the western Afar Rift. Rift structures are only shown for Ethiopia.

From FROSTICK, L.E. *et al.* (eds) 1986, *Sedimentation in the African Rifts*, Geological Society Special Publication No. 25, pp. 241–51.

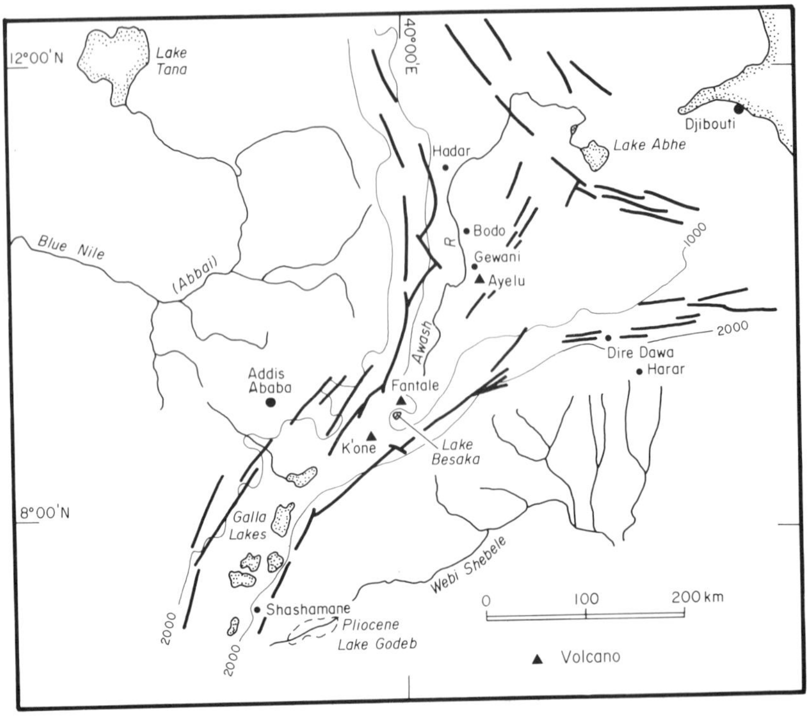

FIG. 2. Junction of the Ethiopian and Afar Rifts. K'one, Fantale and Ayelu are Quaternary central volcanoes of the Ethiopian Rift. Hadar and Bodo prehistoric sites are in a Neogene rift-margin basin bounded by the Ethiopian escarpment and by block-faulted Afar stratoid basalts to the W and E respectively. Lake Abhe is on a microplate of the actively spreading central Afar. The Awash river thus traverses four distinct geotectonic provinces from its source in the Ethiopian Uplands near Addis Ababa to its present terminus in Lake Abhe.

There was also a period of rapid ice build-up in the northern hemisphere during 2.4 Ma (Shackleton & Opdyke 1977; Shackleton *et al.* 1984). In China loess began to accumulate at 2.4 Ma, when the late Pliocene climate became cool and dry (Heller & Liu 1982), and in southern Europe the inception of the present seasonally dry Mediterranean style of climate was at *c.* 2.3 Ma (Suc 1984).

A case therefore exists in support of the view that stone tool-making may have been a cultural adjustment, first evident in the western Afar Rift, in response to climatic and associated environmental changes during the late Pliocene (Assefa *et al.* 1982). This paper considers the nature and sedimentary repercussions of some of these changes, focusing upon the Middle Awash valley between latitudes 10° and 11°N.

Volcano-tectonic setting

Uplift of Ethiopia-Arabia began during the Cretaceous (Mohr 1967). By the mid-Tertiary, depressions had formed along the embryonic Gulf of Aden and Red Seas (Hutchinson & Engels 1972; Mohr 1975). Eruption of the Trap Series flood basalts had begun by 30 Ma, culminating in widespread and voluminous lava extrusion across the uplifted Mesozoic, Palaeozoic and Precambrian basement rocks towards 25–30 Ma (Jones 1976; Mohr 1983).

Late Oligocene–early Miocene opening of the Afar Rift was triggered by the periclinal uplift associated with eruption of the Trap basalts, and resulted in continental rifting of the Afar, with episodic faulting along its western and southern margins (Barberi *et al.* 1975).

The Ethiopian proto-Rift began as a narrow late Miocene trough. In southwestern Ethiopia there may have been shallow proto-rifts before then (Davidson & Rex 1980). The eastern scarp of the Ethiopian Uplands, which forms the western margin of the Afar Rift (Fig. 2), seems to have evolved earliest in the N, moving southwards to reach the Ethiopian proto-Rift during the early Pliocene (Zanettin & Justin-Visentin 1975). Pliocene ash flows erupted along the rift margins, mantling both plateaux and rift floors (Mohr

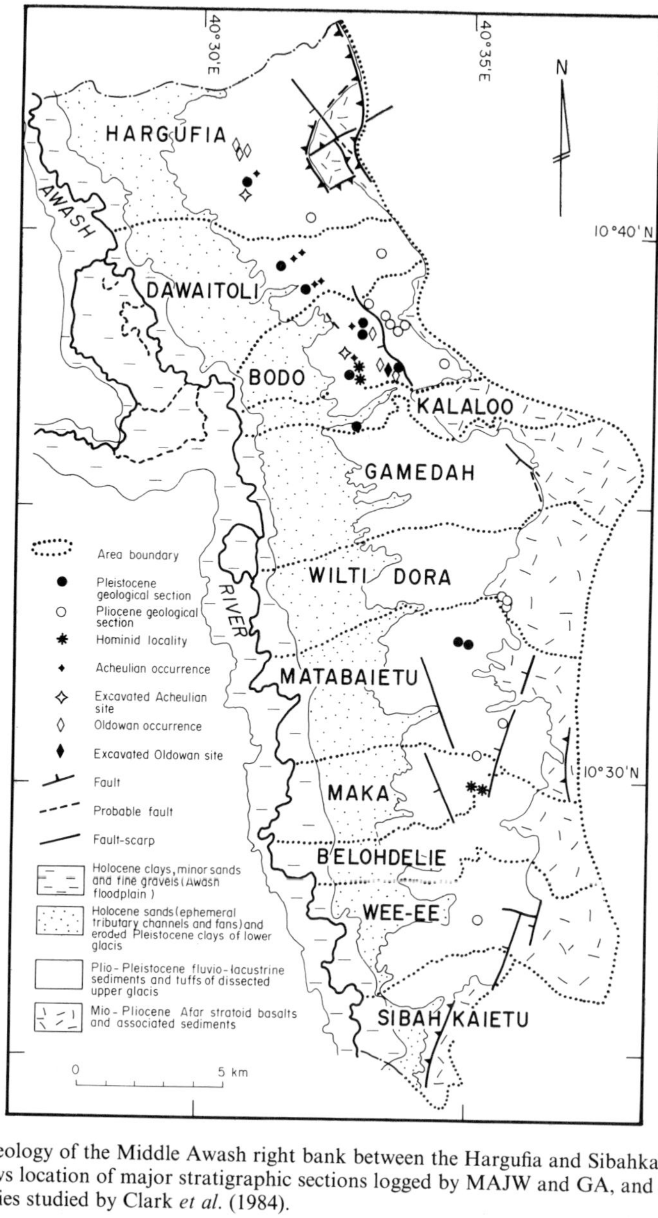

FIG. 3. Simplified geology of the Middle Awash right bank between the Hargufia and Sibahkaietu tributary valleys. Figure shows location of major stratigraphic sections logged by MAJW and GA, and of archaeological and hominid localities studied by Clark *et al.* (1984).

1968). Late Miocene faulting of the Ethiopian Rift margin intensified later in the Pliocene during a renewal of swell uplift, with reactivation of the western Afar margin. Sizeable lakes occupied the floors of downwarped basins which developed along the foot of the escarpments overlooking western and southern Afar. The Upper Miocene fluvio-lacustrine Ch'orora Formation accumulated in one such basin (Tiercelin *et al.* 1979), and the Neogene fluvio-lacustrine deposits of the Middle Awash valley N of Gewani (Fig. 2) were laid down further N, in a basin parallel to the present eastern scarp of the Ethiopian Highlands.

During the Quaternary, continental rifting gave way in central and eastern Afar to sea-floor spreading along discrete axes (Barberi & Varet 1977). Intermittent Quaternary faulting was

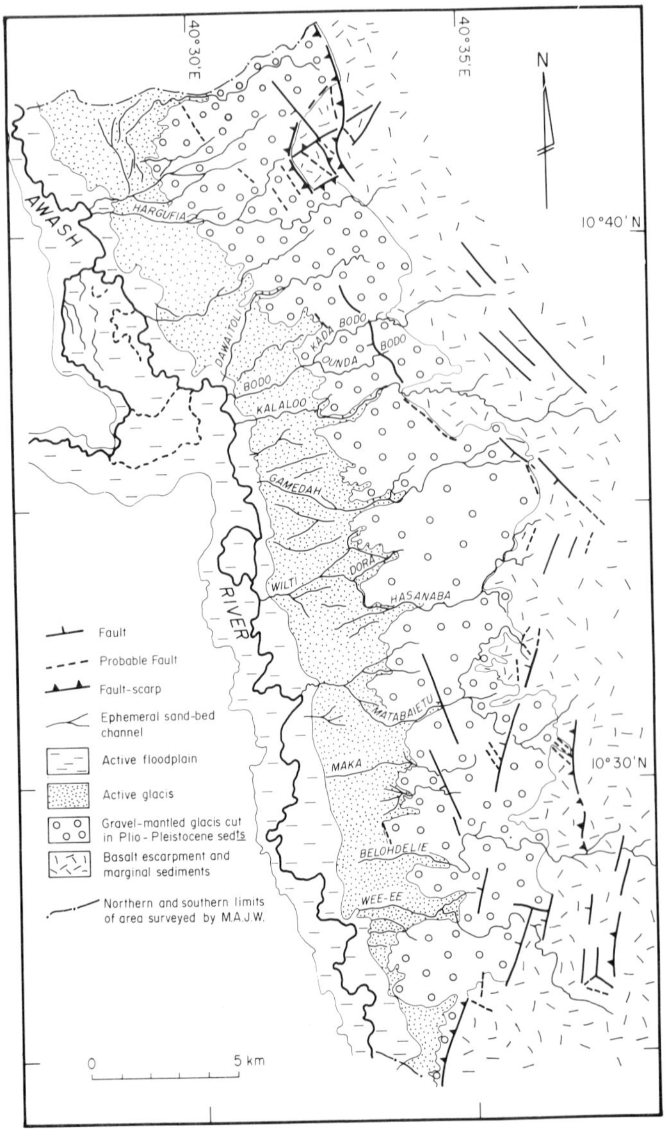

FIG. 4. Geomorphic map of the Middle Awash right bank between latitudes 10°25′ and 10°45′N. Note the inactive glacis mantled by basalt cobbles cut across early Pliocene to Middle Pleistocene formations. In their middle and upper reaches the ephemeral sand-bed channels are incised, but become shallow distributary systems on emerging onto the active glacis flanking the Awash floodplain. Solid lines are faults for which the direction of throw is not known.

accompanied by development of large silicic central volcanoes in the Ethiopian and Afar Rifts (Fig. 2). Fluctuations in the rift lakes reflect regional volcano-tectonic events, as well as global Plio-Pleistocene climatic oscillations (Gasse *et al.* 1980). How did these events influence the pattern and tempo of sedimentation in the Middle Awash valley during the past 4–5 Ma?

Pliocene depositional environments

Taieb's pioneering research (1974) revealed the palaeontological potential of the Middle and Lower Awash, as did later surveys by Kalb and colleagues (Kalb *et al.* 1982a, b), and by Johanson and co-workers at Hadar (Johanson *et al.* 1982). Aronson & Taieb (1980) discussed the possible

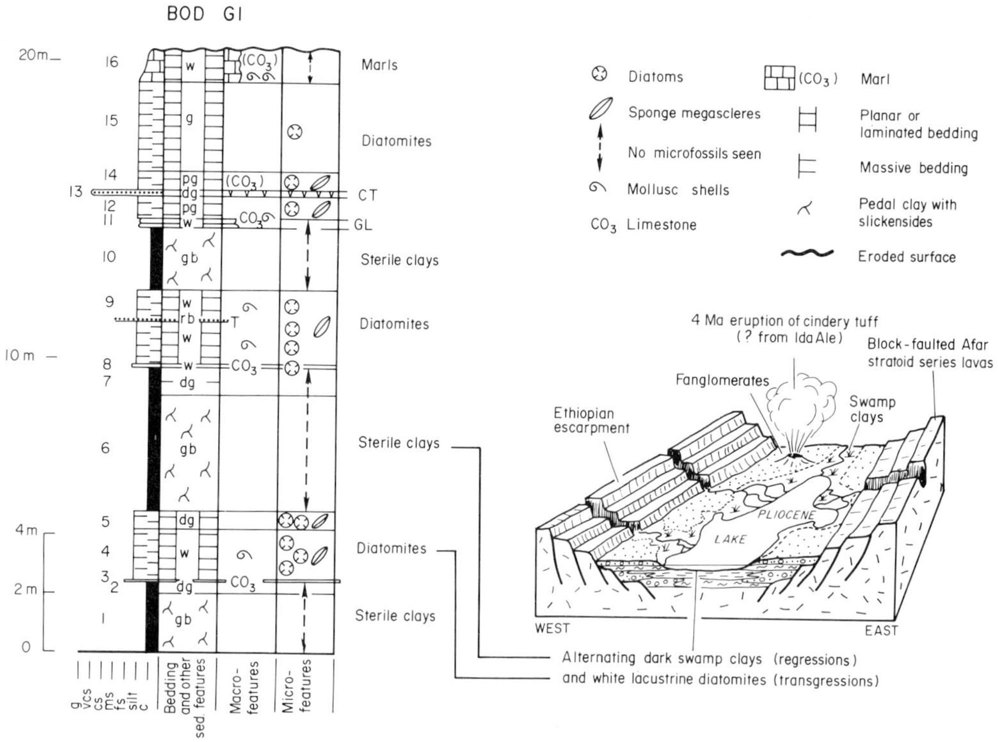

FIG. 5. Stratigraphy and geomorphic setting of the early Pliocene Middle Awash Lake. Capped by the 3.8–4.0 Ma Cindery Tuff, the lacustrine diatomites crop out over a minimum N–S distance of 25–30 km. In Bodo geological section 1 (BOD G1), GL = gastropod limestone, CT = Cindery Tuff. Colour symbols are w = white, g = grey, p = pale, d = dark, b = brown, r = red. The lithological symbols are black = clay, dashes = diatomite, dots = sand. On bottom axis, g = gravel, vcs = very coarse sand, cs = coarse sand, ms = medium sand, fs = fine sand, c = clay. Note alternation of diatomites and sterile clays, devoid of microfossils. The relief and degree of faulting E of the Pliocene lake are considerably exaggerated. (For section location see Fig. 9.)

origins of the fossiliferous Pliocene sands at Hadar, but confusion persists as to whether the non-laminated Hadar clays are predominantly lacustrine (Taieb & Tiercelin 1979; Hillaire-Marcel *et al.* 1982) or fluviatile (Aronson & Taieb 1980), local or allochthonous.

An E–W geological transect through the Bodo catchment (studied in some detail by MAJW and GA in 1981) passes from the Afar stratoid basalts of probable end-Miocene to early Pliocene age, via faulted Pliocene and Pleistocene fluvio-lacustrine clays, diatomites, sands, tuffs and gravels, to the Holocene floodplain of the Awash (Figs. 3 and 4). The youngest sediments are topographically lowest and closest to the river, and are separated from the next youngest block of sediments by faults that run NNW–SSE, parallel to the faulted margin of the basalt scarp.

Geomorphically, the critical contrast is between the lower, active glacis which is devoid of a gravel mantle, and the upper defunct glacis which is capped by a 1–3 m thick layer of basalt cobbles in a matrix of gritty brown clay. This gravel mantle (the MPA or Middle Pleistocene Awash Gravels of Kalb *et al.* 1982b) is of local provenance here, and it thins away from the basalt scarp. It was laid down as low-angle alluvial fans by channels emerging from the horst and flowing W across the eroded surface of the faulted Plio-Pleistocene formations during the mid to late Pleistocene, and possibly at intervals before then. A distal portion of the gravel mantle overlies the fossiliferous Middle Pleistocene beds that yielded the Bodo hominid skull in 1976 (Kalb *et al.* 1980), and a large left parietal fragment of a second hominid in 1981 (Asfaw 1983; Clark *et al.* 1984). Basalt cobbles of the size found in the glacis mantle (15–100 cm) are conspicuously absent from any of the Pliocene sections we examined (Fig. 3), a point relevant to our interpretation of the sections shown in Figs. 5–8.

Although exposure is good, lateral correlation

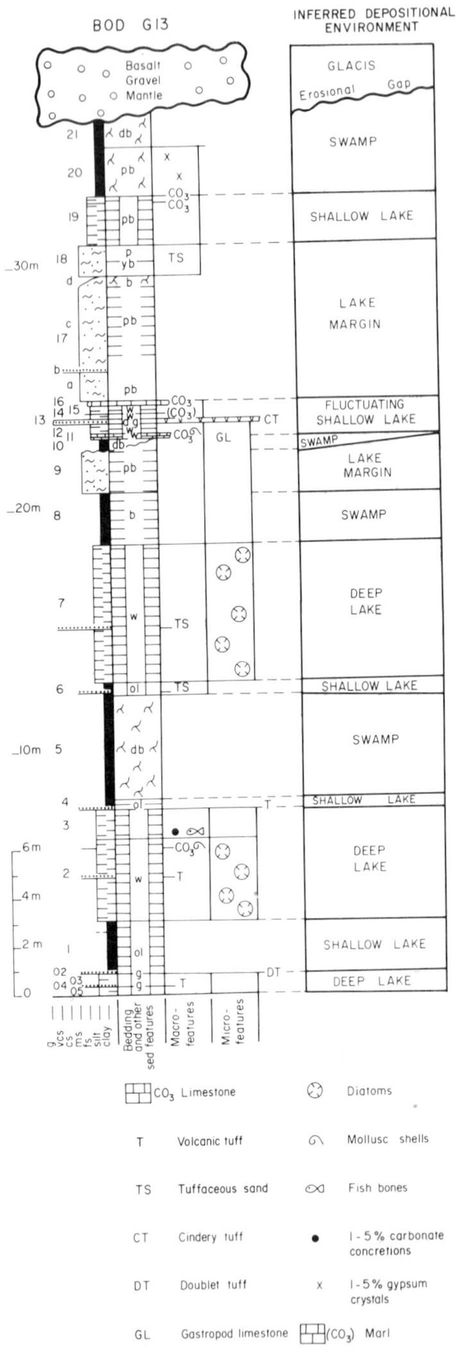

FIG. 6. Early Pliocene geological section Bodo 13 (BOD G13). Note alternation of diatom-rich 'deep' and clay-rich 'shallow' lake or lake margin facies prior to eruption of 3.8–4.0 Ma Cindery Tuff, after which lake margin, swamp or floodplain facies predominate. Lithological symbols as in Fig. 5, plus clay–sand mixtures as dots and wavy lines. Gypsum crystals are authigenic here, and in Fig. 7 and 10. (For section location see Fig. 9.)

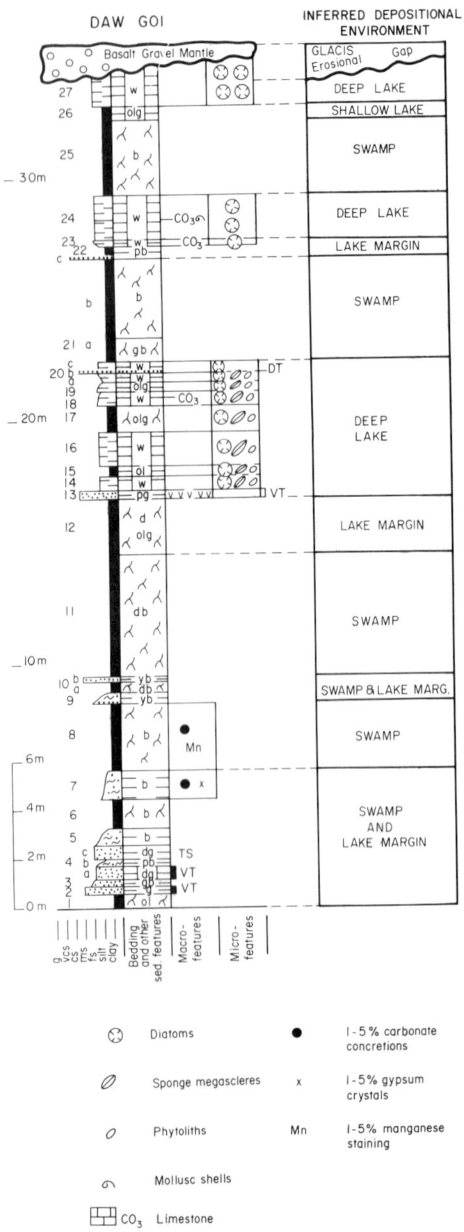

FIG. 7. ?Late Miocene–early Pliocene geological section Dawaitoli 01 (DAW G01). Note alternation of diatom-rich and clay-rich lake facies above the pale grey vitric tuff (VT). Beneath this tuff lake margin, swamp or floodplain facies predominate. Colour symbols as in Fig. 5, plus y = yellow, ol = olive. (For section location see Fig. 9.)

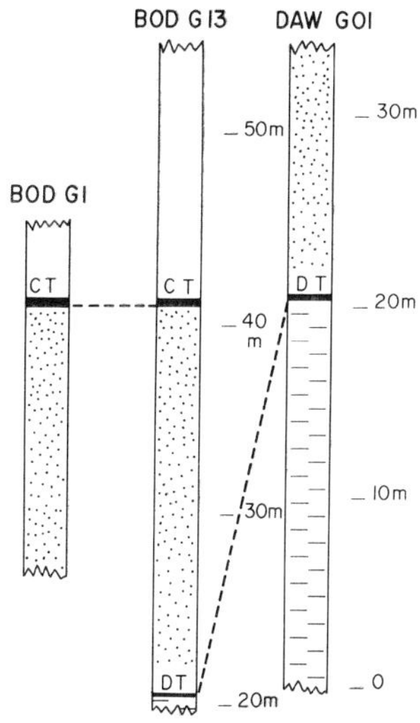

FIG. 8. Stratigraphic correlation between geological sections Bodo 1 (BOD G1), Bodo 13 (BODG 13) and Dawaitoli 01 (DAW G01). CT is the 3.8–4.0 Ma Cindery Tuff. DT is the Doublet Tuff, as yet undated. (For section location see Fig. 9.)

between sections is hindered by the NNW–SSE and NNE–SSW faults that traverse the Plio-Pleistocene deposits (Fig. 4), bringing rocks of different ages into fault contact. Lithologic correlation is further impeded by erosional gaps in the sequence, as well as by the repetitive nature of late Neogene sedimentation in this area. Including Bodo catchment, we have surveyed 300 km^2 of terrain E of the Awash River between 10°25′ and 10°45′N.

An invaluable isochronous marker bed is the 'Cindery Tuff' (Figs. 5 and 6) dated by Walter and colleagues to 3.8–4.0 Ma (Hall *et al.* 1984), that may be traced laterally from Wee-ee catchment to Dawaitoli (Fig. 4), a distance of over 25 km. Other distinctive tephra units, including the 'Doublet Tuff' (Figs. 6 and 7), allow confident correlation between Lower Pliocene sections (Fig. 8).

The oldest end-Miocene to Lower Pliocene sediments (Fig. 7) comprise at least 15 m of alternating brown clays and horizontally-bedded tuffaceous sands and clayey sands, that we interpret as possible swamp and lake margin facies. Capping these is a pale grey vitric tuff overlain by diatomaceous lake clays and diatomites, locally abundant in phytoliths and sponge spicules, and indicative of lake margin facies. Above the doublet Tuff there are some 24 m of alternating brown clays and lake sediments (massive to finely-laminated diatomites). The *c.* 3.9 Ma cindery Tuff caps the sequence (Figs. 5 and 6), which reflects an alternation between lake transgressions (diatomites) and regressions (pedal brown clays devoid of microfossils). Immediately beneath the dark grey Cindery Tuff is a diatomaceous clay, 25–100 cm thick, that also contains sponge megascleres. A hard band of limestone underlies the clay, and usually contains recrystallized shells of *Melanoides* and other freshwater mollusca (Fig. 5). We consider that the Cindery Tuff was laid down towards the close of a major lake regression, when the lake was quite shallow, a conclusion consistent with mammalian hoof impressions on the upper surface of the tuff at Gamedah (Clark *et al.* 1984).

Beneath the Cindery Tuff there are thus at least 45 m of horizontally-bedded silts, clays and diatomites, with minor lenses of fossiliferous gravel and sand, and occasional bands of volcanic ash and limestone. Diatomite is characteristic of the upper 20 m of the sediments beneath the Cindery Tuff. Beds of diatomite 1–6 m thick alternate with brown clays of comparable thickness. The brown clays are highly pedal, with slickensides and manganese staining on the ped faces, and they are invariably devoid of diatom and sponge microfossils. The upward transition from brown clay to diatomite is abrupt, that from diatomite to brown clay more gradual, with diatomite passing upwards into grey clays which may represent diagenetic alteration of diatomite to clay (F. Gasse, pers. comm.). We interpret the diatomite as mainly lacustrine, and the brown clays tentatively as fluviatile overbank and backswamp deposits of a major axial river, possibly the early Pliocene Awash.

Above the Cindery Tuff facies changes are frequent. Much of the sequence of sands, silts and brown clays has been truncated by erosion, and many sections are capped unconformably by the gravel mantle discussed above. The total thickness of Pliocene sediments is not yet known, but it is at least 100 m.

'Floating' sections (not shown here) of Upper Pliocene cross-bedded sands and gravels, alternating with sub-horizontal silts and clays, indicate a major contrast between early and late Pliocene sedimentation—a contrast which persisted during the Pleistocene. Tectonic events were almost certainly responsible for the demise of the early Pliocene lakes in this area, but climatic oscillations may well have caused the repeated early Pliocene lacustrine transgressions and regressions evident in the stratigraphic record (Figs. 6 and 7). Climatically controlled late Pliocene lake fluctuations are well documented in eastern Afar (Gasse *et al.* 1980), as well as in the southeastern Ethiopian Uplands (Williams *et al.*

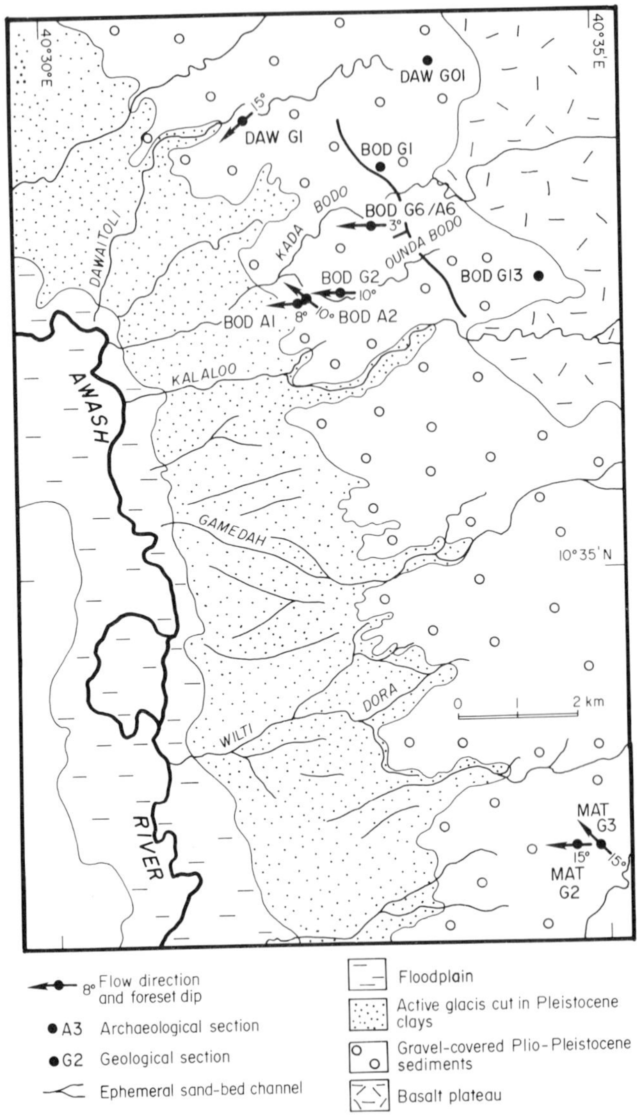

FIG. 9. Palaeocurrent directions and foreset dips in the Middle Pleistocene fossil and artefact-bearing sands. The sands are sandwiched within sterile clays, interpreted here as palaeo-Awash floodplain clays. Flow directions run normal to the modern Awash axial stream, and parallel to present-day ephemeral sand-bed channels. The badland relief and deep gullies between the basalt plateau and the active glacis developed in the late Pleistocene or Holocene, after deposition of the basalt gravel mantle.

1979; Gasse 1980). Until 3.8–4.0 Ma, there was little influx of coarse clastic material from adjacent slopes. This suggests that local relief was subdued, and that the marginal fault scarps (exaggerated in Fig. 5) were low, and possibly did not exist as such E of the early Pliocene lake. After deposition of the Cindery Tuff, fluviatile deposition of the pedal brown clays replaced the early Pliocene accumulation of lacustrine diatomites.

Pleistocene depositional environments

Two distinctive lithofacies are characteristic of the Pleistocene sequence. They are the massive brown clay facies, devoid of macro- and microfossils, and the cross-bedded gravelly sand facies, often rich in vertebrate fossils (Figs. 9 and 10). We interpret the clays as palaeo-Awash overbank

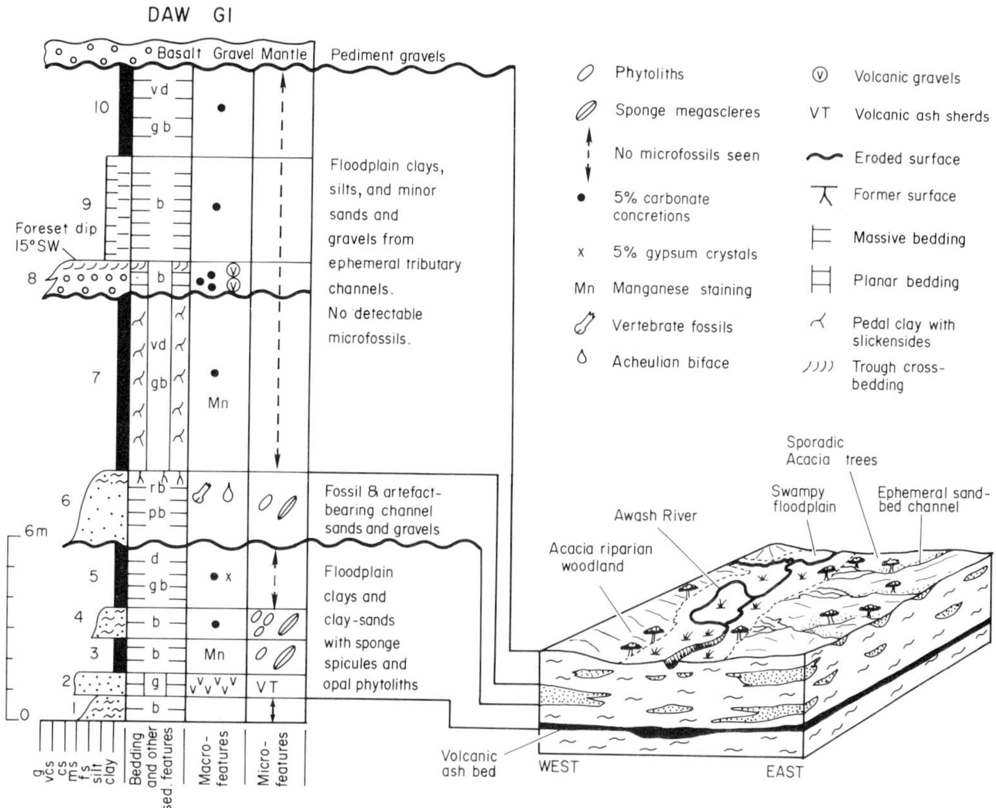

FIG. 10. Stratigraphy and geomorphic setting of the Middle Pleistocene Awash floodplain. Geological section Dawaitoli 1 (DAW G1) (see Fig. 9) shows that the fossil and artefact-bearing channel sands and gravels are intercalated between massive to sterile grey-brown clays. The Middle Pleistocene depositional environments have a modern analogue in the interface between the Awash floodplain and active glacis drained by ephemeral sand-bed channels. Dense acacia woodland covers the main floodplain, and tall scattered acacias grow along the sandy channels.

and backswamp deposits, analogous to those laid down seasonally by the late Quaternary Blue Nile in central Sudan (Adamson *et al.* 1980). When traced eastwards along the Kalaloo stream channel (Fig. 4), the clays wedge out and interfinger with fanglomerates from the basalt scarp.

Palaeocurrent directions indicated by the Pleistocene gravelly sand facies invariably parallel those in the present-day ephemeral sand-bed channels that run W towards the Awash (Fig. 9). The scale of the cross-beds, the often poor sorting, and the equal proportions of rolled carbonate and basalt gravel are also consistent with seasonal or episodic deposition by small local streams with shifting braided or distributing channels in their distal reaches.

We conclude that the sterile brown clays are allochthonous, palaeo-Awash floodplain deposits, in contrast to the fossil-bearing sands and fine gravels, which are mostly of local provenance (Fig. 10). Such an interpretation may help to explain the sudden disappearance of vertebrate fossils on moving laterally from sand to clay. Streams capable of moving and concentrating bones during peak flows could no longer transport such bones when their velocity was abruptly checked on debouching into seasonally-flooded swamps along the Pleistocene Awash. Then, as now, the Awash was highly seasonal. During dry glacial maxima, individual peak flows may have been high, but total duration was doubtless less than today, when flood peaks occur from July to October.

Within the Lower to Middle Pleistocene sequence, archaeological concentrations (Fig. 3) are always absent from the massive brown clays (Fig. 10). The least disturbed of the Early Stone Age occurrences described by Clark *et al.* (1984)

are within the massive silts and clayey fine sands immediately above the sterile clays. Where not disrupted by faulting, sands often merge laterally into coarse channel sands and gravels. Where some artefacts are found within channel sands, they may be imbricated, but have not necessarily been moved very far. The depositional context suggests that the Pleistocene hominids avoided the floodplain and swamps of the palaeo-Awash, preferring their more elevated margins in the vicinity of local sandy tributary channels.

Conclusions

The environmental interpretation offered here to account for major changes in lithofacies over the past 5 Ma in the Middle Awash valley between 10°25′ and 10°45′N may also help in interpreting the Hadar sequence. The Lower Pliocene diatomites at Bodo have no counterpart at Hadar. However, the hominid-bearing sands and clays which were laid down at Hadar towards 3.1–2.8 Ma (or ?3.8–2.8 Ma) may also be mainly fluviatile, reflecting both local and allochthonous sediment influx.

Our reconstruction of Plio-Pleistocene events in the Middle Awash valley, summarized diagrammatically in Figs. 5 and 10, is based on the following assumptions, each of which requires independent testing:

1 Finely laminated diatomites are lacustrine.
2 Clays devoid of microfossils denote seasonal or episodic flooding on low-angle floodplains.
3 Abundant or common microfossils denote permanent swamps or lakes.
4 If phytoliths are the sole microfossils, deposition was non-aquatic.
5 Absent or rare microfossils denote fluvial deposition and a lack of still, permanent water.
6 Sediments rich in diatoms and sponge spicules reflect deposition in permanent water or lakes, river pools or swamps. (Presence, absence, and relative abundance of diatoms, sponge megascleres, and opal phytoliths can usually be determined on site, using a portable biological microscope.)

ACKNOWLEDGEMENTS: We thank the government of Ethiopia and Professor J.D. Clark for the privilege of working on this project, NSF and Macquarie University for support, and Berhane Asfaw, R. Blumenschine, B.C. Clark, J.D. Clark, T.G. Egziabher, J.W.K. Harris, L. Krishtalka, H. Kurashina, M. Tesfaye, R.C. Walter and T.D. White for help and discussion in the field. MAJW thanks J.D. Clark, F. Gasse, M.R. Talbot and J.J. Tiercelin for constructive suggestions.

References

Adamson, D.A., Gasse, F., Street, F.A. & Williams M.A.J. 1980. Late Quaternary history of the Nile. *Nature* **288,** 50–5.

Aronson, J.L. & Taieb, M. 1980. Geology and paleogeography of the Hadar hominid site, Ethiopia. In: Rapp, G. & Vondra, C.F. (eds) *Hominid Sites: Their Geologic Settings*. Westview Press, Boulder, Colorado, pp.165–95.

Asfaw, Berhane, 1983. A new hominid parietal from Bodo, Middle Awash valley, Ethiopia. *Am. J. Phys. Anthrop.* **61,** 367–71.

Assefa, Getaneh, Clark, J.D. & Williams, M.A.J. 1982. Late Cenozoic history and archaeology of the Upper Webi Shebele basin, east-central Ethiopia. *Sinet: Ethiop. J. Sci.* **5,** 27–46.

Barberi, F., Ferrara, G., Santacroce, R. & Varet, J. 1975. Structural evolution of the Afar Triple Junction. In: Pilger, A. & Rösler, A. (eds), *Afar Depression of Ethiopia*. Schweizerbart, Stuttgart, pp.174–8.

—— & Varet, J. 1977. Volcanism of Afar: small-scale plate tectonics implications. *Bull. Geol. Soc. Am.* **88,** 1251–66.

Bonnefille, R. 1983. Evidence for a cooler and drier climate in the Ethiopian uplands towards 2.5 Myr ago. *Nature* **303,** 487–91.

Clark, J.D., Asfaw, Berhane, Assefa, Getaneh, Harris, J.W.K., Kurashina, H., Walter, R.C., White, T.D. & Williams, M.A.J. 1984. Palaeoanthropological discoveries in the Middle Awash valley, Ethiopia. *Nature* **307,** 423–8.

Davidson, A. & Rex, D.C. 1980. Age of volcanism and rifting in southwestern Ethiopia. *Nature* **283,** 657–8.

Gasse, F. 1980. Les diatomées lacustres plio-pléistocènes du Gadeb (Ethiopie): systématique, paléoécologie, biostratigraphie. *Rev. Algol.* **3,** 1–249.

——, Richard, O., Robbe, D., Rognon, P. & Williams, M.A.J. 1980. Évolution tectonique et climatique de l'Afar Central d'après les sédiments plio-pléistocènes. *Bull. Soc. Géol. Fr.* **22,** 987–1001.

Hall, C.M., Walter, R.C., Westgate, J.A. & York, D. 1984. Geochronology, stratigraphy and geochemistry of Cindery Tuff in Pliocene hominid-bearing sediments of the Middle Awash, Ethiopia. *Nature* **308,** 26–31.

Harris, J.W.K. 1983. Cultural beginnings: Plio-Pleistocene archaeological occurrences from the Afar, Ethiopia. *Afr. Arch. Rev.* 1, 2–31.

Heller, F. & Liu, T-S. 1982. Magnetostratigraphical dating of loess deposits in China. *Nature* **300,** 431–3.

Hillaire-Marcel, C., Taieb, M., Tiercelin, J.J. & Page, N. 1982. A 1.2-Myr record of isotopic changes in a late Pliocene rift lake, Ethiopia. *Nature* **296,** 640–2.

HUTCHINSON, R.W. & ENGELS, G.G. 1972. Tectonic evolution in the southern Red Sea and its possible significance to older rifted continental margins. *Bull. Geol. Soc. Am.* **83,** 2989–3002.

JOHANSON, D.C., TAIEB, M. & COPPENS, Y. 1982. Pliocene hominids from the Hadar Formation, Ethiopia (1973–1977): stratigraphic, chronologic, and paleoenvironmental contexts, with notes on hominid morphology and systematics. *Am. J. Phys. Anthropol.* **57,** 373–402.

JONES, P.W. 1976. Age of the lower flood basalts of the Ethiopian Plateau. *Nature* **261,** 567–9.

KALB, J.E., JOLLY, C.J., MEBRATE, ASSEFA, TEBEDGE, SLESHI, SMART, C., OSWALD, E.B., CRAMER, D., WHITEHEAD, P., WOOD, C.B., CONROY, G.C., ADEFRIS, TSRHA, SPERLING, L. & KANA, BERHANE 1982a. Fossil mammals and artefacts from the Middle Awash valley, Ethiopia. *Nature* **298,** 25–29.

—— OSWALD, E.B., TEBEDGE, SLESHI, MEBRATE, ASSEFA, TOLA, E. & PEAK, D. 1982b. Geology and stratigraphy of Neogene deposits, Middle Awash valley, Ethiopia. *Nature* **298,** 17–25.

——,WOOD, C.B., SMART, C., OSWALD, E.B., MABRETE, ASSEFA, TEBEDGE, SLESHI & WHITEHEAD, P. 1980. Preliminary geology and palaeontology of the Bodo hominid site, Afar, Ethiopia. *Palaeogeogr. Palaeoclimatol. Palaeoecol.* **30,** 107–20.

MOHR, P.A. 1967. The Ethiopian Rift System. *Bull. Geophys. Obs. Addis Ababa* **11,** 1–65.

—— 1968. The Cenozoic volcanic succession in Ethiopia. *Bull. Volc.* **32,** 5–14.

—— 1975. Structural setting and evolution of Afar. In: PILGER, A. & RÖSLER, A. (eds) *Afar Depression of Ethiopia.* Schweizerbart, Stuttgart, pp. 27–37.

—— 1983. Ethiopian flood basalt province. *Nature* **303,** 577–84.

ROCHE, H. & TIERCELIN, J.J. 1977. Découverte d'une industrie lithique ancienne *in situ* dans la formation D'Hadar, Afar Central, Ethiopie. *C. R. Acad. Sci. Paris* **284D,** 1871–4.

—— & —— 1980. Industries lithiques de la formation plio-pléistocène d'Hadar, Ethiopie (campagne 1976). In: LEAKEY, R.E. & OGOT, B.A. (eds) *Proc. 8th Panafr. Congr. Préhist. Quat. Studies Nairobi 1977.* TILLMIAP, Nairobi, pp.194–9.

SARICH, V.M. & WILSON, A.C. 1967. Immunological time scale for hominid evolution. *Science* **158,** 1200.

SCHMITT, J.J. & NAIRN, A.E.M. 1984. Interpretation of the magneto-stratigraphy of the Hadar hominid site, Ethiopia. *Nature* **309,** 704–6.

SHACKLETON, N.J. & OPDYKE, N.D. 1977. Oxygen isotope and palaeomagnetic evidence for early Northern Hemisphere glaciation. *Nature* **270,** 216–9.

SHACKLETON, N.J. & 16 OTHERS. 1984. Oxygen isotope calibration of the onset of ice-rafting and history of glaciation in the North Atlantic region. *Nature* **307,** 620–3.

SUC, J-P. 1984. Origin and evolution of the Mediterranean vegetation and climate in Europe. *Nature* **307,** 429–32.

TAIEB, M. 1974. Évolution quaternaire du bassin de l'Awash (Rift éthiopien et Afar). Unpublished PhD thesis, University of Paris, 390pp.

—— & TIERCELIN, J.J. 1979. Sédimentation pliocène et paléoenvironnements de rift: exemple de la formation à Hominidés d'Hadar (Afar, Éthiopie). *Bull. Soc. Géol. Fr.* **21,** 243–53.

TIERCELIN, J.J., MICHAUX, J. & BANDET, Y. 1979. Le miocène supérieur du Sud de la Dépression de l'Afar, Éthiopie: sédiments, faunes, âges isotopiques. *Bull. Soc. Géol. Fr.* **21,** 255–8.

WALTER, R.C. & ARONSON, J.L. 1982. Revisions of K/Ar ages for the Hadar hominid site, Ethiopia. *Nature* **296,** 122–7.

WILLIAMS M.A.J., WILLIAMS, F.M., GASSE, F., CURTIS, G.H. & ADAMSON, D.A. 1979. Plio-Pleistocene environments at Gadeb prehistoric site, Ethiopia. *Nature* **282,** 29–33.

ZANETTIN, B. & JUSTIN-VISENTIN, E. 1975. Tectonic and volcanological evolution of the western Afar margin (Ethiopia). In: PILGER, A. & RÖSLER, A. (eds). *Afar Depression of Ethiopia.* Schweizerbart, Stuttgart, pp. 300–9.

M.A.J. WILLIAMS, Dept of Geography, Monash University, Clayton, Victoria 3168, Australia.

GETANEH ASSEFA, Faculty of Sciences, Addis Ababa University, PO Box 1176, Addis Ababa, Ethiopia.

D.A. ADAMSON, School of Biological Sciences & Quaternary Research Unit, Macquarie University, North Ryde, New South Wales 2113, Australia.

Palaeoenvironment of Lake Abijata, Ethiopia, during the past 2000 years

R. Bonnefille, C. Robert, G. Delibrias, C. Elenga, J.P. Herbin, A.M. Lezine, G. Perinet & J.J. Tiercelin

SUMMARY: Multidisciplinary studies, including the examination of sedimentology, palynology, diatoms and organic matter have been undertaken on a 6 m core from Lake Abijata, Ethiopia. The bottom of this core has been dated at 1720±80 BP. Analyses have shown four distinct stages in lake evolution. Two transgressive phases that have been recognized can be correlated with other lake level evidence. The transgressions result from changing inputs from lowland and highland regions within the catchment. They are most probably related to changes in seasonal rainfall patterns.

Lacustrine fluctuations in East Africa during the Pleistocene and Holocene have been of considerable interest in palaeoclimatic interpretations (Peterson *et al.* 1979). On time scales of thousands of years, several changes in lake level have been described since 12,000 BP (Gillespie *et al.* 1983), suggesting a progressive trend towards aridity over the last 4000 years (Butzer *et al.* 1972; Gasse & Street 1978; Street & Grove 1979; Owen *et al.* 1982; Vincens *et al.*, this volume). Very recent shoreline fluctuations have also been studied, using annual or decennial time scales (Makin *et al.* 1976; Vincent *et al.* 1979).

In 1982, following on from geomorphological studies on lake levels in the Ziway-Shalla basin (Street 1979), as well as the analysis of diatoms (Descourtieux 1979) and pollen (Lezine 1981; Lezine & Bonnefille 1982), several 6 m cores were recovered from Lakes Langeno, Shalla and Abijata using a Mackereth corer (Fig. 1). Carbon-14 dates obtained from these cores have shown sedimentation rates varying from 15 to 300 cm ka^{-1}. Multidisciplinary analysis of one of the cores recovered from Lake Abijata (7°10′N), in 11 m of water, is presented here. This core is 5.46 m long and has been dated at 1720±80 BP close to its base.

Environmental setting

The region of the Ziway-Shalla-Abijata basin is located in the Ethiopian part of the East African Rift System. It is bounded by Cenozoic basaltic plateaux, locally reaching 3600 m. The rift floor (1600 m) is covered by Quaternary rhyolitic flows, ignimbrites and ashes. Acidic or trachytic volcanoes are located on the plateaux margins or within the rift (Mohr 1971). On the rift floor, Lakes Abijata, Langeno, Shalla and Ziway occupy the lower parts of a small endoreic basin. Large rivers that feed the lakes drain the margins of the basaltic platforms, as well as the slopes and the rift floor. More locally, water is added from overland flow and small ephemeral streams (Street 1979). Furthermore, hydrothermal springs are located close to the shoreline of Lakes Langeno and Shalla (Baumann *et al.* 1975). At present, the four lakes form two different hydrological systems: Ziway and Langeno overflow into Abijata, and Shalla is a separate terminal lake (Street 1979).

The climate is characterized by alternating wet and dry seasons following the annual movements of the Intertropical Convergence Zone. The mean annual rainfall is no more than 600 mm around the lakes, and it rises with elevation to a maximum of 1600 mm at the 3000 m contour. Rainfall distribution is bimodal with the main rain from July to September, and a secondary peak in March–April.

From the viewpoint of soil development, vertisols or cambisols are dominant in the rift area, including the uplands bordering the rift. However, only weakly developed soils have formed on the lacustrine sediments, river alluvium and pumice on the floor of the rift. Generally these soils are rich in smectite, but amorphous material is also present. At higher altitudes, reddish latosols reflect the former extent of montane forests. Generally, kaolinite develops in these soils (Millot 1964; Riche & Segalen 1969; Street 1979).

Main features of the sedimentation

A description of the macroscopic stratigraphy is followed by an analysis of the microfacies and mineralogy of the bulk sediment (Fig. 2).

The bottom of the sedimentary column (5.46–2.85 m depth in the core) consists of a dominantly

From Frostick, L.E. *et al.* (eds) 1986, *Sedimentation in the African Rifts*, Geological Society Special Publication No. 25, pp. 253–65.

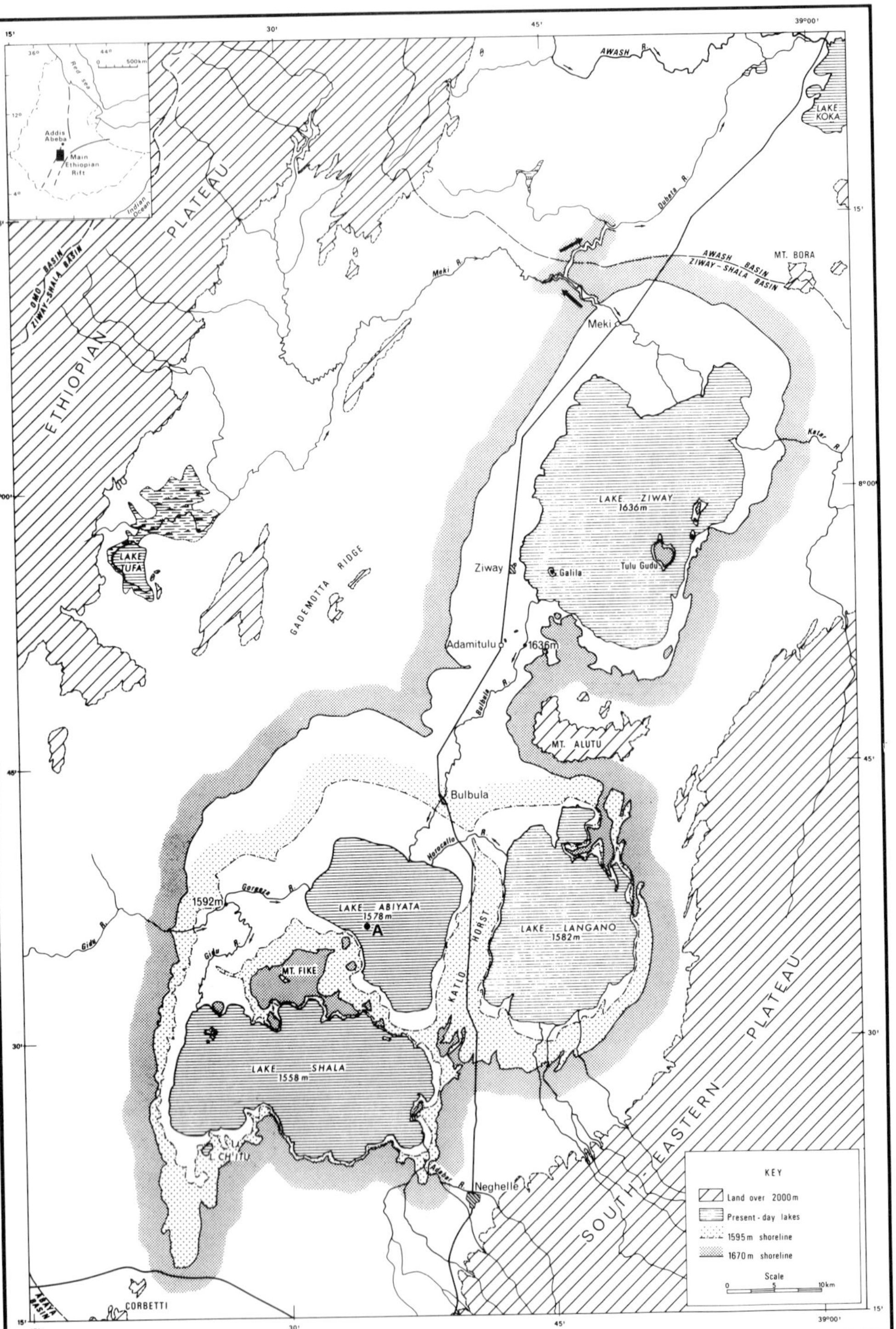

FIG. 1. Location map of Lake Abijata in the Ziway-Shalla basin, Ethiopia (A = core location).

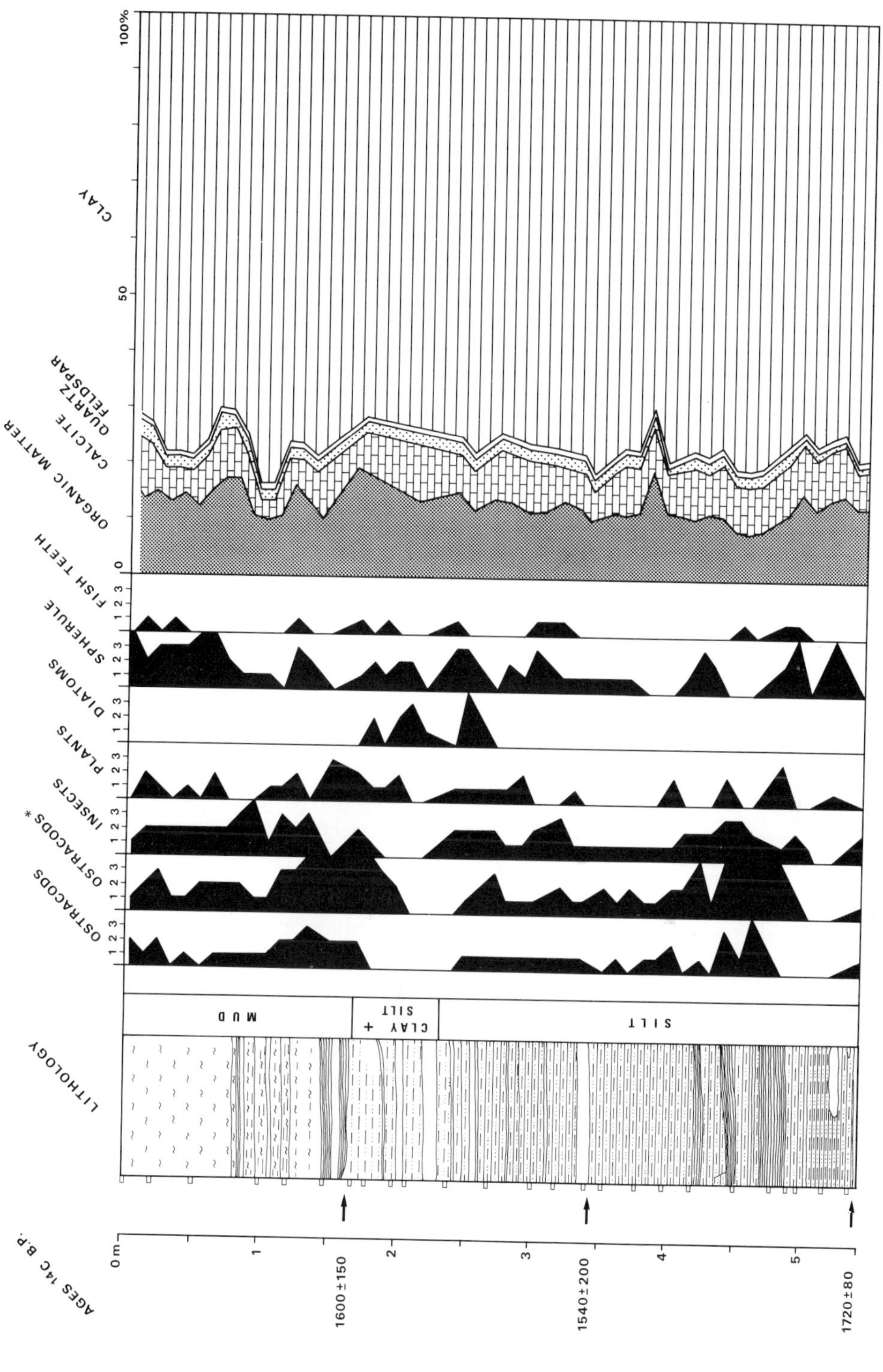

FIG. 2. Main features of the sediments observed on smear slides and in bulk sediment samples. 1 = rare; 2 = abundant; 3 = very abundant. Ostracods* are small; others are large.

grey silt which has black laminae (between 4.92 and 4.20 m) and light mottles (between 4.20 m and 3.50 m). A brown silt (2.85–2.60 m) is followed by a clayey silt (2.60–1.65 m) which varies from grey-red to brown and dark red, and which ends abruptly with a possible discontinuity. Above these silts (1.65–1.25 m), a beige mud shows fine black laminae. Then, alternating brownish black and grey laminae are present up to 0.82 m. The top of the sedimentary column consists of a clayey mud, with a layer of green mottles that progressively becomes brown to beige upwards. Laminae in the core are millimetric, irregularly distributed and more frequent near the bottom.

Analyses of smear slides have revealed the presence of ostracods (both large and small shells), insect fragments, plant remains, diatoms, red spherules and fish teeth. They form the main recognizable particles within a terrigenous matrix that includes volcanic detritus. From the base of the core to 4.80 m, identifiable particles are scarce, with the exception of two levels at 5.17 and 4.90 m, where there are numerous red spherules. From 4.80 to 4.00 m, ostracods are present in great abundance. From 4.00 m to 2.50 m, the various identifiable grains are only present in small quantities, but ostracods, insect fragments, plant remains and spherules dominate in varying proportions. And from 2.50 to 2.00 m, diatoms and spherules are the main components of the sediment. Above this level (at 1 m), ostracods and insect fragments are very abundant. From 0.80 m upwards, the top of the sedimentary column is dominated by the red spherules.

X-ray diffraction of the bulk sediment has demonstrated the importance of clay particles (70–85%) in the sediment (Fig. 2). Total organic matter, expressed as dry weight after calcination at 350°C, generally ranges from 10 to 20%. Calcite and quartz are the only other minerals present in measurable quantities, accompanied by trace amounts of feldspars. The abundance of calcite and quartz shows only minor variation along the core, their mean value being *c.* 10%. With the exception of the calcite, which is possibly authigenic as in Lake Shalla (Baumann *et al.* 1975; Street 1979), no other authigenic mineral was recognized. We consider that most of the mineral fraction of the Lake Abijata sediments are detrital, as proposed by Street (1979). The results from Abijata differ from data obtained in other regions of the East African Rift, such as Magadi in Kenya and Assal in Ethiopia, where authigenic minerals (mainly zeolites), are abundant (Surdam & Eugster 1976; Street 1979). These results also differ from the data obtained from older Pleistocene sediments in the Abijata long core, where zeolites are present. The mineralogical composition of the bulk sediment from Abijata is similar to that of surficial samples from Lake Shalla (Baumann *et al.* 1975) and from other East African lakes (Yuretich 1982).

Aquatic constituents

Organic matter

The organic matter was analysed by pyrolysis, using the method described by Espitalie *et al.* (1977). This method is based on the selective detection of the hydrocarbon and oxygen compounds produced during the heating of the samples in an inert atmosphere. It provides several indices; two indices, called the hydrogen index and the oxygen index, correlate closely with the atomic ratios H:C and O:C obtained from the elemental analysis of the kerogens (Tissot *et al.* 1974; Debyser & Gadel 1977). Data obtained from pyrolysis analysis have been standardized with the different types of organic matter identified during analysis of the kerogen fraction (Herbin 1980; Tiercelin *et al.* 1980). Consequently, a kerogen fraction that contains aquatic plankton is characterized by hydrogen indices >400 mg of hydrocarbon per gram of organic carbon, with oxygen indices <100. A kerogen fraction containing terrestrial material (mostly vegetal fragments) is characterized by hydrogen indices <100 and oxygen indices <200. The temperature reached for the maximum peak of pyrolysis is related to the thermal maturation of the sediment. The zone of weak thermal maturation, called the zone of diagenesis, corresponds with pyrolysis temperatures of between 400 and 430°C.

In the Abijata core (Fig. 3), the amount of organic carbon exceeds 2%, and reaches the exceptionally high value of 9.23% at 4.90 m, with a mean value of *c.* 5%, which is very high for recent sediments. The weak maturation of the organic matter is shown by pyrolysis temperatures of 420–425°C, and by very high oxygen indices (≤400). Hydrogen indices lie between 247 and 643, with a mean value of *c.* 420 (Fig. 3). Thus organic matter of aquatic origin dominates in the sediment. A good correlation exists between the percentages of organic carbon and the hydrogen index. From these results, the following four zones can be distinguished: (1) from the bottom of the core to 4.90 m, high percentages of organic carbon (6.2–9.2%) are coincidental with high values for the hydrogen index (500–650), suggesting that there is abundant aquatic organic matter in the sediment, and/or that the organic matter is well preserved;

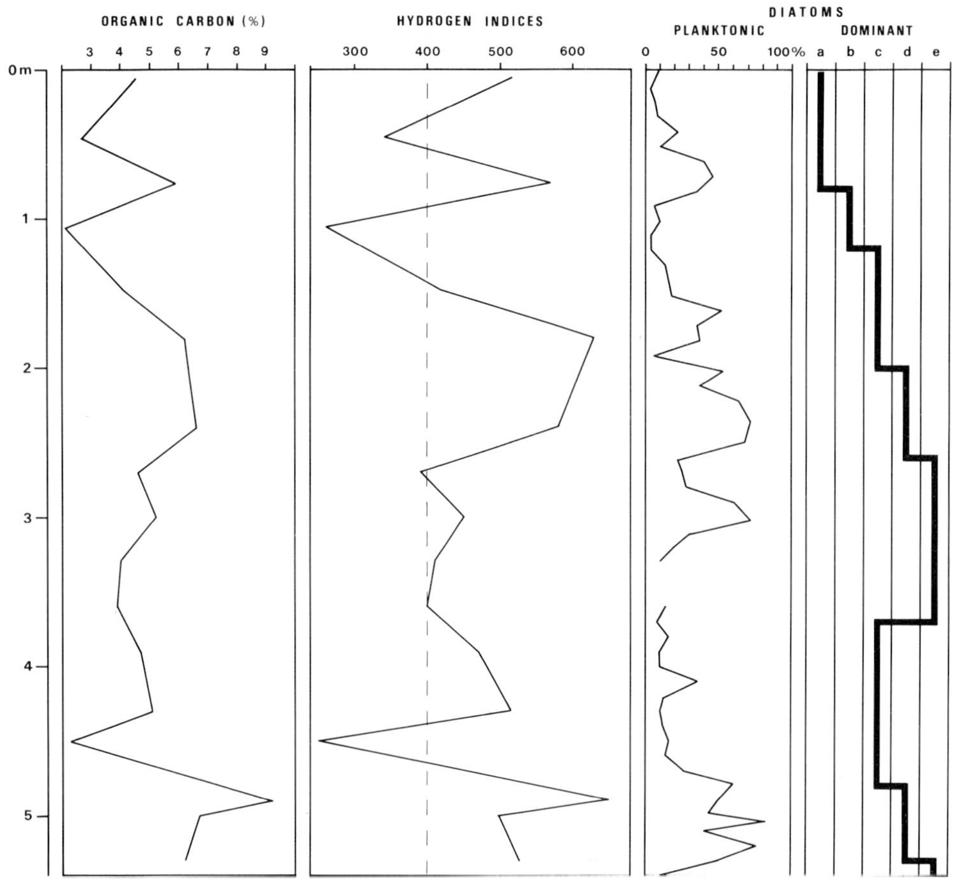

FIG. 3. Relative proportions of aquatic constituents in lake core. a = *Anomoeoneis sphaerophora*; b = *Nitzschia frustulum*; c = *Rhopalodia gibberula*; d = *Nitzscha amphibia*; e = *Nitzschia vitrea*.

(2) from 4.90 to 3.60 m, organic carbon decreases (2–5%), as does the hydrogen index (250–520) due to lack of, or, the poor preservation of, the aquatic organic matter; (3) between 3.60 and 1.80 m, the indices increase progressively (organic carbon: 3.9–6.6%; hydrogen index: 390–630); and (4) from 1.80 m to the top of the core, the organic carbon decreases (2–5%), as does the hydrogen index (260–570), but fluctuations in productivity are still evident (Fig. 3).

Diatoms

The diatom microflora shows variations in planktonic species abundance as well as in the succession of the various associations (Fig. 3). Detailed results have been given by Elenga (1984). A succession of eight associations are present from the bottom to the top of the core:

1 The first association includes *Nitzschia vitrea*, *Navicula halophila* and *Melosira agasizii*. The dominant species, *Nitzschia vitrea*, is common in waters intermediate between concentrated and fresh.

2 The second association comprises *Nitzschia amphibia*, *N. vitrea*, *Navicula halophila* and *Anomoeoneis sphaerophora*.

3 The third association includes *Rhopalodia gibberula*, *Stephanodiscus astrea* and *Anomoeoneis sphaerophora*.

4 The fourth association comprises *Nitzschia vitrea*, *Anomoeoneis sphaerophora*, *Rhopalodia gibberula* and *Nitzschia* sp.

5 The fifth association includes *Nitzschia amphibia*, *Anomoeoneis sphaerophora* and *Rhopalodia gibberula*.

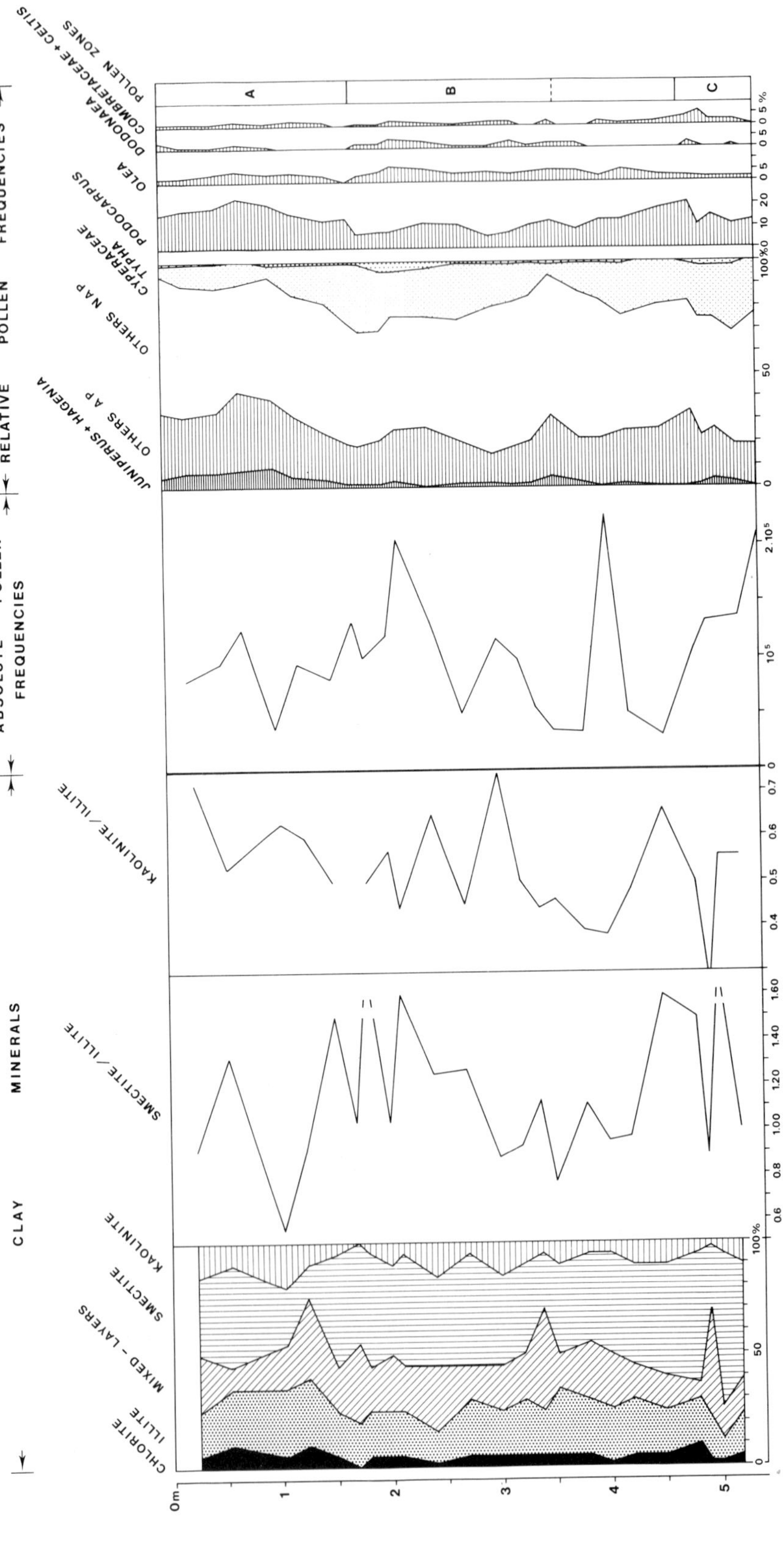

FIG. 4. Absolute and relative abundance of terrestrial constituents in lake core, clay minerals and pollen.

6 The sixth association comprises *Rhopalodia gibberula*, *Cyclotella ocellata*, *Anomoeoneis sphaerophora*, and *Stephanodiscus astrea*.
7 The seventh association comprises *Nitzschia frustulum*, *Anomoeoneis sphaerophora*, and *Rhopalodia gibberula*.
8 The eighth association comprises *Anomoeoneis sphaerophora*, *Rhopalodia gibberula*, *Cyclotella ocellata*, *Nitzschia frustulum* and *Stephanodiscus astrea*.

Previous work has demonstrated that *Nitzschia vitrea*, *N. amphibia* and *Stephanodiscus astrea* are usually more abundant in dilute water, whereas *Anomoeoneis sphaerophora*, *Nitzschia frustulum* and *Rhopalodia gibberula* develop in more saline waters (Gasse *et al.* 1983) (Fig. 3). The sequence containing these associations, mainly characterized by their dominant species, together with variations of the percentages of the planktonic diatoms, allows four zones to be distinguished:
1 From the bottom of the core to 4.80 m, the planktonic species are abundant (40–80%), with *Nitzschia vitrea* and *N. amphibia* dominant. Such high percentages probably indicate a large volume of water in the lake.
2 From 4.80 to 3.90 m, the percentage of the planktonic species decreases (15%) suggesting a decrease in the volume of water. The dominant species, *Rhopalodia gibberula*, is characteristic of more concentrated waters.
3 Between 3.90 and 2.00 m, planktonic species increase (40–70%), with *Nitzschia vitrea* and *N. amphibia* again dominating. These suggest a new expansion of the lake and more dilute water than previously.
4 Finally, from 2.00 m to the top of the core, there is a tendency for the abundance of the planktonic species to decrease, although two minor fluctuations are registered. *Rhopalodia gibberula*, *Nitzschia frustulum* and *Anomoeoneis sphaerophora* dominate, indicating a trend towards increased salinity.

Conclusion

A fairly close correlation between organic carbon content, hydrogen index and the abundance of planktonic diatoms is shown in Fig. 3, despite the fact that the sampling intervals are different. This indicates that the organic carbon curve correlates with lake productivity, rather than with conditions of preservation of organic matter. The Abijata core shows four stages in the aquatic productivity of the lake during the last 2000 years. When the volume of water increases, species of dilute water diatoms appear first, followed by an increase in other planktonic species, that ultimately leads to higher organic matter content.

Terrestrial constituents

Clay minerals

The clay fraction (less than 2 μm) was studied using X-ray diffraction. The percentages of the different minerals were estimated and peak height ratios of smectite:illite (S:I) and kaolinite:illite (K:I) were deduced. The minerals present are chlorite (< 10%), illite (10–35%), irregular mixed layers (10–50%), smectite (15–70%) and kaolinite (0–20%). No significant variation of these values are seen down the core (Fig. 4). The clay mineralogical association is very similar to that recorded from Lake Shalla (Baumann *et al.* 1975), Lake Mobutu (Stoffers & Singer 1979) and other East African lakes (Yuretich 1982).

On the diffractograms, the clay minerals appear very poorly crystalline, and amorphous materials seem abundant. On electron micrographs, amorphous components correspond to aggregated small spherules (Beutelspacher & Van der Marel 1968) and to biogenic fragments probably derived from diatoms (Fig. 5).

Chlorite, illite and irregular mixed-layer clays are derived mainly from poorly weathered bedrocks (Millot 1964). In Ethiopia, trachytes and andesites, rich in potassium, could weather to illite (Riche 1976). Kaolinite forms during the alteration of feldspars, and also as a result of soil genesis in the regions where precipitation and drainage favour leaching (Millot 1964). In Ethiopia, kaolinite is present in temperate zones with areas of high relief (Riche 1976). In Lake Abijata sediments, kaolinite forms small hexagonal particles dispersed in the sediment (Fig. 5). Smectites are the most important clays; they develop in abundance in low-lying plain areas, where poor drainage prevents leaching. Basic volcanic rocks containing abundant cations tend to weather to smectites (Millot 1964; Riche 1976). In the Abijata region, smectites form in the soils of the poorly-drained areas, basaltic platforms and the rift floor, which are mantled with pyroclastics. Electron micrographs of the Abijata samples show the presence of clay flocs characteristic of pedogenic smectites described by Chamley & Bonnot-Courtois (1981) and Robert (1982). Smectites can also form as a result of the recrystallization of amorphous material eroded and washed into the lacustrine basin (cf. New Caledonia, Baltzer & Trescases 1971).

Electron micrographs also show the presence of particles joined in small wisps, or short opaque laths (Fig. 5). These particles, which have a fine and fragile appearance, could be formed *in situ*. Exceptionally long laths resembling smectite particles derived from the subaerial alteration of volcanic products such as those in the Philippine

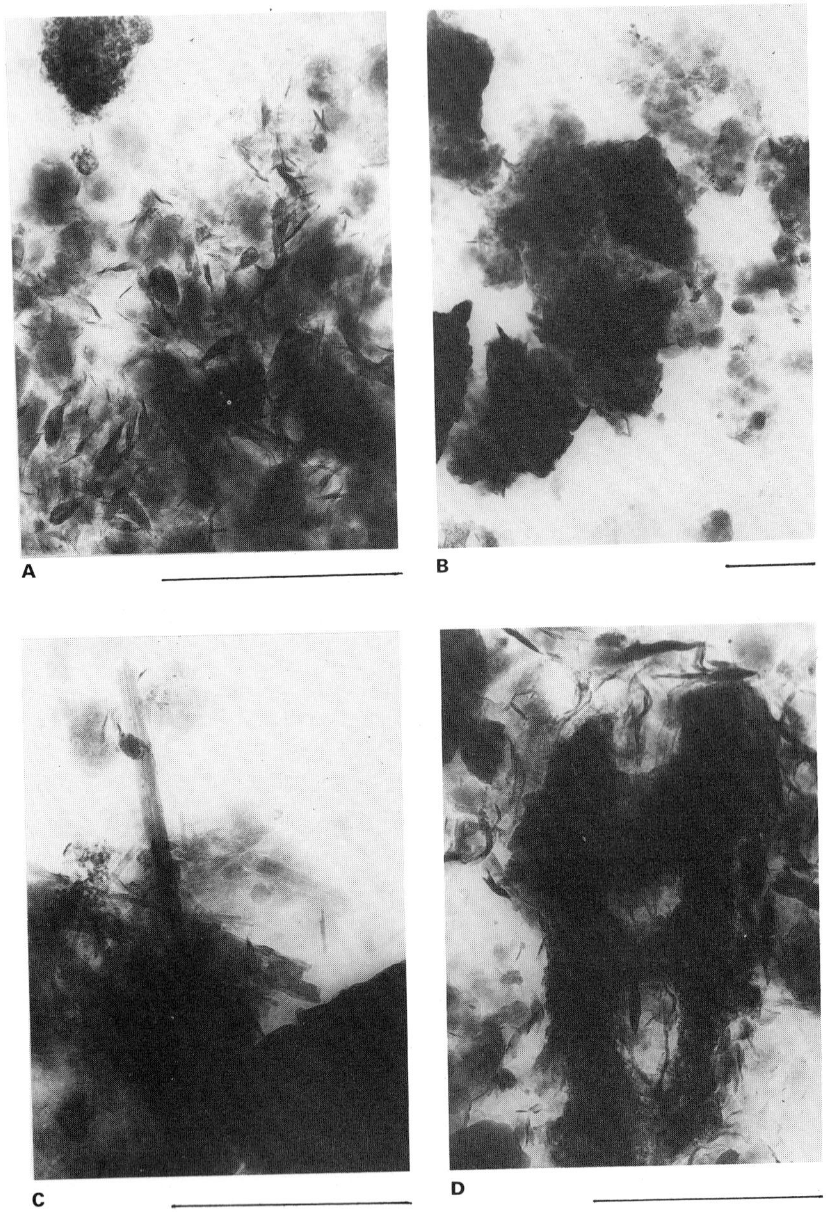

FIG. 5. Electron micrographs of sediments from the Abijata Holocene core (scale bars = 1 μm). A. Aggregated small spherules (top left), small hexagon (top right), abundant wisps and flocs. Core level 1.5 m (× 51,000). B. Hexagonal particle (middle right), aggregated small spherules (top and middle right), small wisps and laths (middle and bottom left), and flocs in the background. Core level 0.15 m (× 20,500). C. Hexagonal particle (middle right), aggregated small spherules (middle left), long laths, and flocs in the background. Core level 0.15 m (× 51,000). D. Biogenic fragment (?diatom) with diffuse sheets and wisps in the cells and at the periphery. Core level 1.5 m (× 51,000).

Sea (Chamley 1980) or on the Walvis Ridge (Chamley *et al.* 1984), are also present. On biogenic fragments that probably originate from diatoms, transparent veils and short laths at the margins and within the frustules are probably smectite particles. Similar features have been recorded from lakes in Bolivia (Badaut *et al.* 1979), but seem less important in Lake Abijata, where most smectites originated from the soils that develop on the volcanic materials of the rift floor.

Since the clay particles are assumed to be mainly detrital in origin if the morphology of the region has remained stable during the last 2000 years, the clay mineral assemblages may be used as climatic indicators. Variations of the smectite:illite and kaolinite:illite ratios in the Abijata core, shown in Fig. 4, have led to the distinction of four different zones:

1 From the bottom of the core to 4.50 m, S:I and K:I ratios show similar fluctuations, due to the simultaneous supply of smectite from the rift, and kaolinite from the mountainous regions to the E.

2 From 4.50 to 3.50 m, low values of S:I and K:I are related to a decrease in the crystallinity of, and/or the supply of smectite and kaolinite.

3 Between 3.50 and 1.80 m, S:I increases while K:I fluctuates, due to the increasing importance of supply from the rift floor.

4 Lastly, from 1.80 m to the top of the core, S:I decreases while K:I increases, detritus from the upland areas being more abundant in the sediment.

Pollen

Pollen analyses were performed on 26 samples taken every 10–25 cm. A great number of pollen taxa (121) were identified. Afro-montane taxa, such as *Juniperus*, *Hagenia abyssinica*, *Podocarpus* and *Olea* pollen have relative frequencies of 1% each. They are associated with taxa such as *Dodonea viscosa*, *Combretaceae* and *Celtis*, and are more akin to Sudano-Zambezian associations. Among herbaceous taxa, *Gramineae* are dominant, associated with *Amaranthaceae*, *Chenopodiaceae* and *Urticaceae*, as well as with the peri-aquatic *Cyperaceae* and *Typha*. The absolute pollen frequency per gram of dry sediment has a mean value of *c.* 1×10^5. However, strong fluctuations are recorded at the bottom of the core with a minimum value of 3×10^4 pollen g^{-1}. Arboreal absolute pollen frequency accounts for 8×10^3 to 8×10^4 pollen g^{-1}. The variations in these quantitative data, compared with the percentages of the different taxa, have allowed three pollen zones to be distinguished:

1 *Pollen zone A*. From the bottom of the core to 4.70 m, peri-aquatic taxa *Typha* and *Cyperaceae* represent 20–31% of the total count. They are associated with arboreal pollen, which include Afro-montane (*Podocarpus*, 5–20%; *Juniperus*, 5%) and Sudano-Zambezian taxa (*Combretaceae*, 2%; *Dodonea viscosa*, 3%; *Celtis*, 3%). *Olea* is present in small amounts (2–5%).

2 *Pollen zone B*. Between 4.70 and 3.50 m, percentages of arboreal pollen >30% are not considered significant because of the bad pollen preservation at this level. From 3.50 to 1.80 m, peri-aquatic taxa such as *Typha* (5%), *Cyperaceae* (29%), and *Gramineae* (60%) increase. Afro-montane arboreal pollen such as *Podocarpus* (5%) and *Juniperus* (1%), decrease dramatically, while *Olea* (7%) and *Dodonea* (4%) show a slight increase, indicating that mid-altitude vegetation expanded at the expense of the montane forest.

3 *Pollen zone C*. From 1.80 m to the top of the core, *Typha* is rare (0.3%) and *Cyperaceae* (11%) decreases. This zone is characterized by a progressive increase of the arboreal pollen, which exceeds 40% at 0.72 m, the dominant taxa being *Podocarpus* (23%), associated with *Juniperus* (7%) and *Hagenia abyssinica* (2%). These are afro-montane taxa and indicate greater pollen input from the highlands.

Conclusion

The terrestrial components of the sediment, including pollen and clay minerals, show fluctuations which correlate well. When kaolinite from high-relief areas is more abundant, afro-montane pollens are supplied in greater quantities. When smectite from the rift increases, peri-aquatic pollens are more abundant. Erosional processes simultaneously supply pollen and clay to the lake.

Evolution of Lake Abijata during the last 2000 years

The comparative studies of organic matter, diatoms, clay minerals and pollen have led to the recognition of four stages in the evolution of the lake during the last 2000 years (Fig. 6). The response of aquatic and terrestrial components to environmental changes are not completely synchronous.

Zone 4 (5.46–4.9 m)

The bottom of the core was dated 1720±80 BP. The abundance of planktonic diatoms, including dilute water species, and the predominance of organic matter of aquatic origin, indicate high

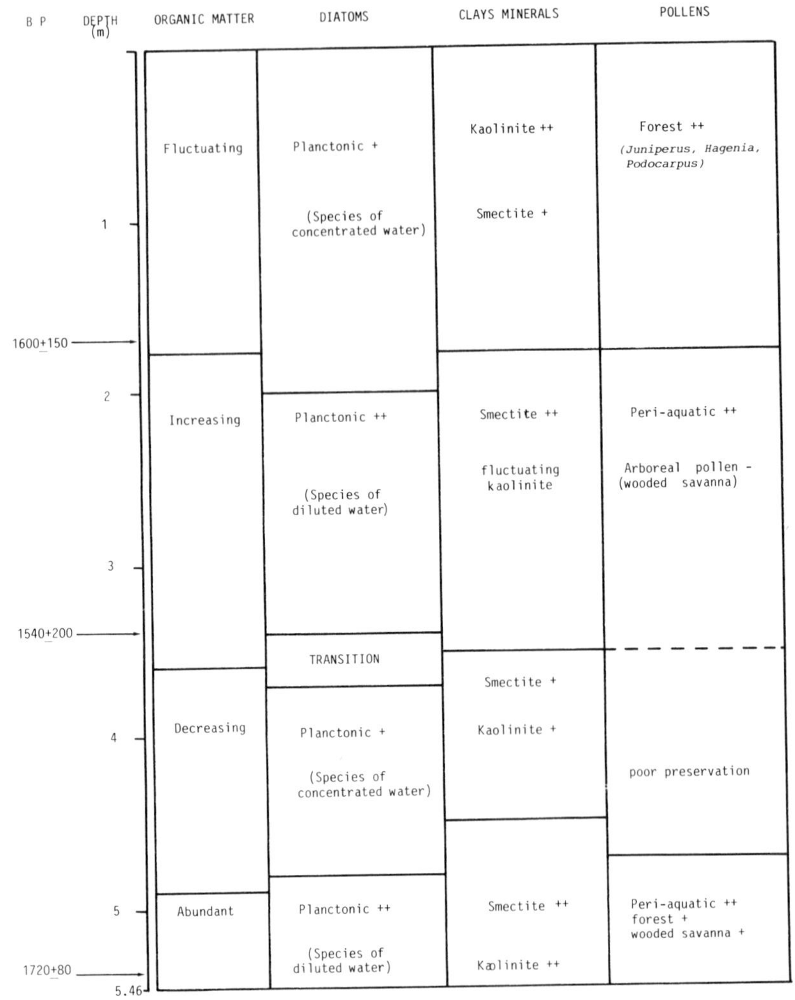

FIG. 6. Evolution of Lake Abijata during the last 2000 years. + = relative abundance; P = precipitation; E = evaporation.

lake levels with high plankton productivity. Of the clay fraction, smectite was delivered by ephemeral streams that had crossed the rift floor, while kaolinite was introduced by perennial rivers draining the mountains. Precipitation exceeded evaporation. Both the lowland savanna and highland forest produced abundant pollen. Lacustrine expansion characterizes Zone 4 and could well correspond to the Ziway-Shalla VII transgression (Street 1979), marked by a shoreline at 42 m above the present level of Lake Shalla. This transgression was previously dated using charcoal at 2510 ± 100 BP (Gillespie *et al.* 1983). It can be correlated with the lacustrine transgression reported from the Northern Rift in Lake Abhé (Fontes & Pouchan 1975) and in the Unda-Dobi (Gasse 1975).

Zone 3 (4.5–3.7 m)

Following a transition zone, Zone 3 was characterized by a decrease in aquatic organic matter. Diatoms were present but less abundant, with species characteristic of more saline water being dominant. Evaporation exceeded precipitation. The organic productivity decreased while the shrinking lake became more saline. Low pollen absolute frequencies indicate climatic conditions less favourable for vegetation. Alteration of pollen grains has been observed, which could be

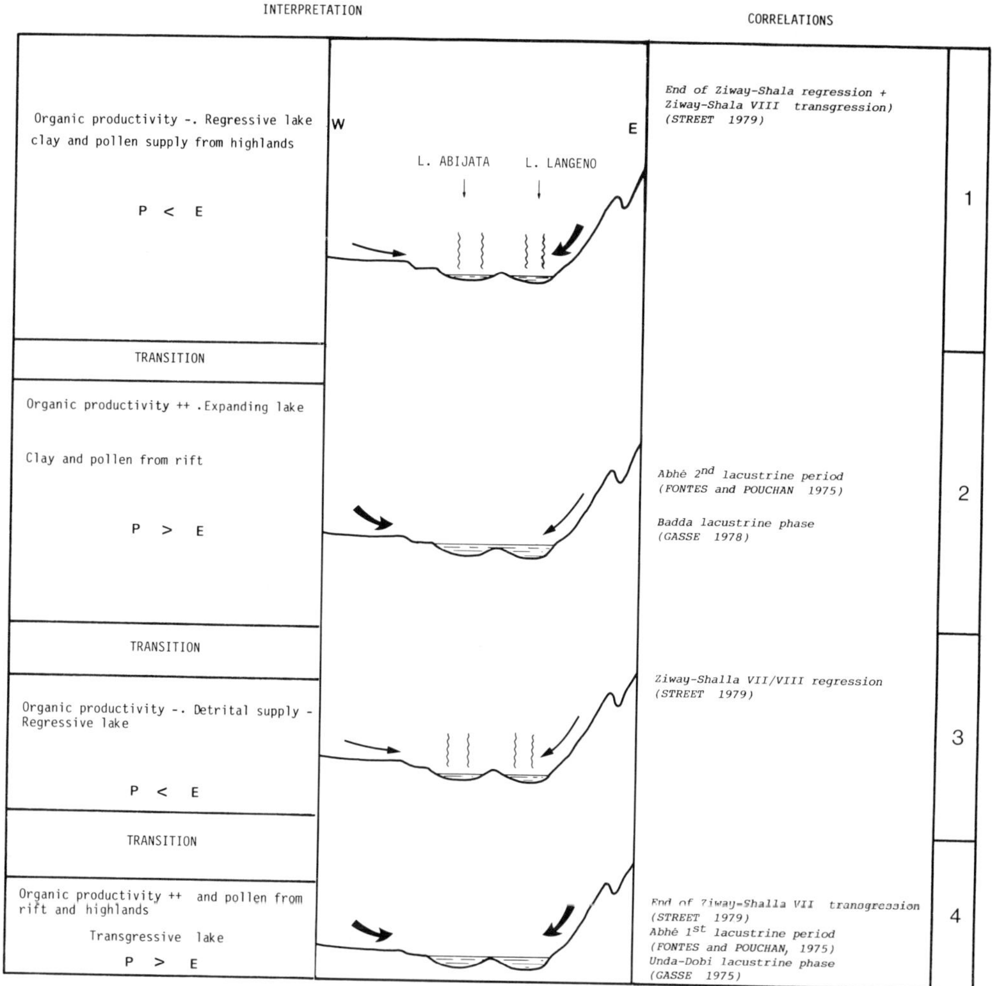

FIG. 6. Continued.

an indicator of their deposition in a more alkaline lake. All the evidence points to a lacustrine regression and regional aridity which could correspond to the beginning of the Ziway-Shalla VII–VIII regression, whose age remains uncertain (Street 1979).

Zone 2 (3.4–2.0 m)

After a short transition period, the tendency was reversed again towards an increasing lake productivity, as diatom species characteristic of more dilute waters became dominant. The increase in the plankton productivity implies a higher volume of water in the lake, lower concentration, relatively more precipitation, and less evaporation. The supply of detrital kaolinite fluctuated, while the smectites increased progressively. Moreover, forest tree pollen decreased while peri-aquatic pollen (*Cyperaceae*, but mainly *Typha*) increased, indicating a strong inflow of freshwater. *Olea* and *Dodonea*, both taxa from the drier, lower montane zone, also increased. This lacustrine phase of Zone 2 in Abijata core A has no outcrop equivalent (Street 1979), but as suggested from diatom evidence (Elenga 1984) it could be correlated with a lacustrine phase recognized on Mount Badda (Gasse 1978) and in Lake Abhé (Fontes & Pouchan 1975). This last high lake level would have occurred at *c.* 1500 BP.

Zone 1 (1.8 m to top)

A short transition is seen between 1.8 and 2.0 m when diatom associations characteristic of dilute waters are replaced by an association that indicates concentrated waters. The decreasing trends in the organic matter, pollen and diatom content, although fluctuating, indicate a regressive phase and a deficit in the water budget. Kaolinite was more abundant than smectites at the same time that Afro-montane tree pollen (*Juniperus*, *Hagenia*) increased. Thus a greater supply of the detrital sediment and pollen from the mountains could be explained by most of the precipitation occurring in the highlands—in other words a rainfall distribution pattern similar to the present one. Runoff occurred on the slopes above 1700 m but the bottom of the rift remained semi-arid (Street 1979). Zone 1 of the Abijata core would correspond to the last Ziway-Shalla regression which is still continuing to the present time (Makin *et al.* 1976; Street 1979).

Conclusions

The multidisciplinary analysis of the Abijata core sediments has demonstrated two transgressive and two regressive phases during the last 2000 years. These phases are recognized by variations in both the aquatic and the terrestrial environments. They are in agreement with lake level fluctuations determined from geomorphological studies. Biological, physical and geochemical indicators have produced complementary evidence. Studies of organic matter and diatoms are linked to the lake productivity, while sedimentology and palynology reflect erosion and climatological processes on a more regional basis. The analyses have shown that the two high lake levels might not have resulted from similar patterns of rainfall distribution.

ACKNOWLEDGEMENTS: We express thanks to the Ethiopian authorities, the Ministry of Mines and Addis Ababa University for their permission to conduct research; and also to Getaneh Assefa and C. Barton for invaluable help in the field. This work was supported by ASP Afrique (CNRS).

References

BADAUT, D., RISACHER, F., PAQUET, H., EBERHART, J.P. & WEBERT, F. 1979. Néoformation de minéraux argileux à partir de frustules de Diatomées: le cas des lacs de l'Altiplano bolivien. *C. R. Acad. Sci. Paris* **289D,** 1191–3.

BALTZER, F. & TRESCASES, J.J. 1971. Première estimation du bilan de l'altération, de l'érosion et de la sédimentation sur péridotites, sous le climat tropical de la Nouvelle-Calédonie. *C. R. Acad. Sci. Paris* **273D,** 2034–7.

BAUMANN, A., FORSTNER, U. & ROHDE, R. 1975. Lake Shala: water chemistry, mineralogy and geochemistry of sediments in an Ethiopian rift lake. *Geol. Rundsch.* **64,** 593–609.

BEUTELSPACHER, H. & VAN DER MAREL, H.W. 1968 *Atlas of Electron Microscopy of Clay Minerals and their Admixtures.* Elsevier, Amsterdam.

BUTZER, K.W., ISAAC, G.L., RICHARDSON, J.L. & WASHBOURN-KAMAU, C. 1972. Radiocarbon dating of East African lake levels. *Science* **175,** 1069–76.

CHAMLEY, H. 1980. Clay sedimentation and paleoenvironment in the area of the Daito Ridge (northwest Philippine Sea) since the early Eocene. In: KLEIN, G. DE V. *et al.* (eds) *Initial Report DSDP, 58.* US Government Printing Office, Washington, pp.683–93.

—— & BONNOT-COURTOIS, C. 1981. Argiles authigènes et terrigènes de l'Atlantique et du Pacifique NW (legs 11 et 58 DSDP): apport des terres rares. *Oceanol. Acta* **4,** 229–38.

——, MAILLOT, H., DUEE, G. & ROBERT, C. 1984. Paleoenvironmental history of the Walvis Ridge at the Cretaceous–Tertiary transition, from mineralogical and geochemical investigations. In: MOORE, T.C. *et al.* (eds) *Inistian Reports DSDP, 74.* U.S Government Printing Office, Washington, pp.685–95.

DEBYSER, Y. & GADEL, F. 1977. Etude géochimique des composés humiques et des kérogènes. In: *Géochimie organique des sédiments marins profonds, Orgon 1, mer de Norvège.* CNRS, Paris, pp.247–68.

DESCOURTIEUX, C. 1979. *Les Diatomées du sondage du lac Abiyata (Ethiopie): systématique et paléoécologie.* Unpublished D. Sp. thesis, Université P. et M. Curie, Paris, 76pp.

ELENGA, C. 1984. *Les sédiments lacustres d'âge Holocène supérieur en Ethiopie et en République de Djibouti. Etude des Diatomées et sédimentologie de quelques séquences d'altitudes différentes.* Unpublished D. Sp. thesis, Université P. et M. Curie, Paris, 177pp.

ESPITALIE, J., LAPORTE, J.L., MADEC, M., MARQUIS, F., LEPLAT, P., PAULET, J. & BOUTEFEU, A. 1977. Méthode rapide de caractérisation des roches mères, de leur potentiel pétrolier et de leur degré d'évolution. *Rev. Inst. Fr. Pét.* **32,** 23–42.

FONTES, J.C. & POUCHAN, P. 1975. Les cheminées du lac Abhé (TFAI): stations hydroclimatiques de l'Holocène. *C. R. Acad. Sci., Paris* **280D,** 383–6.

GASSE, F. 1975. *L'évolution des lacs de l'Afar central (Ethiopie et TFAI) du Plio-Pléistocène à l'Actuel. Reconstitution des paléomilieux lacustres à partir de l'étude des Diatomées.* Unpublished D.Sci. thesis, Université P. et M. Curie, Paris.

—— 1978. Les Diatomées holocènes d'une tourbière (4040 m) d'une montagne éthiopienne: le Mont Badda. *Rev. Algol.* **13,** 105–49.

—— & STREET, A. 1978. Late Quaternary lake-level

fluctuations and environments of the Northern Rift Valley and Afar region (Ethiopia and Djibouti). *Paleogeogr. Paleoclimatol. Paleoecol.* **24,** 279–325.

——, TALLING, J.F. & KILHAM, P. 1983. Diatom assemblages in East Africa: classification, distribution and ecology. *Rev. Hydrobiol. Trop.* **16,** 1–34.

GILLESPIE, R., STREET-PERROTT, F.A. & SWITSUR, R. 1983. Post-glacial arid episodes in Ethiopia have implications for climate prediction. *Nature* **306,** 680–3.

HERBIN, J.P. 1980. Géochimie organique des sédiments récents du lac Bogoria, rift Gregory, Kenya. *Rech. Geol. Afr.* **5,** 140–3.

LEZINE, A.M. 1981. *Le lac Abijata (Ethiopie): palynologie et paléoclimatologie du Quaternaire récent.* Unpublished D. Sp. thesis, Université de Bordeaux I, 125pp.

—— & BONNEFILLE, R. 1982. Diagramme pollinique holocène d'un sondage du lac Abijata (Ethiopie 7°42′N). *Poll. Spores* **24,** 463–80.

MAKIN, H.J., KINGHAM, T.J., WADDAMS, A.E., BIRCHALLE, C.J. & EAVIS, B.W. 1976. *Prospect for Irrigation Development around Lake Ziway, Ethiopia.* London Resources Division UK, Ministry of Overseas Development, Publication No. 26, 407pp.

MILLOT, G. 1964. *Géologie des argiles.* Masson, Paris, 499pp.

MOHR, P.A. 1971. *The Geology of Ethiopia.* University College of Addis-Ababa Press, 268pp.

OWEN, R.B., BARTHELME, J.W., RENAUT, R.W. & VINCENS, A. 1982. Palaeolimnology and archaeology of Holocene deposits north-east of Lake Turkana, Kenya. *Nature* **298,** 523–9.

PETERSON, G.M., WEBB III, T., KUTZBACH, J.E., VAN DER HAMMEN, T., WIJMSTRA, T.A. & STREET, F.A. 1979. The continental record of environmental conditions at 18,000 yr BP. An initial evaluation. *Quat. Res.* **12,** 47–82.

RICHE, G. 1976. Les processus de formations de sols dans le bassin du Wabi Shebelle (Ethiopie). *Cah. Orstom Pédol.* **14,** 227–43.

—— & SEGALEN, P. 1969. La zonalité verticale des sols en Ethiopie du Centre et du Sud-Est. *Cah. Orstom Pédol.* **7,** 5–9.

ROBERT, C. 1982. *Modalité de la sédimentation argileuse en relation avec l'histoire géologique de l'Atlantique Sud.* D. Sci., Universite de Aix-Marseille II.

STREET, F.A. 1979. *Late Quaternary lakes in the Ziway-Shala basin. Southern Ethiopia.* Unpublished PhD thesis, University of Cambridge, 457pp.

—— & GROVE, A.T. 1979. Global maps of lake level fluctuations since 30 000 years BP. *Quat. Res.* **12,** 83–118.

STOFFERS, P. & SINGER, A. 1979. Clay minerals in Lake Mobutu Sese Seko (Lake Albert). Their diagenetic changes as an indicator of the paleoclimate. *Geol. Rundsch.* **68,** 1009–24.

SURDAM, R.C. & EUGSTER, H.P. 1976. Mineral reactions in the sedimentary deposits of the Lake Magadi region, Kenya. *Bull. Geol. Soc. Am.* **87,** 1739–52.

TIERCELIN, J.J., LE FOURNIER, J., HERBIN, J.P. & RICHERT, J.P. 1980. Continental rifts: modern sedimentation, tectonic and volcanic control. Example from the Bogoria-Baringo graben, Gregory Rift, Kenya. In: *Geodynamic Evolution of the Afro-Arabian Rift System.* Accad. Nazionale dei Lincei Rome, pp.143–63.

TISSOT, B., DURAND, B., ESPITALIE, J. & COMBAZ, A. 1974. Influence of the nature and diagenesis of organic matter in formation of petroleum. *Bull. Am. Assoc. Petrol. Geol.* **58,** 499–506.

VINCENT, C.E., DAVIES, T.D. & BERESFORD, A.K.C. 1979. Recent changes in the level of Lake Naivasha, Kenya, as an indicator of equatorial westerlies over East Africa. *Climatic Change* **2,** 175–89.

YURETICH, R.F. 1982. Possible influences upon lake development in the East African Rift valleys. *J. Geol.* **90,** 329–37.

R. BONNEFILLE, C. ROBERT, A.M. LEZINE & G. PERINET, Laboratoire de Géologie du Quaternaire, CNRS—Case 907 Luminy, 13288 Marseille, Cedex 9, France.

G. DELIBRIAS, Centre Faibles Radioactivités—CNRS, 91190 Gif-sur-Yvette, France.

C. ELENGA, Ecole Normale Supérieure, 5 rue Boucicaut, 92260 Fontenay-aux-Roses, France.

J.P. HERBIN, Institut Français du Pétrole, BP18, 92506 Rueil Malmaison, Cedex, France.

J.J. TIERCELIN, GIS Océanologie et Géodynamique, Université de Bretagne Occidentale, 29287 Brest, Cedex, France.

Early rift sedimentation in the Turkana basin, northern Kenya

P.G. Williamson & R.J.G. Savage

SUMMARY: The 'Turkana Grit' deposits of northern Kenya are a thick sequence of immature arkosic sediments of pre-Plio-Pleistocene age. Originally described by Murray-Hughes in the 1930s, these sediments have been variously considered as of either Mesozoic or Miocene age by subsequent authors. These extremely immature, tectonically controlled sediments clearly reflect intense local deformation; their age is of considerable interest, as they clearly signal the onset of significant tectonic activity in the central sector of the East African rift. A regional survey of the Turkana Grits as currently known, coupled with recently available information from the hitherto unmapped Kajong area in the SE Turkana basin, clearly indicates that these sediments represent two distinct episodes of local tectonism and deposition. On the basis of available radiometric and biostratigraphic evidence, the earlier of these episodes appears to be of Cenomanian-basal Palaeocene age, while the later episode dates from the mid-Miocene. These observations suggest that significant tectonic activity in East Africa commenced in the late Mesozoic rather than—as generally assumed—in the mid-Cenozoic. The earliest (late Mesozoic) episode of clastic deposition in the Kajong area was initially controlled by a set of major E–W trending transform faults; these deposits are primarily preserved in associated fault-wedge basins. The documentation of a major transform element in the Kajong area has important implications for the early development of the African Rift as a whole. We suggest here the possibility that the enigmatic distribution of the late Mesozoic rift sediments—which are restricted to the southern sector of the Western Rift and the northern sector of the Eastern Rift—reflects the fact that the original late Mesozoic rift consisted of a single graben. The southern and northern portions of this graben have been subsequently displaced by a major E–W trending dextral transform system, extending some 500 km from the SE Turkana basin to the present northern termination of the Western Rift on the Uganda–Sudan border.

The Turkana depression is a triangular lowland situated between the Kenyan and Ethiopian domal uplifts in the N Kenyan sector of the East African Rift. This depression extends between approximately 2°30′, 4°30′ N and 35°30′, 37°0′ E, is bordered on the W by the Turkwel and Uganda escarpments, and on the E by the Kinu Sogo Fault Zone (see Fig. 1). To the S the Turkana depression merges with a well defined graben, the Baringo-Suguta graben, essentially a southern continuation of the Kino Sogo fault zone; while to the N a distinct graben is again developed on the Ethiopian border, occupied by Chew Bahir (Lake Stephanie). Various views have been expressed regarding the history and development of the structural boundaries of the Turkana depression (e.g. Pulfrey 1960; McCall 1964; Bishop & Trendall 1967). Recent reviews indicate that the Turkwel escarpment is a fault scarp and the Uganda escarpment an eroded monoclinal flexure (e.g. Walsh & Dodson 1969); both structures are considered to be of essentially Lower Miocene age (e.g. Fuchs 1939; Baker *et al.* 1972). The Kino Sogo Fault Zone, like its southern extension, the Suguta graben, is considered to have originated during the Plio-Pleistocene (Baker *et al.* 1972).

Four major lithostratigraphic units have long been recognized from the Turkana depression: (1) Precambrian and ?lowermost Palaeozoic gneisses of the Mozambique fold belt; (2) a thick sequence of coarse, immature clastics (the Turkana Grits) overlying this basement; (3) a sequence of Miocene volcanics, with subsidiary sediments, overlying the Turkana Grits; and (4) a thick, heterogeneous assemblage of Plio-Pleistocene volcanics and sediments (Murray-Hughes 1933; Fuchs 1939; Arambourg 1943; Baker *et al.* 1972).

The thick sequence of clastic sediments immediately overlying the basement complex was initially described by Murray-Hughes (1933). He described these 'Turkana Grits' as 'pebbly conglomerates, feldspathic grits, and sandy shales, all the elements of which appear to have been derived from the Basement Complex on which the beds rest'. A Jurassic age was proposed by Murray-Hughes for these deposits, but the basis for this suggestion is unclear. Because significant tectonic activity in the area has long been considered to have commenced with Lower Miocene rifting (Baker *et al.* 1972), the Turkana Grits—immature arkosic sediments primarily derived from a rapidly uplifted basement terrain—have been assumed by most authors to be of exclusively

From FROSTICK, L.E. *et al.* (eds) 1986, *Sedimentation in the African Rifts*, Geological Society Special Publication No. 25, pp. 267–83.

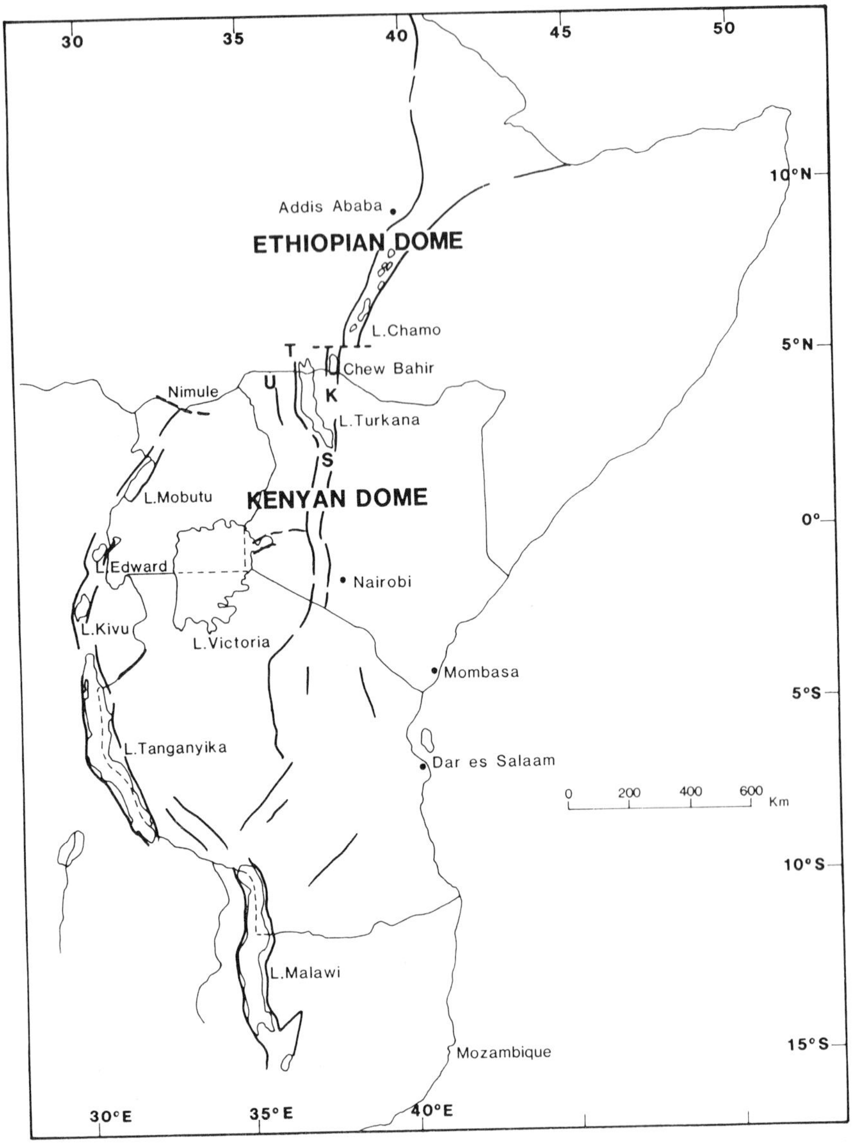

FIG. 1. Structural setting of the Turkana depression, northern Kenya. K = Kino Sogo fault zone; S = Suguta graben; T = Turkwel escarpment; U = Uganda escarpment.

Miocene age. These sediments have indeed yielded Miocene mammal faunas in certain areas and associated volcanic horizons have yielded consonant radiometric ages. However, as early as 1943, Arambourg suggested that the Turkana Grits in fact represent two distinct episodes of sedimentation, one of Miocene and the earlier of Mesozoic age. The precise age and relationships of these texturally and mineralogically immature sediments is of considerable interest, reflecting as they do the onset of significant tectonic activity in the area.

This paper is a discussion of the probable age and relationships of the Turkana Grits, with particular reference to sections recently investigated in the Kajong area of the SE Turkana depression; we also discuss the general implications of these deposits for the early development of the African Rift as a whole.

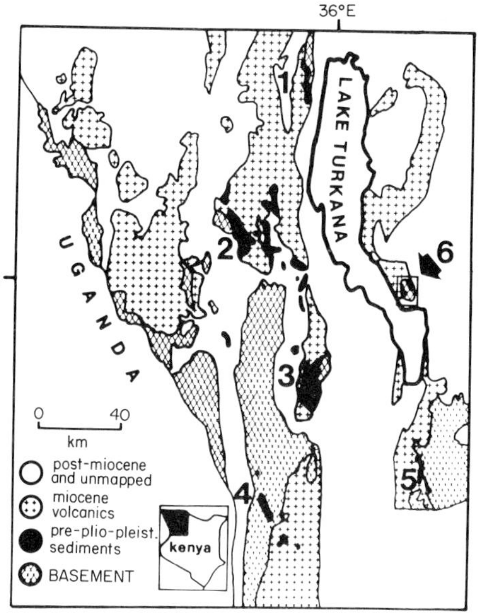

FIG. 2. Generalized lithostratigraphy of the Turkana depression and neighbouring areas in NW Kenya, showing distribution of basement, pre-Plio-Pleistocene sediments, Miocene volcanics, post-Miocene sediments and unmapped areas. Compilation based on data from Fuchs 1939; Arambourg 1943; Mason & Gibson 1957; Baker 1963; Dodson 1963; Joubert 1966; Walsh 1966; Walsh & Dodson 1969; Fairburn & Matheson 1970; Dodson 1971; Vondra & Bowen 1976; Savage & Williamson 1978; Williamson & Savage, this paper. Geological data from NW Uganda is not incorporated in this figure. Numbers refer to localities discussed in the text. 1 = Lapurr Range; 2 = Lodwar area; 3 = Loperot area; 4 = Kalossia-Tiati area; 5 = area N of Baragoi; 6 = Kajong area. Detailed geological map of the Kajong area (indicated by black arrow) is given in Fig. 3.

Regional survey of early clastic sedimentation in the Turkana depression

As indicated in Fig. 2, deposits of the Turkana Grits are known from six principal areas in the general area of the Turkana depression: (1) in the Lapurr Range; (2) in the Lodwar area; (3) in the area around Loperot; (4) in the Kalossia-Tiati area; (5) in the area N of Baragoi; and (6) in the Kajong area. Most of these exposures are located in the vicinity of, and can be demonstrated to nonconformably overlie, the Precambrian basement. A brief survey of the geology of these areas is given below.

The Lapurr range

The Lapurr (Labur; Lubur) Range W of Lake Turkana rises to around 1500 m on Mt Lapurr, some 18 km NNW of Lokitaung. The geology of the area is described by Arambourg (1933b, 1935, 1943); Arambourg & Wolff (1969); Fuchs (1934, 1935, 1939); Murray-Hughes (1933); and Walsh & Dodson (1969). All reports of the stratigraphy of this area confirm a general succession consisting of between 200 and 300 m of sandstones and conglomerates dipping 15° to the SSW, resting nonconformably on basement and unconformably overlain by Miocene lavas. The sedimentary sequence comprises a suite of interbedded clays, siltstones, medium to coarse sandstone, conglomerates, and minor limestone (Arambourg 1943; Walsh & Dodson 1969). Coarse sandstones dominate the succession. The sandstones are generally immature to submature subarkoses or sublitharenites; the larger quartz fractions are generally subrounded to angular, while the smaller fractions are generally subround; cements are generally siliceous or calcareous, with a varying admixture of ferruginous material; the beds vary widely in colour, but are generally maroon, green, or off-white, and are occasionally cross-bedded. The conglomerates range from granule to cobble grade, and consist primarily of well-rounded quartz clasts, although angular quartz and lithic (gneiss and schist) fragments from granule to cobble grade also occur (Arambourg 1943; Walsh & Dodson 1969). Cements in the conglomerates consist of an iron-oxide staining with a matrix of chlorite, muscovite, kaolinite, silica and rare pyrite (Walsh & Dodson 1969). No volcaniclastic elements occur in any of these sediments, and all components appear to be derived from the underlying basement terrain.

Sediments at the basal contact with the basement are variable, ranging from 'soft, shaly bands' to coarse sands or discontinuous conglomerates (Walsh & Dodson 1969).

Fossil evidence from the sediments of the Lapurr range is sparse. In the absence of any identifiable fauna, Arambourg (1933c) compared the succession with the Adigrat Sandstone of Ethiopia, and suggested a Trias-Jurassic age (although the Adigrat is now considered to be Lower Jurassic, Whiteman 1971). Fuchs (1939) reported large silicified tree trunks at Lokitaung gorge, identified as *Dryoxylon*, a genus recorded from the Cenozoic of both India and Africa. He inferred that the sediments were of 'Oligo-Miocene' age. Arambourg (1943) argued that *Dryoxylon* gives a poor indication of age and that Eocene or even Cretaceous was a more likely guess. Subsequently, Arambourg & Wolff (1969) reported F.H. Brown's discovery of the distal end of a sauropod humerus from the eastern end of Lokitaung gorge. More recently, additional dinosaur material, hitherto undescribed, has been discovered in neighbouring areas by a National Museums of Kenya Expedition (J.M. Harris, pers. comm.). On available evidence, the Turkana Grits sequence of the Lapurr range therefore appears to be of Jurassic-Cretaceous age.

Although the clastic sequence of the Lapurr range was originally mapped as Turkana Grits by Murray-Hughes and has been so mapped subsequently by the Kenya Geological Survey (Walsh & Dodson 1969), Arambourg (1933b) introduced the term 'Lubur Series' for these sediments, on the grounds that sediments known elsewhere in the Turkana depression and termed Turkana Grits had a major volcaniclastic component and had yielded a Miocene fauna. However, his definition of the Lubur Series, which stressed the fact that these sediments lack any volcaniclastic component or associated volcanics, was essentially the same as Murray-Hughes' (1933) original description of the Turkana Grits (see above); although Arambourg's adoption of the term Lubur Series for the clastic sediments of the Lapurr Range has certainly contributed to the terminological confusion which surrounds the earliest clastic sediments of the Turkana depression, he deserves credit for being the first to publish the suggestion that these sediments in fact represent two distinct depositional episodes of different age (although this suggestion was apparently made at the same time in an unpublished manuscript by Champion, see Mason & Gibson 1957).

The Lodwar area

The geology of the Lodwar area is described by Arambourg (1933a, c, 1935), Walsh & Dodson (1969) and Dodson (1971). Two distinct sedimentary sequences are exposed in an extensive terrain E of the town of Lodwar; a lower sequence of coarse clastic sediments lacking any volcaniclastic component and referred by Arambourg to his Lubur Series, and a sequence of interbedded basalts, tuffs, sands, silts, and conglomerates which overlie the lower sequence uncomformably. This upper sequence passes up conformably into a thick and continuous sequence of overlying Miocene volcanics.

The lower sedimentary sequence comprises a heterogeneous assemblage of clays, silts, fine to coarse sands, and granule to cobble grade conglomerates; at least 100 m is exposed. The sandstones are primarily immature to submature subarkoses and sublitharenites, intermittently cross-bedded. Minor, discontinuous limestone bands also occur in places. The larger clasts in the conglomerates are primarily subangular to well-rounded quartz; as in the sandstones, cements are generally calcareous or siliceous, with a varying ferruginous admixture. The sediments range in colour from green or maroon to buff or off-white, and are intermittently lateritic. There is no volcaniclastic component, and these deposits appear to be derived exclusively from the underlying basement terrain. The only fossils reported are unidentifiable bone and wood fragments.

The upper sedimentary sequence comprises a heterogeneous assemblage of tuffs, agglomerates, and thin waterlain clastic sediments which are superficially similar to the underlying Lubur Series sediments (Walsh & Dodson 1969). These sediments range from silts to coarse sands, and occasional conglomerates to boulder grade also occur. The sands are generally immature to submature subarkoses and sublitharenites, with a large volcaniclastic component and much quartz; the coarser sands are frequently cross-bedded. The finer-grained sediments in this upper sequence have yielded an extensive but rather poorly preserved fauna, primarily from two separate localities some 11 km apart: Lothidok and Moruarot.

The fauna includes Miocene mammals (Arambourg 1933c, 1947; Clark & Leakey 1951; Deraniyagala 1951; Whitworth 1954, 1958; Savage 1965; Hooijer 1966; Churcher 1970; Madden 1972; Wilkinson 1976), various reptiles, and the freshwater molluscs *Lanistes* and *Etheria* (Verdcourt 1963; Van Damme & Gautier 1972). The Lower-Middle Miocene age indicated by the faunal elements so far known from the upper sedimentary sequence at Lodwar is in broad agreement with available radiometric (K-Ar) age determinations. These determinations, apparently from basalts immediately and conformably overlying this sequence at Lothidok, are in the general region of 14 Ma (Baker *et al.* 1972).

In the extreme S of the Lodwar area, almost midway between the main Lodwar area exposures and those of Loperot (see Fig. 2), Dodson (1971) described a series of clastic sediments which he compared closely to the lower part of the Turkana Grits section described by Joubert (1966) from Lakhapelinyang in the Loperot area (see below).

Loperot area

Murray-Hughes (1933), Champion (1937) and Dixey (1945) all noted the presence of Turkana Grits in the Loperot area; Murray-Hughes provided a preliminary map of the area, and detailed mapping was performed by Joubert (1966). These reports indicate that the sediments described as Turkana Grits in this area occur primarily as a fringe to a basement inlier in the northern areas, and to the S and W are surrounded by Miocene volcanics (see Fig. 2). The sedimentary sequence in the southern part of the area is best known; it rests nonconformably on basement and the contact is marked by a distinctive yellowish 'calcareous grit' or 'basal limestone' (calcrete?) which grades into sandstones laterally and vertically (Joubert 1966) and which is said to occur, on occasion, 'within small fractures in the basement'. Basal conglomerates may also occur at this contact. The deposits appear to be relatively thick; Joubert gives a section E of Lakhapelinyang which has a calcareous contact with the basement, and consists of some 230 m of heterogeneous silts, fine to coarse sands, granule to cobble-grade conglomerates, and calcareous horizons. The lower part of the sequence includes gypsum layers, and the middle part of the succession is characterized by numerous tuff horizons. Occasional freshwater limestones occur. The sandstones appear to consist primarily of immature to submature subarkoses and sublitharenites. Cements are generally calcareous and limonitic, occasionally siliceous. Broadly equivalent sediments elsewhere in the southern areas include substantial interbedded basalt horizons.

The sediments in the southern part of the Loperot area have yielded an extensive series of Miocene faunas (Dixey 1945; Clark & Leakey 1951; Hooijer 1966, 1971, 1973; Maglio 1969; Van Damme 1969; Mead 1975). Of particular note is the presence of a ziphiid whale—no other non-marine whale is known on the African continent. The *Platybelodon* (Maglio 1969) is similarly unique to Africa, and appears to represent a very early member of a stock that diversified into Asia and North America. In addition to the mammals, these deposits have also yielded an extensive freshwater fauna, which includes the fish *Protopterus* and *Polypterus*, the crocodile *Tomistoma*, and the molluscs Mutelidae indet., *Pleiodon* sp., *Lanistes carinatus*, *Pila ovata*, *Saulea lithiodes*, *Limicolaria leakeyi* and *Melanoides tuberculata* (Van Damme 1969). Unfortunately, although these faunal elements clearly indicate a broadly mid-Miocene age, none permit precise biostratigraphic correlation. However, several radiometric (K-Ar) age determinations performed on basalts in the general stratigraphic vicinity of the fossiliferous units, reported by Maglio (1969), Hooijer (1971) and Mead (1975), range in age from 15.8 ± 1.2 Ma to 17.5 m± 0.9 Ma, and therefore confirm a general mid-Miocene age for these sediments.

Kalossia-Tiati area

Mason & Gibson (1957) described an extensive series of coarse-grained clastic sediments resting on basement in this area. No information on the age of these sediments is currently available, but these authors, like Arambourg and Champion (see above), drew an important distinction between these sediments (in the southern part of the area) containing a substantial volcaniclastic component and conformably overlain by Miocene basalts, and sediments (in the northern part of the area) consisting exclusively of basement debris but lacking any volcaniclastic component. The former sedimentary sequence they termed the 'Tiati Grits'; the latter they referred to as the 'Turkana Grits'. In the sense that these authors reserve the term 'Turkana Grits' for sediments lacking a volcaniclastic component, their usage clearly conforms to Murray-Hughes' (1933) original definition (see above).

The Tiati Grits of Mason & Gibson range from coarse sands to cobble-grade conglomerates; the sands are generally immature to submature subarkoses or sublitharenites; in addition to quartz and feldspars, they contain basalt fragments, and apparently also include minor interbedded basalts. Cements are calcareous and limonitic, and occasionally siliceous. The sequence can be demonstrated to nonconformably overlie basement, and in places is immediately and conformably overlain by a sequence of Miocene basalts.

Sediments referred to by Mason & Gibson as the Turkana Grits in the Kalossia-Tiati area comprise a thin (< 20 m), poorly exposed section of fine sands to cobble-grade conglomerates, with coarser grain grades predominating; the sands are generally immature to submature subarkoses. Cements are commonly siliceous or limonitic, more rarely calcareous. Unfortunately, the superpositional relationship of Mason & Gibson's Turkana and Tiati Grit sequences is unknown.

Area N of Baragoi

Baker (1963) describes a series of clastic sediments, which he termed the 'sub-volcanic sediments', in the area N of Baragoi (see Fig. 2). The main sequence consists of some 70 m of coarse sands and conglomerates; the conglomerates consist chiefly of well-rounded quartz clasts to pebble grade. The lowest part of the sequence consists of a 3 m thick conglomerate containing pebbles of quartzite, pegmatite, and biotite gneiss up to small boulder size, clearly derived from the basement; the base of the sequence is not seen. Immediately overlying the sediments, which are

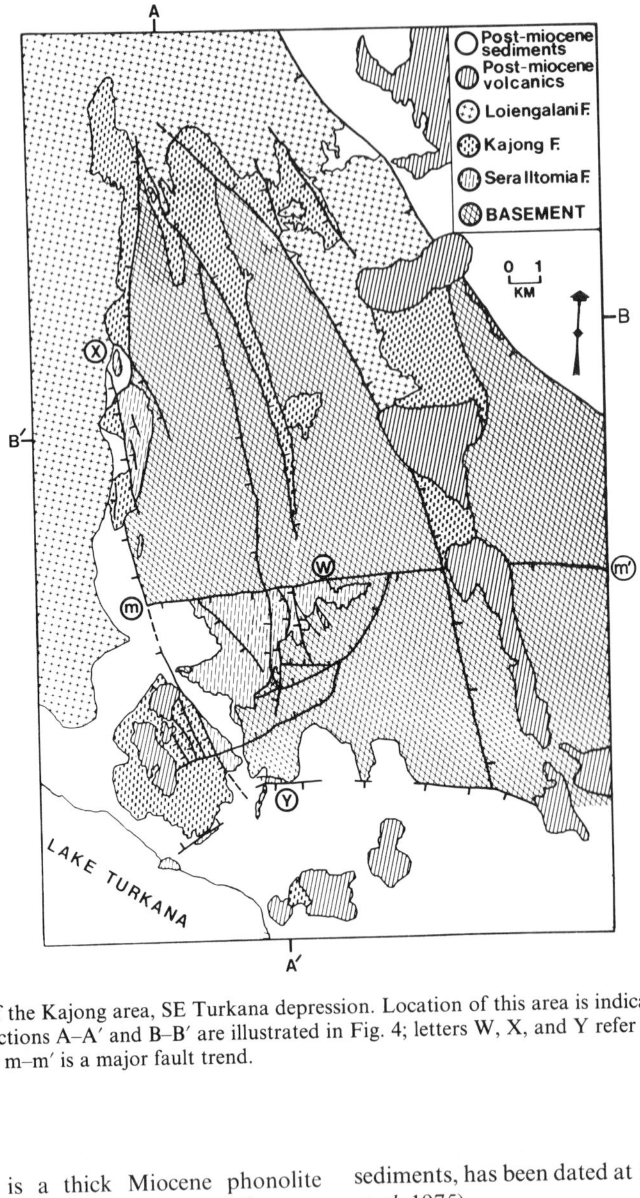

FIG. 3. Geology of the Kajong area, SE Turkana depression. Location of this area is indicated by black arrow in Fig. 2. Cross-sections A–A′ and B–B′ are illustrated in Fig. 4; letters W, X, and Y refer to localities mentioned in text; m–m′ is a major fault trend.

unfossiliferous, is a thick Miocene phonolite sequence. Unfortunately, the nature of the contact between the sediments and the overlying volcanic sequence is unclear. In addition to the principal areas of early clastic sedimentation in the Turkana depression discussed above, a restricted exposure of intravolcanic fossiliferous sediments in the Buluk area, NE of Lake Turkana, has been described recently by Harris & Watkins (1974). The sequence is some 50 m thick, and comprises clays, silts, and fine to coarse immature subarkoses with intercalated tuffs. The base of the conformably overlying basalts, which themselves include later intercalated fossiliferous sediments, has been dated at 17.3 ± 1.4 Ma (Fitch *et al.* 1975).

Geology of the Kajong area

General geological setting

An extensive series of clastic sediments was first reported from the Kajong area in the SE Turkana basin during the explorations of Teleki and von Höhnel in 1888 (Höhnel *et al.* 1891). Subsequent investigation has revealed an extensive series of exposures occupying a rectangular area of *c.* 330 km^2 centred on 3°03′N and 36°32′E (Savage &

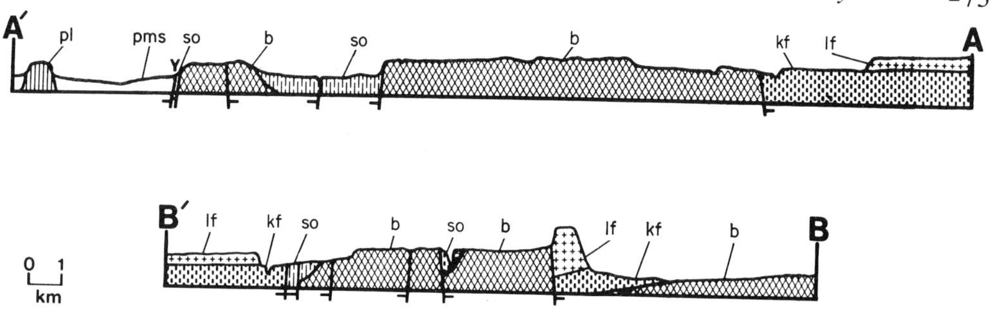

FIG. 4. Generalized geological cross-sections across the Kajong area; the alignment of these sections is indicated in Fig. 3. Key as in Fig. 3; b = basement; so = Sera Iltomia Formation; kf = Kajong Formation; lf = Loiengallani Formation; pl = Plio-Pleistocene volcanics; pms = post-Miocene sediments and alluvium. 'Y' indicates position of locality Y, the geographic position of which is shown in Fig. 3 and which is discussed in the text.

Williamson 1978). Apart from recent alluvium and Holocene lake beds, five principal lithostratigraphic units can be recognized in this area (Savage & Williamson 1978). These units comprise: (1) basement granites and gneisses of the Mozambique fold belt; (2) an earlier sedimentary sequence (the Sera Iltomia Formation), which rests unconformably on basement; (3) a later sedimentary sequence (the Kajong Formation), which also has an unconformable contact with the basement; (4) a thick sequence of Miocene basalts (the Loiengallani Formation) which conformably overlies the Kajong; and (5) a series of Plio-Pleistocene basalts.

The general geology of the area is indicated in Figs. 3 and 4. As indicated in these figures, the central part of the area is occupied by a large basement horst composed of a series of fault bounded blocks. In the northern areas these blocks are defined by a set of NNW–SSE trending normal faults downthrowing to either the E or W; in the southern areas of the horst this fault set intersects a series of broadly E–W trending normal faults downthrowing to either the N or S (see Fig. 4). The NNW–SSE trending fault set closely parallels the local basement trend, but the E–W set is strongly discordant with this trend. The sedimentary sequences and overlying Miocene volcanics have been preserved by these fault sets, either in the angles between adjacent blocks defined by the NNW–SSE fault set, or in small graben defined primarily by the E–W faults in the southern part of the area (see Fig. 4).

The Sera Iltomia Formation

The Sera Iltomia Formation outcrops principally in the southern part of the area, where it primarily occupies an extensive graben in the basement; additional smaller exposures also occur in the western areas, chiefly around locality 'Y' (see Figs. 3 and 4). The sequence comprises a variegated sequence of silts, fine to coarse sands and conglomerates up to large cobble grade. At least 220 m are currently exposed, but as indicated below, the original thickness is likely to have been much greater.

As shown in Fig. 5, the Sera Iltomia can be conveniently divided up into four principal members. Member 1 immediately and unconformably overlies basement. This member comprises

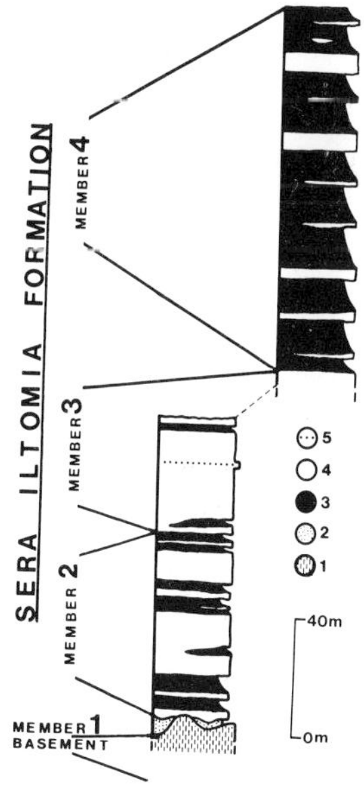

FIG. 5. Generalized stratigraphic section through the Sera Iltomia Formation. 1 = basement; 2 = regolith; 3 = fine sands/silts; 4 = medium/coarse sands; 5 = major conglomerates.

some 4–5 m of silts and fine sands with intercalated large-cobble grade conglomerates. In places these sediments rest directly on basement, with a clean contact, but elsewhere altered basement can be seen breaking down *in situ* to form this deposit. Where the contact is abrupt, the basement is invariably heavily iron stained. The silts and sands are partly lateritic, heavily iron cemented, and of mottled maroon and greenish colour; no sedimentary structures are apparent. The conglomerates may rest directly on the basement, but they usually occur as lensoid intercalations within the sands and silts; they consist exclusively of well-rounded quartz clasts of pebble to large cobble size. Member 1 is interpreted as a regolith deposit which mantled the original basement surface immediately prior to deposition of the remainder of the Sera Iltomia Formation.

Member 2 immediately overlies Member 1 and in places channels down into it. Member 2, some 60 m in thickness, consists of a series of between five and eight laterally extensive, erosion-surface bounded, fining-up sequences (only the major sequences are shown in Fig. 5). These sequences typically begin with granule to pebble-grade conglomerates supported by a matrix of medium to very coarse sand. The clasts are dominated by angular to subangular quartz and basement fragments up to cobble size, and the matrix is generally an immature sublitharenite or subarkose. These basal units, which may be intermittently cross-bedded, grade up rapidly through medium to fine subarkoses and sublitharenites, or even into fine silty mudstones, and are erosively truncated by the overlying fining-up sequence. Cements are generally calcareous, with a varying admixture of ferruginous material. These sediments are generally maroon, red or brown in colour. Contrary to previous interpretations (Savage & Williamson 1978), the deposits of Member 2 are considered here to represent relatively high-energy channel sands and overbank deposits in a rapidly aggrading fluvial system. The few palaeocurrent directions available indicate transport directions in a broadly S–N direction (Fig. 6).

At locality 'W' (see Fig. 3), sediments which are demonstrably at the same stratigraphic level as more typical Member 2 deposits exhibit a very different facies. At this locality a thick series of grain-supported boulder-grade conglomerates, consisting primarily of angular basement clasts, document the western half of a large alluvial fan complex. Occasional very well-rounded quartz cobbles, similar in appearance to those of the conglomerates in Member 1, suggest that the fan materials may be derived in part from this unit as well as from the basement. The eastern half of this

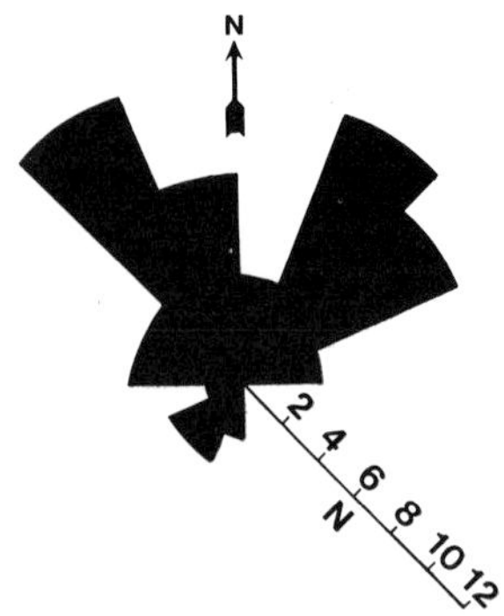

FIG. 6. 'Rose diagram' of palaeocurrent directions from Members 2 and 3 of the Sera Iltomia Formation; N = number of paleocurrent determinations. Note that all determinations between 180° and 270° were obtained from the alluvial fan at locality W, indicated in Fig. 3, and discussed in text.

fan complex has been removed by erosion. The stratigraphic relationships and the few palaeocurrent directions available, taken on coarse sand to granule grade immature sublitharenite intercalations, indicate quite unequivocally that this fan complex fed directly off the major E–W fault defining the northern boundary of the graben in which the majority of the Sera Iltomia sediments are currently preserved (see Figs. 3 and 6). The predominantly SSW palaeocurrent directions obtained from this complex reflect the fact that only the western half of the fan is preserved. This locality is of great importance in demonstrating initial local tectonic control of the Sera Iltomia deposition; this local control is also reflected by the extreme immaturity of more typical Member 2 deposits elsewhere in the area.

Member 3 immediately overlies Member 2 and in places channels down into it. This Member consists of at least 40 m of intercalated trough cross-bedded sandstone lenticles up to 4 m thick and 15 m across. The bases of individual lenticles may be very coarse, consisting of matrix supported conglomeratic layers dominated by angular to subangular quartz and basement clasts up to small cobble size, and including mudstone intraclasts. Member 3 is generally coarser than Member 2, the sand grade comprising medium to very coarse immature to submature subarkoses, sublitharenites, and subsidiary submature orthoquartzites. Occasional silty mudstone lenses occur at the top of some of the sandstone lenticles. Cements are generally calcareous; Member 3 is generally dirty yellow brown to grey in colour. Wood fragments and large drifted trunks occur at several localities. The 40 m section for Member 3 indicated in Fig. 5 is certainly an underestimate; the sections illustrated are separated by a major fault (see Fig. 3), and the contact between Members 3 and 4 is not seen in the lower section. In the western areas, along the Sera Iltomia drainage (at locality 'Y', Fig. 3) deposits of Member 3 appear to reach at least 80 m in thickness, but these exposures are heavily faulted, and a precise estimate of thickness is not possible. Member 3 clearly represents high-energy braided stream conditions; the few palaeocurrent directions available indicate transport directions in a broadly S–N direction (see Fig. 6).

Member 4 is superficially similar to, but finer and more mature than, Member 2, and consists of at least 120 m of medium to coarse sands and silts. The exposed upper surface of this unit everywhere coincides with a recent erosion surface, and its original thickness may well have been much greater. The sequence consists of 20–30 erosion-surface bounded fining-up cycles. The base of these cycles consists of laterally extensive, tabular, intermittently cross-bedded, coarse to very coarse submature subarkoses, occasionally including matrix-supported granule to small pebble-grade conglomerates consisting primarily of round to subround quartz clasts. These basal units are locally thickened into elongated lensoid channel bodies up to 8 m across, which protrude below the general level of the lower surface of the unit and apparently represent major distributary axes. The basal units of the various fining-up sequences pass up into medium to fine, submature subarkoses and occasional coarse silts. Cements are generally calcareous and ferruginous, and the sediments are pinkish or bright red in colour. Like Member 2, Member 4 apparently represents channel and overbank deposition in an aggrading, non-braided, fluvial system.

The Sera Iltomia Formation lacks any volcaniclastic component or associated volcanics, and the only fossils so far known are fossil wood fragments from Member 3. The age of these deposits is therefore unclear. However, it is of note that of 12 samples of fossil wood from this member, all on sectioning proved to be angiosperm.

The Kajong and Loiengallani Formations, and post-Miocene sediments and volcanics

As the main focus of this paper is the earliest rift sedimentation in the African Rift, only a brief summary of the Miocene Kajong and later units in the SE Turkana basin is given here; further details are reported elsewhere (Savage & Williamson 1978). The Kajong Formation outcrops throughout the Kajong area (see Fig. 3) and comprises a varied assemblage of mudstones, silts, fine to coarse sands and conglomerates up to pebble grade. The sediments are primarily white and red in colour, and cements are primarily calcareous. Pedogenetic calcrete horizons occur throughout the sequence. The total aggregate thickness of the formation is 160 m.

As indicated in Savage & Williamson (1978), the Kajong Formation may be conveniently divided into four members. Member 1 unconformably overlies a massively caliched, low-relief basement surface; calichification of the basement extends in most places at least 5 m below the sediment–basement contact. Member 1 is restricted to the northern Kajong area, and consists of up to 55 m of bright red, green or white immature fine to coarse subarkoses, sublitharenites, silts and mudstones. Major calcrete horizons up to 1.5 m thick are common, as are halite lenses, stromatolite accumulations, *Etheria* reefs, and a restricted fauna of other freshwater and terrestrial molluscs including *Melanoides*, *Mutela* and *Cerastua*. Member 1 is interpreted as an extensive playa deposit (Savage & Williamson 1978).

Member 2 of the Kajong conformably overlies Member 1 in the northern Kajong area, but in the S rests on massively caliched basement, often with a major basal conglomerate. Member 2 comprises some 35 m of tabular, laterally extensive, erosion-surface bounded, fining-up sequences up to 4 m thick with occasional calcretes. This unit is interpreted as the overbank and distributary channel deposits of a rapidly aggrading, proximal fluvial system.

Members 3 and 4 of the Kajong Formation are broadly comparable in petrological and palaeoenvironmental terms to Member 2. Member 3, comprising some 50 m of sediment, represents a somewhat higher energy regime than Member 2, whereas Member 4 is quite comparable to Member 2, but includes a number of thin (<0.1 m) tuff horizons which apparently herald the onset of the major volcanic activity documented by the overlying Loiengallani Formation. In the northern Kajong area, Member 3 is conformably overlain by Member 4; in the S, Member 4 rests on Member 3 with a pronounced angular unconformity (Savage & Williamson 1978).

A restricted mammal fauna occurs in all four members of the Kajong Formation. This fauna comprises *Paraphiomys* sp., Hyracoidea indet., *Prodeinotherium hobleyi*, *Gomphotherium kisumuensis*, *Dorcatherium* cf. *pigotti*, and indeterminate rhinocerotids, suids, and anthracotheriids (see Savage & Williamson 1978, Table 1).

Member 4 of the Kajong Formation is everywhere conformably overlain by the Loiengallani Formation (Savage & Williamson 1978); this contact is marked by a reddened 'baked zone' up to 1 m thick. The Loiengallani Formation comprises at least 200 m of fine-grained olivine basalts and subsidiary interbedded pyroclastics. A recent K-Ar date on the base of the Loiengallani Formation (Fitch & Miller Associates, Cambridge, unpub. data) has yielded a date of 18.5 ± 0.45 Ma, consonant with the broadly mid-Miocene mammal fauna of the underlying Kajong. This date indicates that, contrary to the recent suggestion of Brotzu *et al.* (1984), the Loiengallani Formation is quite distinct from the claimed episode of latest Miocene (*c.* 6.4 Ma) fissure volcanism in other areas E of Lake Turkana.

The Miocene sediments and volcanics of the Kajong area are immediately overlain by a series of areally restricted Plio-Pleistocene volcanics, not formally named here (see Fig. 3). These fine-grained olivine basalts have recently yielded a K-Ar date of 3.38 ± 0.25 Ma (Fitch & Miller Associates, Cambridge, unpub. data). Coastal areas of the Kajong area are mantled by up to 20 m of fine lacustrine silts apparently equivalent to the sub-recent Galana Boi Beds known from the Koobi Fora area to the N (Vondra & Bowen 1976), and Recent alluvium occurs in the main river valleys.

Summary

In summary, a regional survey of the Turkana Grit deposits of the Lapurr, Lodwar, Loperot, Kalossia-Tiati, Baragoi, and Kajong areas of the Turkana depression strongly supports Arambourg's original (1943) contention that these clastic sequences comprise two distinct chronostratigraphic units. Available evidence suggests that these deposits fall into two distinct categories: (1) the upper clastic sequences at Lodwar, the sequence at Loperot (Arambourg 1933b), and the upper clastic sequence at Kajong (the 'Kajong Formation', Savage & Williamson 1978); and (2) the lower clastic sequence at Lodwar, the sequence at Lapurr (both originally allocated to the 'Lubur Series', Arambourg 1933b), and the lower sedimentary sequence at Kajong (the Sera Iltomia Formation, Savage & Williamson 1978).

The upper sequence at Lodwar, the sequence at Loperot, and the Kajong Formation are petrologically similar; all of these deposits include a significant volcaniclastic component. All rest nonconformably on basement, and all are, on biostratigraphic and/or radiometric evidence, of mid-Miocene age. The lower sequence at Lodwar, the sequence at Lapurr, and the Sera Iltomia Formation at Kajong are petrologically comparable; they lack any volcaniclastic component, lie nonconformably on basement, and in the case of the Lodwar and Kajong sections, unconformably underlie deposits of proven Miocene age. As indicated above, the 'Lubur Series' at Lapurr has yielded dinosaur material and angiosperm wood, while the Sera Iltomia Formation has yielded a large amount of (exclusively) angiosperm material. Available evidence suggests a Cenomanian–basal Palaeocene age for these deposits (see below). Although the relative ages of the Turkana Grit deposits at Lapurr, Lodwar, Loperot and Kajong are clear, the relationship of these sections to those in the Kalossia-Tiati area and N of Baragoi remain obscure. The stratigraphic settings of the later two sections are unclear, and no biostratigraphic or radiometric evidence for their age is currently available.

Geological history of the Kajong area

An understanding of the early depositional history of the Kajong area clearly requires the establishment of the relative ages of the two major depositional episodes (the Sera Iltomia and Kajong Formations) and the two generations of faulting (the NNW–SSE and E–W sets, see above). Contrary to previous reports (Savage & Williamson 1978), additional work in the Kajong area has demonstrated that it is not possible to demonstrate unequivocably the superpositional relationship of the Kajong and Sera Iltomia Formations in any single section. But the Sera Iltomia clearly predates the Kajong: in several places, notably at locality 'X' (see Fig. 3), the eroded top of the Sera Iltomia is massively caliched, and this zone of calichification can be traced laterally into the caliched basement surface which everywhere underlies the basal contact of the Kajong. With the exception of this massive secondary calichification at its eroded upper surface, the Sera Iltomia, unlike the Kajong, is completely devoid of calcrete.

So far as the age of the fault sets in the Kajong area is concerned, it is clearly the case that the E–W set predates the NNW–SSE series. The latter fault set commonly terminates the E–W trending fault series, displaces the Kajong and Loiengallani Formations, and—clearly still active—has displaced the sub-Recent Galana Boi lake bed deposits by up to 2 m. In contrast, the E–W fault set has nowhere displaced the Kajong and Loiengallani Formations (or indeed the overlying Plio-Pleistocene volcanics, see Fig. 3).

Three major erosion levels are generally recognized in East Africa, of late Mesozoic, mid-Cenozoic and late Cenozoic age (Baker *et al.* 1972). In the sense that the formation overlying the Sera Iltomia—the Kajong— is demonstrably of mid-Miocene age, the Sera Iltomia was presumably deposited on the earliest of these erosion surfaces. This late Mesozoic basement surface was apparently of low relief; the lateritic regolith mantle overlying it (Member 1 of the Sera Iltomia) is often thick and well developed. The basement topography is subdued in the limited series of exposures in which the surface of nonconformity can be observed, and the conglomeratic channel lenses developed in places within this regolith consist exclusively of well-rounded quartz clasts. The absence of any lithic fragments from these conglomerates indicates that little local erosion of basement was taking place.

In contrast, the coarse immature sediments of Member 2 of the Sera Iltomia, which are dominated by angular quartz and gneiss fragments, clearly signal the onset of profound local tectonism and erosion of basement. Local tectonic control of these Member 2 deposits is unambiguously indicated by the observation of at least one major alluvial fan system feeding directly S off the major E–W fault trend at locality 'W' (see above and Figs. 3 and 6). Local uplift along this trend is apparently responsible for the observation that further N, at locality 'X' (see Fig. 3), Members 1 and 2 are absent and Member 3 rests directly on basement.

By Member 3 times, dominant palaeocurrent directions were firmly established in a principally N–S orientation throughout the area, indicating that by this time the aggrading fluvial surface had completely buried the underlying basement topography, and that the E–W fault scarps in the southern Kajong area had no further influence on local drainage patterns. Insofar as the underlying basement surface was clearly buried by Member 3 times, the establishment of this broadly N–S drainage pattern was clearly not immediately controlled by the local NNW–SSE basement trend; indeed, the appreciable scatter of palaeocurrent directions between the NNE and SSW reinforces this suggestion (see Fig. 6). The establishment of this palaeocurrent pattern implies the development of a generally N–S oriented topography comparable to that presently pertaining in the area. The relatively coarse braided river deposits of Member 3 indicate a significant acceleration in the degree of regional tectonism, whereas the finer-grained channel and overbank deposits of Member 4 imply a waning of this activity.

Deposition of the Sera Iltomia was terminated by a period of profound fault movement along both NNW–SSE and E–W fault trends. In places the Sera Iltomia was completely removed, elsewhere it was preserved by downfaulting. A section at locality 'Y' (see Fig. 3) has preserved evidence of this erosive interval; a small downfaulted section consists of large (up to 2 m) tumbled angular blocks of Mesozoic regolith and sandstones, as well as large basement boulders, in a coarse sand matrix.

The Kajong Formation constitutes the second cycle of clastic sedimentation in the Kajong area. As indicated above, the mammal fauna of the Kajong and a radiometric date on the base of the conformably overlying Loiengallani Formation indicate a mid-Miocene age, suggesting that the heavily calcretized erosion surface on which the Kajong was deposited represents the mid-Cenozoic erosion level widely recognized in eastern Africa (Baker *et al.* 1972). Deposition of the Kajong sediments and emplacement of the overlying Loiengallani basalts was followed by major fault movement, although some movement was apparently in progress during the deposition of the Kajong, as indicated by the significant intraformational unconformities documented within this succession. Post-Loiengallani fault movement was restricted to the NNW–SSE fault set; no disruption of either the Kajong or Loiengallani Formations along the E–W fault trend is apparent. Block faulting along the NNW–SSE fault set led to removal of the Miocene section over much of the area, although in places the Miocene deposits are preserved by downfaulting, or as erosional remnants on uplifted basement blocks (see Figs. 3 and 4). The Plio-Pleistocene basalt sequence overlying the earlier units clearly post-dates this erosive episode by a considerable interval; these basalts frequently lap directly onto uplifted basement surfaces from which the Loiengallani and Kajong (and Sera Iltomia) Formations had been completely removed.

Late Mesozoic syntectonic deposition in the Kajong area

The early sedimentary record of the Kajong area is of particular significance in that there is clear evidence that deposition of the earliest clastic sequence (the late Mesozoic Sera Iltomia Formation) was, locally, tectonically controlled. This control is primarily evidenced by the major fan complex documented in Member 2 of the Sera Iltomia, which feeds directly off a major element (trace m–m′, Fig. 7) of the broadly E–W fault trending system developed in the southern Kajong area. Although there has clearly been significant vertical movement on this and the

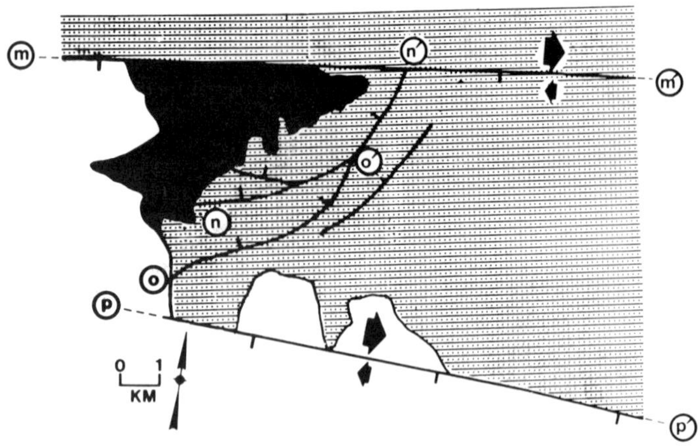

FIG. 7. Generalized map of pre-Miocene faults in the southern Kajong area of the SE Turkana depression; only major fault traces are indicated. Area of map corresponds to the southern portion of the generalized geological map of the Kajong area given in Fig. 3; orientation is indicated by the major fault trend m–m′. Stipple indicates basement; dark area indicates Sera Iltomia Formation; blank areas indicate Miocene and post-Miocene sediments and volcanics. The Miocene and post-Miocene NNW–SSE trending fault system (indicated in Fig. 3) is not plotted. Arrows indicate sense of (dextral) transform movement. Note that apparent truncation of the major fault trends m–m′ and p–p′ is due to Miocene and post-Miocene faulting or results from cover by recent alluvium. Further discussion in text.

other local E–W faults, two lines of evidence indicate that there has also been significant lateral movement.

Firstly, it has proved impossible to match basement lithologies across the major fault trace m–m′, despite the fact that certain marker beds—notably two distinctive series of marble bands present N of this fault—can be traced several kilometres to the N. Secondly, the obvious dragging of basement lineation against certain of the fault traces (e.g. trace n–n′, Fig. 7) testifies to significant dextral strike-slip movement. Given that basement cannot anywhere be matched across the major fault trace m–m′, despite the fact that this trace can be followed for at least 15 km and even then is terminated either by later faulting or by recent cover, it is not possible to estimate the total lateral displacement (although this must obviously be greater than 15 km). This fault trend clearly constitutes a major transform system. In this context it is significant that major transform elements have been postulated previously in other areas of the northern sector of the Eastern Rift; in particular, the 50 km lateral offset of the main rift trend some 300 km N of the Kajong area, along the Sagan River drainage in southern Ethiopia, has been suggested by Baker *et al.* (1972) to be the result of a major dextral transform system.

Our interpretation of the E–W fault trend in the southern Kajong area as a major transform system suggests intriguing parallels with the transform fault systems of the California Borderland as reported by Crowell (1974). Crowell indicates that the San Andreas Fault Zone in this area is composed of a series of anastomosing subparallel faults, all with the same direction of offset. As these split and rejoin, movement along them leads to either convergence, compression and uplift of the tip of the fault-wedge, or to divergence, extension and subsidence of the fault-wedge tip. These fault-wedge basins and highs are typically rhomboidal in form.

The braided, anastomosing geometry of the fault sets in the southern Kajong area, and the rhomboidal basins and uplifts these define, are clearly comparable to the transform terrain described by Crowell (1974; see Fig. 7). It is characteristic of such braided transform systems that faults lying parallel to the main direction of relative motion extend for long distances, are relatively straight, and are characterized by near vertical dips. Faults lying at an angle to the main direction of relative motion are generally shorter, have dips departing significantly from the vertical, and because they frequently rotate progressively further out of alignment with continued movement on the main faults, are often strongly curved (Crowell 1974). The two major fault traces in the southern Kajong area (m–m′ and p–p′, Fig. 7) extend for long distances, are relatively straight, and have near vertical dips; these traces are basically oriented E–W. Additional subsidiary faults in the area, which often depart considerably from an E–W orientation (e.g. n–n′, Fig. 7) are generally shorter, more curved, and have low-angle dips. Following Crowell's analysis, the main direction of (dextral) strike-slip movement in the Kajong area was broadly E–W.

The marked asymmetry of the major fault-wedge basin in which the majority of the Sera Iltomia sediments are currently preserved is readily explicable in terms of Crowell's analysis. The southern margin of this basin preserves the Sera Iltomia basement contact, whereas on the northern margin the Sera Iltomia is abruptly faulted out against the basement (see Figs. 3, 4 and 7). This asymmetry is apparently due to the fact that the southern boundary is defined by a subsidiary, low-dip fault lying at a high angle to the main direction of relative motion, whereas the northern margin is defined by a high-angle normal fault parallel to the main direction of motion. Although it is certainly the case that the fault scarp bounding the northern margin of this fault-wedge basin controlled initial deposition of the Sera Iltomia, as shown by the large alluvial fan system feeding off the northern bounding fault, evidence for direct local control of sedimentation is limited to the earliest phases of the Sera Iltomia. Deposition was widespread over the Kajong area by Member 3 times, and there is no evidence of immediate local tectonic control of deposition. Evidence for the precise timing of movement on this transform system is scanty. Insofar as initial deposition of the Sera Iltomia Formation, of late Mesozoic age (see below), was controlled by this fault system, these faults were obviously active at this time. In addition, the presence of stripped basement and basal Sera Iltomia debris at locality 'Y' (see above and Fig. 3), preserved between two elements of this E–W fault system, suggest significant movement during the period of uplift which terminated deposition of the Sera Iltomia. However, none of these E–W fault traces dislocate the stratiform basalts of the Loiengallani Formation, indicating that there has been little or no movement on this fault series during the last 18 Ma.

Discussion

Age and significance of the Turkana Grits

Murray-Hughes (1933) originally mapped all clastic sediments lying above the basement complex and below the main Miocene volcanic sequence in the Turkana depression as a single lithostratigraphic unit, which he termed the 'Turkana Grits' and regarded as of Jurassic age. Several subsequent authors have similarly considered these sediments to represent a single lithostratigraphic unit, but given the presence of undeniably Miocene faunas at certain localities, and consonant radiometric age determinations on associated or conformably overlying basalts, have suggested that these sediments are exclusively of Miocene age. This view was clearly encouraged by the fact that these extremely immature sediments, in large part locally derived from the underlying basement terrain and often associated with volcanics, clearly reflect vigorous local tectonism. Given the assumption that significant regional tectonic activity commenced during the earliest Miocene rifting episodes, a Miocene age for all these sediments seemed probable.

However, the brief regional survey given above, and the evidence presented from the Kajong area, clearly indicates that the Turkana Grits comprise two superficially similar but distinct sequences, as Arambourg (1933b) originally suggested. The upper clastic sequence at Lodwar (Arambourg 1933a; 'Turkana Grits' of Murray Hughes 1933; Walsh & Dodson 1969), deposits S of Lodwar ('Turkana Grits' of Walsh & Dodson 1969; Dodson 1971), deposits at Loperot ('Turkana Grits' of Murray-Hughes 1933; Champion 1937; Dixey 1945; Joubert 1966; Walsh & Dodson 1969), and the upper clastic sequence in the Kajong area (Kajong Formation of Savage & Williamson 1978) are all, on the basis of either radiometric or biostratigraphic age determinations, of broadly mid-Miocene age. These deposits are generally coarse-grained arkoses and subarkoses which include a significant fraction of detrital volcaniclastic material and interbedded volcanic horizons. The mid-Cenozoic erosion level on which these deposits rest is cut either in basement or in an earlier clastic sequence; where well-exposed this basal contact is heavily caliched (as at Loperot and Kajong). Subsidiary basal conglomerates, consisting of angular quartz and gneiss clasts, also occur. Where overlain by Miocene volcanics, as at Kajong and Lodwar, the contact is conformable.

The mid-Miocene sediments described above clearly post-date an earlier clastic sequence in the Turkana depression. These deposits include the clastic deposits in the Lapurr Range ('Lubur Series' of Arambourg 1933b; 'Turkana Grits' of Murray-Hughes 1933; Walsh & Dodson 1969), the lower clastic sequence in the Lodwar area ('Lubur Series' of Arambourg 1933b; 'Turkana Grits' of Murray-Hughes 1933; Walsh & Dodson 1969), and the lower clastic sequence in the Kajong area (Sera Iltomia Formation of Savage & Williamson 1978). These deposits are generally coarse, immature arkoses and subarkoses which superficially resemble the mid-Miocene deposits of the area but which are distinguished, as

Arambourg (1933b) originally pointed out, by the fact that they lack any detrital volcaniclastic component or associated volcanics. These deposits are relatively unfossiliferous; the sequence at Lapurr has, however, yielded dinosaur material (Arambourg & Wolff 1969; J.M. Harris, pers. comm.) and the angiosperm '*Dryoxylon*' (Fuchs 1939; Arambourg 1943). In addition, the Sera Iltomia Formation in the Kajong area has yielded exclusively angiosperm wood. The combination of dinosaur material and a dominantly angiosperm flora strongly suggests a Cenomanian-basal Palaeocene age for these early clastic deposits. The (presumably) late Mesozoic erosion level on which these deposits rest is characterized by the development of a thick, lateritic regolith (as in the Kajong area) or by thick basal conglomerates consisting of well-rounded quartz clasts (as in the Kajong area and at Lapurr). At Lapurr this early clastic sequence is overlain unconformably by Miocene volcanics, and at Lodwar by Miocene clastics (Arambourg 1933b). In the Kajong area, the early clastic sequence (the Sera Iltomia Formation) can be inferred to underlie the Miocene clastics of the Kajong Formation due to the fact that the thick caliched zone on which the latter rests can be demonstrated to extend across the upper surface of the downfaulted Sera Iltomia.

The relationship of the two major clastic sequences in the Lapurr, Lodwar, Loperot and Kajong areas seems relatively clear. But unfortunately, due to the lack of any biostratigraphic or radiometric evidence from the early clastic deposits of the Kalossia-Tiati area ('Turkana' and 'Tiati Grits' of Mason & Gibson 1957) and the Baragoi area ('sub-volcanic sediments' of Baker 1963), and uncertainty as to the precise stratigraphic setting of these sequences, they cannot be readily related to other early clastic deposits in the Turkana depression.

The coarse, immature, arkosic character of clastic sequences hitherto described as the Turkana Grits clearly implies that these are the product of significant local tectonism. Moreover, as discussed above, deposition of the earlier clastic sequence (the Sera Iltomia Formation) in the Kajong area can be demonstrated to be syntectonically controlled by a major E–W trending transform fault system. This observation implies that significant rifting activity commenced in the Turkana depression area as early as the late Mesozoic, rather than during the mid-Miocene, as is commonly assumed. In this context it is worth noting the recent observation that significant tectonic activity is evident by earliest Cenozoic times in the southern sector of the Red Sea Rift (Richardson & Harrison 1976).

General implications for the early development of the African Rift

While the demonstration of a major transform system in the S Kajong area of the SE Turkana depression is clearly of intrinsic interest, particularly given that this system at least initially controlled late Mesozoic sedimentation in the area, the presence of this transform system also prompts speculations concerning the early development of the rift as a whole. It may shed light on two outstanding puzzles of general rift geology: the curiously circumscribed distribution of late Mesozoic sediments in the Eastern and Western Rifts, and the enigmatically abrupt northern termination of the Western Rift at Nimule on the Uganda–Sudan border.

There is a significant contrast between the distribution of Karroo (Carboniferous through early Jurassic) and late Jurassic–Cretaceous sediments in East and Central Africa. Continental and marine deposits of Karroo age are distributed throughout the area, although deposits of definite Karroo age are not known N of approximately 2°S. By contrast, deposits of definitely late Jurassic and Cretaceous age in East and Central Africa are restricted to a narrow zone paralleling the East African coast, or—inland—are restricted to areas in the immediate vicinity of the present African Rift (Fourmarier 1914; Dixey 1930; Bond *et al.* 1957; Brandt *et al.* 1957; Cahen & Lepersonne 1957; Cooper & Dixey 1957; Pallister 1957; Pulfrey 1957a, b; Nunes 1966). Inland, these late Mesozoic deposits extend between approximately 16°30′S and 4°30′N, and are exclusively non-marine. They include the Lupata Series of the lower Shire valley in Malawi (Dixey & Campbell-Smith 1929), the equivalent 'Dinosaur Beds' on the NW coast of Lake Malawi (Dixey 1928), the 'Red Sandstone Group' of Aitken (1950) (equivalent to the 'Red Sandstones' of Spence 1954) in the Rukwa graben, Tanzania, and the 'Lubur Series' at Lapurr and Lodwar (Arambourg 1933b) and the Sera Iltomia Formation (Savage & Williamson 1978), all in the northern sector of the Eastern Rift.

In short, although Karroo age sediments are distributed throughout East and Central Africa, deposits of late Jurassic and Cretaceous age are limited, outside of the narrow coastal belt, to the immediate vicinity of the southern portion of the Western Rift (S of 8°30′S) and the northern portion of the Eastern Rift (N of 2°45′N). All of these deposits appear comparable insofar as, on the basis of published accounts, they all comprise primarily immature arkoses.

The late Mesozoic clastic deposits of the southern sector of the Western Rift have long been considered to signal an initial phase of rift formation (Pulfrey 1957b; Logatchev *et al.* 1972), implying that early rifting was confined to the southern extremity of the present rift system. However, the episode of late Mesozoic syntectonic deposition reported here from the Turkana depression—in the northern sector of the Eastern Rift—suggests that significant rifting activity took place simultaneously in this area. In this context, the abrupt northern termination of the Western Rift is of considerable significance. It is in marked contrast to the southern extremities of both the Eastern and Western Rifts, which 'degenerate' gradually into a series of splayed, fault-bounded blocks (e.g. King 1970).

The abrupt northern termination of the Western Rift has been ascribed by some authors to the pronounced NE–SW trend of the local basement (the Madi Series and Aswa fold belt of southern Sudan and northern Uganda). However, an appeal to basement structure as an explanation for this phenomenon is questionable; as Baker *et al.* (1972) have pointed out with respect to the Eastern Rift, 'rift faults have followed Precambrian structures where the older structures happened to be favourably located and oriented, but the rift system as a whole was formed by imposition of new structural lines across the heterogeneous pre-existing fabric'. Similar conclusions have been drawn regarding the relationship between the Rhine and Baikal Rifts and their respective basement foundations (Baker *et al.* 1972). The abrupt northern termination of the Western Rift is effected by a series of major E–W or NE–SW trending faults (e.g. King 1970). Although basement structures in northern Uganda and the extreme southern Sudan remain unmapped in any detail (e.g Whiteman 1971), an E–W or NW–SE structural trend—variously described as 'joints' or 'faults'— is known from the Kitgum and Kaabong areas of northern Uganda (Almond 1962; MacGregor 1962). These areas are essentially aligned between the transform fault trend we report here from the Kajong area in the SE Turkana depression and the northern termination of the Western Rift at Nimule on the Uganda–Sudan border.

Clearly, a comprehensive understanding of the early development of the African Rift must await detailed mapping of basement structures in northern Uganda and southern Sudan. We raise here the possibility that the enigmatic restriction of late Mesozoic rifting activity to the southern sector of the Western Rift and the northern sector of the Eastern Rift may reflect the fact that the initial late Mesozoic Rift consisted of a single graben incorporating these portions of the present rift system. They have been subsequently displaced some 500 km along the major transform system we document here from the Kajong area in the SE Turkana depression.

Conclusions

Arambourg's original contention (1933b) that the Turkana Grits represent two distinct chronostratigraphic units is supported by our observation that these deposits in the Kajong area and elsewhere in the Turkana depression can be demonstrated to unequivocably represent two depositional episodes, one of Miocene, the other of Cenomanian-basal Palaeocene age. In the Kajong area, late Mesozoic deposition can be demonstrated to have been, at least initially, controlled by a major, broadly E–W trending transform fault system; these Mesozoic sediments are preserved in associated fault-wedge basins.

Our observations in the Kajong area clearly support Arambourg's original suggestion that the Turkana Grits comprise deposits of both Mesozoic and mid-Cenozoic age, and indicate, as he originally implied, that significant rifting activity in the Turkana depression commenced in the late Mesozoic rather than the mid-Cenozoic. The recognition that late Mesozoic sedimentation in the Kajong area was controlled by a major E–W transform fault system may have more general implications for the early development of the African Rift as a whole. In this context we note that: (1) proven late Mesozoic rift sediments are exclusively restricted to the N sector of the Eastern Rift and the S sector of the Western Rift; and (2) that the western segment of the rift is abruptly terminated to the N—for reasons still obscure—by a broadly E–W trending fault system at Nimule on the Uganda Sudan border (King 1970). Our observation that a major E–W transform fault system controlled late Mesozoic clastic sedimentation in the Kajong area, in combination with the limited evidence currently available for a significant, broadly E–W or SSE–NNW structural trend at various localities in northern Uganda between Nimule and the SE Turkana depression, raises the following intriguing possibility. That is that the original late Mesozoic rift may have consisted of a single graben system incorporating what is now the northern sector of the Eastern Rift and the southern sector of the Western Rift, and that the southern section of this original rift was subsequently displaced some 500 km to the W by a major dextral transform system aligned between the Kajong area in the SE Turkana depression (and presumably extending further E) to what is now the northern termination of the Western Rift near Nimule on the Uganda–Sudan border. Detailed mapping of northern Uganda and the southern portions of Sudan will be required to test this tentative hypothesis.

ACKNOWLEDGEMENTS: The authors wish to acknowledge the Natural Environment of Research Council of Great Britain (GR 3/2382A) and the Shaler Fund of Harvard University for their generous financial support. We particularly thank F.H. Brown for his comments and criticism, and R.E.F. Leakey and the National Museums of Kenya for their invaluable assistance in the field.

References

AITKEN, W.G. 1950. Songwe-Rukwa. *Ann. Rep. Geol. Surv. Tanganyika* **1948,** 15–18.

ALMOND, D.C. 1962. Explanation of the geology of sheet 15 (Kitgum). *Rep. Geol. Surv. of Uganda* **8,** 55 pp.

ARAMBOURG, C. 1933a. Decouverte d'un gisement de Mammifères burdigaliens dans le Bassin du Lac Rodolphe (Afrique Orientale). *C. R. Soc. Géol. Fr.* **14,** 221–2.

—— 1933b. Les formations pretertiares de la bordure occidentale du Lac Rodolphe (Afrique Orientale). *C. R. Acad. Sci. Paris* **197,** 1663–5.

—— 1933c. Mammiferes Miocene du Turkana (Afrique Orientale). *Ann. Paleont.* **22,** 123–46.

—— 1935. Esquisse geologique de la bordure occidentale du Lac Rodolphe. *Mission Scient. Omo* (Mus. Natn. Hist. Nat. Paris) **X(1),** 59pp.

—— 1943. Contribution a l'etude geologique et paleontologique du Bassin du Lac Rodolphe et de la Basse Vallee de L'Omo. *Mission Scient. Omo* (Mus. Natn. Hist. Nat. Paris) **1,I(2),** 157–230.

—— 1947. Contribution a l'etude geologique et paleontologique du Bassin du Lac Rodolphe et de la Basse Vallee de L'Omo. *Mission Scient. Omo* (Mus. Natn. Hist. Nat. Paris) **2,I(3),** 231–562.

—— & WOLFF, R.G. 1969. Nouvelles donnees paleontologique sur l'ages des 'gres du Lubur' (Turkana grits) a l'ouest du Lac Rodolphe. *C. R. Soc. Géol. Fr.* **6,** 190–202.

BAKER, B.H. 1963. Geology of the Baragoi area. *Rep. Geol. Surv. Kenya* **53,** 74pp.

——, MOHR, P.A. & WILLIAMS, L.A.J. 1972. Geology of the Eastern Rift System. *Spec. Pap. Geol. Soc. Am.* **136,** 66pp.

BISHOP, W.W. & TRENDALL, A.F. 1967. Erosion surfaces, tectonics and volcanic activity in Uganda. *J. Geol. Soc.* **122,** 385–420.

BOND, G., MAUFE, H.B. & TYNDALE-BISCOE, R. 1957. *Rhodesie du Sud—Southern Rhodesia.* Lexique Stratigraphique Internationale: Afrique Fasc. No. 9c, CNRS, Paris, pp. 29–78.

BRANDT, R.T., MAUFE, H.B., PHILLIPS, K.A. & REEVE, W.H. 1957. *Rhodesie du Nord—Northern Rhodesia.* Lexique Stratigraphique Internationale: Afrique Fasc. No. 9b, CNRS, Paris, pp.11–28.

BROTZU, P., MORBIDELLI, L., NICOLETTI, M., PICCIRILLO, E.M. & TRAVERSA, G. 1984. Miocene to Quaternary volcanism in eastern Kenya: sequence and geochronology. *Tectonophysics* **101,** 75–86.

CAHEN, L. & LEPERSONNE, J. 1957. *Congo Belge.* Lexique Stratigraphique Internationale: Afrique Fasc. No. 7a, CNRS, Paris, pp. 1–129.

CHAMPION, A.M. 1937. Physiography of the region to the west and southwest of Lake Rudolf. *Geogr. J.* **89,** 97–118.

CHURCHER, C.S. 1970. Two new Upper Miocene Giraffids from Fort Ternan, Kenya, East Africa: *Palaeotragus primaevus* n. sp. and *Samotherium africanum* n. sp. In: LEAKEY, L.S.B. & SAVAGE, R.J.G. (eds) *Fossil Vertebrates of Africa. 2.* Academic Press, London, pp. 1–105.

CLARK, W.E. LE GROS & LEAKEY, L.S.B. 1951. *The Miocene Hominoidea of East Africa. Fossil Mammals of Africa, 1.* British Museum (N.H.) London, pp. 1–117.

COOPER, W.G.G. & DIXEY, F. 1957. *Nyassaland-Nyasaland.* Lexique Stratigraphique Internationale: Afrique Fasc. No. 9a, CNRS, Paris, pp. 1–10.

CROWELL, J.C. 1974. Origin of late Cenozoic basins in southern California. In: DICKINSON, W.R. (ed) *Tectonics and sedimentation.* Spec. Publ. Soc. Econ. Paleont. Miner. No. 22, Tulsa, pp. 190–204.

DERANIYAGALA, P.E.P. 1951. A hornless rhinoceros from the Mio-Pliocene deposits of East Africa. *Spolia Zeylan* **26(2),** 133–5.

DIXEY, F. 1928. The dinosaur beds of Lake Nyasa. *Trans. R. Soc. South Afr.* **16(1),** 55–6.

—— 1930. The Karroo of the lower Shire-Zambezi area. *C. R. 2 XVth Geol. Internat. Congr.* 120–42.

—— 1945. Miocene sediments in South Turkana. *J. E. Afr. Nat. Hist. Soc. Nairobi* **18,** 13–14.

—— & CAMPBELL-SMITH, W. 1929. The rocks of the Lupata Gorge and the north side of the lower Zambezi. *Geol. Mag.* **66,** 241–59.

DODSON, R.G. 1963. Geology of the South Horr area. *Rep. Geol. Surv. Kenya* **60,** 53pp.

—— 1971. Geology of the area south of Lodwar. *Rep. Geol. Surv. Kenya* **87,** 36pp.

FAIRBURN, W.W. & MATHESON, F.J. 1970. Geology of the Loiya-Lorugumu area. *Rep. Geol. Surv. Kenya* **85,** 52 pp.

FITCH, F.J., WATKINS, R.T. & MILLER J.A. 1975. Age of new carbonatite locality in North Kenya. *Nature* **254,** 581–3.

FOURMARIER, P. 1914. Le bassin charboniere d'age permo-triassique de la Lukuga. Etude geologique de la region de la Lukuga et de la Lubumba au voisinage du lac Tanganyika. *Ann. Soc. Geol. Belg.* **41,** C77–227.

FUCHS, V.E. 1934. The geological work of the Cambridge Expedition to the East African Lakes 1930–1931. *Geol. Mag.* **71,** 97–112.

—— 1935. The Lake Rudolf Rift Valley Expedition, 1934. *Geogr. J.* **86,** 111–42.

—— 1939. The geological history of the Lake Rudolf Basin, Kenya Colony. *Phil. Trans. R. Soc. Lond.* **229,** 219–74.

HARRIS, J.M. & WATKINS, R. 1974. New early Miocene vertebrate locality near Lake Rudolf, Kenya. *Nature* **252,** 576–7.

HÖHNEL, L.R. VON, ROSIWAL, A., TOULA, F. & SUESS, E. 1891. Beiträge zur geologischen Kenntnis des östlichen Afrika. *Denkschr. Akad. Wiss., Wien. Math-nat. Kl.* **58,** 447–584.

HOOIJER, D.A. 1966. Miocene Rhinoceroses of East Africa. *Fossil Mammals of Africa* **21.** British Museum (N.H.) London, pp. 117–90.

—— 1968. A note on the mandible of *Aceratherium acutirostratum* (Deraniyagala) from Moruraret Hill, Turkana District, Kenya. *Zool. Med. Museum Leiden* **42(21),** 231–5.

—— 1971. A new rhinoceros from the Late Miocene of Loperot, Turkana District, Kenya. *Bull. Mus. Comp. Zool.* **142,** 339–92.

—— 1973. Additional Miocene to Pleistocene Rhinoceroses of Africa. *Zool. Med. Museum Leiden* **46(11),** 149–77.

JOUBERT, P. 1966. Geology of the Loperot area. *Rep. Geol. Surv. Kenya* **74,** 52pp.

KING, B.C. 1970. Volcanicity and rift tectonics in East Africa. In: CLIFFORD, T.N. & GASS, I.G. (eds) *African Magmatism and Tectonics.* Hafner Publishing Co., Darien, Conn, pp. 263–83.

LOGATCHEV, N.A., BELOUSSOV, V.V. & MILANOVSKY, E.E. 1972. East African rift development. In: GIRDLER, R.W. (ed) *Developments in Geotectonics 7: East African Rifts.* Elsevier, pp. 71–81.

MACGREGOR, J.P. 1962. Explanation of the geology of sheet 10 (Kaabong). *Rep. Geol. Surv. Uganda* **9,** 42 pp.

MADDEN, C.T. 1972. Miocene mammals, stratigraphy and environment of Muruarot Hill, Kenya. *Paleobios* **14,** 1–12.

MAGLIO, V.J. 1969. A shovel-tusked Gomphothere from the Miocene of Kenya. *Mus. Comp. Zool. Breviora* **310,** 1–10.

MASON, P. & GIBSON, A.B. 1957. Geology of the Kalossia-Tiati area. *Rep. Geol. Surv. Kenya* **41,** 34pp.

MCCALL, G.J.H. 1964. Geology of the Sekerr area. *Rep. Geol. Surv. Kenya* **65,** 84 pp.

MEAD, J.G. 1975. A fossil beaked whale (Cetacea: Ziphiidae) from the Miocene of Kenya. *J. Palaeo.* **49(4),** 745–51.

MURRAY-HUGHES, R. 1933. Notes on the geological succession, tectonics and economic geology of the western half of Kenya colony. *Rep. Geol. Surv. Kenya* **3,** 8pp.

NUNES, A. DE F. 1966. *Mocambique.* Lexique Stratigraphique Internationale: Afrique Fasc. No. 10a, CNRS, Paris, pp. 1–86.

PALLISTER, J.W. 1957. *Uganda.* Lexique Stratigraphique Internationale: Afrique Fasc. No. 8b, CNRS, Paris, pp. 95–123.

PULFREY, W. 1957a. *Kenya.* Lexique Stratigraphique Internationale: Afrique Fasc. No. 8a, CNRS, Paris, pp. 1–94.

—— 1957b. *Tanganyika.* Lexique Stratigraphique Internationale: Afrique Fasc. No. 8c, CNRS, Paris, 1–171.

—— 1960. Shape of the sub-Miocene erosion bevel in Kenya. *Bull. Geol. Surv. Kenya* **3,** 18 pp.

RICHARDSON, E.S. & HARRISON, C.G. 1976. Opening of the Red Sea with two poles of rotation. *Earth Planet. Sci. Lett.* **30,** 135–42.

SAVAGE, R.J.G. 1965. *The Miocene Carnivora of East Africa. Fossil Mammals of Africa 19.* British Museum (N.H.) London, pp.239–316.

—— & WILLIAMSON, P.G. 1978. The early history of the Turkana Depression. In: BISHOP, W.W. (ed) *Geological Background to Early Man.* Scottish Academic Press, Edinburgh, pp. 375–94.

VAN DAMME, D. 1969. Fossile Zoetwatermollusken van de Omorivier, Ethiopia. Licentiaat Wetenschappen (Dierkunde), Rijksuniversiteit Gent.

—— & Gautier, A. 1972. Some fossil Mollusca from Moruarot Hill (Turkana District, Kenya). *J. Conch.* **27,** 423–6.

VERDCOURT, B. 1963. The Miocene non-marine Mollusca of Rusinga Island, Lake Victoria and other localities in Kenya (Abstract). *Palaeontographica* **121A,** 1–37.

VONDRA, C.F. & BOWEN, B.E. 1976. Plio-Pleistocene Deposits and Environments, East Rudolf, Kenya. In: COPPENS, Y., CLARK HOWELL, F., ISAAC, G.L.L. & LEAKEY, R.E.F. (eds) *Earliest Man and Environments in the Lake Rudolf Basin.* University of Chicago Press, pp. 79–93.

WALSH, J. & DODSON, R.G. 1969. Geology of northern Turkana. *Rep. Geol. Surv. Kenya* **82,** 42 pp.

WHITEMAN, A.J. 1971. *The Geology of the Sudan Republic.* Clarendon Press, Oxford, 290pp.

WHITWORTH, T. 1954. *The Miocene Hyracoids of East Africa. Fossil Mammals of Africa 7.* British Museum (N.H.) London, 58pp.

—— 1958. *Miocene Ruminants of East Africa. Fossil Mammals of Africa 15.* British Museum (N.H.) London, 50pp.

WILKINSON, A.F. 1976. The Lower Miocene Suidae of Africa. In: SAVAGE, R.J.G. & CORYNDON, S.C. (eds) *Fossil Vertebrates of Africa, 4.* Academic Press, London, pp.174–282.

P.G. WILLIAMSON, Dept of Geological Sciences & Museum of Comparative Zoology, Harvard University, Cambridge, Massachusetts, USA.

R.J.G. SAVAGE, Dept of Geology, University of Bristol, Bristol BS8 1SS, UK.

Sedimentary stratigraphy of the Tugen Hills, Baringo, Kenya

A. Hill, G. Curtis & R. Drake

SUMMARY: The Tugen Hills sequence within the northern Kenya Rift Valley W of Lake Baringo is unusual in providing sedimentary and faunal information over a relatively long period of time, from 14 Ma to the Pleistocene, and including some periods not yet noted elsewhere in sub-Saharan Africa. The sediments record changing depositional environments in an evolving rift system, and the enclosed fossils permit the evolutionary lineages of a number of families of animals important in the modern African fauna to be elucidated. We document the history of investigation of sedimentary units belonging to the Tugen Hills succession, and describe their current status. Information is given concerning the general sedimentary character of the units and their age is determined by faunal, isotopic and palaeomagnetic methods. Faunal change through the sequence is summarized.

The Tugen Hills

The Tugen Hills, sometimes known as the Kamasia Range, is a complex block tilted to the W and extending about 75 km N–S within the northern Kenya Rift Valley, W of Lake Baringo (Fig. 1). The succession of over 3000 m of rocks exposed in its fault scarps and eastern foothills, includes the major recorded outcrops of sedimentary deposits in sub-Saharan Africa which span the time period from 14 Ma to less than 4 Ma.

The first European with geological interests to visit the area was Thomson (1884), who mistakenly described the Tugen Hills as being composed of metamorphic rocks. This impression was rectified by Gregory (1921; see also Prior 1903), and more information was subsequently provided by Bailey Willis (1936) and Shackleton (1951). From the mid-1960s to the mid-1970s work was carried out by the East African Geological Research Unit (EAGRU) under the direction of Basil King, and by a smaller group working with Bill Bishop. They produced the framework upon which later studies in the region have been based. More recently, investigation has been resumed by the Baringo Palaeontological Research Project (BPRP), based at Harvard and Yale Universities.

Cenozoic rocks in the northern rift valley have been divided into five groups of formations (Bishop *et al.* 1971; King & Chapman 1972). Williams & Truckle (1980) point out that N of 1°30′N this simple stratigraphic sequence becomes difficult to apply, but it remains valid for that part of the rift which includes the Tugen Hills. Group I, consisting of basalts, is not represented in the Tugen Hills sequence. Group II is predominantly flood phonolites, and in the Tugen Hills rests directly on metamorphic basement or on the Kamego Formation. A major unconformity, during which major faulting took place, separates these from the rocks of Group III, which is composed of flood trachytes and alkali basalts. Group IV is principally composed of large trachyte shield volcanoes, but is also represented (*contra* Bishop *et al.* 1971) by clastic rocks in the Tugen Hills sequence. Group V are lavas of diverse composition.

A number of sedimentary units have been described within the sequence (Bishop *et al.* 1971; Chapman *et al.* 1978), most defined and separated from each other by volcanic lavas (Fig. 2). The Sidekh Phonolite Formation, belonging to Group II, contains five sedimentary members. The Muruyur Beds are sediments found between flows of the overlying Tiim phonolites, also belonging to Group II. Above the Tiim phonolites is the extensive Ngorora Formation, also within Group II, bounded at the top by the Ewalel phonolites. The Mpesida Beds occur between flows of the Kabarnet Trachyte Formation of Group III. Also part of Group III, above the Kabarnet trachytes, and beneath the Kaparaina basalts, is the Lukeino Formation. And above the Kaparaina basalts is the Chemeron Formation, which probably represents Group IV in the Tugen Hills succession. The Kapthurin Beds, of Group V, overstep from the Chemeron Formation to the Kaparaina basalts. Another unit, the Kaperyon Formation, is a clastic representative of Group IV, but one interpretation allocates rocks in this unit to the Lukeino and Chemeron Formations.

In the following account isotopic dates have been corrected for new constants where necessary, using the formula provided by Ness *et al.* (1980).

Kamego Formation

The lowest sedimentary unit in the Tugen Hills sequence is the Kamego Formation (Martyn 1969; Chapman *et al.* 1978), lying directly on metamorphic basement. It is composed of quartz-

From Frostick, L.E. *et al.* (eds) 1986, *Sedimentation in the African Rifts*, Geological Society Special Publication No. 25, pp. 285–95.

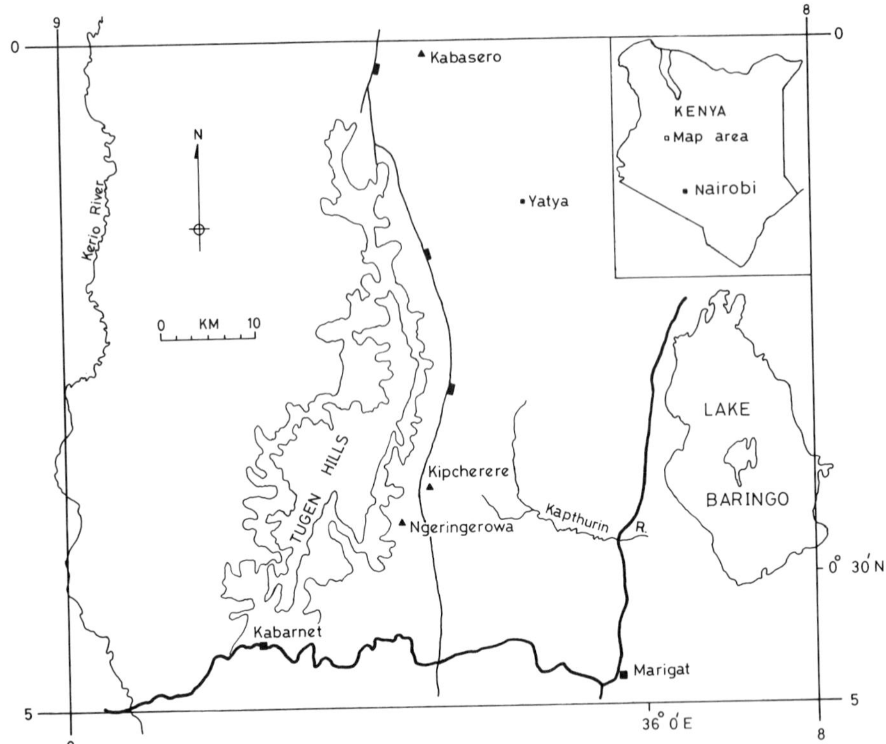

FIG. 1. Map of the Tugen Hills, Baringo District, Kenya, to show the location of the area and some localities mentioned in the text. The line of principal faults forming the eastern scarp is indicated.

ites and siltstones up to 80 m in thickness. Its exact age is not known, but pollen found within it is Tertiary in character. No vertebrate fossils have been recorded from these sediments.

Sedimentary members of the Sidekh Phonolite Formation

Martyn (1969) records a number of sedimentary units within his Saimo Phonolite Formation, subsequently renamed the Sidekh Phonolite Formation by Chapman (1971). This belongs to Group II, and unconformably overlies the Kamego Formation and basement rocks. The lowest of the sedimentary units is the Chepkuno grit and breccia, composed of thin pyroclastics. Above these are the Tinyerinyer Beds, fine-grained purple and red mixed lacustrine and fluviatile sediments, which incorporate locally-derived phonolite pyroclastics, and lead into fully lacustrine green shales. The Aimeno shales and marls are also lacustrine, grey and green shales showing local evidence of anaerobic conditions. They are followed by the Kapkiai tephrites, which include evidence of sedimentary episodes in the form of dark shales. The black Surkai shales are stratigraphically above the Kapkiai tephrites.

Chapman & Brook (1978) quote two isotopic dates for the Sidekh phonolites: one, of 16.4 ± 0.6 Ma on a lower flow, and another of 14.8 ± 0.5 Ma on an upper flow. Apart from Martyn's original mapping and description these units have received no attention. No fossils are reported from them.

Muruyur Beds

The Muruyur Beds occur within the Tiim Formation, which is composed of phonolites belonging to Group II. They were originally described by Chapman (1971), and some information was published in Bishop *et al.* (1971). Pickford (1975a) reported additional work on the unit. He discovered new fossil sites, and expanded the faunal list considerably.

The unit crops out at the N end of the Tugen Hills, along and to the W of its crest, and is up to 275 m thick. The sediments are varied, mainly tuffaceous, lacustrine, and show traces of intermittently emergent conditions, with some local

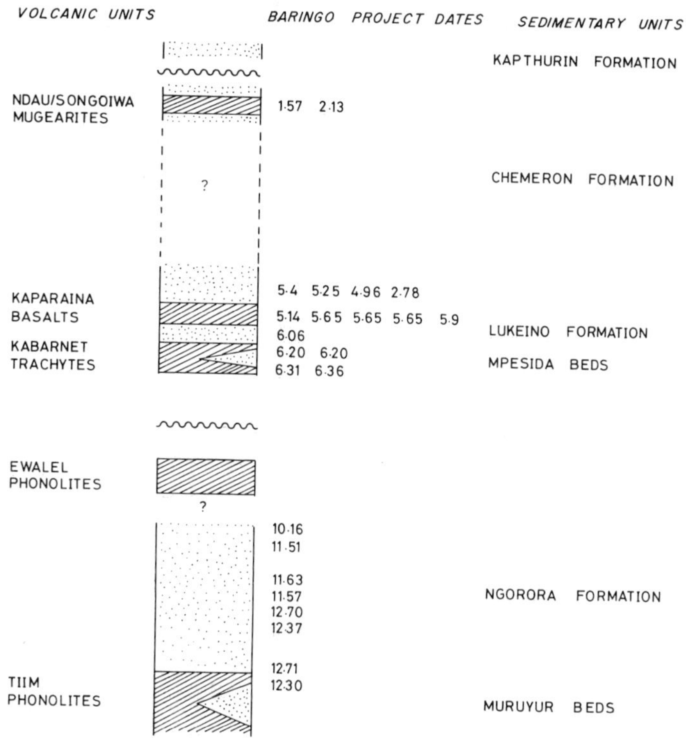

FIG. 2. Schematic stratigraphical succession to show the relationship of the sedimentary units in the upper part of the Tugen Hills sequence. The vertical scale corresponds neither to sedimentary thickness nor exactly to time. Baringo Palaeontological Research Project isotopic dates are given (in Ma).

palaeosols. Correlation between the separate outcrops by which the unit is represented is not yet established.

As the unit is defined as lying between flows of the Tiim Formation, its age is estimated by determinations on the Tiim phonolites. The Tiim Formation is equivalent to the upper part of the more extensive plateau phonolites of the Uasin Guishu Plateau. Dates on the Tiim phonolites themselves are 13.3 ± 0.4 Ma, 14.7 ± 0.5 Ma and 15.3 ± 0.4 Ma for lower units, and 11.7 ± 1.2 Ma, 13.1 ± 0.2 Ma and 13.5 ± 1.4 Ma for upper flows. Two dates of 10.2 ± 0.5 Ma and 7.4 ± 0.5 Ma found further S are regarded as aberrant (but see below) (Bishop *et al.* 1969; Baker *et al.* 1971; Chapman & Brook 1978). A more recent BPRP date, determined by Curtis & Drake (Hill *et al.* 1985), is 12.30 ± 0.50 Ma for the phonolite just beneath the Ngorora type section at Kabasero, which is slightly in conflict with a date of 12.71 ± 0.09 Ma for sediments just above it.

There is a quite diverse fauna from the Muruyur Beds (see Table 1, taken largely from Pickford 1975a). The main problem with the Muruyur Beds is simply that they are so little known, and there are difficulties with the correlation of discrete outcrops. Yet they are important,

TABLE 1. *Mammal fauna from the Muruyur Beds*

- Primates
  - Hominoidea
- Proboscidea
  - Gomphotheriidae
    - sp. indet.
  - Deinotheriidae
    - *Deinotherium* sp.
- Perissodactyla
  - Rhinocerotidae
    - *Aceratherium acutirostratum*
- Artiodactyla
  - Suidae
    - *Listriodon*
  - Bovidae
    - cf. *Protragocerus*
    - medium sized sp.
  - Giraffidae
    - sp. indet.

being among the oldest fossiliferous sediments within the rift, and should provide information regarding the early history of the rift and its fauna. It is also important to understand the faunal links between this unit and the better known fossil assemblages of equivalent age in western Kenya.

Ngorora Formation

The Ngorora Formation is a relatively widespread and thick unit of sediments (370–400 m in the type area) bounded by the Tiim phonolites below and the Ewalel phonolites above. The sediments are exposed in a number of fault-bounded blocks, and crop out along the whole length of the Tugen Hills. Within each of these, sedimentation has been largely fault controlled, and considerable lateral facies variation makes it difficult to correlate between them accurately. The unit is formed of a great variety of sediments dominated by volcanics, ranging from agglomerates to finely laminated lacustrine shales. There are also both primary and reworked tuffs.

The type section was defined at Kabasero, in the N, by Bishop & Chapman (1970), and more details about the Formation are given in Chapman (1971). Chapman (1971) included in the formation the Poi Tuffaceous Sandstones and Kapkiamu Shales of Martyn (1969). A supplemented faunal list appears in Bishop & Pickford (1975) and Pickford (1975a, 1978a).

Chapman & Brook (1978) provide a review of relevant dates prior to our more recent work. The maximum age of the unit is determined by dates on upper flows of the Tiim phonolites, which were discussed in connection with the Muruyur Beds, and which have most recently been dated by Curtis & Drake, beneath the Ngorora type section at 12.30±0.50 Ma (Hill *et al.* 1985). However, this may be too young. Until recently only two dates had been obtained on sanidines from within the sequence, of 12.6±0.7 Ma and 12.2±0.7 Ma (Chapman & Brook 1978). Dates on the overlying Ewalel phonolites, which in the N are represented by a single flow of up to 230 m, are 6.9±0.1 Ma, 7.5±0.3 Ma, 9.0±0.3 Ma and 9.0±0.8 Ma.

BPRP has performed a more extensive series of determinations in the type section at Kabasero in connection with dense palaeomagnetic sampling (Hill *et al.* 1985). The stratigraphically lowest date is 12.71±0.09 Ma from just above the Tiim phonolite. Determinations on samples up the section, in stratigraphic order, are 12.37±0.10 Ma, 12.70±0.38 Ma, 11.57±0.14 Ma, 11.63±0.11 Ma, 11.51±0.22 Ma and the topmost sample, 270 m above the base, is 10.16±0.38 Ma. The section continues for another 100–130 m above this level.

The suggested correlation with the geomagnetic reversal time scale provides estimates of ages from the base in chron 13, from older than 12.74 Ma, into chron 9 at younger than 10 Ma. The isotopic dates and the palaeomagnetic calibration exhibit a very good fit, and show that the sediments in the type section represent at least 2.5 Ma, and maybe much more. This is a very long sequence of relatively continuous sedimentation for continental sediments in Africa.

One of the best collections of fossil macroflora from sub-Saharan Africa has been made from a tuff in the type section dated at 12.37 Ma. Preliminary analysis by Fine Jacobs (pers. comm.) suggests that the fossil leaves reflect forested conditions, and not necessarily forest confined just to riverine margins (Hill *et al.* 1985). The fossil fauna (Table 2) is different in different parts of the formation, which is no doubt a function of age. A series of productive sites in the type area probably fall between 11.0 Ma and 12.4 Ma, and show many similarities with earlier Middle Miocene faunas in the W of Kenya. The collected fauna is much more abundant than that of the Muruyur Beds, so it is difficult to know whether the appearance of new species is due simply to better sampling or to the evolution of new taxa.

Sites elsewhere in the formation, particularly those such as Ngeringerowa in fault blocks to the S, have a fauna with elements of more modern aspect. These include equids which may be the earliest in sub-Saharan Africa. Dates on these and other sites outside the type area are not yet established, but they could be only a little older than the Ewalel phonolites.

Additional work on isotopic dating and palaeomagnetics is being carried out to solve this problem, but it is not yet complete. One date, on what is regarded as Tiim phonolite beneath the sedimentary sequenced at Ngeringerowa, is 7.4±0.5 Ma (Chapman & Brook 1978). This has usually been regarded as aberrant, but it is possible that it is not actually Tiim phonolite at all. We may eventually find evidence for the rocks at Ngeringerowa and their equivalents being a separate and younger formation.

Mpesida Beds

The Mpesida Beds were described by Chapman (1971), and comprise sediments that occur between flows of the Kabarnet trachytes, but their detailed stratigraphy has been relatively little worked on. At Kasigoryan there is *c.* 30 m of

TABLE 2. *Mammal fauna from the Ngorora formation*

- Insectivora
  - ?Soricidae, gen. & sp. indet.
- Rodentia*
  - *paraulacodus*, n. sp.
  - Thryonomyidae, n. gen. & sp.
  - Bathyergidae, gen. & sp. indet.
  - Myocricetodontinae, gen. & sp. indet.
  - Cricetidae, gen. & sp. indet. (3 spp.)
- Primates
  - Cercopithecoidea, gen. & sp. indet.
  - Hominoidea, gen. & sp. indet. (2 spp.)
- Creodonta
  - cf. *Masrasector* sp.
  - *Dissopsalis* cf. *pyroclasticus*
  - Gigantic hyaenodontine
- Carnivora
  - ?*Pseudocyon macrodon*
  - Arctoidea, gen. & sp. indet.
  - *Vishnuonyx* n. sp.
  - Mellivorinae, n. gen. & sp.
  - *'Percrocuta' tobieni*
  - cf. *Protictitherium* sp.
  - cf. *Thalassictis robusta*
- Tubulidentata
  - Oryteropodidae
    - ?*Orycteropus chemeldoi*
- Hyracoidea
  - *Parapliohyrax* sp.
- Proboscidea
  - Deinotheriidae, gen. & sp. indet.
  - Gomphotheriidae
    - *Choerolophodon ngorora*
    - 'Tetralophodont gomphothere'
- Perissodactyla
  - Rhinocerotidae
    - *Brachypotherium* cf. *lewisi*
    - *Chilotheridium pattersoni*
    - Aceratherini, gen. & sp. indet.
  - Equidae
    - Hipparioninae, gen. & sp. indet.**
- Artiodactyla
  - Suidae
    - *Lopholistriodon kidogosana*
  - Suidae, gen. & sp. indet.
  - Hippopotamidae
    - *Kenyapotamus coryndoni*
  - Tragulidae
    - *Dorcatherium* cf. *minus*
    - *Dorcatherium* cf. *songhorensis*
  - Climacoceridae
    - *Climacoceras gentryi*
  - Giraffidae
    - *'Palaeotragus' primaevus*
  - Giraffidae, gen. & sp. indet. (large)
  - Boividae
    - *Caprotragoides gentryi*
    - *Pachytragus* aff. *solignaci*
    - *'Protragocerus' labidotus*
    - *Sivoreas eremita*
    - ?Reduncini, gen. & sp. indet.
    - *Homoiodorcas tugenium*
    - *Antidorcas*(?) sp.

* Bishop & Pickford (1975) report a ?Peditidae, a ?Sciuridae, a Phyomyidae, a Dendromurinae, and two Cricetidae from the Ngorara Formation. These specimens remain undescribed and are not available for evaluation.
** Not known from sites in the type section.

silts, sandstones, feldspathic tuffs, argillaceous pumice tuffs and lenses of fine conglomerate.

The Mpesida Beds are bracketed by flows of the Kabarnet trachyte, for which there is a series of relatively consistent isotopic dates. There is a considerable angular unconformity between the Ewalel phonolites and the Kabarnet trachytes. This, together with the formation of a deep weathering horizon on the phonolites, suggests that a significant period of time is unrepresented in the sequence. This has also led to the suggestion that young dates on the Ewalel phonolites, which overlap those on Kabarnet trachytes, are likely to be inaccurate, and to more confidence being placed on the older results (Chapman & Brook 1978). Three determinations on one specimen of Kabarnet trachyte from the base of the sequence in the Tugen Hills near Kabarnet, give dates of 7.3 ± 0.3 Ma, 7.5 ± 0.3 Ma and 7.5 ± 0.3 Ma (Walsh 1969; Baker *et al.* 1971). Chapman & Brook (1978) record another date of 6.9 ± 0.3 Ma from near Morop, and one on a lower flow near Yatya, just beneath an outcrop of the Mpesida Beds, of 7.0 ± 0.4 Ma.

More recent BPRP determinations on upper flows in the Yatya area above the sediments, provide dates of $6.36\pm0{\cdot}08$ Ma, 6.31 ± 0.20 Ma, 6.20 ± 0.19 Ma and 6.20 ± 0.26 Ma (Hill *et al.* 1985).

In terms of fauna the Mpesida Beds are very important within the Tugen Hills succession, and in sub-Saharan Africa generally. This is because they provide the earliest evidence of a major change in the African fauna from the typical mid-Miocene character of the Ngorora assemblages, to one which shows the broad characteristics of the modern Ethiopian condition (Table 3). It is unfortunate that the collections from this unit are

TABLE 3. *Mammal fauna from the Mpesida Beds*

Lagomorpha	Artiodactyla
Leporidae, gen. & sp. indet.	Hippopotamidae, gen. & spp.
Proboscidea	Giraffidae
Gomphotheriidae	?*Giraffa* sp.
Anancus kenyensis	Bovidae
Elephantidae	*Tragelaphus* sp.
Stegotetrabelodon orbus	*Kobus* aff. *porrecticornis*
Perissodactyla	*Madoqua* sp.
Calicotheriidae	*Gazella* or *Raphicerus* sp.
Chemositia tugenensis	
Rhinocerotidae	
Brachypotherium cf. *lewisi*	
Ceratotherium praecox	
Equidae	
Hipparioninae, gen. & sp. indet.	

so far relatively limited. Some new families of animals appear, the Elephantidae, and the first Leporidae in Africa. Within earlier families there are new genera, such as *Anancus* within the Gomphotheriidae, and a modern genus—*Ceratotherium*—within the Rhinocerotidae. There are new species in already established genera such as *Brachypotherium lewisi*, and *Hipparion turkanense*.

Apart from the desirability of getting a more extensive fauna from this important time period, one that is particularly important from the point of view of hominoid evolution, further work is needed to establish the relations between separate outcrops. Chapman (1971) noted that correlation between the different exposures was dubious.

Lukeino Formation

The Lukeino Formation rests on Kabarnet trachytes, and is usually separated from the overlying Chemeron Formation by flows of Kaparaina basalt. Rocks belonging to it were mapped and named in 1967 as the Lukeino Member of the Kaparaina Basalts Formation by Chapman (1971), with subsequent work being done by Martyn (1969) and McClenaghan (1971). It was Pickford (1975b) who published a type section, after he had performed more extensive mapping and fossil collecting in the area of its outcrop. He describes this in more detail in Pickford (1978b). Part of the Lukeino outcrop figured by Pickford was originally mapped as the Kaperyon Formation by Chapman (1971).

The Lukeino Formation consists of a variety of mainly fine-grained epiclastic and pyroclastic sediments. There are red beds and a number of levels of diatomaceous and tuffaceous shales. The sediments are separated by a number of lava units within the succession that have been interpreted as sills (Pickford 1975b).

Bounding ages are given by dates on upper flows of the Kabarnet trachyte and on Kaparaina basalts. Dates on the Kabarnet trachytes have been discussed in connection with the Mpesida Beds, our recent work suggesting an age of 6.20 Ma for the uppermost flows. There is one date on sediments from within the formation of 6.06 ± 0.13 Ma on sanidines from the basal part, and another of 5.62 ± 0.5 Ma on a basaltic unit within the sediments (Hill *et al.* 1985).

Dates on the Kaparaina basalts, which separates the formation from sediments of the Chemeron Formation over much of its outcrop, suggest a minimum age for the unit. Dates quoted by Chapman & Brook (1978) cover a wide range; 4.0 ± 0.1 Ma, 4.8 ± 0.1 Ma (see also Dagley *et al.* 1978), 5.4 ± 0.2 Ma, 5.5 ± 0.2 Ma and 8.4 ± 0.4 Ma from sites in the N. The last date, at least, is justifiably regarded as unreliable in view of its xenocrystic mineralogy. Chapman & Brook (1978) also quote two determinations of 7.1 ± 0.2 Ma and 6.9 ± 0.3 Ma from a site to the S near Sabor.

Recent determinations by the BPRP have produced rather more consistent results, possibly because the specimens come from a more restricted area. There are dates of 5.65 ± 0.07 Ma and 5.65 ± 0.13 Ma on lower flows, two others of 5.65 ± 0.07 Ma and 5.9 ± 0.9 Ma, and one of 5.14 ± 0.34 Ma on an upper flow (Hill *et al.* 1985).

The fauna from the Lukeino sites is more diverse than that from Mpesida, this probably simply reflects the greater abundance of material rather than a real change with time (Table 4). New taxa include *Hysterix*, a number of carnivores, a new elephantid—*Primelephas gomphotheroides*, *Ancylotherium*, and the suid *Sivachoerus*

TABLE 4. *Mammal fauna from the Lukeino Formation*

Lagomorpha	Perissodactyla
Leporidae, gen. & sp. indet.	Calichotheriidae
Rodentia	*Ancylotherium hennigi*
Hystricidae	Rhinocerotidae
Hystrix sp.	Equidae
Primates	Hipparioninae, gen. & sp. indet.
Cercopithecoidea, gen. & sp. indet.	Artiodactyla
Hominoidea, gen. & sp. indet	Suidae
Carnivora	*Sivachoerus tulotos*
Arctoidea, gen. & sp. indet.	Hippopotamidae, gen. & spp.
Enhydriodon cf. *iluecai*	Giraffidae
Atilax n. sp.	*Giraffa* sp.
?cf. *Pachycrocuta* sp.	Bovidae
Homotherium sp.	*Tragelaphus* cf. *spekei*
Tubulidentata	*Ugandax* cf. *gautieri*
Orycteropodidae	*Cephalophus* sp.
Orycteropus spp.	*Aepyceros* sp.
Proboscidea	
Deinotheriidae, gen. & sp. indet.	
Gomphotheriidae	
Anancus kenyensis	
Elephantidae	
Stegotetrabelodon orbus	
Primelephas gomphotheroides	

tulotus. These probably do not mark a great change between Mpesida and Lukeino, but are the effects of sampling. The greatest faunal change in the sequence possibly correlates with what may be the greatest gap in the record, between the top of the Ngorora Formation and the Mpesida Beds.

Kaperyon Formation

The Kaperyon Formation was described by Chapman (1971), and referred to by Bishop *et al.* (1971) and Chapman *et al.* (1978). It consists mainly of pyroclastics, well-bedded and waterlaid in the SW of its outcrop, becoming more massive and subaerial to the N. MacInnes (in Fuchs 1950) remarked on these sediments, which were included in the Kamasia Lake Beds. The type section is at Chesoton where about 80 m of sediments are exposed.

Re-mapping by Pickford (1975a) led him to the view that rocks belonging to the unit should be included variously in the Mpesida Beds, Lukeino Formation, and Chemeron Formation, and that the Kaperyon Formation was superfluous. Some of his criteria appear to be faunal rather than lithostratigraphical (see footnote in Chapman & Brook 1978, p. 217), and Chapman still disputes aspects of this re-interpretation (Chapman, pers. comm.). Further work is necessary to settle the question.

Chemeron Formation

Sediments exposed in river valleys to the W of Lake Baringo were the first in the Tugen Hills succession to be noted scientifically. Gregory (1921) described them as the Kamasia sediments and placed them in his Nyasan division, which he thought was Oligocene in age. The sediments became the basis of the notion of a Kamasian pluvial stage, which Solomon (in Leakey 1931) believed to be Pleistocene. Fuchs collected vertebrate fossils from these sediments, which Hopwood (in Fuchs 1950) dated as Pleistocene.

McCall *et al.* (1967) discerned two sedimentary units in these Kamasian sediments, separated by a lava and an angular unconformity. They proposed the name Chemeron Beds to designate the lower of the two formations, which lay on basalts subsequently named the Kaparaina Basalt Formation. The upper sedimentary formation was named the Kapthurin Formation. Martyn (1967, 1969) carried out more extensive geological work in the area, describing sedimentary successions and recording new fossil sites. In the Chemeron Formation, Martyn recognized two distinct areas of outcrop, the Chemeron basin, and the Kipcherere basin, between which the successions were tentatively and broadly equated lithologically. He also collected from them one of the earliest recognized pre-*Homo sapiens* hominids from Kenya (Tobias 1967).

TABLE 5. *Mammal fauna from the Chemeron Formation*

Rodentia	Perissodactyla
Hystricidae	Calicotheriidae
cf. *Xenohystrix crassidens*	*Ancylotherium hennigi*
Primates	Rhinocerotidae
Cercopithecidae	*Ceratotherium praecox*
*Papio baringensis**	*Ceratotherium simum**
*Paracolobus chemeroni**	Equidae
Cercopithecoidea, gen. & sp. indet.	Hipparioninae, gen. & sp. indet.
Hominidae	Artiodactyla
Australopithecus sp	Suidae
Hominoidea, gen. & sp. indet.	*Nyanzachoerus kanamensis*
Hominidae, gen. & sp. indet.	*Nyanzachoerus jaegeri*
Carnivora	*Potamochoerus porcus*
Enhydriodon cf. *campani*	Hippopotamidae, gen. & spp.
cf. *Pachycrocuta* sp.	Bovidae
Crocuta sp.	*Tragelaphus* cf. *spekei*
Proboscidea	*Kobus* aff. *porrecticornis*
Deinotheriidae	*Aepyceros* sp.
Deinotherium bozasi	
Gomphotheriidae	
Anancus kenyensis	
Elephantidae	
Primelephas gomphotheroides	
Loxodonta adaurora	
Elephas recki	

* Only from site JM 90 in the Chemeron basin.

As a result of his investigations of sediments further N, Pickford described new exposures of the Chemeron Formation, and re-allocated to this formation sediments that had previously been attributed to the upper part of the Kaperyon Formation (Pickford 1975a, 1978b; Pickford *et al.* 1983). Part of this northern extension lies on Kaparaina basalts, and from the point of view of homotaxy and internal lithological characters the equation of this part with Chemeron exposures in the Kipcherere area is little disputed. Where no Kaparaina basalts separate the northern extension sediments from the underlying Lukeino Formation the situation is more controversial (Chapman *et al.* 1978). Some additional details concerning the northern extension are given in Hill (1985).

Sediments consist predominantly of relatively finely stratified silts and tuffs. There are brown sediments at the base, possibly derived from the weathering of underlying lavas, and horizons of ferruginous gravels which are often fossiliferous. Some diatomaceous units also occur.

Isotopic dates relevant to the age of the formation were referred to in Bishop *et al.* (1971), and more fully published in Chapman & Brook (1978). Maximum dates for the Chemeron Formation are given by a variety of isotopic determinations on the Kaparaina basalts (Chapman & Brook 1978) which have been discussed in connection with the Lukeino Formation. Published dates on these underlying basalts range from 4.0 Ma to 8.4 Ma, but our recent dates on lavas which occur very near the top of the unit in some places ranges more tightly from 5.14 Ma to 5.9 Ma.

There are a few dates on tuffs within the formation. Two, recorded by Chapman & Brook (1978) at 8.2 ± 0.2 Ma on sanidines, clearly do not accurately date the formation. Some have more recently been carried out by the BPRP. At Tabarin (BPRP KO77) two determinations on a tuff 20 m above the Kaparaina basalt and 20 m below the main fossiliferous level give dates of 4.96 ± 0.03 Ma and 5.25 ± 0.04 Ma (Hill 1985). Others at sites further N have given dates of 5.4 ± 0.23 Ma, and one of 2.78 ± 0.16 Ma which appears to be anomalous. There are two dates, of 1.57 Ma and 2.13 Ma, on lavas in the Chemeron basin of the formation, which were originally thought to overlie the formation but actually are within it, very near the top.

A faunal list for the Chemeron Formation is given in Bishop *et al.* (1971) and in Table 5. The fauna from the majority of sites, particularly those not in the Kipcherere basin, are consistent with a date of from 5 Ma to 4 Ma, but there is at least one anomaly. One main change from the Lukeino fauna is in the suids, where *Nyanzachoerus jaegeri* and *Nyanzachoerus kanamensis* replace *Sivachoerus tulotos*, and are in fact prob-

TABLE 6. *Mammal fauna from the Kapthurin Formation*

Rodentia*	Perissodactyla
Muridae	Rhinocerotidae
Oenomys sp.	*Ceratotherium simum*
Thryonomyidae	Artiodactyla
Thryonomys sp.	Suidae
Primates	*Mesochoerus majus*
Cercopithecidae	*Phacochoerus aethiopicus*
Colobus sp.	Hippopotamidae
Cercopithecus sp.	*Hippopotamus amphibius*
Papio sp.	*Bovidae*
Hominidae	*Tragelaphus* cf *scriptus*
Homo sp.	*Tragelapohus* sp.
Carnivora	*Redunca* sp.
Felidae	*Kobus ellipsiprymnus*
Panthera leo	*Alcelaphus* sp.
Hyaenidae	*Cephalophus* sp.
Crocuta crocuta	*Sylvicapra* sp.
Proboscidea	*Antidorcas* sp.
Elephantidae	Tubulidentata
sp. indet.	Orycteropodidae
Hyracoidea	*Orycteropus* sp.
Procaviidae	
Heterohyrax sp.	

* Jacobs (pers. comm.) reports a total of at least eight rodent species.

ably derived from it. Another is in the Elephantidae, where *Stegotetrabeledon orbus* disappears and *Loxodonta adaurora* is present. There are new species of Cercopithecidae, *Papio baringensis* and *Paracolobus chemeroni*, both of which come from a site which may be younger than the rest.

Also from this site (JM 90/91, Kapthurin River section) in the Chemeron basin, are specimens of *Elephas recki* and *Ceratotherium simum*. Maglio (1970, 1973) suggested on the basis of his study of fossil elephants, and particularly the specimen from this site, that the Chemeron Formation as a whole (*sensu* Martyn) represented a long time span. Although, fossils from most of the sites indicated an age of between 5 Ma and 4 Ma, *Elephas recki* Stage 2 from JM 90 suggested a much more recent age, of just over 2 Ma. This was supported by the presence at that site of a specimen of *Ceratotherium simum* (Hooijer 1969).

This faunal discrepancy highlights one of the major problems concerning the formation. At present the stratigraphical relations are obscure between much of the outcrop, which is early Pliocene in age, and other sediments such as some in the Chemeron basin, which appear on faunal grounds to be considerably younger.

Kapthurin Formation

The Kapthurin Formation overlies the Chemeron Formation with a pronounced unconformity to the W of Lake Baringo, overstepping onto Kaparaina basalts. It originally formed, with the Chemeron sediments, the Kamasian unit of Gregory (1921) and later workers, until it was defined by McCall *et al.* (1967). Martyn did much additional work (Martyn 1967, 1969; Leakey *et al.* 1969), and more detailed mapping was carried out by Tallon (1976, 1978). At present various aspects of the formation are under investigation by a research team directed by Francis van Noten.

The unit ranges from coarsely-bedded boulder conglomerates interpreted as flash flood deposits to fine-grained well-bedded pumiceous tuffs, thought to have been deposited lacustrine conditions.

The Kapthurin Formation is younger than the Chemeron Formation, and the lower unit of pumice tuff may be equivalent in age to the Lake Baringo trachyte, dated at 0.27 Ma (Tallon 1976, Chapman & Brook 1978). There are two series of dates on sanidines from a pumice tuff within the formation; one gives dates of 0.67 ± 0.10 Ma and 0.70 ± 0.10 Ma and the second series 0.89 ± 0.26 Ma, 0.87 ± 0.28 Ma and 0.84 ± 0.28 Ma. A bedded tuff has produced dates of 0.25 ± 0.12 Ma and 0.24 ± 0.08 Ma (Tallon 1978). Dagley *et al.* (1978) have shown that the pumice tuff is of normal polarity, and hence younger than 0.7 Ma. This favours the younger dates for this unit.

The first faunal list for the formation appeared in Leakey *et al.* (1969), and was revised in Bishop *et al.* (1971). The list provided here (Table 6) is based on the one by Harris (in Tallon 1978) with a few revisions.

Conclusions

The extensive sediments in the Tugen Hills sequence are well calibrated by isotopic means and also, over a large part of the succession, by palaeomagnetic stratigraphy. The sediments cover a substantial amount of the period from older than 14 Ma to younger than 4 Ma, which is not available elsewhere in Africa. One formation, the Ngorora Formation, does this more or less continuously for at least 2.5 Ma, and possibly much longer. This provides an opportunity to examine changing conditions and environments in the rift valley situation from its inception. The sediments, some of which occupy separate fault-bounded basins, permit the elucidation of tectonic development in this part of the rift.

Vertebrate fossils from these sediments document important biological change over this period when major shifts were taking place in Old World faunas. During this time many modern families of animals originated, including the Hominidae. Fossil sites in this succession, particularly when correlated with other well-calibrated Old World fossiliferous sequences, make it possible to investigate palaeobiogeographical history, and the causes and dynamics of Neogene faunal change.

ACKNOWLEDGEMENTS: This work has been financed by grants from the National Science Foundation to David Pilbeam, John Barry and Kay Behrensmeyer (No. BNS 81-40818) and to David Pilbeam, John Barry and Andrew Hill (No. BNS 84-07575), and by grants from the Louise Brown Foundation to David Pilbeam and Andrew Hill. We would like to thank Greg Chapman, Marc Monaghan, Lisa Tauxe, John Barry, Louis Jacobs, Bonnie Fine Jacobs and David Pilbeam for their help. We are very grateful to the National Museums of Kenya, and to Richard Leakey the Director, for logistical help. The government of the Republic of Kenya provided research permission.

References

BAKER, B.H., WILLIAMS, L.A.J., MILLER, J.A. & FITCH, F.J. 1971. Sequence and geochronology of the Kenya Rift volcanics. *Tectonophysics* **11,** 191–215.

BISHOP, W.W. & CHAPMAN, G.R. 1970. Early Pliocene sediments and fossils from the northern Kenya Rift Valley. *Nature* **226,** 914–18.

——,——, HILL, A. & MILLER, J.A. 1971. Succession of Cainozoic vertebrate assemblages from the northern Kenya Rift Valley. *Nature* **233,** 389–94.

——, MILLER, J.A. & FITCH, F.J. 1969. New potassium-argon age determinations relevant to the Miocene fossil mammal sequence in East Africa. *Am. J. Sci.* **267,** 669–99.

—— & PICKFORD, M. 1975. Geology, fauna and palaeoenvironment of the Ngorora Formation, Kenya Rift Valley. *Nature* **254,** 185–92.

CHAPMAN, G.R. 1971. *The geological evolution of the northern Kamasia Hills, Baringo District, Kenya.* Unpublished PhD thesis, University of London.

—— & BROOK, M. 1978. Chronostratigraphy of the Baringo basin, Kenya Rift Valley. In: BISHOP, W.W. (ed) *Geological Background to Fossil Man.* Scottish Academic Press, Edinburgh, pp.207–23.

——, LIPPARD, S.J. & MARTYN, J.E. 1978. The stratigraphy and structure of the Kamasia Range, Kenya Rift Valley. *J. Geol. Soc.* **135,** 265–81.

DAGLEY, P., MUSSETT, A.E. & PALMER, H.C. 1978. Preliminary observations on the palaeomagnetic stratigraphy of the area west of Lake Baringo, Kenya. In: BISHOP, W.W. (ed) *Geological Background to Fossil Man.* Scottish Academic Press, Edinbourgh, pp. 225–35.

FUCHS, V.E. 1950. Pleistocene events in the Baringo Basin. *Geol. Mag.* **87,** 149–74.

GREGORY, J.W. 1921. *The Rift Valleys and Geology of East Africa.* Seeley Service, London.

HILL, A. 1985. Early hominid from Baringo, Kenya. *Nature* **315,** 222–4.

—— DRAKE, R., TAUXE, L., MONAGHAN, M., BARRY, J.C., BEHRENSMEYER, A.K., CURTIS, G., FINE JACOBS, B., JACOBS, L., JOHNSON, N. & PILBEAM, D. 1985. Neogene palaeontology and geochronology of the Baringo Basin, Kenya. *J. Hum. Evol.* **14,** 759–73.

HOOIJER, D.A. 1969. Pleistocene East African rhinoceroses. In: LEAKEY, L.S.B. (ed) *Fossil Vertebrates of Africa 1.* Academic Press, London, pp.71–98.

KING, B.C. & CHAPMAN, G.R. 1972. Volcanism of the Kenya Rift Valley. *Phil. Trans. R. Soc. Lond.* **271A,** 185–208.

LEAKEY, L.S.B. 1931. *The Stone Age Cultures of Kenya Colony.* Cambridge University Press, Cambridge.

LEAKEY, M., TOBIAS, P.V., MARTYN J.E. & LEAKEY, R.E.F. 1969. An Acheulian industry with prepared core technique and the discovery of a contemporary hominid at Lake Baringo, Kenya. *Proc. Prehist. Soc.* **35,** 48–76.

MAGLIO, V.J. 1970. Early Elephantidae of Africa and a tentative correlation of African Plio-Pleistocene deposits. *Nature* **225,** 328–32.

——1973. Origin and evolution of the Elephantidae. *Trans. Am. Phil. Soc.* **63,** 1–149.

MARTYN, J.E. 1967. Pleistocene deposits and new fossil localities in Kenya. *Nature* **215,** 476–7.

—— 1969. *The geological history of the country between*

Lake Baringo and the Kerio River, Baringo District, Kenya. Unpublished PhD thesis, University of London.

McCall, G.J.H., Baker, B.H. & Walsh, J. 1967. Late Tertiary and Quaternary sediments of the Kenya Rift Valley. In: Bishop, W.W. & Clark, J.D. (eds) *Background to Evolution in Africa.* University of Chicago Press, Chicago, pp.191–220.

McClenaghan, M.P. 1971. *Geology of the Ribkwo area, Baringo District, Kenya.* Unpublished PhD thesis, University of London.

Ness, G., Levi, S. & Crouch, R. 1980. Marine magnetic anomaly time scales for the Cenozoic and late Cretaceous: a precis, critique, and synthesis. *Rev. Geophys. Space Phys.* **18(4),** 753–70.

Pickford, M. 1975a. *Stratigraphy and palaeoecology of five late Cainozoic formations in the Kenya Rift Valley.* Unpublished PhD thesis, University of London.

—— 1975b. Late Miocene sediments and fossils from the northern Kenya Rift Valley. *Nature* **256,** 279–84.

—— 1978a. Geology, palaeoenvironments and vertebrate faunas of the mid-Miocene Ngorora Formation, Kenya. In: Bishop, W.W. (ed) *Geological Background to Fossil Man.* Scottish Academic Press, Edinburgh, pp. 237–62.

—— 1978b. Stratigraphy and mammalian palaeontology of the late-Miocene Lukeino Formation, Kenya. In: Bishop, W.W. (ed) *Geological Background to Fossil Man.* Scottish Academic Press, Edinburgh, pp.263–78.

——, Johanson, D.C., Lovejoy, C.O., White, T.D. & Aronson, J.L. 1983. A hominoid humeral fragment from the Pliocene of Kenya. *Am. J. Phys. Anthrop.* **60,** 337–46.

Prior, G.T. 1903. Contributions to the petrology of British East Africa. *Mineral. Mag.* **13,** 228–62.

Shackleton, R.M. 1951. A review of some recent work in the rift valleys of Kenya. *18th Internatinal Geological Congress* **14,** 213.

Tallon, P.W.J. 1976. *The stratigraphy, palaeoenvironments and geomorphology of the Pleistocene Kapthurin Formation, Kenya.* Unpublished PhD thesis, University of London.

—— 1978. Geological setting of the hominid fossils and Acheulian artefacts from the Kapthurin Formation, Baringo District, Kenya. In: Bishop, W.W. (ed) *Geological Background to Fossil Man.* Scottish Academic Press, Edinburgh, pp. 361–78.

Thomson, J. 1884. *Through Masailand.* Samson Low, London, 364pp.

Tobias, P.V. 1967. Pleistocene deposits and new fossil localities in Kenya. *Nature* **215,** 478–80.

Walsh, J. 1969. Geology of the Eldama Ravine—Kabarnet area. *Rep. Geol. Surv. Kenya* **83.**

Williams, L.A.J. & Truckle, P.H. 1980. Volcanic sequences in the Kenya Rift. In: *Geodynamic Evolution of the Afro-Arabian Rift System.* Accad. Nazionale dei Lincei, Rome, pp.123–32.

Willis, B. 1936. *East African Plateaus and Rift Valleys.* Carnegie Institution, Washington.

A. Hill, Dept of Anthropology, Yale University, New Haven, Connecticut 06520, USA.

G. Curtis & R. Drake, Dept of Geology & Geophysics, University of California, Berkeley, California 94720, USA.

Quaternary environments of the Chalbi basin, Kenya: sedimentary and geomorphological evidence

C.K. Nyamweru

SUMMARY: The Chalbi basin is an inland drainage basin surrounded by lava uplands in arid northern Kenya. In its centre lies the Chalbi playa, a bare surface of mud subject to occasional flooding, the surface of which is at *c.* 370 m above sea-level. Along the edges of the basin, basalt overlies sediments. At one site (Algas) sediments contain the fossil bones of aquatic and terrestrial species which suggest an age within the range 1.6 to 2.0 Ma. A K-Ar date of 2.5 ± 0.3 Ma has been obtained for basalt overlying the bone-bearing sediment. Between the basalt front and the Chalbi playa there is a belt of younger sediments that include sands and horizontally-bedded silts. At several localities, fossil freshwater gastropods have given ^{14}C dates ranging from 9530 to 11,080 BP. The gastropods may have originated within a single Lake Chalbi with a surface water level above 410 m, or, alternatively, in a number of separate ponds or swamps. The evidence of early Holocene humid conditions in the Chalbi basin agrees with that from the Lake Turkana basin and from other lake basins in East Africa.

Regional setting

The Chalbi basin is an inland drainage basin in the Marsabit District of northern Kenya (Fig. 1). At its centre lies the Chalbi playa, a surface of mud and salt subject to occasional flooding. The long axis of the Chalbi basin (and of the playa) runs approximately NW–SE; in this direction the playa is *c.* 110 km long. Its width varies from 10 to 20 km. The surface area of the playa is slightly under 1500 km^2, and it lies at an elevation of *c.* 370 m above sea-level.

At North Horr, rainfall records over 24 years give a mean annual figure of 153 mm, and at Maikona 10 years of records yield a mean annual rainfall of 206 mm. At times, rainfall on the Chalbi catchment area brings water to the playa via a number of intermittent drainage lines (lagas), as shown on Fig. 2. In some years, the inflow is enough to create a shallow playa lake a few tens of cm deep. In 1978, for example, parts of the playa were flooded for several months.

The Chalbi playa is largely surrounded by volcanic terrain which includes the basaltic lavas of Kulal, Chari Ache, the Huri Hills and Marsabit. For tens of km the lava forms the top of prominent cliffs that rise 20–30 m above the playa. Elsewhere, a belt of sediments of gently rolling relief separates the lava and the playa. Fig. 3 shows the broad topographical and geological divisions of the study area.

The primary research objective was to describe past environmental conditions in the Chalbi basin, largely through the interpretation of landforms and sediments. A search was carried out for evidence of a former 'Lake Chalbi', such as shoreline cliffs, beach gravels, sand spits, sand beaches and lacustrine sediments. Attention was paid to other landforms and sediments which might provide evidence of environmental conditions that were different from those in the area today. A number of sites yielded faunal and floral remains including fossil bone, fossil wood and fossil gastropod shells. These have given indications of past environmental conditions and have also allowed dating to be carried out. Another objective of the fieldwork in the Chalbi basin was topographic mapping, to supply height data that is missing from the maps of the area. By the use of barometric levelling, heights were obtained for some of the topographic and sedimentary features relevant to the Quaternary history of this basin (see Table 1 for heights of gastropod sites).

Evolution of the Chalbi basin

Fuchs (1939) presents a geological map which shows the Chalbi basin (though not by name) as an area of 'undifferentiated alluvial deposits' between Miocene and Pleistocene lavas. Dixey (1948) describes the 'Chelbi' basin and postulates the existence of a lake there during part of the Lower Pleistocene. The Chalbi basin is not included in the mapping by the Geological Survey of Kenya at the scale of 1:125,000, but relevant information can be found in reports and maps on other parts of northern Kenya, for example, Baker (1963), Dodson (1963, 1971), Joubert (1966), Dodson & Matheson (1969), Walsh & Dodson (1969) and Randel (1970). Reconnaissance mapping at the scale of 1:250,000 is currently being carried out in this area as part of the Samburu-Marsabit Geological Mapping and Mineral Exploration Project of the British Geological Survey and the Department of Mines and Geology, Ministry of Environment and Natural Resources, Kenya.

From Frostick, L.E. *et al.* (eds) 1986, *Sedimentation in the African Rifts*, Geological Society Special Publication No. 25, pp. 297–310.

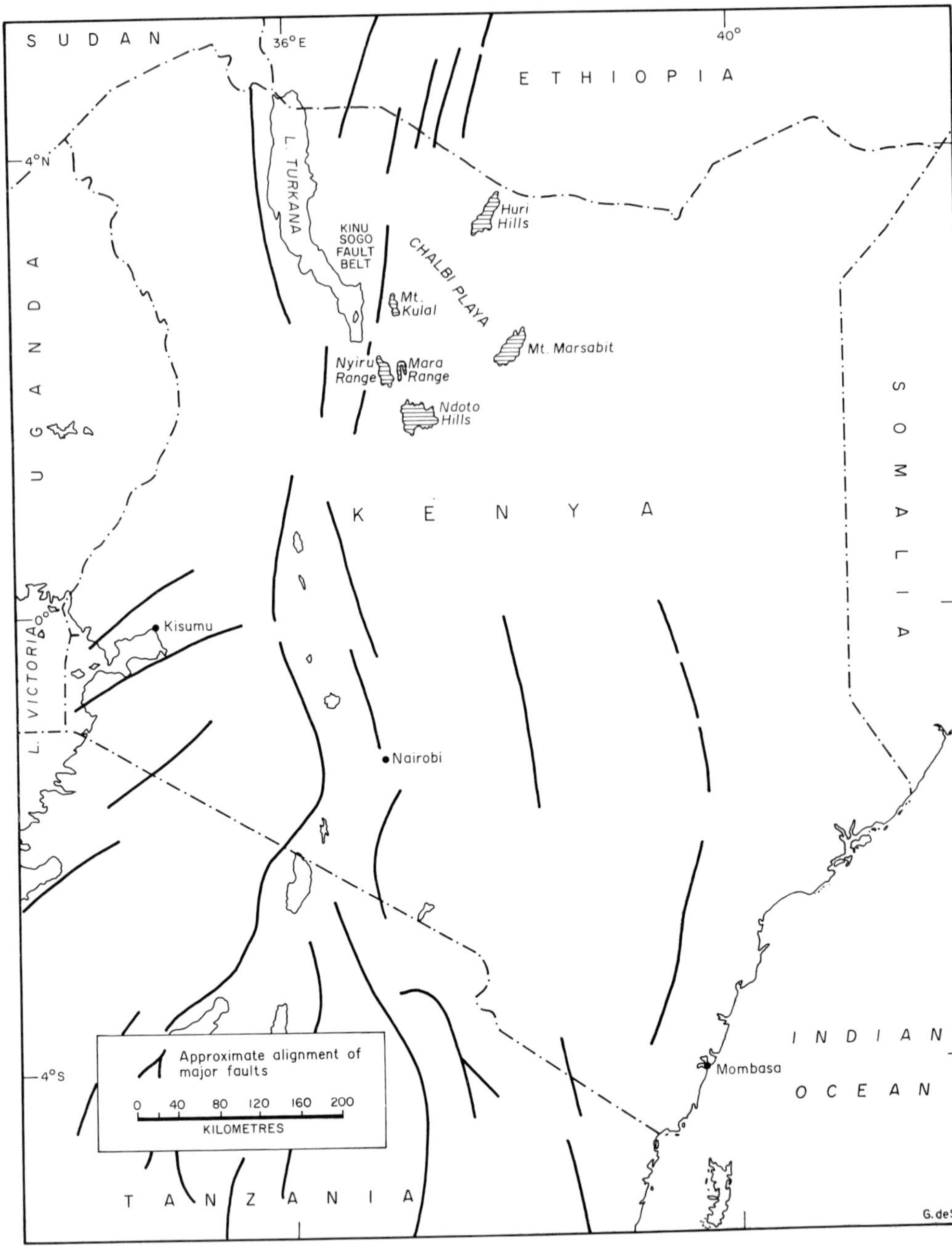

FIG. 1. Location of the Chalbi playa in Kenya.

Research into aspects of the physical environment of this area has been carried out by IPAL (Integrated Project in Arid Lands; UNESCO/UNEP Man and Biosphere Project 3). Reports have been written on geomorphology (Sinda 1981; Bake 1982), vegetation (Herlocker 1979), climatology (Edwards *et al.* 1979; Bake 1983) and hydrology (Ongweny *et al.* 1983).

Understanding of the evolution of the Chalbi basin benefits from the detailed work done in the East Lake Turkana (Koobi Fora) area to the W, for example, Bowen (1974), Vondra & Bowen

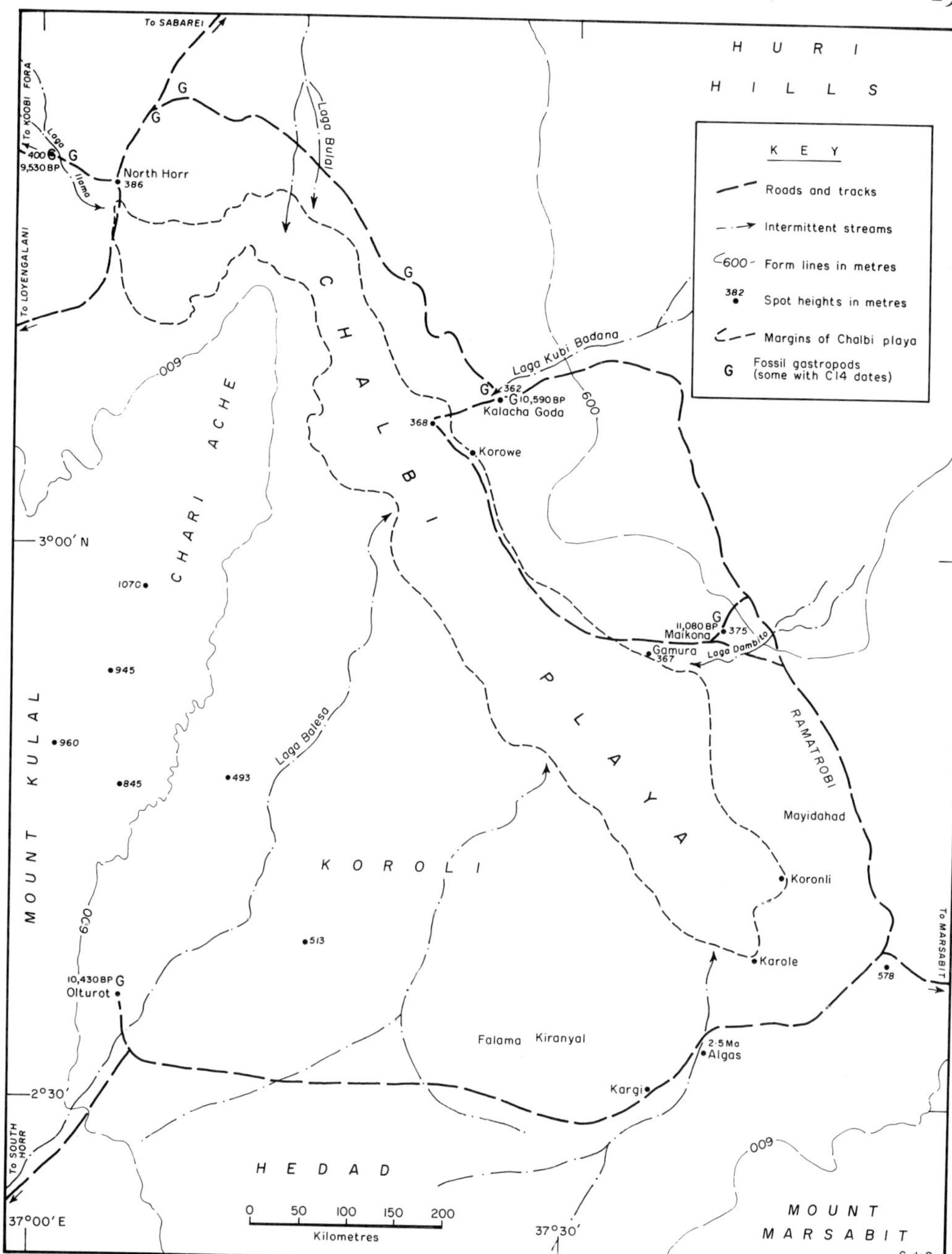

FIG. 2. Topography and place names of the Chalbi playa and its margins.

(1978), Owen *et al.* (1982) and Watkins (1982). Important information on the sequence and geochronology of the volcanic rocks of north-central Kenya is presented by Brotzu *et al.* (1984).

The oldest rocks of this part of N Kenya are Precambrian and metamorphic. They outcrop to the N, S and W of the Chalbi basin, for example in the Nyiru Range and the Ndoto mountains (Baker 1963; Dodson 1963; and Fig. 1). In the area of the Chalbi basin these rocks are overlain by an undetermined thickness of younger sediments and volcanic rocks. However, structures

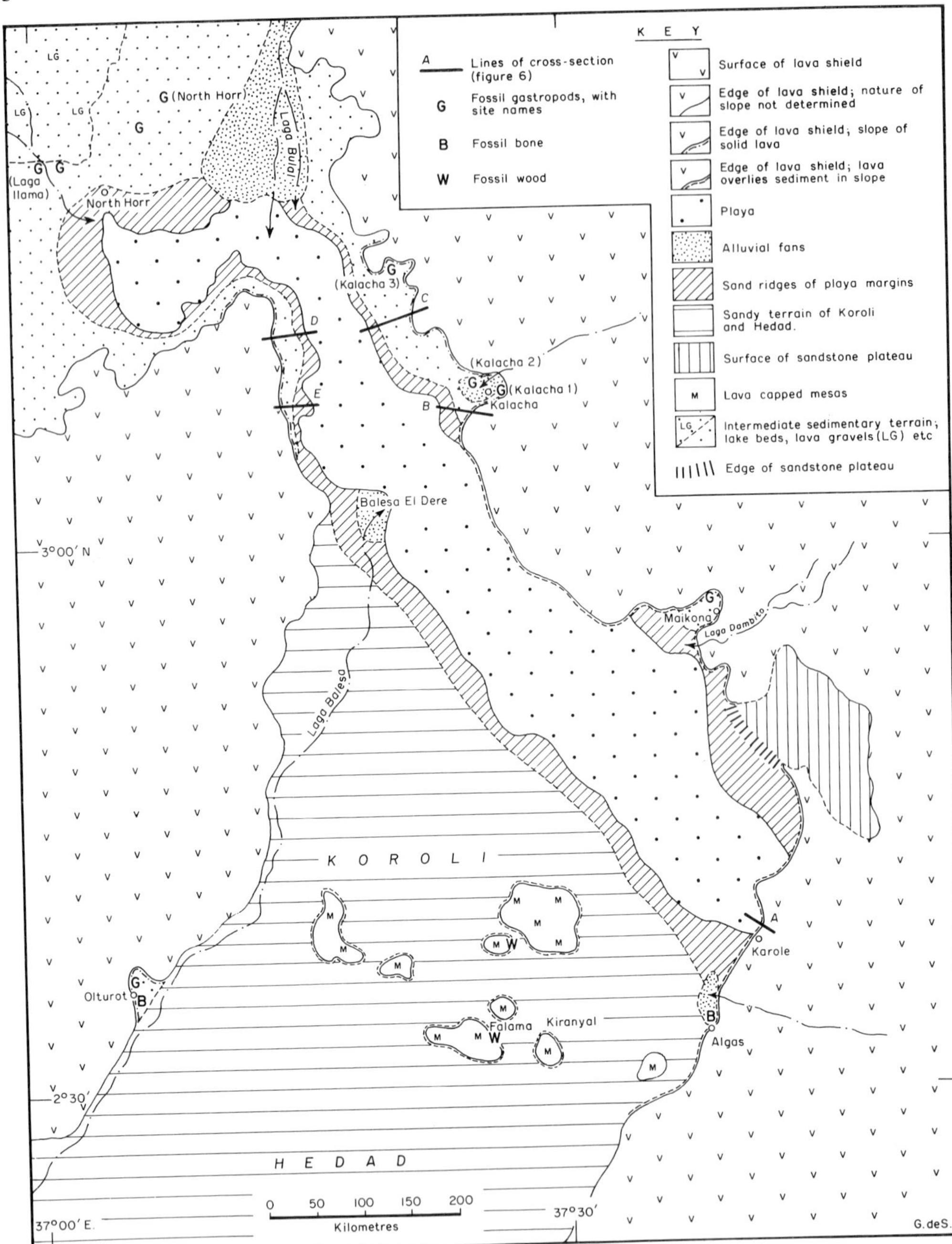

FIG. 3. Geomorphology of the Chalbi playa and its margins.

within the Precambrian basement may have influenced the alignment of the Chalbi basin margins. The Precambrian rocks underwent a long period of erosion, warping and possible faulting, resulting in a surface of rather low relief on which the younger sediments and volcanic material accumulated.

Some of the sediments have been grouped together in the literature (e.g. Joubert 1966; Walsh & Dodson 1969) under the broad heading

TABLE 1. *Altitude and* ^{14}C *dates of gastropod sites on the margins of the Chalbi basin*

Gastropod site*	Altitude† in m a.s.l.	Radiocarbon date, years BP
Maikona	375	11,080 ± 270 (GX-9790)
Kalacha 1	381	10,590 ± 255 (GX-9765)
Kalacha 2	385	—
Kalacha 3	383	—
North Horr	409	—
Laga Ilama	400	9530 ± 220 (GX-9792)
Olturot	541–542.5	10,430 ± 250 (GX-9791)

* See Fig. 3 for location.
† Altitude of gastropod site at Olturot obtained by estimation from form lines on the 1 : 100,000 map, sheet 42. Altitudes of other sites obtained by barometric levelling Dec. 1983 to Jan. 1984. Datum is the Survey of Kenya datum, obtained by levelling from the survey beacons at Iyole (19 PR 1) and Maihadad (SKT-A). Expected error by this method under the conditions that prevailed is ± 7.5 m.

of 'Turkana Grits', which are said to range in age from Cretaceous to Pleistocene. Sediments on the eastern and southern margins of the Chalbi basin may be correlated with some of the Turkana Grits. There is an exposure of sediments covering several tens of km^2 in the Ramatrobi area, E and SE of Maikona (see Fig. 3). The surface of Ramatrobi is covered by drifts of orange sand and patches of smooth rounded quartz pebbles, and it is surrounded by lava on all sides except the W. This western margin of Ramatrobi slopes towards the Chalbi playa. Over a distance of *c*. 2 km (in an E–W direction) there is a terrain in which cliffs (up to 5 m high) of red and purple sandstone and grit rise from debris slopes covered with rounded quartz pebbles, mostly in the size range 2–6 cm.

For over 5 km, lava is absent from the Chalbi playa margin, which here runs broadly NNW–SSE. It seems that lava has never overlain Ramatrobi; there is no evidence of even a thin lava cover removed by later erosion. This surface may thus represent part of a pre-volcanic landscape. Air photographs and satellite imagery indicate that NW–SE trending structures within the rocks of Ramatrobi may have influenced the alignment of the basin margin in this area. To the N of Ramatrobi, exposures in the walls of Laga Dambito (see Fig. 3) show sandstone overlain by basalt. The indication from these exposures (Fig. 4A–C) is that the sandstone had undergone considerable dissection before the lava flowed across and covered the landscape. Fig. 5A and B show other sections from the eastern margins of the Chalbi basin. The red sandstones appear to dip southwards and are lost from the base of the section close to Mayidahad. At Koronli and Karole, pale grey and cream coloured clays and silts are visible; no fossils were found at the localities visited. Further S, fossil bone and fossil wood have washed out of sediments that underlie the volcanic blanket at Algas and at Falama, respectively. The vertebrate fossils (see Table 2) are of particular interest in that they indicate the presence of a large body of deep water in this area at the time the sediments were laid down. The sediments may span a considerable age. R. Key (pers. comm. 1984) considers that the Ramatrobi sandstones may be as old as Jurassic. On the evidence of the fossil wood, it has been suggested that the Falama sediments may be Mio-Pliocene, while the Algas sediments underlie basalt dated at 2.5 Ma (see Appendix).

It is likely that the volcanic rocks of this part of northern Kenya also cover a considerable age range. According to Watkins (1982) volcanism began to the NW of the Chalbi basin in the late Oligocene; basalts of the Suregei-Asille district are dated at 28.2 Ma. Dates quoted by Brotzu *et al.* (1984) suggest that the NW and N margins of the Chalbi basin may have been defined as early as the Miocene. Lavas to the N of North Horr gives dates of 6.1 and 4.5 Ma and, on the Huri Hills, plateau basalts date from 7.3 and 5.4 Ma. In fact, in the Huri Hills, the volcanic foundation is overlain by younger multicentre volcanic rocks which are dated between 2.3 and 1.4 Ma.

Brotzu *et al.* (1984) have no rocks dated as older than 0.6 Ma from Marsabit; the rest of their samples had ^{40}Ar levels below the limit of detection and it is assumed that these rocks are probably less than 1 Ma (Brotzu *et al.* 1984, p.78). However, a basalt from Algas, near Kargi on the western margin of the Marsabit lava, has been dated for the present study at 2.5 ± 0.3 Ma (sample R-6920; see Appendix). Thus the eastern margin of the Chalbi basin may not have come into existence until the Pleistocene.

The southwestern and western margins of the basin are formed by the dissected volcanic ridge Kulal and its northwards extension, Chari Ache. Earlier authors (e.g. Fuchs 1939) dated Kulal as Miocene, but it also includes younger rock. Dodson (1963) shows Pleistocene basalts covering a low plateau on the southern slopes of Kulal, while Brown (pers. comm. 1984) has dates of 2.0 and 2.4 Ma, also for the southern slopes.

On the southern margin of the Chalbi basin lie

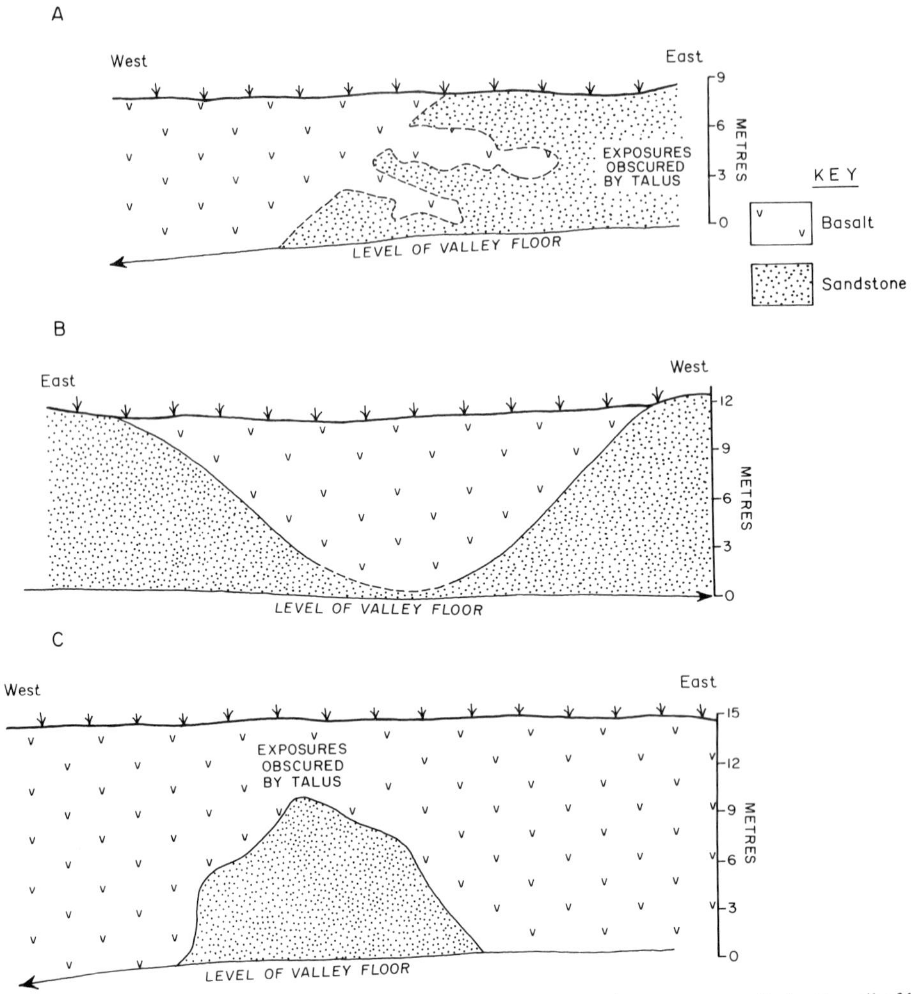

FIG. 4. Rock sections from the walls of Laga Dambito. A. North wall of laga, upstream. B. South wall of laga. C. North wall of laga, downstream.

the Koroli Plateaux (Fig. 3). There are about ten mesas whose approximately horizontal tops stand at between 100 and 150 m above the surrounding surface. Exposures in the mesa side slopes show a maximum of *c.* 10 m of basalt overlying sediments (see Fig. 5C). The mesas probably originated as a continuous sheet of lava, which has been considerably reduced by erosion. No radiometric dates are yet available for these lavas, but the degree of dissection which has taken place (some of the plateaux are separated from their nearest neighbours by over 4 km) suggests that they may be among the older volcanic rocks of this area.

The Chalbi basin appears to be a feature of compound origin. In part it exists as a result of the building of lava mountains to the N, E and W, leaving the Chalbi as an interior lowland between them. However there are indications of tectonic influences on the landscape. The western margin lies less than 60 km from the Gregory Rift System, which at this latitude runs from the southern section of Lake Turkana through the Kinu Sogo fault belt, with an alignment slightly E of N (Fig. 1). Landsat images show similar alignments in the Precambrian terrain to the S (Mara, Nyiru and the Ndoto Ranges), as well as along the western margin of the Marsabit lavas, the SW edge of the Huri Hills, and the E edge of Chari Ache.

The eastern margin of the Chalbi basin does not owe its existence entirely to the accumulation of the lavas of Mt Marsabit. Along much of this margin, lava only forms a cap of 2–3 m that

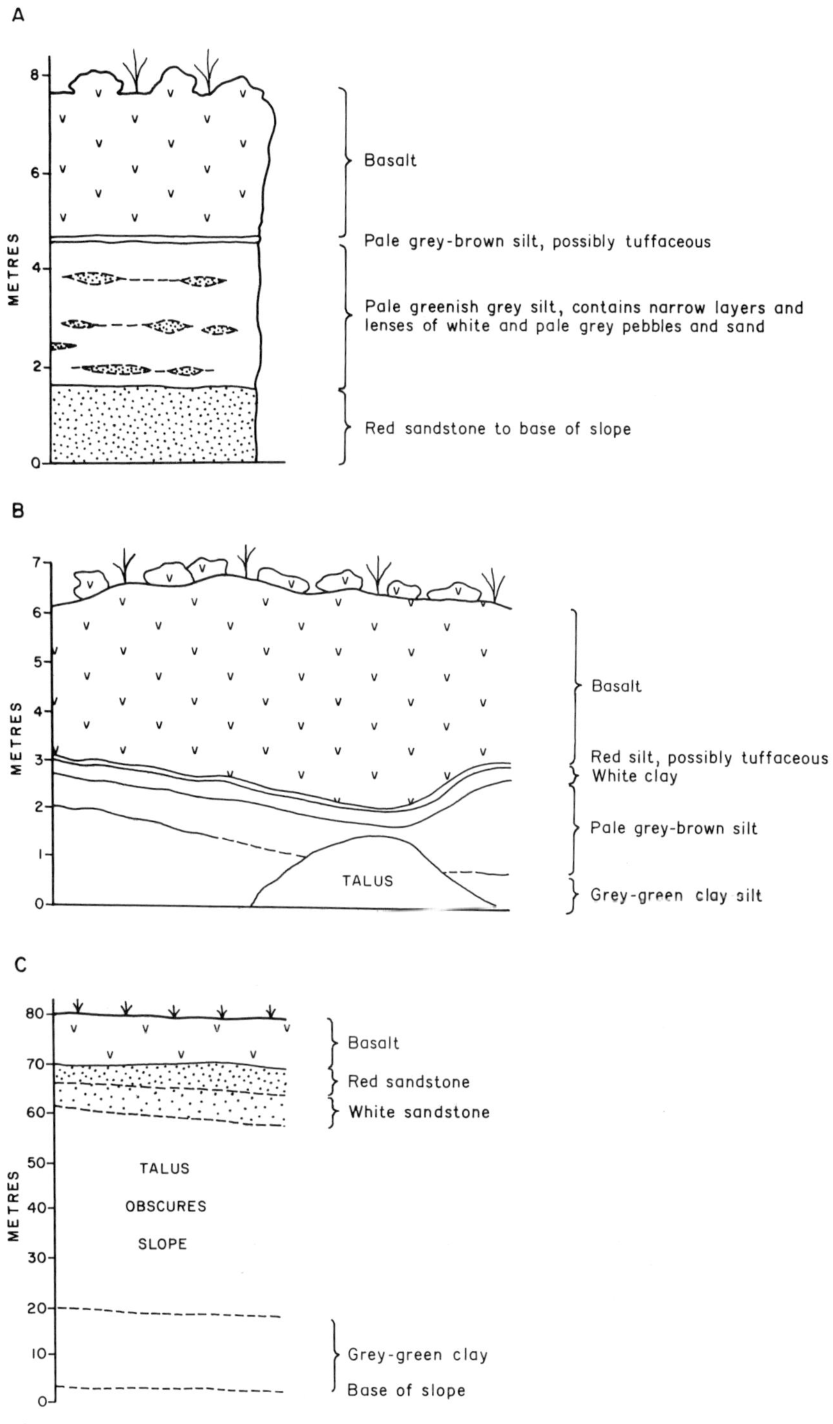

FIG. 5. Rock sections from the playa margins. A. Edge of lava near Mayidahad. B. Edge of lava near Karole. C. Northern slope of Kiranyal mesa.

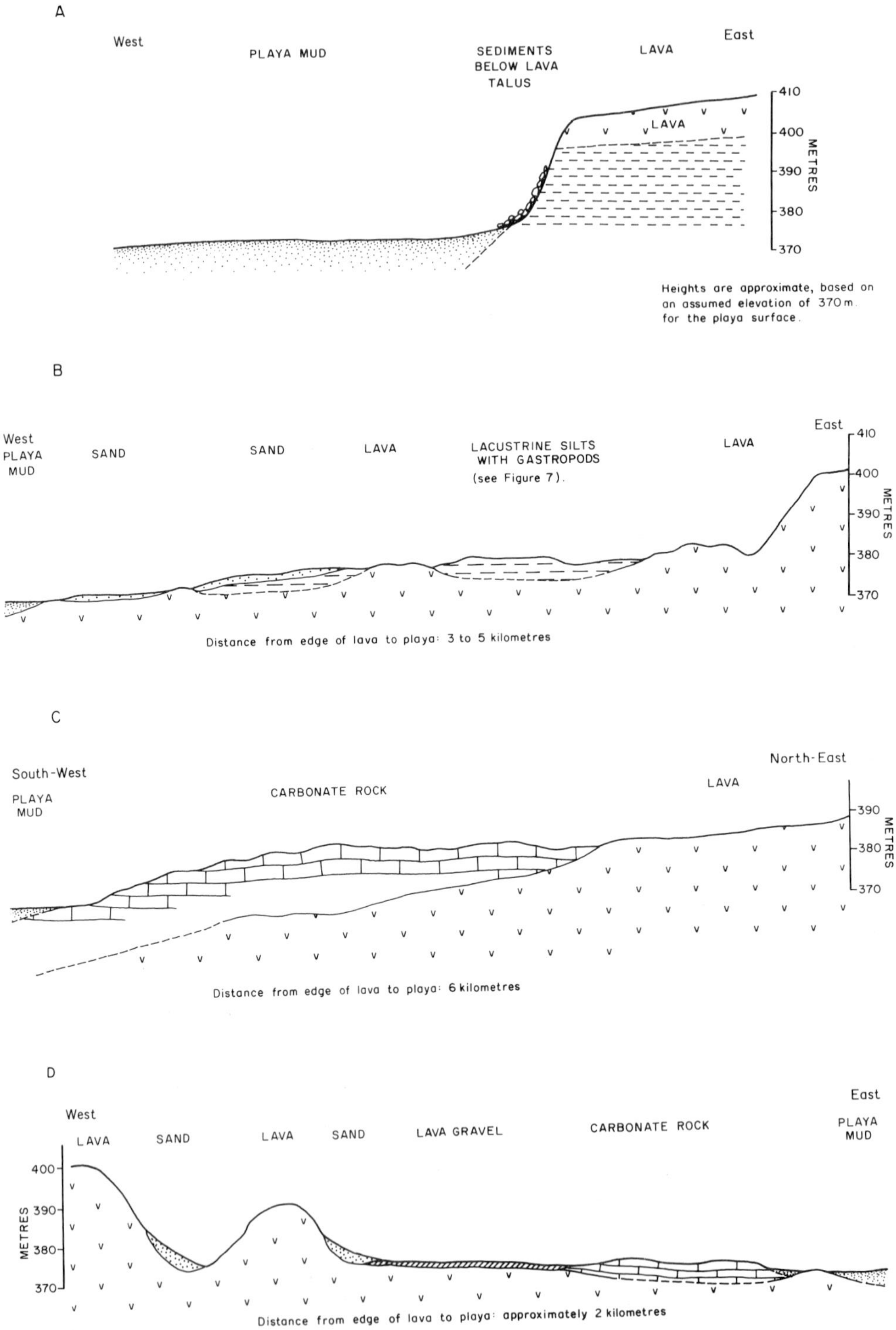
A
West
PLAYA MUD
SEDIMENTS
BELOW LAVA
TALUS
LAVA
East
LAVA
410
400
390
380
370
METRES
Heights are approximate, based on an assumed elevation of 370 m. for the playa surface.
B
West
PLAYA
MUD
SAND
SAND
LAVA
LACUSTRINE SILTS
WITH GASTROPODS
(see Figure 7).
LAVA
East
410
400
390
380
370
METRES
Distance from edge of lava to playa: 3 to 5 kilometres
C
South-West
PLAYA
MUD
CARBONATE ROCK
LAVA
North-East
390
380
370
METRES
Distance from edge of lava to playa: 6 kilometres
D
West
LAVA
SAND
LAVA
SAND
LAVA GRAVEL
CARBONATE ROCK
East
PLAYA
MUD
400
390
380
370
METRES
Distance from edge of lava to playa: approximately 2 kilometres

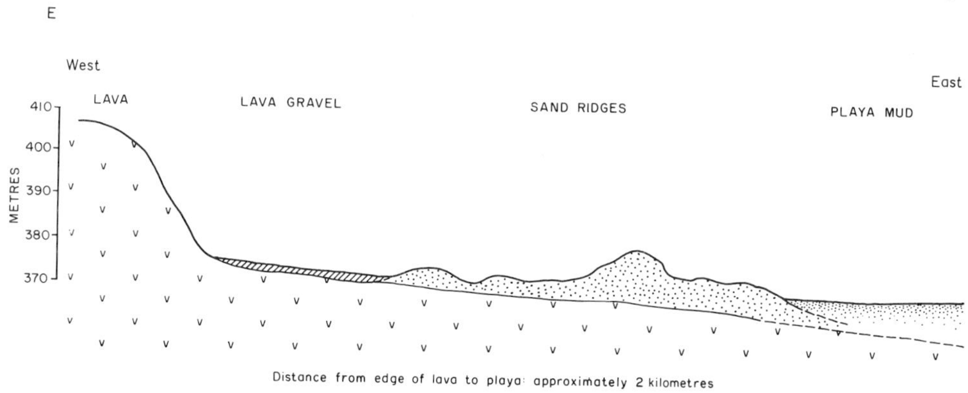

FIG. 6. (*This and facing page.*) Cross-sections of the playa margins. A. Near Karole. B. Huri Hills, S of Kalacha Goda. C. Huri Hills, N of Kalacha Goda. D. Northern end of Chari Ache. E. Central part of Chari Ache.

overlies 5–10 m of exposed sediment. In the Ramatrobi area, lava falls back from the basin margin for a distance of more than 5 km. The W edge of Ramatrobi may represent the margin of a depression that existed before the formation of Marsabit and the Huri Hills. Alternatively it is possible that the eastern margin of the Chalbi basin proper has been formed by downfaulting to the W *after* the eruption of the Marsabit lavas. In this context it is of interest to note that close to the Karole waterhole there is lava exposed in the Chalbi playa floor. A dark grey vesicular basalt occupies an area of several hundred m^2 between the waterholes and the base of the marginal slope.

Along the E edge of Chari Ache there is evidence of relatively recent faulting. Air photographs show a clear lineation that runs for more than 10 km with an alignment varying between N and *c.* 10° E of N. Where it is crossed by ephemeral drainage lines running broadly W–E off Chari Ache, it has caused several of them to be deflected southward for some hundreds of metres.

The rift system lies less than 100 km W of the Chalbi. Swain (pers. comm.) reported that, in the course of gravity surveys across the Chalbi basin in the 1960s and 1970s, he found the area to be seismically active. Loupekine (1971) mentions an earthquake felt at Allia Bay on the E shore of Lake Turkana in 1896, and another (estimated at VII on the Modified Mercalli scale) in the Suguta valley, S of Lake Turkana, in 1924. The Chalbi basin may be, at least in part, a structural basin formed by downfaulting, and owing its shape to the alignments in the basement rocks.

Geomorphology of the Chalbi basin

The geomorphology of the area is shown in Fig. 3. During the course of the present study, particular attention was paid to the margins of the playa, as being the most likely areas to show evidence of former lake levels in the basin. Cross-sections of the playa margins are given in Fig. 6.

The Chalbi playa

Most of the playa surface is composed of brown mud which, when dry, forms a polygonal pattern of cracks. In some places the surface is covered by a white crust of halite (NaCl) and thenardite (Na_2SO_4) (K. Pye, pers. comm. 1984). This crust appears to occur where the groundwater is particularly close to the playa surface, for example, near Karole and Korowe. In some other places the playa surface is loose and powdery, composed of pale grey and pale brown silty clay. No information about the thickness of sediment below the playa is at present available.

The playa margins

In places, as shown on Fig. 3, the Chalbi playa abuts directly onto the margins of the basalt. It is felt that these margins cannot strictly be termed 'lava fronts' since it is apparent from their topography that there has been a certain amount of dissection and erosional retreat since deposition. The lava–playa junction forms quite a spectacular feature, and is illustrated in Fig. 6A. The angle of slope may be over 30°, and the height of the slope >30 m. On most of these slopes, a few metres of *in situ* lava are visible at the top of the slope; the rest of the slope is heavily mantled with coarse, angular lava talus. Within this talus it is possible in places to distinguish fine-grained white to pale grey sediments which apparently underlie the lava. At the slope base there is a sudden transition (within less than 5 m) between the talus-mantled slope and the brown mud of the playa. Scattered fragments of lava extend for only a few metres out onto the playa surface. At one locality close to Gamura, lava boulders on the playa itself show a distinct and horizontal change in the colour of the weathering patina, which appears to indicate a waterline *c.* 30 cm above the playa surface, evidence of the occasional flooding of the playa. The material on the slope and at its base is all angular to subangular in shape. No well-rounded material suggestive of shingle beach or river gravel was seen.

North of Kalacha the lava slope of the Huri Hills is separated from the playa by a belt of sediments that in places is over 5 km wide. Cross-sections through this terrain are shown in Fig. 6B and C. A few km to the S of Kalacha Goda village, the lava of the Huri Hills appears to dip beneath horizontally bedded sediments; a section through these sediments is given in Fig. 7. They are visible in ten little buttes that rise 2–4 m above the surrounding terrain, and with surface areas ranging from a few m^2 to a few hundred m^2. These buttes are at present undergoing rapid erosion. The fine grain size and horizontal bedding of these sediments indicates deposition in a low-energy environment, a lake or pond. This is also indicated by the included gastropod shells, of which *Melanoides tuberculata* is dominant. Diatoms are present in some of the sediments. An analysis of these (currently in progress) should provide further indications of the conditions at the time of deposition. These sediments appear to be overlain by sands that occur in a narrow belt along the playa margins, and are presumably also overlain by the modern playa muds. However, a core through the playa mud is needed to establish whether these lake sediments extend below the modern playa.

North of Kalacha (Fig. 6C) more outcrops of fossil gastropods occur in several places (Fig. 3), but the greater part of this terrain between basalt and playa is composed of ridges and mounds of carbonate rock, which form a rolling terrain of 5–10 m of relief. Despite this relief, no vertical sections are exposed, except in the walls of Laga Kubi Badana, where a thickness of *c.* 5 m of a carbonate rock appears to overlie lava. This rock has been studied in thin section by K. Pye and identified as a calcrete, formed as a result of deep weathering of the underlying basalt. Other similar rock specimens from the terrain N of Kalacha have proved difficult to identify, even following microscopic and X-ray diffraction analyses. According to Pye, some samples show a rounded

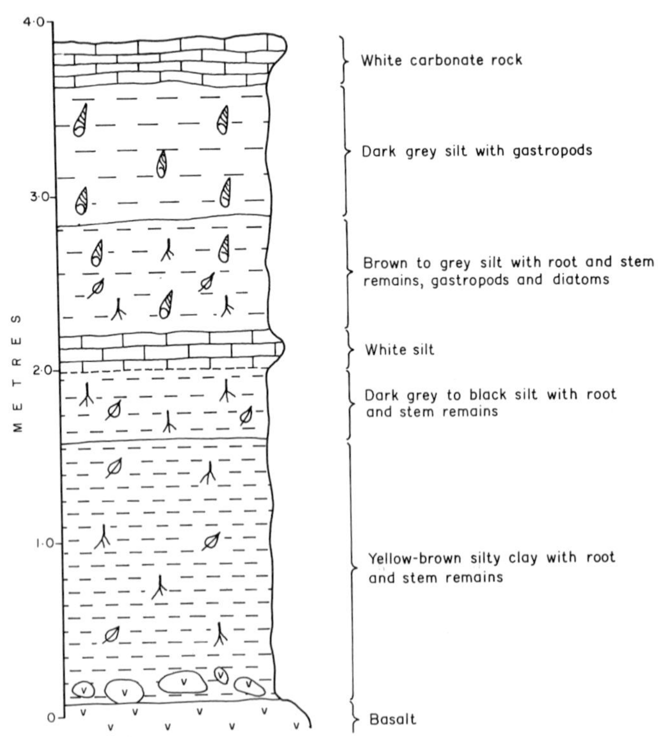

FIG. 7. Cross-section of sediments S of Kalacha Goda.

(pisoid) structure, which may indicate an algal origin, precipitation of the carbonate occurring in a shallow lake or around a spring. There again, two samples from low mounds on the playa margins close to North Horr may be reworked calcareous lake sediments (marls). A search for diatoms within these sediments has been unrewarding. It seems probable, though, that most of these carbonate sediments originated at times when conditions were wetter than at present. It is unlikely, however, that any of these sediments formed in a deep lake environment. According to Pye (pers. comm. 1984) they are more likely to have formed in a series of shallow basin-margin pools of different salinity. This is likely to be particularly true of the sediments at Olturot (see Fig. 3), which lie at an elevation of *c.* 540 m on the SW slopes of Kulal. Here fossil gastropods (again predominantly *Melanoides tuberculata*) occur in pale grey silts which form the top of a sedimentary sequence of *c.* 2–3 m, whose base is not visible. These sediments outcrop in two buttes whose plan dimensions are each *c.* 1 km by 0.5 km.

Comparable sediments occur at a number of other localities, as shown on Fig. 3. North of North Horr there is a surface of low relief that rises very gently northward. Fossil gastropods were found within the sediments that underlie this surface, to a maximum elevation of *c.* 409 m. The surface continues northwards for several tens of km but with no sign of a former lake shoreline. Buttes of horizontally-bedded sediments occur near Laga Ilama, where fossil gastropods are found at an elevation of *c.* 400 m. Sandy terrain (such as around the village of North Horr) separates the old lacustrine sediments from the modern playa.

Other features of local importance on the basin margins are the alluvial fans of the major drainage lines. These fall into two classes: the gravel fans (e.g. of Laga Bulal and Laga Kubi Badana) and sand fans (e.g. of Laga Balesa). The gravel fans, now partially dissected, are recognizable in the field by their gently convex relief and surface of smooth, rounded basalt pebbles. The Laga Bulal fan consists of a series of near-parallel ridges with pebble-covered surfaces, separated by broad flat-floored depressions containing fine-grained brown sediments. Within these depressions, some erosion and transportation of sediments by running water appears to take place, but the greater part of the fan (i.e. the gravel ridges) seems to be fossil, no longer forming under present-day conditions. It is possible that these fans developed during a past, more humid period, when greater amounts of runoff flowed into the Chalbi basin. Frostick & Reid (1980) describe a comparable feature from the NE margins of the Lake Turkana basin, the Koobi Fora Plateau Gravel. They consider this to represent the last significant deposit of overspill from Chew Bahir to the Lake Turkana basin, and date it as late Pleistocene.

Sequence of events

Based on the features described above, the following tentative sequence of events can be outlined for the study area. In the period (pre-Pliocene and Pliocene) before the Chalbi basin came into existence in its present form, there were basins of sedimentation in this area including some which contained relatively deep water supporting a freshwater aquatic fauna. The Chalbi basin probably took its present form during the early Pleistocene through a combination of volcanic and tectonic events. These events included downwarping and downfaulting along an axis aligned approximately NW–SE, and the emplacement of lava along the E and W margins of the basin.

During the late Pleistocene and early Holocene wet phase that is documented for most of Africa, the Chalbi basin contained either a single body or several small bodies of freshwater. A single lake extended to an elevation of *c.* 410 m above sea-level, as indicated by the gastropods at Maikona, Kalacha, Laga Ilama and to the N of North Horr. The gastropods at Olturot lived in a separate, higher body of water ponded up for a time on the SW slope of Kulal. Lake Chalbi would have had shorelines along the lava slopes to the E and W, and extended some distance S across the Koroli sand plain. Its limits to the N and NW are not recognizable. Sand beaches, bars and spits probably formed along parts of the lake shoreline. Remains of these features may be recognizable along the southern and western margins of the playa, though evidence is inconclusive especially given the present erosional and constructional activity of the easterly and southeasterly Trade Winds in shifting sand.

Since the fall of the lake level during the last 9000 years, the sediments have been covered in places by alluvial gravels and by a more ubiquitous fine-grained material that is still being brought down and deposited in the modern playa.

Fossil fauna and flora

Gastropods

Nine collections of fossil gastropods from sediments on the playa margins were identified by M.

Pickford. Some of these were re-examined later by P. Kat. Two other collections (both from Olturot) were identified at the Department of Macropalaeontology, The Netherlands Government Geological Service. In all these collections *Melanoides tuberculata* was dominant, with *Biomphalaria* of secondary importance. Present in much smaller numbers were *Lymnaea*, *Gyraulus*, *Segmentorbis* and some other genera. The indication from these gastropods is of shallow freshwater conditions, and possibly of a perennial spring feeding a series of pools and streams. Living *Melanoides* have not so far been recorded from this area. Two collections of modern gastropods were made from the lower slopes of Chari Ache. According to Pickford (pers. comm. 1984) *Zootecus insularis* was present in both of them; this is a desert terrestrial species, which reflects the extremely arid conditions on the sand ridges where it occurs. *Bloyetia* is adapted to semi-arid conditions typical of much of the Chalbi basin. The third species, *Bellamya unicolor*, is a freshwater gastropod. Its shells were found widely scattered over the brown mud of the playa, at the base of the N–W slope of Chari Ache. The snails presumably flourish and breed rapidly during the short periods when this area is flooded with freshwater. No *Melanoides*, either modern or fossil, were found in the area where the *Bellamya* occur.

Fossil bone

Bone from two sites, Olturot and Algas (see Fig. 3), was identified by M. Pickford. At Olturot, fossil bone is washing out from a slope of pale grey silt, less than 1 m high. The fauna, listed in Table 2, indicates an environment similar to that of the area today. The modern vegetation around Olturot is described by Herlocker (1979) as annual grassland, with a strip of woodland (*Acacia tortilis* with *Duosperma indigofera*) along the course of Laga Balesa a few km E of the site. No inferences about the age of this bone can be made, except to note that these are all modern species.

TABLE 2. *Faunal list for Olturot site*

Equus grevyi	common
Diceros bicornis	1 specimen
Gazella granti	3 or 4 specimens
Gonnochaetes taurinus	3 or 4 specimens
Bovidae spp.	moderately common
Giraffa spp.	1 specimen or more
Phacochoerus aethiopicus	4 specimens

The site at Algas may be that described by Dixey (1948) as the Marsabit Road site, and said to contain 'lacustrine limestones and clays yielding a rich mammalian fauna referred to the uppermost Lower Pleistocene'. Bone collected in the 1940s has been identified; Table 3 summarizes the results of the analysis of this bone as well as that collected in 1983. Pickford (pers. comm. 1984) suggested a date between 1.6 and 2.0 Ma for the fauna, possibly closer to the older limit of this range. K-Ar dating of the lava overlying the bone-bearing sediments yielded an age of 2.5 ± 0.3 Ma. The fauna from the Algas/Marsabit Road site comprises both lacustrine and terrestrial elements. Among the lacustrine elements are *Lates niloticus*, crocodile and hippopotamus. The large size of the *Lates* vertebrae suggests an extensive and deep body of water at or near the site. Terrestrial mammals include bovids and *Equus*, suggesting the presence nearby of open country. It is possible that different strata are yielding discrete aquatic and terrestrial faunas which are becoming mixed by present-day erosional processes in a manner described by Frostick & Reid (1982).

Evidence from just one site does not allow a palaeoenvironmental reconstruction to be carried

TABLE 3. *Faunal list for Marsabit Road/Algas site*

	Year of collection	
	1940s	1983
Lates niloticus	c	c
Clariidae	c	c
Trionyx	p	p
Pelusios	p	p
Crocodylus niloticus	c	c
Euthecodon	—	p
Cercopithecinae	—	p
Elephas recki	p	p
Deinotherium bozasi	p	p
Rhinocerotidae	p	—
Equus oldowayensis	c	c
Mesochoerus limnetes	p	—
Metridiochoerus andrewsi (II)	p	p
Hippopotamus cf. *imaguncula*	c	c
Camelus	p	—
Sivatherium maurusium	c	—
Tragelaphini	p	—
Alcelaphini	p	p
Menelikia lyrocera	c	p

c = common; p = present

out, but the indication is that at the time of deposition of the Algas sediments (predating at least some of the Marsabit volcanicity) there was a large, deep body of water in this area. However it is not possible to say how extensive this body of water was, or how long it existed, or indeed what kind of climatic conditions brought it into existence. The presence of *Lates niloticus* so far to the E is of interest, as it suggests faunal links with the Nile system to the W. By 2.5 Ma it would seem likely that Kulal and Chari Ache were already in existence, as well as the Chalbi–Turkana divide further N. It is possible that the link with the Turkana basin and thereby the Nile was to the S of Mt Kulal.

Fossil wood

Fragments of fossil wood were collected from the slopes of Falama, one of the lava-capped mesas (see Fig. 3). Five pieces have been identified by M. Crawley of the Fossil Plants Section of the British Museum (Natural History). Four of them show similarity with the living genera *Afzelia*, *Dichrostachys*, *Macrolobium* and *Piptadenia*, and are assignable to the fossil genera *Pahudioxylon*, *Dichrostachyoxylon*, *Berlinioxylon* and *Mimosoxylon* respectively. The fifth specimen shows similarities to *Copaifera* and *Guibourtia*, but also to *Chukrasia*, and is tentatively assigned to the *Copaiferoxylon* genus. The age of these specimens is estimated as Cenozoic, more precisely Mio-Pliocene. The fossils are broadly indicative of tropical forests or woodlands, but it is impossible to assign them specific habitats on the evidence of the wood alone.

Although it is necessary to be cautious in drawing inferences from a few small pieces of fossil wood, it is interesting to note that Crawley (pers. comm.) assigns the wood to a significantly earlier date than has been obtained for the Algas fossil bone—Mio-Pliocene as opposed to terminal Pliocene. This tends to support the suggestion that the lavas capping Falama and the other mesas are older than those of Marsabit. The modern vegetation in the area around the mesas is described by Herlocker (1979) as shrubland. The fossil assemblage of tropical forest genotypes indicates much more humid conditions than at present.

Conclusions

A large deep body of water existed near Kargi, on the southwestern margin of the Marsabit shield at *c.* 2 Ma. This was within the confines of a depression that pre-dated the formation of the modern Chalbi basin. Faunal evidence indicates that this depression has been linked with water bodies further to the W. The Chalbi basin owes its existence largely to the outpouring of volcanic rock that built up the basin margins. However, faulting, guided by earlier structural alignments, may have influenced the form of the basin margins.

Conditions in the Chalbi basin were influenced by the general climatic fluctuations of the terminal Pleistocene and the Holocene that affected East Africa. Patches of lake silts containing freshwater gastropods indicate the existence of either a Lake Chalbi whose surface level stood close to 410 m, or of a number of small bodies of freshwater on the basin margins, both of these occurring between 9500 and 11,000 years ago. It is not possible from evidence so far available to recognize later Holocene moist phases known to occur elsewhere in the same broad geographical province.

ACKNOWLEDGEMENTS: This work was supported by Kenyatta University Nairobi, Kenya, and by the UN Environment Programme—Global Environment Monitoring System. The research was carried out under Research Permit No. RDST.13/001/C856/33 of the government of Kenya. I am very grateful to Martin Pickford for reading a first draft of this paper, and for analysing the gastropods and fossil bone. Tim Charsley and Roger Key commented critically on the final draft of the paper, and gave me access to the results of their own work in this area. Adrian Lewis identified some of the volcanic rocks, and Ken Pye some of the sediments. Sarah Metcalfe helped me with the analysis of diatomaceous sediments, and Arie van Kekem gave me the report on the gastropods studied at the Department of Macro-palaeontology, Netherlands Government Geological Service. Gastropods were also identified by Pieter Kat. Mark Crawley identified the fossil wood. Help with equipment and advice in connection with the topographic survey work was given by Peter Opie-Smith and Bob Caukwell. I am very grateful to all of the above for their help, and also to the many other people, in Marsabit, Nairobi and Great Britain who contributed towards the completion of this project.

Appendix

Potassium-argon age determination was carried out by Geochron Laboratories, sample number R-6920. The material submitted was whole rock (basalt lava, my sample number 301283–0) which was treated with dilute HF and HNO_3. The results of the analyses are as follows:

Radiogenic ^{40}Ar, ppm	Radiogenic ^{40}Ar/total ^{40}Ar	Ave. radiogenic ^{40}Ar, ppm
0.000089	0.105	0.000077
0.000065	0.092	—

%K	Ave. %K	^{40}K, ppm
0.456	—	—
0.449	0.453	0.540

Radiogenic $^{40}Ar/^{40}K = 0.000143$
Age $= 2.5 \pm 0.3$ Ma.

Constants used:
$\lambda_\beta = 4.962 \times 10^{-10}\ a^{-1}$
$(\lambda_\varepsilon + \lambda_\varepsilon') = 0.581 \times 10^{-10}\ a^{-1}$
$^{40}K/K = 1.193 \times 10^{-4} g\ g^{-1}$

Radiocarbon age determinations were carried out by Geochron Laboratories, sample numbers GX-9765, 9790, 9791, 9792. The material submitted was freshwater gastropod shells which were ultrasonically cleaned and then etched with dilute HCl to remove altered surficial material. The shells were hydrolysed with HCl under vacuum and the carbon dioxide recovered for analysis. The dates are ^{13}C corrected and are based on the Libby half-life (5570 years) for ^{14}C. The age is referenced to the year AD 1950. Values of $\delta^{13}C_{PDB}$ are:
GX-9765 = −6.4‰
GX-9790 = −4.8‰
GX-9791 = −2.4‰
GX-9792 = −4.0‰.

References

Bake, G. 1982. *Abiotic components of the IPAL study area in Northern Kenya.* IPAL unpublished report, UNESCO, Nairobi.

—— 1983. An analysis of climatological data from the Marsabit District of Northern Kenya. *IPAL Technical Rep.* **B3,** UNESCO, Nairobi, 89 pp.

Baker, B.H. 1963. Geology of the Baragoi area. *Rep. Geol. Surv. Kenya* **53,** 74 pp.

Bowen, B.E. 1974. *The geology of the Upper Cenozoic sediments in the East Rudolf embayment of the Lake Rudolf basin, Kenya.* Unpublished PhD thesis, Iowa State University, Ames.

Brotzu, P. Morbidelli, L., Nicoletti, M., Piccirillo, E.M. & Traversa, G. 1984. Miocene to Quaternary volcanism in eastern Kenya: sequence and geochronology. *Tectonophysics* **101,** 75–86.

Dixey, F. 1948. Geology of Northern Kenya. *Rep. Geol. Surv. Kenya* **15,** 43 pp.

Dodson, R.G. 1963. Geology of the South Horr area. *Rep. Geol. Surv. Kenya* **60,** 53 pp.

—— 1971. Geology of the area south of Lodwar. *Rep. Geol. Surv. Kenya* **87.**

—— & Matheson, F.J. 1969. Geological map of the area NE of Marsabit (Map). *Rep. Geol. Surv. Kenya* **94** (map only).

Edwards, K.A., Field, C.R. & Hogg, I.G.G. 1979. A preliminary analysis of climatological data from the Marsabit District of Northern Kenya. *IPAL Technical Rep.* **B1,** UNESCO, Nairobi, 44 pp.

Frostick, L.E. & Reid, I. 1980. Sorting mechanisms in coarse-grained alluvial sediments: fresh evidence from a basalt plateau gravel, Kenya. *J. Geol. Soc.* **137,** 431–41.

—— & —— 1982. Taphonomic significance of subaerial transport of vertebrate fossils on steep semi-arid slopes. *Lethaia* **16,** 157–64.

Fuchs, V.E. 1939. The geological history of the Lake Rudolf basin, Kenya Colony. *Phil. Trans. R. Soc. Lond.* **229B,** 219–74.

Herlocker, D. 1979. Vegetation of southwestern Marsabit District, Kenya. *IPAL Technical Rep.* **D1,** UNESCO, Nairobi, 68 pp.

Joubert, P. 1966. Geology of the Loperot area. *Rep. Geol. Surv. Kenya* **74,** 52 pp.

Loupekine, I.S. 1971. *A Catalogue of Felt Earthquakes in Kenya.* Seismology Unit, Geol. Dept, University of Nairobi, Kenya.

Ongweny, G.S., Baur, P.J., Borgeson, R.H. & Ndombi, J.M. 1983. The hydrology of southern-west Marsabit District, Kenya (Consultancy reports). *IPAL Technical Rep.* **B2,** UNESCO, Nairobi, 359 pp.

Owen, R.B., Barthelme, J.W., Renaut, R.W. & Vincens, A. 1982. Palaeolimnology and archaeology of Holocene deposits northeast of Lake Turkana, Kenya. *Nature* **298,** 523–9.

Randel, R.P. 1970. Geology of the Laisamis area. *Rep. Geol. Surv. Kenya* **84,** 24 pp.

Sinda, P.M. 1981. *Provisional Geomorphological Map of Southwestern Marsabit District, Northern Kenya.* UNESCO, Nairobi.

Vondra, C.F. & Bowen, B.E. 1978. Stratigraphy, sedimentary facies and paleoenvironments, East Lake Turkana, Kenya. In: Bishop, W.W. (ed) *Geological Background to Fossil Man.* Scottish Academic Press, Edinburgh, pp.395–414.

Walsh, J. & Dodson, R.G. 1969. Geology of Northern Turkana. *Rep. Geol. Surv. Kenya* **82,** 42 pp.

Watkins, R.T. 1982. *The geology of the Suregei-Asille District and the Upper Bakate valley, Nothern Kenya.* Unpublished PhD thesis, Birkbeck College, University of London.

C.K. Nyamweru, Dept of Geography, Kenyatta University, PO Box 43844, Nairobi, Kenya.

Sedimentology, stratigraphy and palaeoenvironments of the Holocene Galana Boi Formation, NE Lake Turkana, Kenya

R.B. Owen & R.W. Renaut

SUMMARY: Lake Turkana lies in a closed drainage basin at the northern end of the Kenya Rift Valley. During the early and middle Holocene the lake was fresh, much expanded and subject to many fluctuations in surface level. Periodically it formed one linkage in a greater Nile drainage network. Late Holocene, climatically-induced regression left a ring of sediments stranded up to 80 m above the modern lake. To the NE, these deposits are known as the Galana Boi Beds and here we propose raising them to formation status. They consist of diatomaceous silts, silty diatomites, quartz-arenites, arkoses, litharenites, coquinas and at least one air-fall tuff. Depositional environments ranged from low-energy offshore, through littoral, lagoonal and deltaic to alluvial and aeolian. A rich, diverse microfossil assemblage of diatoms, pollen and ostracods has been recorded and used in palaeoenvironmental analysis. Macrofossils include molluscs, fish debris and other vertebrates. With their excellent outcrop and diverse environments represented, the Galana Boi Formation can serve as a model for sedimentation of the margins of a large rift lake, and aid in the interpretation of similar, but poorly exposed graben sequences in the geological record.

Some of the most distinctive lacustrine sediments in the Kenya Rift Valley are light grey to pure white diatomaceous silts and diatomites. They occur throughout the late Cenozoic geological record from the Miocene of the Baringo District to the Holocene of many of the rift lakes (Owen 1981) and are commonly associated with periods of relatively high lake level. Studies of the diatom floras can often provide a wealth of palaeoenvironmental information (e.g. Richardson & Richardson 1972; Owen & Renaut 1981; Gasse *et al.* 1983).

Diatomaceous silts and associated lake margin sands of Holocene age are well exposed around the margins of Lake Turkana, Kenya, recording former higher lake levels. We describe here their stratigraphy and sedimentology in the Koobi Fora region of northeastern Lake Turkana. Aided by diatom analysis, we interpret the palaeogeography of the region during the early Holocene lake maxima. Detailed discussion of radiocarbon dates and Holocene lake level fluctuations are presented elsewhere (Owen *et al.* 1982). Full details of field and analytical methods are given in Owen (1981).

General setting

Modern limnology

Lake Turkana is situated in the semi-arid region of NW Kenya (Fig. 1). With an average depth of *c.* 35 m (maximum: 125 m) and a surface area of 7500 km^2, it ranks as the fourth largest natural lake and largest 'closed-basin' lake in Africa (Yuretich 1979). More than 80% of the water originates from the Omo River (Fig. 1), much of which is supplied during the July flood season when rains fall in the Ethiopian Highlands. The remaining inflow derives from the Turkwel and Kerio Rivers, draining high ground to the SW. Very little direct precipitation falls on to the lake, and all water input is approximately balanced by evaporation, the surface level lying at *c.* 375 m a.s.l.

The lake is mildly saline with total dissolved solids of 2500 ppm (principal ions: Na^+, HCO_3^- and Cl^-); carbonate alkalinity ranges from 17 to 23 mEq l^{-1} and the pH is 9.2 (Yuretich & Cerling 1983). In the N, salinity is seasonally reduced through mixing with dilute Omo floodwaters.

Regional geology

Lake Turkana lies to the W of the main Eastern Rift System within a broad half-graben. The main line of the Kenya Rift cuts through and bounds the S end of the lake. The present structural pattern became defined in the Miocene, though the structural evolution can be traced back to the late Cretaceous. Miocene volcanism produced extensive basic and acidic rocks. During the early Pliocene, major faulting cut the depression into a series of N–S trending graben, half-graben and horsts (Vondra & Bowen 1976; Cerling & Powers 1977).

Plio-Pleistocene units crop out sporadically around the lake margins. To the NE, such sediments were deposited in and on the margins of a former large embayment. Up to 325 m of fluvial, deltaic and lacustrine sediments were laid down over an area of 5000–6000 km^2 (Vondra &

From FROSTICK, L.E. *et al.* (eds) 1986, *Sedimentation in the African Rifts*, Geological Society Special Publication No. 25, pp. 311–22.

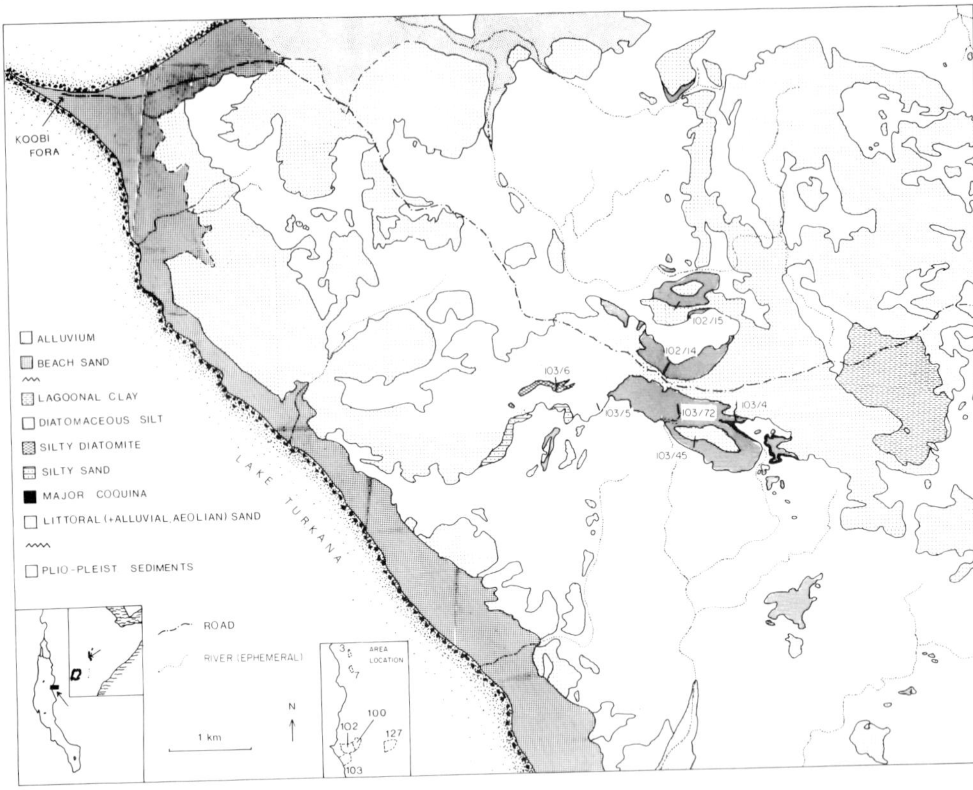

FIG. 1. Distribution of Holocene sediments E of Koobi Fora.

Bowen 1978; White *et al.* 1981; Brown & Cerling 1982).

Previous studies of the Holocene sediments

Investigations of Holocene deposits have been slight in comparison with the Plio-Pleistocene units, despite their excellent exposure. Lake Turkana is ringed by Holocene sediments. The lacustrine units occur up to *c.* 80 m above the modern lake—the level of an overflow linkage to the Nile drainage system (Harvey & Grove 1982). Hemming & Trapnell (1957), Walsh & Dodson (1969), and Dodson (1971) have made observations W of the lake, Robbins (1980) to the SW, and Phillipson (1977) to the SE. Butzer (1971, 1980) has provided valuable data for the Omo region to the N.

The Holocene sediments of NE Lake Turkana were first described briefly by Vondra *et al.* (1971) and designated 'Galana Boi Beds' by Bowen & Vondra (1973). Raynolds (1973) described their occurrence E of Koobi Fora. More extensive studies of the beach-ridges were conducted by Barthelme (1977, 1981), essentially in an archaeological context. Frostick & Reid (1980) have described beaches to the S of the Koobi Fora Ridge. Owen (1981) and Owen *et al.* (1982) have discussed aspects of the stratigraphy, chronology and micropalaeontology.

Stratigraphy and dating

Definition

The exposed Holocene deposits of NE Lake Turkana have been informally referred to as the Galana Boi Beds. Here, we suggest that they be elevated to formation status, in view of their distinctive lithological features. However, as no single type-section can be assigned that adequately documents the entire spectrum of lithologies present, a series of type-sections is presented to illustrate the Holocene lithostratigraphy of the region from near Ileret in the N to Allia Bay in the S (Fig. 2). Isaac (1978) encountered similar problems in defining the predominantly lacustrine Olorgesailie Formation in southern Kenya.

Sediment relationships and chronology

Fig. 2 shows a series of type-sections that illustrate the Holocene stratigraphy of the region. Sections 103/4, 5, 6 and 11 as well as section 102/14 all lie E of Koobi Fora. There, extensive deposits have been preserved and deeply dissected to provide excellent exposures.

Fluctuating lake levels during sedimentation produced complex, interdigitating units with primary dips of up to 15°. The Galana Boi Formation has a maximum aggregate thickness of *c.* 32 m in Area 103 (Fig. 1 inset) (Bowen 1974). However, throughout most of its outcrop it is less than 10 m thick. The sediments are mostly unaffected by tectonics, but are locally uplifted near the Kokoi Uplands (Fig. 2).

The succession commences with a sublitharenite that rests on a deeply eroded surface (sections 103/4 and 11, Fig. 2). This widespread unit, up to 3 m thick, is locally rich in pebbles and in places shows small channel-fill sequences. It records a late Pleistocene to early Holocene phase of fluvial and colluvial activity when lake levels were low.

Diatomaceous silts succeed the basal sands sharply and signify a lacustrine transgression. In section 103/4 (Fig. 2), they are *c.* 13 m thick, but locally exceed 20 m. Fish debris, ostracods and rootmarks occur at many levels. Molluscs are scattered throughout the diatomaceous silts and are dominated by *Melanoides tuberculata*, *Corbicula africana*, *Cleopatra pirothii* and *Bellamya unicolor*. Molluscs are also present in thin (< 10 cm) laterally extensive concentrations. One such coquina of *Corbicula africana* near the base of section 103/4 has been dated at 9880 ± 670 BP and may represent a storm accumulation formed at heights of 50–55 m above the modern lake.

Pollen collected from below the coquina (Fig. 2) are dominated by montane and thicket taxa. These are rare in the lake today (Bonnefille & Vincens 1977) and contrast with the Graminae-dominated spectra of younger sediments. This flora may reflect former overflow linkages from

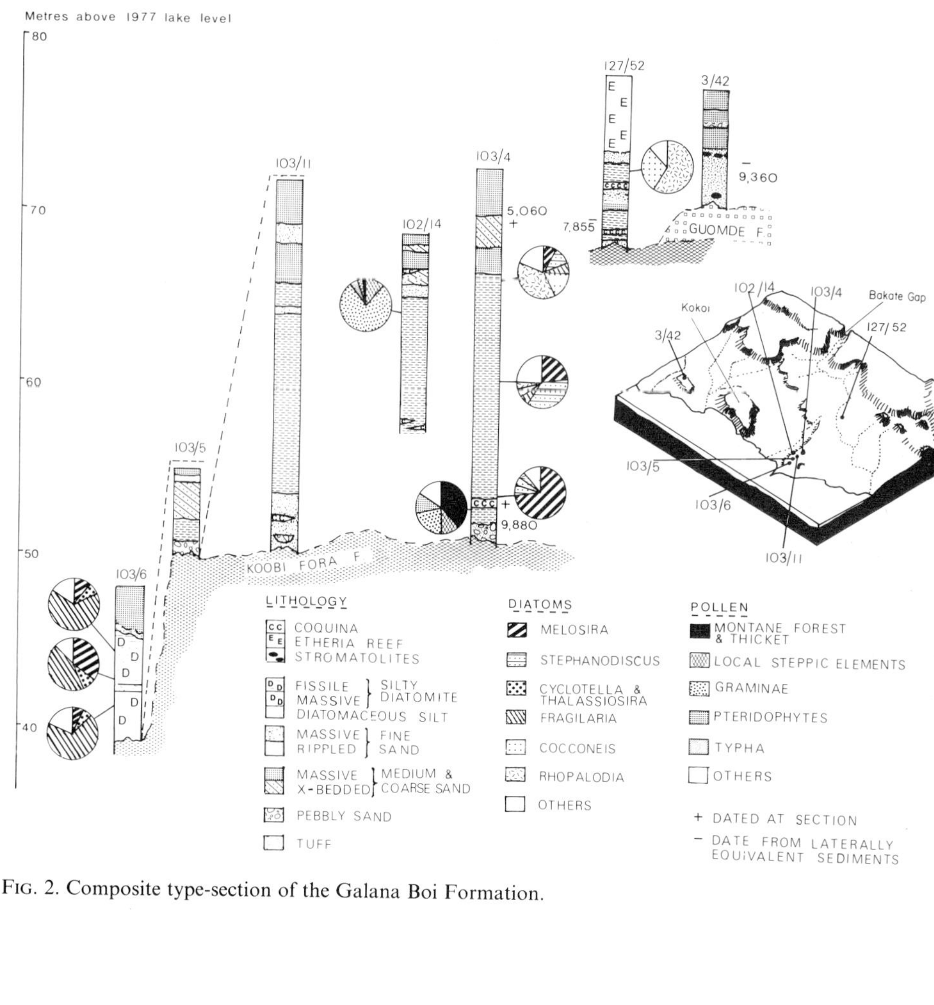

FIG. 2. Composite type-section of the Galana Boi Formation.

the Ethiopian Highlands. The diatom flora of the silts suggests many lake level fluctuations. However, a broad pattern is evident. An early Holocene high lake phase was dominated by *Melosira* and *Stephanodiscus*. This was followed by intermediate and high lake levels, reflected by the varying percentages of benthonic types such as *Rhopalodia* and *Cocconeis*. While several minor disconformities occur, evidence for a major fall in lake level during the timespan represented by the silts is lacking.

Subarkosic and quartz sands with locally developed tabular cross stratification and climbing ripples rest unconformably on the diatomaceous silts. Mollusc shells from the sands have given a date of 5060 ± 245 BP at section 103/4. These capping sands become younger to the W, where dates between 4580 and 3945 BP have been obtained at various levels between 43 and 47 m above lake level (Owen *et al.* 1982). These arenites probably represent a diachronous series of coalesced, regressive lake margin sand bodies. Minor transgressions within the sand sequence are indicated by localized interbedded diatomaceous silts (e.g. section 103/5, Fig. 2).

Although undated by ^{14}C, section 103/6 (Fig. 2) post-dates the deposits described above. The sequence consists of a lower massive diatomite and an upper fissile silty diatomite, separated by an 8 cm thick, white vitric tuff. One possible source for the ash is North Island (Fig. 1), where trachytic lavas overstep wave-cut benches of probably Holocene age. Eroded lithoclasts of silty diatomite occur in sections to the SW, where the deposits are silty sands and sands, commonly with ostracods and fish bones. When traced W towards the lake, they pass into linear beach sands and lagoonal silts laid down in the latest Holocene and most recent times.

Although the sediments are best preserved E of Koobi Fora, evidence has been obtained from elsewhere within the region. South of the Kokoi Uplands (Fig. 2), dates of 9540 ± 260 and 9260 ± 235 BP have been obtained from molluscs in high-level spits. In the same region, mammal bones from littoral sands are dated at 8395 ± 270 and 8355 ± 235 BP. Near Ileret to the N, beach sands at +70–80 m have given a shell date of 9360 ± 135 BP. Locally they are rich in stromatolites (section 3/42, Fig. 2). Subarkosic sands, arkoses, diatomaceous silts, silty diatomites and claystones in area 127 (Fig. 1) represent littoral and lagoonal palaeoenvironments. Shells from these units have given ages of 8520 ± 130 and 8710 ± 130 BP. Locally these deposits are overlain by bioherms of *Etheria elliptica* (section 127/52, Fig. 2).

Facies and environments

Nine facies of varying importance can be recognized within the Galana Boi Formation. Their complex distribution and interfingering relationships reflect the numerous transgression/regression cycles and consequent lateral migration of depositional environments experienced by the palaeolake. The main sedimentary features that characterize each facies are summarized in Table 1.

Cross-stratified subarkose/sublitharenite/quartz arenite facies

This lithofacies typically forms linear bodies < 2 m thick. Cross-stratification is common and consists mostly of low-angle tabular cross-sets. Well washed, medium and coarse-grained subarkoses and sublitharenites predominate; heavy mineral bands are commonly well developed. Calcified root tubules and fish-bone debris are locally present. Molluscs are common and include *Mutela emini*, *Etheria elliptica*, *Pila ovata* and *Melanoides tuberculata*. Microfossils are generally absent. Very similar deposits (though lacking modern molluscs) are forming today at Koobi Fora spit. This facies is thus interpreted as representing a beach or beach bar environment which formed along an exposed shoreline.

Structureless sublitharenite/subarkose facies

This facies is characterized by sheet-like bodies of well sorted fine and medium sands. Although appearing mostly massive, ripple laminae and horizontal bedding are found. Sublitharenites and subarkoses are abundant, although arkosic arenites also occur. Calcified root tubules, ostracods, fish debris and molluscs are all present to varying degrees. Some diatoms occur in the finer units. From the sedimentological characteristics and location relative to other better defined facies, the environment of deposition was probably sub-shoreface to proximal offshore, adjacent to a high wave-energy shore.

Litharenite/arkosic arenite facies

This heterogeneous sediment type encompasses a range of lithologies generally exhibiting poor sorting. Litharenites and arkosic arenites of fine to coarse grade predominate. Individual units are < 2 m thick and stratification is poorly developed, if at all recognizable. Calcareous concretions, rhizoliths, root-marked palaeosols, molluscs, fish and other vertebrates are variously

TABLE 1. *Summary of the principal sedimentological and palaeontological features that characterize the main depositional environments of the Galana Boi Formation*

	Lacustrine	Beach shoreface	Lagoon	Beach bar	Channel	Floodplain
Gross features						
Major lithology	Silt and clayey silt	Sand	Silt, clay	Sand	Sand, gravel	Silt
Sorting	Good	Good	Good	Good	Poor	Moderate
Bed form geometry	Extensive sheets	Extensive sheets	Ribbon; pods	Linear; shoestring	Linear; shoestring	Sheets
Bedding						
Massive	+	+	+	+	+	+
Horizontal	+	+	+	–	+	+
Cross-stratification	–	RT	R	P	RPT	RT
Horizontal lamination	+	+/–	+	–	–	+
Other features						
Mudcracks	–	–	+	–	–	++
Mud clasts	–	–	–	–	+	+?
Root casts	–	++	++	+	+	++
Limonitic						
Ca concretions	+	–	–	–	–	–
Palaeosols	–	–	–	–	–	+
Calcrete; soil-$CaCO_3$	–	–	–	–	–	+
Bioturbation	+/–	+	+	+	+	+
Diatoms	++	+	+	–	–	–
Ostracods	+	+	–	–	–	–
Oncolites	–	+	–	–	–	–
Molluscs	+/–	+	+	+	–	+
Fish bones	++	++	+	+	–	+

++ = abundant; + = present; +/– = uncommon, locally present; – = absent; T = trough cross-stratification; P = planar cross-stratification; R = climbing ripple stratification.

represented. This lithofacies reflects a series of environments ranging from distal alluvial and delta plain settings, to littoral regions adjacent to river mouths. In general, lower energies and less exposed settings are envisaged than for facies 1 and 2, except where sediment input was sufficient to overcome wave sorting.

Diatomaceous silt and fine arenite facies

Diatomaceous silts and fine sands are very common. The sediments often show well-developed, horizontal lamination (laminae up to 1 cm thick) and are locally fissile. A maximum thickness of 20 m has been recorded. Diatom content ranges up to 10^8 individuals g^{-1} of sediment. Nitratine ($NaNO_3$) is common locally, as are calcareous, limonite-stained siltstone nodules. Carbonaceous root marks, molluscs, fish bones, pollen, sponge spicules and ostracods are all widespread. These sediments were laid down under low-energy conditions. The water depth varied, but from the difference in height between the highest recorded lake level and the underlying topography, a maximum water depth of *c.* 35 m can be demonstrated within this area.

Silty diatomite facies

Large numbers of benthonic diatoms are found together with some mineral silt in this facies. In several outcrops it is calcareous, laminated and/or fissile. Locally it is massive, containing abundant rootmarks. An environment with fairly low clastic input that was shallow enough for vegetation periodically to take root is indicated. Examples might include lagoons or protected embayments.

Black shale facies

This rare facies is restricted to Area 127 (Fig. 1). There, black clays consisting mostly of smectite (*c.* 85%) are mixed with fine to medium feldspathic silts. The deposits are generally thin (4–5 cm) sheets, although one unit reaches 1 m in thickness. Fossils are absent; some lamination is present. These shales were laid down in restricted lagoonal environments, given their association with littoral sediments. The absence of fauna and the colour suggest anoxicity.

Etheria 'reef' facies

Highly localized outcrops of this biogenic facies occur in Area 127. It consists of high concentrations of *Etheria elliptica* in biohermal units up to 4 m thick, tightly cemented by calcite. Clastic material is almost entirely absent. This freshwater mollusc is found today in high-energy environments, and has been reported from interdistributary channels of the Omo delta (Butzer 1971).

Coquina facies

Molluscs are commonly concentrated in lensoid bodies less than 1 m wide and only 4–5 cm high. However, there are also locally extensive coquinas traceable laterally for up to 400 m. The molluscs mostly occur in silt or poorly sorted fine to medium sand matrices. Three common types of coquina occur: (1) *Melanoides tuberculata*-dominated; (2) *Corbicula africana*-dominated; and (3) *M. tuberculata/C. africana* co-dominated. A wide variety of other species also tend to occur, including *Mutela emini*, *Cleopatra* spp., *Caelatura* sp. and *Bellamya unicolor*.

Stromatolite facies

This facies is best developed in the extreme N, around the Chari Ridge. There, sub-spheroidal stromatolites (oncolites) have nuclei of molluscs and rock fragments. Algal carbonate cortices range from 1 cm to > 50 cm. They commonly rest on a planar erosion surface cut into arkosic arenites, and range from small lensoid beds to laterally extensive horizons traceable for > 100 m (Abell *et al.* 1982). The stromatolites probably formed under fairly high-energy conditions in littoral zones of the palaeolake.

Mineralogy and petrology

Excluding diatoms, which may form up to 75% by volume, feldspars are the dominant constituents of the silts. Albite forms up to 50%, with microcline and orthoclase also common. Quartz accounts for a maximum of 25% of the mineral fraction. Smectite, showing variable degrees of crystallinity on X-ray diffractograms, is the only common clay mineral. Biotite and muscovite are locally common. Other minerals in the finest fractions include: chlorite, illite, illite-smectite, hornblende and calcite (comminuted shell debris).

The sands are variously dominated by detrital plagioclase (mainly albite), orthoclase, microcline and quartz (of both volcanic and metamorphic affinities). The quartz ranges from angular to subrounded, the latter commonly reflecting secondary overgrowths of silica. Lithic fragments are locally abundant and are predominantly lava fragments and reworked sandstones. Hornblende, augite, hypersthene, chlorite, biotite, muscovite, magnetite and ilmenite are all common. The heavy minerals are commonly concentrated in distinct bands by wave action. Less common are sphene, garnet, zircon, haematite, diopside, and alkali pyroxenes and amphiboles. The provenance of heavy minerals in the region is discussed by Mathisen & Vondra (1983).

Although most sands are poorly lithified, three early cements have been recognized: secondary quartz overgrowths, calcite and iron oxides (haematite, goethite).

Calcareous concretions up to 10 cm in diameter occur (commonly in layers) within the silts and fine sands. They are normally limonite-stained. Molluscs commonly form the nucleus of the concretion. Thin sections reveal fine-grained calcite cement. Limonite has been reported from the modern Omo delta by Butzer (1971) who observed that concretions might develop given longer periods of stability.

Authigenic gypsum is present on several fracture surfaces and permeability barriers within the silts and sands, and may originate from oxidation of pyrite within older sediments (cf. Cerling 1979). Nitratine is widely distributed through the silts of Area 103 (up to 8%; Owen, 1981). It decreases in fine sands and is absent from the coarser units. Although the source of the interstitial nitrate has not been determined it is thought to be a secondary product, and testifies to the aridity of the present climate.

Syndepositional precipitated carbonate appears to be rare, in contrast to the modern lake (Yuretich 1979). Calcite and aragonite found in the silts mostly originates from fragmented shells. Secondary deposits occur in various nodular forms, as micritic and sparite cements, and as several types of calcrete. Traces of dolomite have been found in some cements from Area 103 sands.

Diatom assemblages

The Galana Boi Formation contains an abundant and diverse diatom flora. Some 125 species and varieties have been recognized (Owen 1981). However, this diversity can be condensed into a series of distinctive planktonic and littoral assemblages that form the basis for palaeoenvironmental interpretation.

Planktonic assemblages

Assemblage 1: Melosira granulata

M. granulata dominates (up to 95%), although *M. agassizi* may contribute 0–50%. Several intermediate forms occur, although separation of the two 'end-member' species was attempted. *Stephanodiscus minutula* and *S. rotula* constitute a maximum of 10%. Benthonic taxa are rare. The dominant is a planktonic form commonly found in moderately eutrophic lakes of East Africa. Usually, alkalinities are <8 mEq^{-1} and pH is <9. Talling & Talling (1965) observed that *Melosira* spp. occur in waters with conductivities of <600 $\mu S\ cm^{-1}$.

Assemblage 2: Melosira agassizi

M. agassizi is dominant (up to 80%). *M. granulata* is also present together with *Stephanodiscus rotula* and *S. minutula*. Benthonic taxa are rare (<5%). *M. agassizi* occurs today in Shiwa Ngandu—a very dilute water body (Richardson 1968).

Assemblage 3: Melosira nyassensis var. victoriae

The dominant taxon is *M. nyassensis* var. *victoriae* (50–70%). This flora is much less common than the latter two, only occurring in sediments of mid-Holocene age. *S. minutula* occurs also (<10%). This association has not previously been observed in other deposits in Kenya, although similar floras occur in modern Lake Malawi and its Holocene sediments.

Assemblage 4: Stephanodiscus spp.

In this common assemblage, *S. minutula* contributes up to 90% of the flora, with *S. rotula* varying from 0–50%. *Melosira agassizi* is usually present. Benthonic species occur at their lowest frequencies suggesting an offshore, relatively deep water setting.

Assemblage 5: Melosira/Synedra

M. agassizi var. *malayensis* and *S. dorsiventralis* co-dominate (forming *c.* 50% of the total). Other diatoms include *Rhopalodia vermicularis*, *R. gracilis* and *Cocconeis placentula*. The presence of the latter species suggests somewhat shallower water than those represented by assemblages 1–4. This flora is uncommon. These diatoms are not reported as a dominant association in Kenyan lake sediments, although floras dominated by *M. granulata* and *S. acus* are known.

Assemblage 6: Nitzschia palea var. tenuirostris

This highly restricted assemblage occurs in only two thin layers from one section. *N. palea* var. *tenuirostris* dominates (*c.* 50%) and is associated with *N. fonticola*, *Stephanodiscus* spp. and *Melosira* spp. The dominant is often cited from littoral habitats, but here occurs with a number of planktonic taxa. Interpreting water depth is therefore uncertain.

Littoral floras

Assemblage 7: Rhopalodia/Cocconeis/Synedra

This assemblage is dominated by *R. vermicularis*, *C. placentula* and *S. ulna*, in varying percentages. It is extremely common and perhaps should be further subdivided. However, as a group it represents shallow water conditions. Other species associated include *R. gracilis*, *R. hirundiformis*, *Navicula pupula*, *Cymbella* spp., *Epithemia* spp and *Gomphonema* spp. Planktonic taxa are rare (<5%). Floral changes within this assemblage occur rapidly in both vertical and lateral analyses of sections. The assemblage represents a complex intermingling of benthonic microhabitats from a simple silty substrate to epiphytic, reed-dominated settings. Equivalent floras are common in other Kenyan lacustrine sediments.

Assemblage 8: Fragilaria spp.

This assemblage is co-dominated by *F. construens* and its varieties *venter* and *triundulata*, *F. pinnata*, *F. brevistriata* and *F. lapponica*. Together they account for 50–60% of the total diatoms. Also common are *Nitzschia* spp., *Navicula* spp. and *Rhopalodia* spp. *Fragilaria* commonly occur in both planktonic and littoral settings in tropical lakes. However, the association with several benthonic species suggests shallow water. Rare *Thalassiosira rudolfii* and *Cyclotella meneghiniana* also occur. They are present in the modern lake and may indicate slightly more alkaline conditions than those represented by assemblages 1–7.

Diatom stratigraphy

Over the last decade, many cores have been taken from the central, usually deeper parts of the East African lakes (e.g. Richardson & Richardson 1972; Harvey 1976; Holdship 1976). These have demonstrated an apparently simple succession of diatom floras, with perhaps only two or three major species. This reflects the siting of the cores. Only the major environmental changes will affect the flora. This simple picture breaks down in littoral regions where even minor lake level fluctuations can exert a major impact on floral composition.

Littoral complexity is well illustrated by the 10 km^2 region of Holocene diatomaceous silts to the SE of Koobi Fora (Fig. 1). There, a series of detailed diatom analyses have shown that numerous minor lake level changes were superimposed on major fluctuations (Owen 1981). Seasonal oscillations may also be reflected by diatom-rich/diatom-poor laminae in parts of some outcrops.

A detailed study has been published for section 103/72 (Owen *et al.* 1982). This section, located in Fig. 1, suggests two main high-level phases during the early and mid-Holocene. These were followed by a regression, interrupted by brief transgressions during the late stages of the mid-Holocene.

A similar story emerges from section 103/45 (Fig. 3), which contains the thickest accumulation of diatomaceous silts in the area. Although undated, lithological correlation suggests that it too represents early and mid-Holocene sedimentation. A basal litharenite that lacks diatoms indicates fluvial/colluvial environments. The succeeding 3–8 m of silts are dominated by planktonic diatoms (assemblages 1, 2 and 4). Periodic increases in littoral taxa suggest that the levels were not stable. Only fragmentary diatoms occur between 8 and 9 m, probably due to low levels and reworking. A second high stage (at the section site) is represented between 9 and 14 m, with a recurrence of assemblages 1, 2 and 4. Again minor oscillations are probable. From 14 to 18.5 m, assemblage 7 is dominant indicating shallow conditions. However, at least one return to deep waters can be inferred from the presence of assemblage 3 at 15 m. Diatoms are absent from

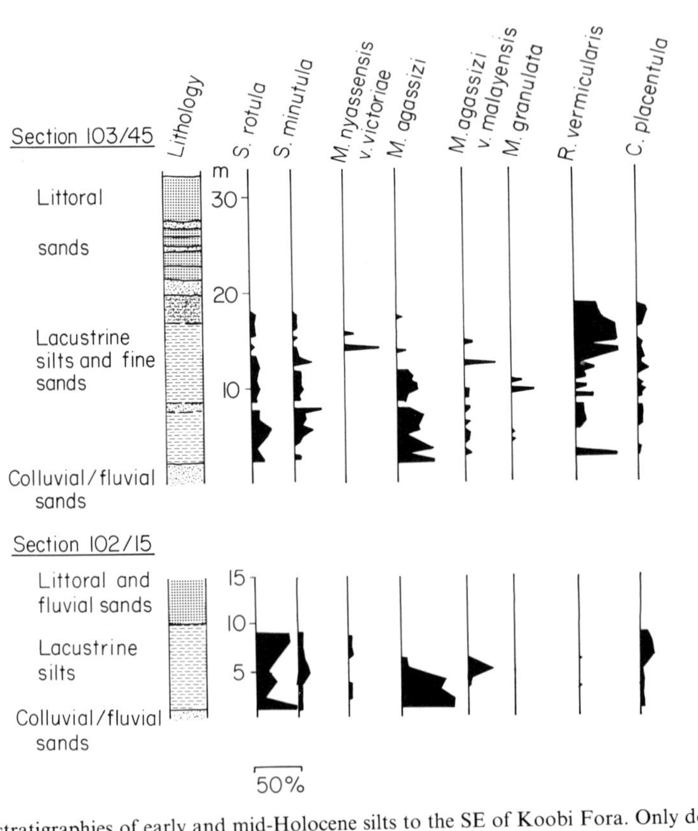

FIG. 3. Diatom stratigraphies of early and mid-Holocene silts to the SE of Koobi Fora. Only dominant species are shown (31 taxa in section 102/15; 40 in section 103/45).

the succeeding arenites which are marginal lacustrine in origin.

This broad pattern is not represented everywhere. Section 102/15 (Fig. 4) is isolated and difficult to correlate, although it is thought to encompass at least the early Holocene. Planktonic *Melosira* and *Stephanodiscus* dominate throughout, but with an increase in the littoral diatom *Cocconeis placentula* above 5 m. Sampling intervals are coarser (1 m) compared to sections 103/45 (40 cm) and 103/72 (10 cm), resulting in a somewhat smoothed appearance to the percentage curves.

The youngest diatomaceous sequence examined occurs at section 103/6 (Fig. 4) and is of late Holocene age. This is dominated by *Fragilaria* spp. (assemblage 7). An interesting feature is the temporary increase in *Melosira granulata* above an ash fall horizon. Perhaps the diatom bloom was influenced by chemical changes in the aftermath of the eruption (e.g. increased dissolved SiO_2, changes in nutrient status, elimination of competitors, etc.). This sequence is the only one to contain *Thalassiosira rudolfii* and *Cyclotella meneghiniana*—diatoms present in the modern alkaline lake.

Early Holocene depositional environments and regional setting

Depositional environments E of Koobi Fora

A full discussion of Holocene lake level changes has been presented elsewhere (Owen *et al.* 1982). Here we concentrate on the early Holocene (10,000 to *c.* 7000 BP) and on the distribution of facies at that time. Fig. 5 shows the inferred patterns of sedimentation E of Koobi Fora for the early Holocene lake maxima. The dominant controlling feature was the Koobi Fora Ridge, around which several depositional sectors can be recognized.

To the N, micaceous sands of facies 1 and 3 probably reflect local sediment input from eroding scarps, and reworking into beaches. Two major coquinas developed and silts were deposited in sheltered, flooded valleys.

Another major coquina formed at the western terminus of the ridge at *c.* 9800 BP, possibly due to storm concentration. However, transgression followed and resulted in a long period of deeper waters. Diatomaceous silts were laid down in protected valleys. Along the narrow shoreline itself, sands of facies 1 and 2 were forming.

The S side of the Koobi Fora Ridge is composed of gently shelving slopes (in contrast to the N and W). Protected embayments gave rise to lagoons in the SW. Linear shelly beach ridges and reedy lagoons prevailed over much of the southern shoreline.

In the E, broad shallow-water sand sheets formed and were broken by a river overflowing from Lake Chew Bahir (Fig. 6). A complex variety of environments developed in the vicinity of the former overflow, ranging from lagoons and beaches to swamps. However, there is no major sediment wedge. This may reflect: (1) post-depositional erosion; (2) removal of sediment by longshore drift, possibly supplying sand and silt to areas further W; (3) low sediment load in the palaeo-river, perhaps due to upstream trapping by Chew Bahir; and (4) only intermittent overflow.

Also common in this region are 'reefs' of *Etheria elliptica*. This species is favoured by freshwater high-energy settings (White *et al.* 1981; Pickford, this volume); their local occurrence may also be related to the site of former inflow. Dating attempts have proved unsuccessful, yielding modern ages (Barthelme, pers. comm.) and we cannot therefore confirm an early Holocene age for the bioherms.

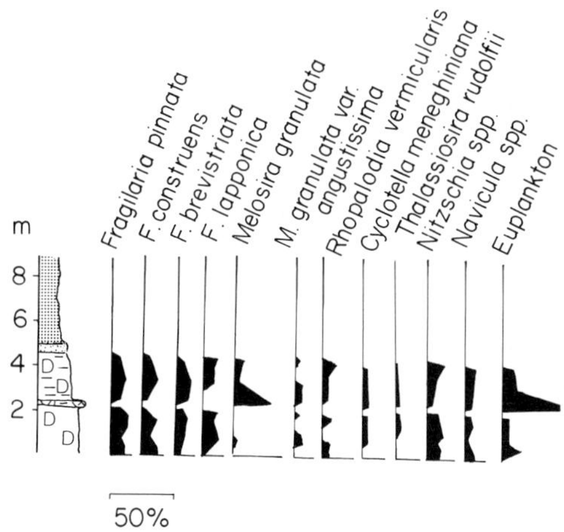

FIG. 4. Diatom stratigraphy of a late Holocene sequence. Key to lithologies given in Fig. 2.

Regional setting of Lake Turkana

The early Holocene high phase of Lake Turkana was only one example of expanded waters in E Africa. Fig. 6 summarizes the inferred palaeogeography for this period and also indicates the heights of inferred overflows. At its maximum, Lake Turkana extended *c.* 340 km from N to S and had a width of *c.* 70 km. Most of this expansion was accommodated by a northwards

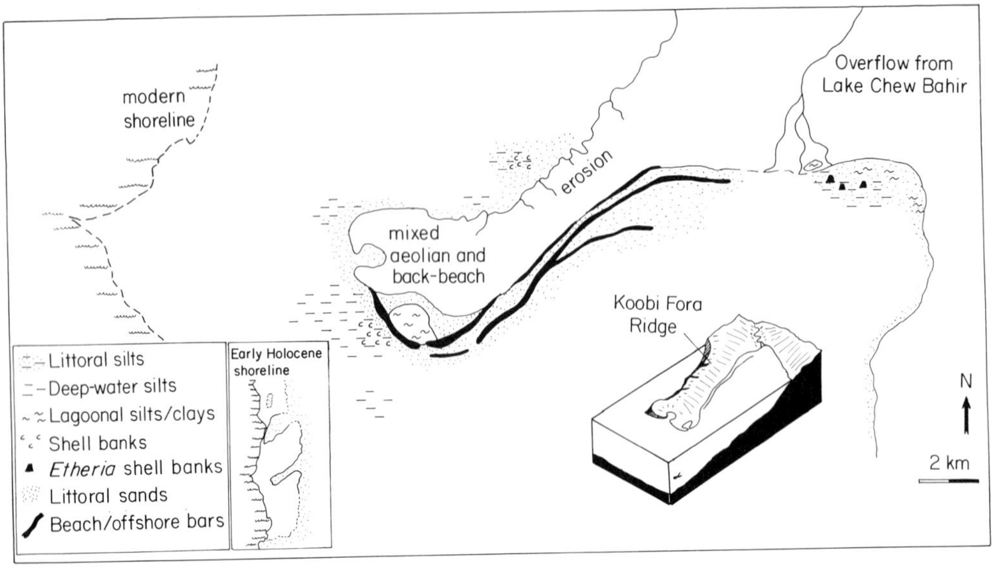

FIG. 5. Inferred depositional environments to the E of Koobi Fora during the early Holocene.

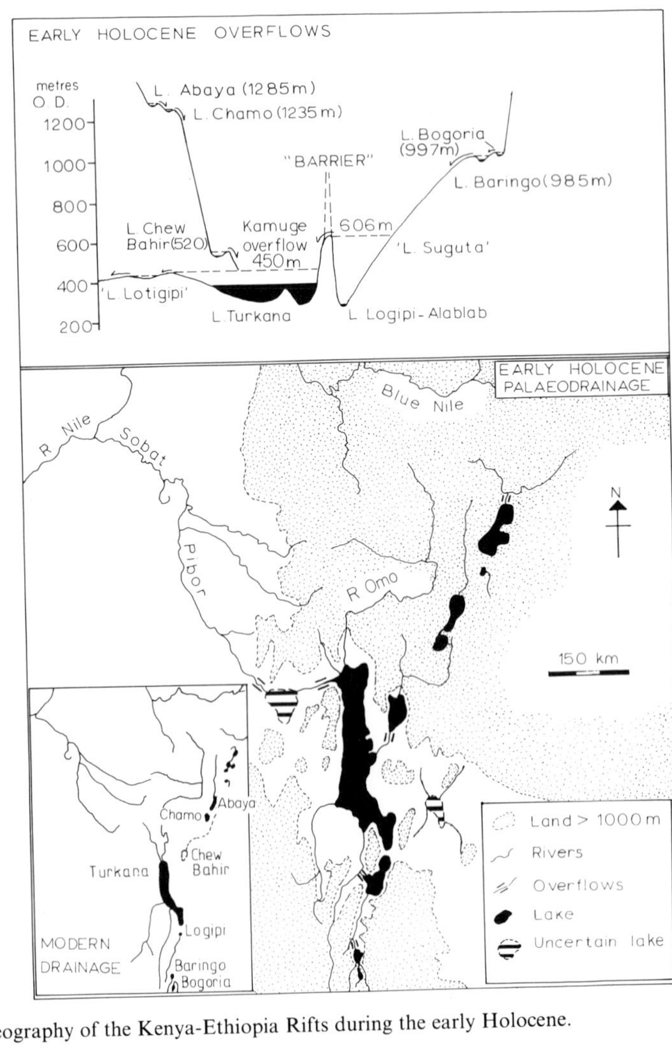

FIG. 6. Palaeogeography of the Kenya-Ethiopia Rifts during the early Holocene.

transgression over the gentle slopes of the Omo valley.

Water input may have been enhanced by the addition of two drainage systems that are fragmented into a series of closed basins today. To the S, overflow may have occurred from a lake occupying the present, arid Suguta basin. There, lacustrine sediments lie at up to 600 m a.s.l., only *c.* 6 m below the level of overflow into the Kerio River, which in turn flows into Lake Turkana (Truckle 1976; Owen & Renaut 1980). The former Lake Suguta may, in turn, have been augmented by periodic overflow from Lake Baringo and possibly also Lake Bogoria (Renaut & Owen 1980; Renaut 1982).

A second probable drainage network derives from the Ethiopian Highlands, where a series of overflows may have connected Lakes Abaya, Chamo and Chew Bahir (Grove *et al.* 1975). Although the chronology is as yet uncertain, Chew Bahir may have flowed into Turkana via the Bakate Gap.

Lake Turkana would overspill its basin today at *c.* 455 m a.s.l., which closely corresponds to the maximum height of early Holocene beach ridges. When the lake reached that level, during the early Holocene, it may have drained into the White Nile via the Lotigipi mudflats (Butzer 1980; Harvey & Grove 1982).

ACKNOWLEDGEMENTS: This research was supported by NERC and the Royal Society (UK). We thank Glynn Isaac, Ian Findlater, Richard Leakey and Peter Williamson for their assistance and discussions in Kenya. We are grateful to John Barthelme for making much of his dating evidence available to us. We also would like to thank the late Bill Bishop, Bill French and the technicians of Queen Mary College (University of London) for their help.

References

ABELL, P.I., AWRAMIK, S.M., OSBORNE, R.H. & TOMELLINI, S. 1982. Plio-Pleistocene lacustrine stromatolites from Lake Turkana, Kenya: morphology, stratigraphy and stable isotopes. *Sed. Geol.* **32,** 1–26.

BARTHELME, J. 1977. Holocene sites north-east of Lake Turkana: a preliminary report. *Azania* **12,** 33–41.

—— 1981. Unpublished PhD thesis, University of California.

BONNEFILLE, R. & VINCENS, A. 1977. Representation pollinique d'environnements arides a l'est du lac Turkana (Kenya). *Suppl. Bull. Ass. Fr. Etud. Quat.* **50,** 235–47.

BOWEN, B.E. 1974. *The geology of the Upper Cenozoic sediments in the East Rudolf embayment of the Lake Rudolf basin, Kenya.* Unpublished PhD thesis, Iowa State University, 164pp.

—— & VONDRA, C.F. 1973. Stratigraphical relationships of the Plio-Pleistocene deposits, East Rudolf, Kenya. *Nature* **242,** 391–3.

BROWN, F.H. & CERLING, T.E. 1982. Stratigraphical significance of the Tulu Bor Tuff of the Koobi Fora Formation. *Nature* **299,** 212–15.

BUTZER, K.W. 1971. Recent history of an Ethiopian delta. *Univ. Chicago, Res. Pap. Dept. Geogr.* **136,** 184pp.

—— 1980. The Holocene lake plain of north Rudolph, East Africa. *Phys. Geogr.* **1,** 42–58.

CERLING, T.E. 1979. Paleochemistry of Plio-Pleistocene Lake Turkana, Kenya. *Palaeogeogr. Palaeoclimatol. Palaeoecol.* **27,** 247–85.

—— & POWERS, D.W. 1977. Paleorifting between the Gregory and Ethiopian Rifts. *Geology* **5,** 441–4.

DODSON, R.G. 1971. Geology of the area south of Lodwar. *Rep. Geol. Surv. Kenya* **87,** 28–29.

FROSTICK, L.E. & REID I. 1980. Sorting mechanisms in coarse-grained alluvial sediments: fresh evidence from a basalt plateau gravel, Kenya. *J. Geol. Soc.* **137,** 431–41.

GASSE, F., TALLING, J.F. & KILHAM, P. 1983. Diatom assemblages in East Africa: classification, distribution and ecology. *Rev. Hydrobiol. Trop.* **16,** 1–34.

GROVE, A.T., STREET, F.A. & GOUDIE, A.S. 1975. Former lake levels and climatic change in the Rift Valley of southern Ethiopia. *Geogr. J.* **141,** 177–202.

HARVEY, T.J. 1976. *The palaeolimnology of Lake Mobutu Sese Seko, Uganda-Zaire: the last 28000 years.* Unpublished PhD thesis, Duke University, Durham, NC.

—— & GROVE, A.T. 1982. A prehistoric source of the Nile. *Geogr. J.* **148,** 327–36.

HEMMING, C.F. & TRAPNELL, C.G. 1957. A reconnaissance classification of the soils of the South Turkana desert. *J. Soil Sci.* **8,** 167–83.

HOLDSHIP, S.A. 1976. *The palaeolimnology of Lake Manyara, Tanzania: a diatom analysis of a 56 meter sediment core.* Unpublished PhD thesis, Duke University, Durham, NC.

ISAAC, G.L. 1978. The Olorgesailie Formation: stratigraphy, tectonics and the palaeogeographic context of the Middle Pleistocene archaeological sites. In: BISHOP, W.W. (ed) *Geological Background to Fossil Man.* Scottish Academic Press, Edinburgh, pp.173–206.

MATHISEN, M.E. & VONDRA, C.F. 1983. Provenance of the Plio-Pleistocene sediments in the East Turkana basin, northern Kenya. *Palaeogeogr. Palaeoclimatol. Palaeoecol.* **44,** 141–68.

OWEN, R.B. 1981. *Quaternary diatomaceous sediments and the geological evolution of lakes Turkana, Baringo and Bogoria, Kenya Rift Valley.* Unpublished PhD thesis, University of London, 465pp.

——, BARTHELME, J.W., RENAUT, R.W. & VINCENS, A. 1982. Palaeolimnology and archaeology of Holocene deposits north-east of Lake Turkana, Kenya. *Nature* **298,** 523–9.

—— & RENAUT, R.W. 1980. The early Holocene palaeogeography of the northern Kenya Rift Valley. *Rech. Geol. Afr.* **5,** 143–6.

—— & —— 1981. Palaeoenvironments and sedimentology of the Middle Pleistocene Olorgesailie Formation, southern Kenya Rift Valley. *Palaeoecol. Afr.* **13,** 147–74.

PHILLIPSON, D.W. 1977. Lowasera. *Azania* **12,** 1–32.

RAYNOLDS, G.H. 1973. *Summary of mapping—Area 103.* Unpublished East Turkana Stratigraphic Report.

RENAUT, R.W. 1982. *Late Quaternary geology of the Lake Bogoria fault-trough, Kenya Rift Valley.* Unpublished PhD thesis, University of London, 498pp.

—— & OWEN, R.B. 1980. Late Quaternary fluviolacustrine sedimentation and lake levels in the Baringo basin, northern Kenya Rift Valley. *Rech. Geol. Afr.* **5,** 130–3.

RICHARDSON, J.L. 1968. Diatoms and lake typology in East and Central Africa. *Int. Rev. Gesamt. Hydrobiol.* **53,** 299–338.

—— & RICHARDSON, A.E. 1972. History of an East African rift lake and its climatic implications. *Ecol. Monogr.* **42,** 359–66.

ROBBINS, L.H. 1974. The Lothagam site: a Late Stone Age fishing settlement in the Lake Rudolph basin, Kenya. *Publ. Mus. Mich. St. Univ.* **1(2),** 149–216.

TALLING, J.F. & TALLING, I.B. 1965. The chemical composition of African lake waters. *Int. Rev. Gesamt. Hydrobiol.* **50,** 421–63.

TRUCKLE, P.H. 1976. Geology and Cainozoic lake sediments of the Suguta trough, Kenya. *Nature* **263,** 380–3.

VONDRA, C.F. & BOWEN, B.E. 1976. Plio-Pleistocene environments and deposits, East Rudolph, Kenya. In: COPPENS, Y., HOWELL, F.C., ISAAC, G.L. & LEAKEY, R.E.F. (eds) *Earliest Man and Environments in the Lake Rudolph Basin.* University of Chicago Press, Chicago, pp.79–93.

—— & —— 1978. Plio-Pleistocene deposits and environments, East Rudolph, Kenya. In: BISHOP, W.W. (ed.) *Geological Background to Fossil Man.* Scottish Academic Press, Edinburgh, pp.395–414.

——, JOHNSON, G.D., BOWEN, B.E. & BEHRENSMEYER, A.K. 1971. Preliminary stratigraphic studies of the east Rudolf basin, Kenya. *Nature* **231,** 245–8.

WALSH, J. & DODSON, R.G. 1969. Geology of northern Turkana. *Rep. Geol. Surv. Kenya* **82,** 26–30.

WHITE, H.J., BURGGRAF, D.R., BAINBRIDGE, R.B. & VONDRA, C.F. 1981. Hominid habitats in the rift valley: part 1. In: RAPP, G.R. & VONDRA, C.F. (eds) *Hominid Sites: Their Geologic Settings.* Westview Press, Boulder, Colorado, pp. 57–113.

YURETICH, R.F. 1979. Modern sediments and sedimentary processes in Lake Rudolf (Lake Turkana), Eastern Rift Valley, Kenya. *Sedimentology* **26,** 313–31.

—— & CERLING, T.E. 1983. Hydrogeochemistry of Lake Turkana, Kenya: mass balance and mineral reactions in an alkaline lake. *Geochim. Cosmochim. Acta* **47,** 1099–109.

R.B. OWEN, Dept of Geography & Earth Science, Chancellor College, University of Malawi, PO Box 280, Zomba, Malawi.

R.W. RENAUT, Dept of Geological Sciences, University of Saskatchewan, Saskatoon, Saskatchewan S7N OWO, Canada.

Palaeolimnology of Lake Bogoria (Kenya) during the 4500 BP high lacustrine phase

A. Vincens, J. Casanova & J.J. Tiercelin

SUMMARY: Lake Bogoria (altitude 990 m) lies within the Kenya Rift Valley (0°15′N, 36°10′E). A high lake stage is marked by a double stromatolite belt: the first, occurring at 999 m, is radiocarbon dated 4140 ± 60 BP and the second, at 995 m, is dated 3880 ± 60 and 3750 ± 180 BP. Palynological data from shoreline stromatolites and lacustrine sediments recovered in three 15 m cores permit a reconstruction of former limnological conditions. They show that the lacustrine extension resulted from a positive water balance brought about by runoff under climatic conditions more humid than today but drier than those prevailing during the early and mid-Holocene.

Studies of the Bogoria lake basin (formerly Lake Hannington) have been mainly limnological (Worthington 1932) or geological, with little emphasis on elucidating changes in the sedimentary environment (McCall 1967). Since 1976, stratigraphic and sedimentological research has been undertaken throughout the basin. The study of modern sediments (Tiercelin *et al.* 1980; Tiercelin 1981), and an examination of three long sediment cores (Tiercelin *et al.* 1981, 1982) linked to a stratigraphic study of lake margin deposits (Renaut 1982), has allowed the deciphering of late Quaternary fluctuations of lake level and a tentative correlation with the major lake basins of the rift valley.

This paper explores a major late Holocene transgression of Lake Bogoria *c.* 4500 BP. Observations made on stromatolites, lacustrine sediments and associated pollen are compared, and their significance discussed in terms of palaeohydrological and palaeoclimatological environments.

General setting

Lake Bogoria lies at an altitude of 990 m a.s.l. (July–August 1977 lake level). It occupies the southern part of a N–S half-graben at the foot of the eastern flank of the Kenya Rift Valley, just N of the equator (0°15′N, 36°10′E). This half-graben is about 60 km long. At its northern end lies the freshwater Lake Baringo (967 m). The two lakes are separated by the low, marshy Loboi plain (999 m; Fig. 1).

Lake Bogoria is small (length 17 km, width 3.5 km, depth 10 m) and endoreic. It is divided into three sub-basins (southern, central and northern) by sills named Nyalilbuch and Mwanasis (Fig. 1). The lake waters have a salinity >40‰ (Tiercelin *et al.* 1980) due to high concentrations of sodium carbonate-bicarbonate, and a pH of >10.3. The lake is fed by ephemeral streams (Kipsirian-Parkirichai and Emsos-Ngiriki), by the intermittent Sandai-Wasages River and by more than 200 hydrothermal springs which contribute to the meromictic regime.

The region of Lake Bogoria is characterized by a mean annual rainfall of 700 mm with two rainy seasons (March–May and July–August), a mean temperature of 24°C and a potential evaporation of >2500 mm a^{-1}. Because of this, it is included in the semi-arid ecoclimatic zone IV as defined by Pratt & Gwynne (1977) and Le Houerou & Popov (1981). The present vegetation has been classified as part of the Somalia-Masai region by White (1983). It is characterized by wooded grassland and mosaic evergreen bushland. Dominant trees are *Acacia*, *Combretum molle*, *Terminalia brownii*, *Tarchonanthus camphoratus* and *Dodonaea viscosa*. Afro-montane evergreen forests with *Podocarpus*, *Juniperus* and *Olea* occur, but only in the eastern part of the basin, in the Marmanet forest.

The high level lake stage

Evidence

The high level lake stage is marked by a double belt of stromatolites that occurs all around the lake, but is particularly well developed along the western shore (Mwanasis peninsula to Loburu delta; Fig. 2). The earlier stromatolites occur at up to 999 m, and the later ones at up to 995 m. At the beginning of this century, lake levels reached similar altitudes (997–998 m). Besides these, several fluctuations of an ephemeral nature and related to exceptional rainy seasons have been recorded more recently (Renaut 1982).

The late Holocene lacustrine expansion resulted from a positive water balance. Significant runoff is suggested by the presence of conglomerates consisting of well-rounded pebbles which

From FROSTICK, L.E. *et al.* (eds) 1986, *Sedimentation in the African Rifts*, Geological Society Special Publication No. 25, pp. 323–30.

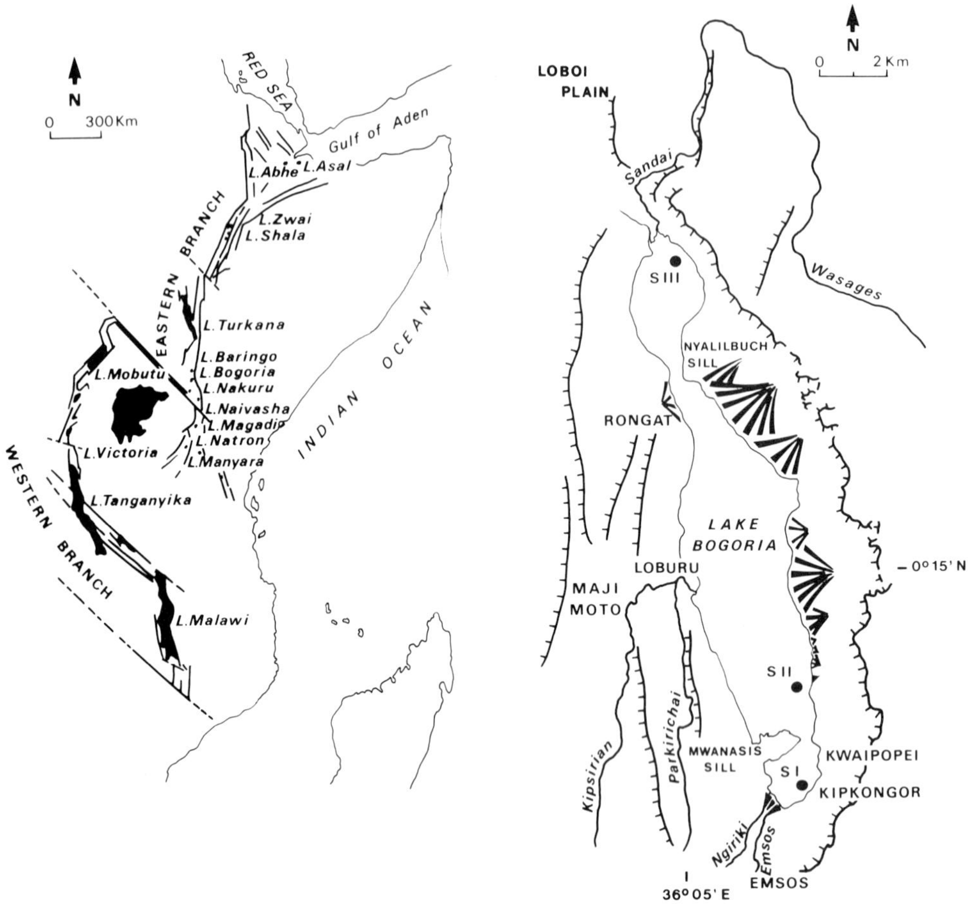

FIG. 1. Location of Lake Bogoria in the East African Rift System, and position of stromatolites, travertines and lake sediment core sites (SI–SIII).

have been individually encrusted and then cemented together by secondary encrustation. These detrital accumulations occur in the lower reaches of the channels that feed the lake. Upstream, there are travertine formations (e.g. at Rongat, Maji Moto, Emsos, Kipkongor, Kwaipopei; Fig. 1). The occurrence of travertine has considerable significance; it indicates a high level of groundwater feeding springs enriched in calcium after passage underground (Casanova 1981). Unfortunately, a lack of radiometric dates makes it impossible to relate the runoff episodes responsible for the travertines to the pebble conglomerates.

^{14}C dating of stromatolites

Three dates have been obtained for the carbonate of the stromatolites using ^{14}C dating. The apparent ages are 4140 ± 60 BP (GIF-5157) for the first and higher generation, and 3880 ± 60 (GIF-5158) and 3750 ± 180 BP (LY-1971) for the second and lower one (Table 1).

Lacustrine carbonates often present significant anomalies in their ^{14}C content. These anomalies arise because of the possibility of hard water effects due to ancient CO_2 in the lake water. This is derived both from hot spring waters and from alkaline lavas (Young & Renaut 1979). As a consequence, the total inorganic dissolved carbon (TIDC) of the palaeolake may not have been in isotopic equilibrium with atmospheric CO_2. In order to argue the validity of ^{14}C dates obtained from stromatolites, we have to consider the stable isotope ratios.

The δPDB ^{13}C of Bogoria stromatolites ranges from +4.21 to +7.79‰. These values are all higher than +2.6‰, i.e. higher than the isotopic composition of a calcite deposited in equilibrium with atmospheric CO_2 at a temperature of *c.* 25°C

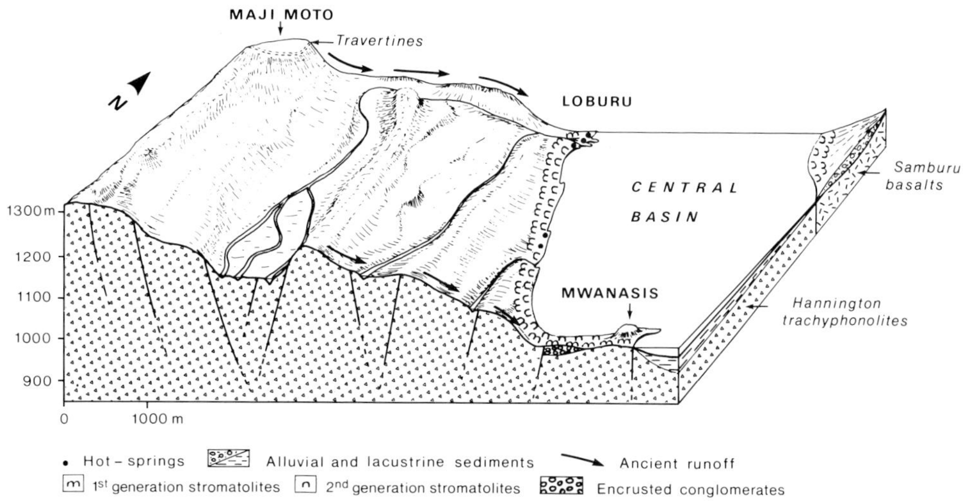

FIG. 2. Schematic diagram of the central western shore of Lake Bogoria.

TABLE 1. *Isotopic composition and dates of Lake Bogoria stromatolites*

Stromatolite age	Lab No.	$\delta PDB^{18}O$ (‰)	$\delta PDB^{13}C$ (‰)	Apparent ^{14}C age (a BP)	Normalized ^{14}C (a BP)
First generation	GIF-5157	+4.83	+5.70	4140±60	4640
Second generation	GIF-5158	+5.03	+7.09	3880±60	4380
Second generation	LY-1971	+4.87	+7.20	3750±180	4250

($\delta^{13}C = c.\ -7$‰; Craig & Keeling 1963). The high $\delta^{13}C$ values can be explained by preferential use of $^{12}CO_2$ during algal photosynthesis. This would lead to enrichment of the levels of ^{13}C contributing to TIDC. However, besides a consideration of the carbon isotopes, the PDB ^{18}O of Bogoria stromatolites ranges from +2.41 to +8.28‰, suggesting long residence-time for the water and a possibility of long-term exchange between dissolved and atmospheric CO_2. For these reasons, we postulate that palaeolake TIDC was in equilibrium with atmospheric CO_2 during stromatolite growth. As a corollary we consider the ^{14}C dates obtained for Bogoria stromatolites to be credible. A complete discussion of East African stromatolite ^{14}C dating is given in Hillaire-Marcel *et al.* (in press).

In order to compare the dates obtained for carbonate with those derived for lake core organic matter, the dates have been normalized to the same $\delta^{13}C$ reference value of −25‰ (Broecker & Olson 1961). This represents a correction of *c.* 500 years. The normalized ages are given in Table 1.

Sedimentological and ecological data

Stromatolites

Lacustrine stromatolites have been observed on all hard substrates from 999 m to 989 m a.s.l., the September 1984 lake level. They take a variety of forms which include flat encrustations, cylindrical encrustations around vegetal remains, and pebble encrustations (Fig. 3A).

The first and higher generation of stromatolites at 999 m are characteristically small with a size that is measurable in millimetres and only rarely in centimetres. The ornamentation consists of small flattened granules (Fig. 3B). In contrast, the second generation at 995 m is composed of encrustations ranging from 2 to 20 cm in thickness. Ornamentation consists of large hemispherical granules (Fig. 3C). The stromatolites of this second generation usually encrust those of the first one. They can accumulate locally reaching thicknesses of up to 1 m, as encrustations on branches that have floated into position and bundles of twigs. Microstructures are identical in

FIG. 3. Lake Bogoria highstand stromatolites. A. Block encrustations along the shore where first generation stromatolites are encrusted by those of the second generation; scale is 30 cm. B. Ornamentation of the first generation stromatolites; scale bar is 2.5 cm. C. Ornamentation of the second generation stromatolites; scale bar is 2.5 cm. D. Stromatolite columnar structure results from competition of the algae during growth; scale bar is 1 cm. E. Sparitic laminae generated by erect bundles of cyanophyte filaments; scale bar is 500 μm. F. Criss-crossed filaments surrounded by micrite; scale bar is 100 μm.

the two stromatolite levels. They are produced by columnar algal growth emanating from a common basal layer (Fig. 3D) that indicates biological competition. The carbonates are composed of low-Mg calcite precipitated around algal filaments. Filaments of 2–2.5 μm in diameter, grouped in erect bundles, represent the dominant cyanophyte species. These generate microsparitic laminae (Fig. 3E). Criss-crossed filaments of between 4.5 and 6 μm in diameter and surrounded by micrite are less abundant (Fig. 3F). Trapping and binding of detrital particles by algae does not occur in these types of stromatolite. It is, therefore, possible to conclude that the carbonates are of biological origin. This is generally the case in East Africa (Abell *et al.* 1982; Casanova, this volume). The laminae range from 50 to 400 μm in thickness, presumably reflecting fluctuations in the growth rate of the algal-bacterial mat. However, the number of laminae in the lower altitude second generation is always two or three times that found in the first generation. If we assume that seasonal rhythms of growth are responsible for the lamination, it is possible to speculate that the younger shoreline was maintained for a longer period than its higher and older counterpart.

Despite their very similar texture, the two stromatolite generations clearly differ in morphology. They characterize two distinct lacustrine stages. The oldest shoreline shows a stabilization of the lake at +9 m above the present average lake level, close to the overflow height into Lake Baringo via the Loboi plain. The second lacustrine stabilization at +5 m was lower in height but longer in duration as suggested by the stromatolite size.

When they are superposed, the two stromatolite generations are separated by a disconformity which is never erosional in character. This fact, together with a lack of encrustation between the two lake levels, suggests a well-defined regression from one level to another. The still water level remained stable during each of these highstands as the stromatolites grew. Each probably corresponds to a hydrological equilibrium within a broad transgressive phase of the Holocene.

Lake bottom sediments

The high lake level is recorded in two of the three cores recovered from the southern, central and northern sub-basins (SI, SII and SIII respectively in Fig. 1).

In the southern sub-basin, the sediments of *c.* 4500 BP are fine grained, essentially massive smectitic brown clays, with low total organic carbon (0.2–0.5%). They have an abundant micro-fauna and micro-flora including diatoms, Spongiae, fish and insect debris. There is a rich ostracod fauna composed of *Limnocythere africana* (the most common species), *Gomphocythere* and *Sclerocypris clavularis*, and *Darwinula* gr. *stevensoni* (abundant). This represents episode b and ecozone 4 of Carbonel *et al.* (1983). The last species is characteristic of calm, clear waters with low salt concentrations. Coarse-grained sediments containing the ostracod *Plesiocypridopsis* or species which proliferate in rough and turbid waters are dated at 4790±280 BP (GIF-5302) and have been taken to represent the increase in runoff that opened the transgressive phase.

In the central sub-basin, homogeneous greyish-green to greyish-black clays, rich in organic matter (1–3% organic carbon) are dated 4060±260 BP (GIF-5080). These indicate the existence of a surface water environment which favours the blooming of planktonic populations. These are encouraged by the development of a less saline upper stratum in the lake waters, the mixolimnion (Tiercelin *et al.* 1980; Tiercelin 1981), which overrides an anoxic saline stratum, the monimolimnion.

The core taken from the northern sub-basin came from the vicinity of the Sandai River delta. Because of this, the core exhibits numerous phases of erosion. Isotopic dating was impossible due to low organic content. It is, therefore, not possible to identify the 4500 BP high lake level.

Pollen

Palynological studies on stromatolite crusts have been carried out for the first time in East Africa. Three samples from the Mwanasis peninsula were analysed, but only two contained enough pollen to allow satisfactory interpretation. They represent the two stromatolite generations. On a normalized chronological radiocarbon scale the sample identified as Bogoria Str. 1 corresponds to an age of *c.* 4640 BP and that identified as Bogoria Str. 2 to an age of between *c.* 4380 and 4250 BP.

The two spectra (Table 2) are very similar and are both well diversified. They are characterized by a large representation of regional afro-montane arboreal pollen. This accounts for >35% of the total count and has three dominant components: *Podocarpus*, *Juniperus* and *Olea*. Taxa which belong to the local Somalia-Masai bushland are present, but as a whole they form <4% of the total. Best represented in this group are *Euclea*, *Dodonaea viscosa* and *Tarchonanthus camphoratus*. Among the herbaceous pollen *Gramineae* are very abundant, but, in addition, only *Cyperaceae*, *Typha* and *Compositae tubuliflorae* have percentages of 1%.

TABLE 2. *Pollen spectra from two stromatolites of Lake Bogoria compared with modern data (means of 17 surface bottom-sediment samples)*

Taxa	Bogoria Str. 1 (%)*	Bogoria Str. 2 (%)*	Modern lake sediments (%)*
Afro-montane arboreal taxa			
Cupressaceae: *Juniperus*	14.9	16.4	4.2
Oleaceae: *Olea africana*	0.8	—	0.3
Oleaceae: *Olea hochstetteri* type	1.8	1.7	2.0
Podocarpaceae: *Podocarpus*	18.0	18.1	12.1
Rosaceae: *Hagenia abyssinica*	—	0.2	0.4
Others	—	—	0.4
sub-total	35.6	36.5	19.4
Somalia-Masai arboreal taxa			
Anacardiaceae: *Lannea* type	0.3	0.1	0.1
Anacardiaceae: *Rhus* type	—	0.1	0.6
Burseraceae: *Commiphora africana* type	—	0.1	0.02
Combretaceae: *Combretum molle* type	0.3	0.1	3.3
Compositae: *Tarchonanthus camphoratus*	0.5	—	2.2
Ebenaceae: *Diospyros*	0.5	—	—
Ebenaceae: *Euclea*	1.0	0.2	0.9
Mimosoideae: *Acacia* group I	—	0.1	1.4
Mimosoideae: *Acacia* group III	0.3	0.1	1.5
Myrtaceae: *Syzygium* type	—	0.1	0.1
Proteaceae: *Protea*	0.3	—	—
Rutaceae: *Fagaropsis* type	0.3	—	0.3
Sapindaceae: *Dodonaea viscosa*	0.5	0.2	2.6
Others	—	—	1.8
sub-total	3.9	1.3	14.8
Non-arboreal taxa			
Acanthaceae: *Justicia anselliana* type	0.3	—	2.4
Amaranthaceae: *Digera muricata* type	0.3	—	0.3
Amaranthaceae/Chenopodiaceae	0.3	0.5	1.7
Caryophyllaceae	0.3	—	0.1
Chenopodiaceae: *Chenopodium* type	0.3	—	1.2
Compositae: *Artemisia*	—	0.1	0.05
Compositae: Tubuliflorae	1.0	0.5	0.7
Cyperaceae	4.9	3.1	11.8
Gramineae	50.0	55.8	35.9
Labiatae: *Satureja* type	—	0.1	—
Rubiaceae: *Anthospermum*	0.3	—	0.07
Typhaceae: *Typha*	1.0	0.7	1.2
Urticaceae	0.3	0.2	0.02
Others	—	—	4.3
sub-total	58.8	61.1	59.7
Undifferentiated taxa			
Compositae: *Vernoniae*	—	0.1	0.1
Loranthaceae: *Plicosepalus* type	0.3	—	—
Papilionoideae	0.3	—	0.07
Papilionoideae: *Rhynchosia* type	0.3	—	—
Rubiaceae	—	0.1	0.2
Others	—	—	1.7
sub-total	0.8	0.2	2.1
Pteridophyta	0.8	0.4	2.5
Unknown	0.3	0.5	1.9

* Total pollen count is Bogoria Str. 1 = 388; Bogoria Str. 2 = 828; modern lake sediments = 273.

A comparison has been made with palynological data from the surface bottom-sediments of Lake Bogoria. There are important differences, especially in the relative frequency of regional and local arboreal pollen (Table 2). The abundance of afro-montane taxa and the low representation of Somalia-Masai vegetation, so characteristic of the fossil spectra, are not mirrored in the modern pollen rain. Because the same taxa are identified in both modern and fossil spectra, such differences cannot be explained in terms of changing large-scale dispersal patterns. It also seems unlikely that there was a selective trapping of pollen by the algae during encrustation. So, a different geographical distribution of the two main vegetation types would seem the best explanation for such differences in composition of the ancient and modern spectra. The abundance of afro-montane pollen and the particular dominance of *Podocarpus*, *Juniperus* and *Olea* suggest an extension of the coniferous forests (*c.* 4500 BP) beyond the area they occupy at present. This implies more humid conditions at that time. However, the presence of Somalia-Masai taxa in the ancient spectra might be taken to indicate localized semi-evergreen bushland around the lake and a colonization of the shore zone by grasses and herbaceous swamp vegetation.

The palynological record preserved in the stromatolite crusts can be correlated with that of the lake deposits. In core SI, the extension of coniferous forest is reflected by an abundance of *Podocarpus* pollen in the massive smectitic clays at 60 cm above the level dated 4790 ± 280 BP. In core SIII, coniferous forests are represented as being extensive later than 4500 BP, a date inferred from ostracod correlations (Carbonel *et al.* 1983). The discontinuous nature of the palynological data for the late Holocene period in this core, however, does not allow us to be more precise about the onset of forest development (Vincens 1986).

The palynological data from Lake Bogoria provides a similar picture to that from high altitude sites for the same time period. The development of afro-montane coniferous forests during the late Holocene has been shown to occur on Mount Kenya by Coetzee (1967) and Perrott (1982), on the Ruwenzori Range by Livingstone (1967), on Mount Elgon by Hamilton & Perrott (in Hamilton 1982), and on the Aberdare Range by Perrott (1982). An increase in *Podocarpus* is also observed in the pollen diagrams for the mid-altitude deposits of Lake Victoria (Kendall 1969) and Lake Abijata (Lezine & Bonnefille 1983). With the exception of this last site, where the age of the increase in *Podocarpus* has been estimated at *c.* 6000 BP, the numerous ^{14}C dates place this episode more or less soon after 4000 BP (Perrott 1982). However, the dating for increases in *Podocarpus* ranges from poor to excellent. As a result, they cannot be synchronized precisely. The *Podocarpus* extension in the Bogoria basin *c.* 4500 BP seems to be contemporary with that recorded at mid-altitude in the Victoria basin, but it is generally older than the increases shown in pollen diagrams from high altitude sites.

In East Africa, all palynological data including that obtained in the two cores of Lake Bogoria (SI and SIII) show that the development of coniferous forests occurred soon after the disappearance of early and mid-Holocene vegetation characteristic of moister climatic regimes. This floristic change suggests a shift towards a drier climate, or at least one with a more seasonal rainfall, in the Bogoria basin after 4500 BP.

Conclusions

High lake levels in the Bogoria basin are shown clearly by two belts of stromatolites. Assuming isotopic equilibrium between TIDC and atmospheric CO_2, as suggested by the carbon and oxygen stable isotope ratios, we can date them at *c.* 4500 BP. Two distinct episodes of stabilized high lake level, each marked by stromatolites, were separated by a short regressive phase. Each high stage can be attributed to greater runoff than at present, and each saw the maintenance of a freshwater surface layer in Lake Bogoria. The incidence of high lake levels correlates well with an extension of the afro-montane forest as suggested by pollen spectra in the lake bottom sediments.

The period around 4500 BP is one of major climatic change over East Africa, and marks the onset of aridification in this part of the African continent. In high altitude pollen diagrams, this manifests itself in the substitution of humid evergreen forests by those more tolerant of drier conditions. In most East African lakes, including Lakes Victoria (Kendall 1969), Nakuru (Richardson 1972), Magadi (Hillaire-Marcel *et al.*, in press) and Manyara (Holdship 1976), the period between *c.* 5000 and 4000 BP was one of falling water levels. In some basins, notably Abhé (Gasse 1977), Ziway-Shala (Gillespie *et al.* 1983) and Turkana (Butzer 1980; Owen *et al.* 1982), this regression was not simple but was accompanied by large oscillations in lake level.

The extension of the coniferous forests, together with the long residence-time of water that is implied by the stromatolite $\delta^{18}O$ ratios suggests the occurrence of a strongly contrasted seasonal climate that includes at least one significant dry season. However, the fact that lake levels were higher than at present implies that rainfall around 4500 BP, though probably less than during the early and mid-Holocene, was greater than it is today.

ACKNOWLEDGEMENTS: This research was possible thanks to the assistance of the Société Nationale Elf-Aquitaine, the Compagnie Française des Pétroles and the Centre National de la Recherche Scientifique (Laboratoire de Géologie du Quaternaire and the Centre des Faibles Radioactivités). The authors would also like to thank the management of Elf-Aquitaine for authorizing this publication and the Kenya Government for a research permit.

References

ABELL, P.I., AWRAMIK, S.M., OSBORNE, R.H. & TOMELLINI, S. 1982. Plio-Pleistocene lacustrine stromatolites from Lake Turkana, Kenya: morphology, stratigraphy and stable isotopes. *Sed. Geol.* **32,** 1–26.

BROECKER, W.S. & OLSON, E.A. 1961. Lamont radiocarbon measurements VIII. *Radiocarbon* **3,** 176–204.

BUTZER, K.W. 1980. The Holocene Plain of North Rudolf, East Africa. *Phys. Geogr.* **1,** 42–58.

CARBONEL, P., GROSDIDIER, E., PEYPOUQUET, J.P. & TIERCELIN, J.J. 1983. Les Ostracodes, témoins de l'évolution hydrologique d'un lac de Rift. Exemple du lac Bogoria, Rift Gregory, Kenya. *Bull. Centres Rech. Explor. Prod. Elf-Aquitaine* **7(1),** 301–13.

CASANOVA, J. 1981. *Etude d'un milieu stromatolitique continental: les travertins plio-pléistocènes du Var (France).* Unpublished thesis 3° Cycle, Université Aix-Marseille II. 130pp.

COETZEE, J.A. 1967. Pollen analytical studies in East and South Africa. *Palaeoecol. Afr.* **3,** 146pp.

CRAIG, H. & KEELING, C.D. 1963. The effects of N_2O on the measured isotopic composition of atmospheric CO_2. *Geochim. Cosmochim. Acta* **27,** 549–51.

GASSE, F. 1977. Evolution of Lake Abhe (Ethiopia and TFAI), from 70,000 BP. *Nature* **265,** 42–5.

GILLESPIE, R., STREET-PERROTT, F.A. & SWITSUR, R. 1983. Post-glacial arid episodes in Ethiopia have implications for climate prediction. *Nature* **306,** 680–3.

HAMILTON, A.C. 1982. *Environmental History of East Africa. A Study of the Quaternary.* Academic Press, London, 328pp.

HILLAIRE-MARCEL, C., CARRO, O. & CASANOVA, J. (in press). ^{14}C and Th/U dating of Pleistocene and Holocene stromatolites from East African palaeolakes. *Quat. Res.*

HOLDSHIP, S.A. 1976. *The paleolimnology of Lake Manyara, Tanzania: a diatom analysis of a 56 meter sediment core.* Unpublished PhD thesis, Duke University, Durham, NC, 121pp.

KENDALL, R.L. 1969. An ecological history of the Lake Victoria Basin. *Ecol. Monogr.* **39,** 121–76.

LE HOUEROU, H.N. & POPOV, G.F. 1981. An eco-climatic classification of intertropical Africa. *FAO Plant Prod. Protect. Pap.* **31,** 40pp.

LEZINE, A.M. & BONNEFILLE, R. 1983. Diagramme pollinique holocène d'un sondage du lac Abijata (Ethiopie, 7°42′ Nord). *Poll. Spores* **24,** 463–80.

LIVINGSTONE, D.A. 1967. Post-glacial vegetation of the Ruwenzori mountains in equatorial Africa. *Ecol. Monogr.* **37,** 25–52.

McCALL, G.J.H. 1967. Geology of the Nakuru–Thomson's Falls–Lake Hannington area. *Rep. Geol. Surv. Kenya* **78,** 122pp.

OWEN, R.B., BARTHELME, J.W., RENAUT, R.W. & VINCENS, A. 1982. Palaeolimnology and archaeology of Holocene deposits north-east of Lake Turkana, Kenya. *Nature* **298,** 523–9.

PERROTT, R.A. 1982. A high altitude pollen diagram from Mount Kenya: its implications for the history of glaciation. *Palaeoecol. Afr.* **14,** 77–83.

PRATT, D.J. & GWYNNE, M.D. 1977. *Rangeland Management and Ecology in East Africa.* Hodder & Stoughton, London 310pp.

RENAUT, R.W. 1982. *Late Quaternary geology of the Lake Bogoria fault-trough, Kenya Rift Valley.* Unpublished PhD thesis, University of London, 498pp.

RICHARDSON, J.L. 1972. Palaeolimnological records from rift lakes in central Kenya. *Palaeoecol. Afr.* **6,** 131–6.

TIERCELIN, J.J. 1981. *Rifts continentaux: tectonique, climats, sédiments. Exemples: la sédimentation dans le Nord du Rift Gregory (Kenya) et dans le Rift de l'Afar (Ethiopie) depuis le Miocène.* Unpublished D.Sc. thesis, Université Aix-Marseille II, 260pp.

——, LE FOURNIER, J., HERBIN, J.P. & RICHERT, J.P. 1980. Continental rifts: modern sedimentation, tectonic and volcanic control: example from the Bogoria-Baringo graben, Gregory Rift, Kenya. In: *Geodynamic Evolution of the Afro-Arabic Rift System.* Acad. Nazionale dei Lincei, Rome, pp.143–64.

——, PERINET, G., LE FOURNIER, J., BIEDA, S. & ROBERT, P. 1982. Lacs du rift est-africain, exemples de transition eaux douces-eaux salées: le lac Bogoria, Rift Gregory, Kenya. *Mém. Soc. Géol. Fr.* **144,** 217–30.

——, RENAUT, R.W., DELIBRIAS, G., LE FOURNIER, J. & BIEDA, S. 1981. Late Pleistocene and Holocene lake level fluctuations in the Lake Bogoria basin, Northern Kenya Rift Valley. *Palaeoecol. Afr.* **13,** 105–20.

VINCENS, A. 1986. Diagramme pollinique d'un sondage pleistocène supérieur-Holocène du lac Bogotia (Kenya). *Rev. Palaeobot. Palynal.* **47,** 169–92.

WHITE, F. 1983. The vegetation of Africa. A descriptive memoir to accompany the UNESCO/AETFAT/UNSO vegetation map of Africa. *Nat. Resources Res.* **20,** UNESCO, Paris, 356pp.

WORTHINGTON, M.A. 1932. Scientific results of the Cambridge expedition to the East African lakes, 1930–1. 1: General introduction and station list. *J. Linnean. Soc.* **38,** 99–118.

YOUNG, J.A.T. & RENAUT, R.W. 1979. A radiocarbon date from Lake Bogoria, Kenya Rift Valley. *Nature* **278,** 243–5.

A. VINCENS & J. CASANOVA, Laboratoire de Géologie du Quaternaire, CNRS—Case 907 Luminy, 13288 Marseille, Cedex 9, France.

J.J. TIERCELIN, GIS Océanologie et Géodynamique, Université de Bretagne Occidentale, 29287 Brest, Cedex, France.

Preliminary magnetostratigraphic record of Pleistocene deposits, Lake Natron Basin, Tanzania

N. Thouveny & M. Taieb

SUMMARY: The declination of characteristic magnetization has been used to determine the polarity changes in two volcano-sedimentary sequences exposed in the Lake Natron basin. Radiometric dating and biostratigraphic data allow us to link the magnetostratigraphy to the standard geomagnetic polarity time scale. In the Lake Natron palaeomagnetic record, the Matuyama Epoch is well defined, possibly including the Olduvai Event (1.87–1.67 Ma). Other normal polarity zones have to be considered cautiously.

The Pleistocene formations of the Lake Natron basin (3°S, 37°E; Fig. 1) were studied for the first time in 1964 by Isaac and Leakey after the discovery of an *Australopithecus robustus* mandible (Leakey & Leakey 1964). Since 1981, stratigraphical, sedimentological, palaeontological and geochemical investigations have been carried out by an international team led by Mturi, Isaac and Taieb in order to reconstruct palaeoenvironmental changes of the last 2 Ma in the Tanzanian Rift.

Magnetostratigraphic studies were made in order to facilitate correlations and chronology. Two main sections in the western part of the basin (Peninj area; Fig. 1) have been sampled: the Kamare section (Moinik Formation) and the Kipalagu section (Humbu and Moinik Formations). These formations have been described by Isaac (1967) as two successive lacustrine deposits corresponding to two different chemical states of the palaeo-Lake Natron. The lower one (Humbu) is a freshwater deposit, the upper one (Moinik) represents saline conditions. The correlation of the two sections is based on two white tuffs and a double thin-layered tuff (Fig. 2).

Sampling and laboratory treatments

Sampling was done at 22 and 56 levels in the Kamare and Kipalagu sections, respectively (Fig. 2). Sampling gaps correspond to poor outcrop exposure. Samples were carefully referenced to the magnetic N and to the horizontal plane before being collected for cutting in the laboratory. At least three samples were obtained from each level.

The study was made on 200 samples while 190 were used for calibration purposes. The criteria

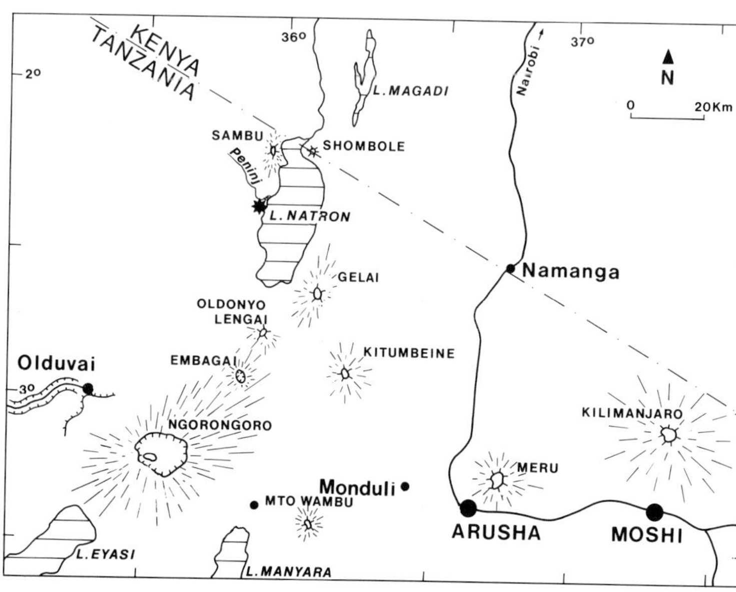

FIG. 1. Location map of the Lake Natron basin. Star = study area.

From FROSTICK, L.E. *et al.* (eds) 1986, *Sedimentation in the African Rifts*, Geological Society Special Publication No. 25, pp. 331–6.

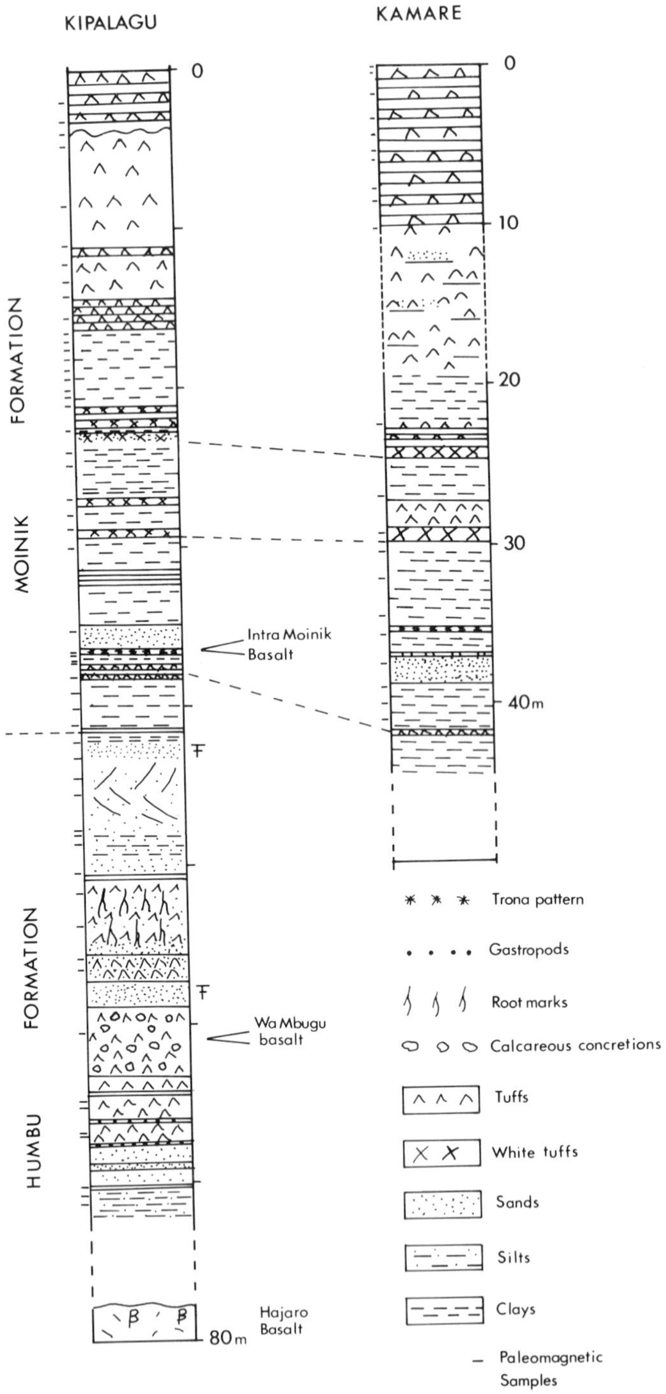

FIG. 2. Location of samples taken from the Kamare and Kipalagu stratigraphic sections, showing also the lithological variation of the Humbu and Moinik Formations.

used for rejection of a sample were either a high discrepancy between direction of magnetization and the mean direction established for the same level, or erratic behaviour during the demagnetization process, which is often due to the fact that intensities are lower than the noise signal of the magnetometer. Measurement was carried out with a spin magnetometer (Molyneux 1971) that was linked to a microcomputer.

The intensity of natural remanent magnetization (NRM) varied between 2×10^{-4} and 1.2×10^{-1} A m^{-1}. The mean value of the NRM intensity was 9×10^{-3} A m^{-1}. NRM is dominated by a component sub-parallel to the present-day field (declination 359°, inclination −30°), with a mean declination of 6°, and a mean inclination of −20°.

Two samples from each stratigraphic level were demagnetized in alternating fields from 0 to 100 mT. Samples still showing normal magnetization after the complete alternating field treatment were submitted to thermal treatments from 0 to 600°C. Analysis of the orthogonal diagrams (Zijderveld 1967) allowed us to choose appropriate alternating field or thermal steps for each particular case. The characteristic magnetization can be isolated by alternating field demagnetization ranging from 10 to 60 mT or by thermal demagnetization ranging from 100 to 600°C. The high coercivity of particular stratigraphic levels shows that one of the carriers of remanence could be haematite or goethite.

Results

The directions retained after treatment were individually compared to the normal (declination 360°, inclination −6°) and the reverse (180°, 6°) dipole fields (Fig. 3). Vectors lying in a 50° arc around the normal or reverse directions are considered as normal or reverse, respectively. Vectors lying outside the 50° arc are considered to be intermediate directions. The data was then classified and statistical parameters were derived for each class. These are given in Table 1.

Intermediate polarity often indicates an incomplete removal of the normal secondary component. The mean value of inclination for the normal group is higher than would be expected, while the mean value of inclination for the reverse group should be positive. These anomalies may be associated with incomplete removal of the normal secondary components acquired in the present-day field, or they may be due to the addition of anhysteretic components acquired during high alternating field treatments. What is clear is that the increase in inclination values during stepwise treatments shows that this anomaly is more likely to be due to the demagnetization process than to errors associated with the sampling or cutting. Since the sign of the inclination value makes their use difficult, the polarity determinations have had to be inferred mainly from the declination data.

Magnetostratigraphy

The Kamare section corresponds stratigraphically to the upper part of the Kipalagu section. The lithological variations in both sections are thus approximately the same, although bed thicknesses can be slightly different. Because the correlation of the sites is based on marker beds, the Kamare plot has been adjusted to fit in with Kipalagu.

The variation in mean declination of the stable magnetization in the two sections is presented in Fig. 4. We define five normal polarity zones (N1–N5) and five reverse polarity zones (R1–R5) in the Kipalagu section. The top of the Kamare record shows the same downward succession of zones N5, R5 and N4. The sampling gap at 10–20 m corresponds to zones R4, N3 and R3 of the Kipalagu section. However, the normal polarity in the middle part of the Kamare record must be considered carefully. NRM intensity was low and the limit of the magnetometer was reached after 30–40 mT treatments. In fact, the changing status of magnetization could not be observed with further treatment, and the mean directions obtained are, in some cases, closer to intermediate than to normal values. Considering the reverse polarity of the same levels in the Kipalagu section, we have to acknowledge the possibility of a normal overprint in this middle part of the Kamare section.

A synthetic magnetostratigraphic column is derived from the polarity values derived for the Kipalagu section and for the upper and lower parts of the Kamare section. This is given in Fig. 5.

Stratigraphic interpretation

Radiometric dating

The Hajaro basalt, which underlies the Humbu Formation (Fig. 1), is reversely magnetized (Reilly, unpub. data, cited in Isaac 1967). Its K-Ar age is 2.0 Ma (Curtis, unpub. data, cited in Isaac 1967). Other K-Ar ages have been obtained for two intercalated basalt flows described by Isaac in the Humbu and Moinik Formations at, respectively, 20 km S and 8 km N of the Kipalagu and Kamare area (Isaac & Curtis 1974). The Wa

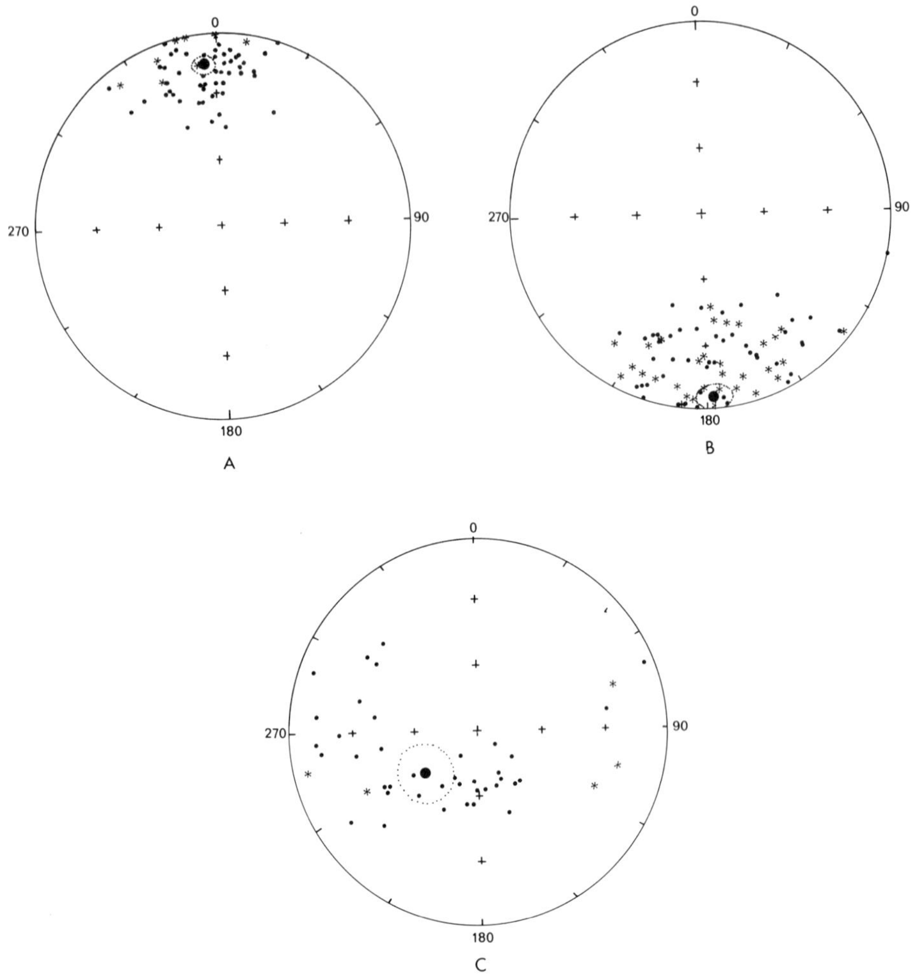

FIG. 3. Stereographic projections of sample characteristic magnetization after treatment. A. Normal polarity. B. Reverse polarity. C. Indeterminate polarity. Dots = negative inclination; stars = positive inclination; dotted enclosures = intersection of cone of confidence with sphere.

TABLE 1. *Parameters of Natron cleaned magnetization*

Polarity groups	Mean declination (deg)	Mean inclination (deg)	K*	α†deg	No. of samples
Normal	4.4	−16.0	15.2	4.3	68
Reverse	177.5	−6.0	6.7	6.0	79
Intermediate	233.0	−59.0	—	—	43

* Fisher's precision parameter.
† Radius of base of cone of confidence at 0.05 significance level.

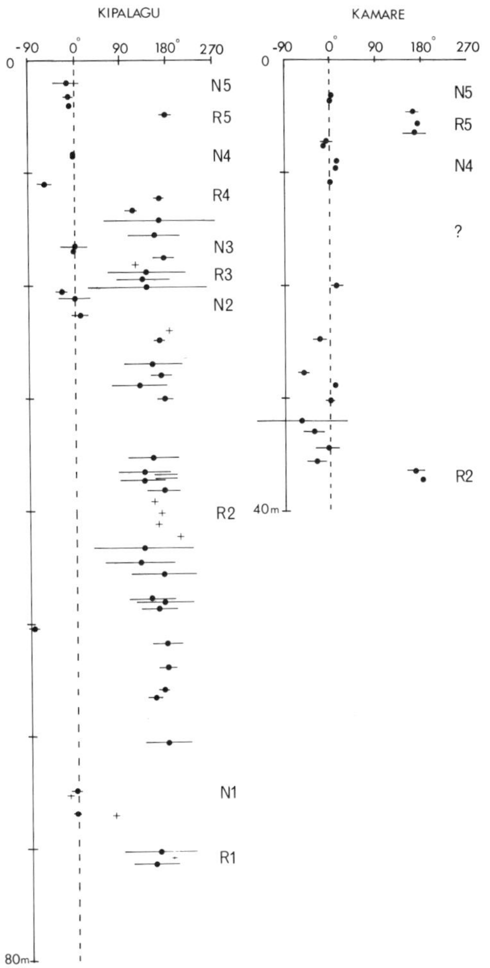

FIG. 4. Mean declination of cleaned magnetization (horizontal axis) in relation to stratigraphic position (vertical axis) in the Kipalagu and Kamare sections. Dots = mean values for each level; horizontal lines = error bars; crosses = single values. Normal (N) and reverse (R) polarity zones are indicated by N1–N5 and R1–R5.

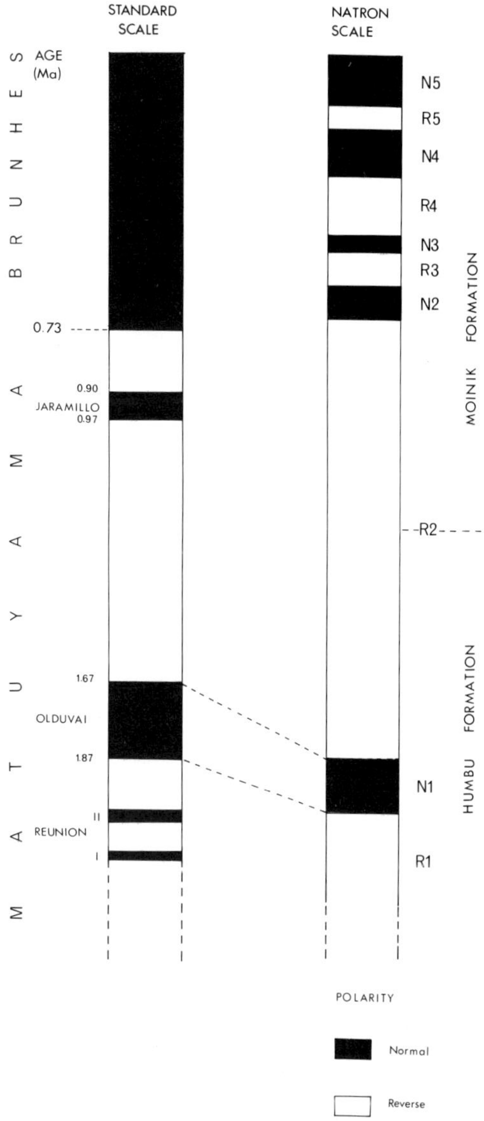

FIG. 5. Synthetic magnetostratigraphy of the Natron Pleistocene Formations correlated with the standard geomagnetic polarity time scale.

Mbugu basalt which is intercalated in the Humbu basaltic tuff in the middle part of the formation, is normally magnetized; the least altered samples give ages between 1.55 ± 0.03 Ma and 2.27 ± 0.06 Ma. The mean age is 1.91 Ma. It was probably laid down during the Olduvai or Reunion Events. The Intra-Moinik basalt found at the base of the Moinik Formation gave very consistant ages: 1.33 ± 0.05 Ma and 1.38 ± 0.09 Ma.

Biostratigraphy

The most significant data obtained on the macrovertebrates come from the study of the bovids collected in the middle part of the Humbu Formation (Gentry & Gentry 1978) and from the following species collected in the upper part of the same formation: *Elephas recki*, *Tragelaphus strepticeros grandis*, *Pelorovis* sp., *Megalotragus* sp., *Kobus* and *Parmularius angusticornis* (Geraads 1986). All three authors propose a correlation of the middle and upper parts of the Humbu Formation, respectively, with the middle or upper part of bed II and the upper part of bed II or lower part of bed III at the neighbouring Olduvai site (Fig. 1). Since the upper limit of the Olduvai

Event (1.67 Ma) lies at the bottom of bed II (Hay 1976), this would imply an age younger than 1.67 Ma for the complete Natron sequence. On the other hand, microvertebrates from the middle part of the Humbu Formation, such as *Arvicanthis primaevus*, *Thallomys quadrilobatus* and *Mastomys minor*, show affinity with those of bed I at Olduvai (Denys 1986). They imply that the Humbu Formation is older.

Despite conflicting evidence all the arguments allow us to assign an age younger than 2 Ma to the Natron record. Our proposed magnetostratigraphy can be referred to the standard geomagnetic time scale (Mankinen & Dalrymple 1979; Fig. 5). The age given by the Wa Mbugu basalt and the microvertebrate evidence support the attribution of N1 to the Olduvai Event. The age of the Intra-Moinik basalt is in good agreement with the reverse polarity of R2 and is assignable to the middle part of the Matuyama Epoch. Unfortunately, the succession of four normal zones, N2, N3, N4 and N5, does not fit the simple pattern of the standard scale.

The stability of normal magnetization is well established. An overprint (chemical or crystalline) might have masked the primary magnetization of the Natron deposits. The same problems have been encountered in many of the sedimentary sections of the Koobi Fora Formation leading Hillhouse *et al.* (1977) to consider the thinness of the normal zones to be as a result of chemical overprints in a normal field. On the contrary, if normal polarity were to be considered as the primary depositional magnetization, N5 and N4 might be correlated with the beginning of the Brunhes epoch and the Jaramillo Event, respectively. If this were the case, N3 and N2 might be identified as short normal events preceding Jaramillo, in a similar manner to the way in which the Cobb mountain event, dated *c.* 1.1 Ma, is interpreted (Mankinen *et al.* 1978).

Conclusions

The normal polarity zone N1 of the Humbu Formation is a manifestation of the Olduvai Event in the Matuyama Epoch. This correlation implies that the 5 m thick sequence of basaltic tuffs at Natron must be the stratigraphic equivalent of the 25–30 m thick sequence at Olduvai bounded by tuff Ia and tuff IIa. These are dated *c.* 1.7 and 1.8 Ma (Curtis & Hay 1972). Since no discontinuity of sedimentation appears in the stratigraphy, there are obvious and important differences in sedimentation rates for the early Pleistocene in neighbouring basins in the southern part of the East African Rift.

ACKNOWLEDGEMENTS: We wish to thank J.P. Valet and P. Tucholka for useful discussions and criticism of the results and for technical assistance; I. MacDougall for useful comments; and G. Isaac who initiated the Lake Natron project. We are grateful to A. Mturi, Director of the Antiquities Department, Ministry of Culture, Dar es Salaam, and to the Tanzanian authorities for research permits.

References

CURTIS, G.H. & HAY, R.L. 1972. Further geological studies and K/Ar dating at Olduvai Gorge and Ngorongoro crater. In: BISHOP, W.W. & MILLER, J.A. (eds) *Calibration of Hominid Evolution*. Scottish Academic Press, Edinburgh, 289–302.

DENYS, C. 1985. Micromammals from the west Natron Pleistocene deposits (Tanzania). Biostratigraphy and palaeoecology. In: FRITZ, B. & TAIEB, M. (eds) *Lac Natron*. Sci. Géol. Mém., Inst. Géol. Univ. Strasbourg (in press).

GENTRY, A.W. & GENTRY, A. 1978. Fossil Bovidae (Mammalia) of Olduvai Gorge, Tanzania. *Bull. Br. Mus. (N.H.)* **29,** 289–446; **30,** 1–83.

GERAADS, D. 1986. La faune des dépôts pleistocènes de l'ouest du Lac Natron (Tanzanie), interprétation biostratigraphique. In: FRITZ, B. & TAIEB, M. (eds) *Lac Natron*. Sci. Géol. Mém., Inst. Géol. Univ. Strasbourg (in press).

HAY, R.L. 1976. *Geology of the Olduvai Gorge*. University of California Press, Berkeley, 203pp.

HILLHOUSE, J.W., NDOMBI, J.W.M., COX, A. & BROCK, A. 1977. Additional results on palaeomagnetic stratigraphy of the Koobi Fora Formation, east lake Turkana (Lake Rudolf) Kenya. *Nature* **265,** 411–15.

ISAAC, G. 1967. The stratigraphy of the Peninj Group. Early and middle Pleistocene formations, west of Lake Natron, Tanzania. In: BISHOP, W.W. & CLARK, J.D. (eds) *Background to Evolution in Africa*. University of Chicago Press, Chicago, pp.229–57.

——& CURTIS, G.H. 1974. Age of early Acheulian industries from the Peninj group, Tanzania. *Nature* **249,** 624–7.

LEAKEY, L.S.B. & LEAKEY, M.D. 1964. Recent discoveries of fossil hominids in Tanganyika at Olduvai and near Lake Natron. *Nature* **209,** 3–9.

MANKINEN, E.A. & DALRYMPLE, G.B. 1979. Revised geomagnetic polarity time scale for the interval 0–5 My BP. *J. Geophys. Res.* **84,** 615–25.

——, DONNELLY, J.M. & GROMME, C.S. 1978. Geomagnetic polarity event recorded at 1.1 My BP on Cobb Mountain Clear Lake volcanic field, California. *Geology* **6,** 653–6.

MOLYNEUX, L. 1971. A complete result magnetometer for measuring the remanent magnetization of rocks. *Geophys. J. R. Astr. Soc.* **24,** 429–33.

ZIJDERVELD, J.D.A. 1967. AC demagnetization of rocks: analysis of results. In: COLLINSON, D.W., CREER, K.M. & RUNCORN, S.K. (eds) *Methods in Palaeomagnetism*. Elsevier, Amsterdam, pp.254–86.

N. THOUVENY & M. TAIEB, Laboratoire de Géologie du Quaternaire, CNRS—Case 907 Luminy, 13288 Marseille, Cedex 9, France.

Sedimentation and the Preservation of Fossil Faunas

Role of tephra in the preservation of fossils in Cenozoic deposits of East Africa

R.L. Hay

SUMMARY: Volcanism has been particularly important in preservation of fossils at some East African sites. It has supplied sediment for burial, and carbonatite ash which has chemically aided fossilization. Moreover, subaerial tuffs have preserved the remains of faunas unlike those of riverine and lacustrine environments. Examples include faunal sites at Napak, Uganda; Songhor and Koru, Kenya; and Laetoli, Tanzania. Some, and perhaps most, of the carbonatite ash erupted from East African volcanoes was natrocarbonatite, alkali-rich and broadly similar to the modern natrocarbonatite. The pre-modern natrocarbonatites have been leached of their alkalies by meteoric water. In some examples the original natrocarbonatite composition can be reliably inferred from textural and chemical evidence; in other examples only the possibility or probability of former natrocarbonatite ash can be demonstrated. Fossiliferous tuffs at Napak, Koru, Songhor, and the Footprint Tuff at Laetoli contain good evidence of natrocarbonatite ash; evidence is suggestive for Rusinga, and for the fossiliferous aeolian tuffs of the Laetolil Beds at Laetoli.

One of the themes of Bill Bishop's work was that volcanism has been particularly important in preserving faunal remains in East Africa (Bishop 1963, 1968). Volcanic activity was a source of sediment for burial, and in East Africa it supplied carbonatite ash, which chemically aided fossilization. Moreover, subaerial ash falls have preserved faunal samples of a type not generally represented in riverine or lacustrine sediment. He based his theme on observations at Napak, Uganda (Bishop 1958, 1962), then extended it to Miocene sites in Kenya, including those at Songhor, Koru, and Fort Ternan (1963, 1968). Burial of faunal remains was relatively easy to demonstrate, as was the distinctive nature of faunas in subaerial tuffs of Napak, Songhor, and Koru. The former presence of carbonatite ash was inferred on the basis of abundant calcite in fossiliferous tuffs erupted from volcanoes with bodies of carbonatite, and trace-element analyses of tuffaceous limestones and calcareous tuffs (Bishop 1968).

The present paper will focus on the nature of carbonatite tephra in East Africa, on methods for documenting its former presence, on its possible role in fossilization, and on fossil sites where carbonatite ash can be demonstrated or inferred.

Eruptive carbonatite

Carbonatite eruptive rocks have been widely recognized in East Africa, generally associated with alkaline volcanism (e.g. Le Bas 1977). The only known modern carbonatite eruptions were from Ol Doinyo Lengai, in northern Tanzania (Fig. 1), which has discharged both lavas and ash (Dawson 1962). These carbonatites are of alkali carbonatite, termed natrocarbonatite, which are about 25% calcium carbonate and 75% alkali (principally sodium) carbonate. Mineralogically the lavas consist of nyerereite, $Na_2Ca(CO_3)_2$, with minor fairchildite $[K_2Ca(CO_3)_2]$ in solid solution, and gregoryite, which is Na_2CO_3 with some $CaCO_3$ in solid solution. By contrast, all known

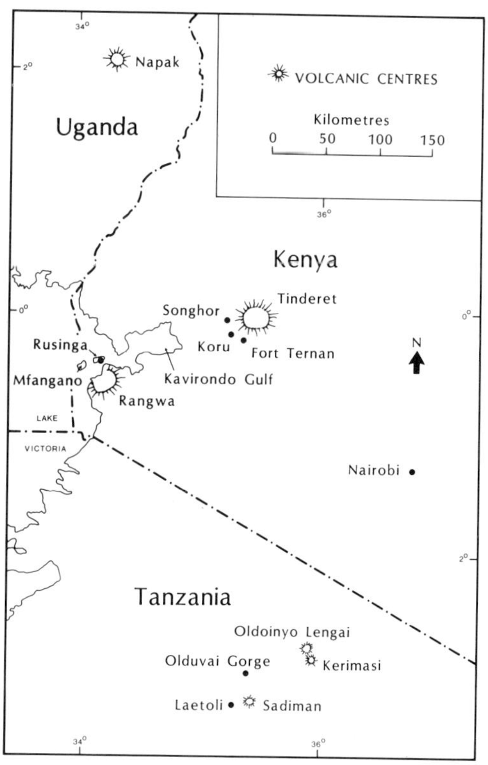

FIG. 1. Map showing volcanoes and localities referred to in text.

From FROSTICK, L.E. *et al.* (eds) 1986, *Sedimentation in the African Rifts*, Geological Society Special Publication No. 25, pp. 339–44.

pre-modern examples are of calcitic, or rarely dolomitic, composition. Recent work on pre-modern carbonatite volcanic rocks of Kerimasi volcano in Tanzania (Hay 1983) and the Tinderet centre in Kenya (Deans & Roberts 1984) has shown that these calcitic carbonatites were originally natrocarbonatite and have been leached of their alkalies by meteoric water. These carbonatites originally consisted of nyerereite with subordinate calcite.

Nyerereite reacts incongruently with water, releasing alkali carbonates into solution and precipitating fine-grained calcite. This reaction can assist fossilization in several ways. The dissolved sodium carbonate will crystallize on evaporation, providing an 'instant' cement. Dissolved sodium carbonate creates an alkaline environment in which bone apatite and calcium carbonate fossils are relatively insoluble. The precipitated calcium carbonate can form a matrix suitable for the preservation of delicate materials, and it can, under appropriate conditions, be redeposited to replace or impregnate organic remains and fill fossil moulds.

Carbonatite tephra can be identified to varying degrees of probability by mineralogical, textural and chemical evidence. Crystals of primary, carbonatitic calcite are coarse, clear, can be mechanically twinned, and commonly contain apatite inclusions. Nyerereite occurs as tabular crystals with a single prominent cleavage, which may be represented by sheets of calcite in porous pseudomorphs (Fig. 2). Calcite replacing nyerereite in the Kerimasi natrocarbonatites is commonly cloudy in thin section, and white and chalklike in reflected light. Pseudomorphs after nyerereite are easy to recognize in rock fragments preserving the primary igneous texture, but they can be difficult or impossible to recognize in tuffs where the original nyerereite crystals were broken and anhedral.

Both sovites and natrocarbonatite are isotopically light in ^{18}O. Most sovites have $\delta^{18}O$ between -6 and -9‰ (SMOW) (Deines & Gold 1973), and natrocarbonatite lava of Oldoinyo Lengai has $\delta^{18}O$ of *c.* $+7$‰ (O'Neil & Hay 1973). Isotopic exchange with meteoric water may be incomplete in the alteration of natrocarbonatite to form calcite (Hay & O'Neil 1983), and the secondary calcite will be isotopically intermediate between unaltered carbonatite and calcite precipitated in equilibrium with meteoric water. Incomplete exchange of $\delta^{18}O$ is demonstrated by carbonatite ejecta of Kerimasi (Table 1). Unaltered carbonatitic calcite has $\delta^{18}O$ values of $+7.2$ to 7.5‰, and a sample of post-volcanic pedogenic calcite deposited from meteoric water has $\delta^{18}O$ of $+29.4$‰. Calcite replacing nyerereite is intermediate between the igneous and pedogenic calcites. Significant amounts of diagnostic trace elements (e.g. Sr, Ba) may be preserved in the replacement of nyerereite by calcite or in early-deposited calcite cement derived from the alteration of natrocarbonatite ash (Hay & O'Neil 1983).

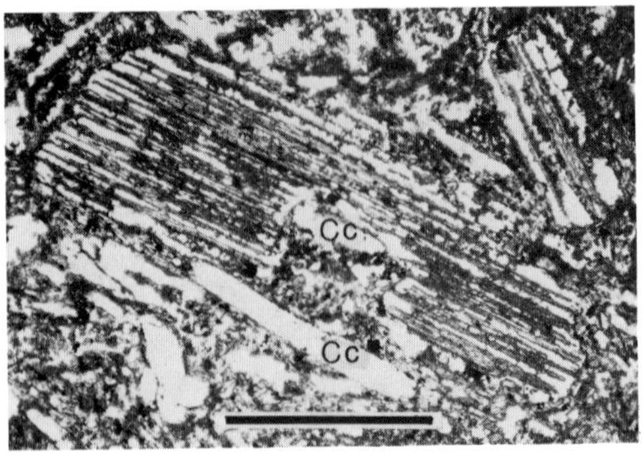

FIG. 2. Photomicrograph of altered porphyritic natrocarbonatite of the volcano Kerimasi. From lower right to upper left is a large phenocryst of nyerereite now represented by a porous pseudomorph with sheets of calcite that parallel a prominent cleavage plane of nyerereite. Crystals of primary carbonatitic calcite are indicated by Cc. Bar scale is 1 mm long.

TABLE 1. *Isotopic composition of carbonatitic calcite, calcite replacing nyerereite, and pedogenic calcrete of Kerimasi volcano, Tanzania**

	$\delta^{18}O$ (SMOW) ‰	$\delta^{13}C$ (PDB) ‰
Calcite of sovite clast	+7.4	−4.3
Calcite of sovite clast	+7.5	−4.2
Calcite phenocrysts of natrocarbonatite lava clast†	+7.2	−4.3
Calcite replacing nyerereite phenocrysts†	+21.2	−1.4
Calcite replacing nyerereite in natrocarbonatite lava clast	+26.4	−3.2
Calcite replacing nyerereite in coarse natrocarbonatite	+22.2	−3.0
Calcite replacing nyerereite in coarse natrocarbonatite	+25.2	−3.3
Pedogenic calcrete overlying Kerimasi tephra deposits	+29.4	−3.3

* Analyses are by J.R. O'Neil, US Geological Survey, Menlo Park, California.
† From same sample.

Miocene of Napak, Uganda

Several rich faunal sites are in tuffs that were deposited on the slopes of Napak, a deeply eroded volcano in eastern Uganda (Bishop 1958, 1962). Lavas are chiefly nephelinite, and carbonatite forms a central plug (King 1949). The fossiliferous tuffs are 30–40 m thick and are subaerial though perhaps locally sorted by water. The tuffs are composed largely of rock fragments and crystals cemented by calcite.

The sites are rich in rodents and contain an unusually high proportion of primates. Non-mammalian fossil materials comprise wood, seeds, fruits and land gastropods. Bishop envisaged the fossilization process as follows (1968, p. 35): 'Natural deaths allowed gradual accumulation of potential fossils over all the slopes of the volcano as bone became scuffed into and covered by the calcareous soils. The next layer of wind-borne ash effectively buried the bone and allowed mineralization to take place. The chemistry of the environment of burial seems to have been crucial and it is likely that the fine matrix was rich in calcite as the ash came from a carbonatitic centre. The primary calcareous nature of the tuff is difficult to prove owing to recrystallization of the calcite.'

A sample collected by the writer on a visit to Napak site I in 1964 documents the occurrence of carbonatite ash. The sampled bed, considered a tuffaceous limestone by Bishop, is a natrocarbonatite-bearing tuff. The sample is *c.* 80% calcite and 20% non-carbonate materials, principally volcanic rock fragments, sodic augite, plagioclase and biotite. The calcitic fraction is *c.* 50% cement, 40% clasts of altered natrocarbonatite, and 10% sovite fragments and crystals of carbonatitic calcite. The natrocarbonatite consists largely of nyerereite replaced by cloudy calcite (see Hay 1983); the sovite and carbonatitic calcite are clear and unaltered.

Kavirondo region of Kenya

Many Miocene fossil sites in the Kavirondo region of western Kenya are in tuffs. Among the best known are Songhor, Koru and Rusinga. The fossiliferous tuffs at Songhor and Koru are dominantly subaerial (Bishop 1963) and originated in eruptions of the Tinderet volcanic complex. Carbonatite is among the eruptive rocks of Tinderet (Pickford 1982). The vertebrate fauna, like that of Napak, is rich in rodents and primates. The trace-element composition of some highly calcareous tuffs and tuffaceous limestones falls within the range of carbonatites (Bishop 1968), and Bishop suggested that 'primary falls of carbonatitic ash seem likely to have been an important factor in the preservation of bone at several fossiliferous localities' (1968, p.38). Bishop's conclusion is fully supported by Deans & Roberts (1984), who convincingly demonstrate the occurrence of calcified natrocarbonatite ash and lava clasts in, and associated with, the fossiliferous deposits. These are similar in composition to the Kerimasi natrocarbonatites, consisting largely of nyerereite replaced by calcite. Primary calcite is also present. At Koru, mammalian fossils are found mainly in soil horizons which separate beds of natrocarbonatite tuff. At Songhor, mammal fossils are in clayey redbeds intercalated within volcaniclastic carbonatite deposits. Fossil leaves are in some of the carbonatitic accretionary lapilli tuffs. Thin natrocarbonatite tuffs are characterized by polygonal cracks, as at Laetoli (see below).

Miocene strata of Rusinga Island are chiefly of volcaniclastic materials from the volcano Kisingiri, or Rangwa, which contains a plug of carbonatite (Le Bas 1977). Fossils are in both subaerial and subaqueous deposits (Bishop 1963). Of special interest from the standpoint of preservation is a 'fauna of soft-bodied animals preserved as solid objects, including caterpillars preserved in the round; animals with skeletons are also preserved. Lizards and birds are fossilized complete with their muscles, which are transformed into calcium carbonate. One lizard can be identified as belonging to a present-day group because its forked tongue is preserved' (Halstead 1982, p.43). Either fossilization was rapid, occurring before decomposition or compaction of the soft-bodied forms; or else moulds were quickly formed in fine-grained material sufficiently cohesive to preserve the shapes of the animals until they were filled by calcite.

The nature of fossilization and of the enclosing matrix have not been described, as far as the writer is aware. As a first step to understanding this unusual occurrence, two samples of matrix and one fossil caterpillar were examined in thin section. The samples of matrix were generously provided by Dr Martin Pickford. Sample R-106 is from the basal layers of Kaswanga Point Tuffs, yielding fossil insects, fruit and leaves. Sample R-126 is from a new site in the Kiahera Formation that yielded a beautifully preserved spider, beetles, fruit and leaves. Both samples are relatively fine-grained calcareous melilititic tuffs, and R-106 contains accretionary lapilli (ash pisoliths). Clay clasts 20–120 μm in diameter are a major non-volcanic component of both samples. The clay clasts are considered an aeolian component, perhaps eroded from a saline mudflat. Calcite forms an estimated 50–60% of both samples and occurs as a cement, a replacement of melilitite and unidentified minerals, and as rare grains of carbonatitic calcite. The latter were identified on the basis of apatite inclusion and mechanical twinning. The tuffs are poorly sorted, and are very likely subaerial deposits. The caterpillar, supplied by Dr M.D. Leakey, consists of clear, coarse calcite which grew inward from the margin of the fossil. Thus the fossilizing calcite filled a mould rather than replaced the tissue of the caterpillar.

Natrocarbonatite ash could have contributed to the rapid formation of a matrix suitable for retaining the moulds of soft-bodied forms. It could also have supplied the calcium carbonate that was later deposited in the mould. Textural evidence of former natrocarbonatite ash is lacking in these samples, as it is in many tuffs of inferred natrocarbonatite origin in the Laetolil Beds; and further work is needed to determine whether or not natrocarbonatite was originally present in these tuffs of Rusinga Island. It should be noted that Deans & Roberts (1984) found natrocarbonatite lava clasts in deposits of Mfanganu Island, only 11 km from Rusinga.

Laetolil Beds at Laetoli, Tanzania

Numerous fossils representing many taxa have been collected from the mid-Pliocene Laetolil Beds at Laetoli, northern Tanzania (Dietrich 1942; Leakey *et al.* 1976). The Laetolil Beds are composed of tuffs erupted from the volcano Sadiman, 15–20 km distant. Only the Upper Unit of the Laetolil Beds, 45–60 m thick, will be considered here, as it has yielded nearly all of the vertebrate fossils at Laetoli.

The Upper Unit is *c.* 80% reworked tuff and 20% air-fall tuff. Nearly all of the reworked tuffs are of air-fall ash redeposited by wind, and are termed aeolian tuff. Air-fall tuffs comprise a thick phonolitic tuff at the top of the formation, and numerous thinner tuffs of inferred melilitite-carbonatite and natrocarbonatite composition which occur at many levels. Tuffs of the Laetolil Beds were deposited on an upland savannah, relatively distant from lakes and major rivers (Hay 1981).

Air-fall tuffs

Melilitite-carbonatite tuffs are of two types: medium to coarse-grained tuffs consisting largely of lava globules and crystals, and fine-grained tuffs consisting largely of calcite (Hay 1978). The tuffs mantled a flat to gently undulating land surface and were reworked by streams in a few places. Curiously, a polygonal fracture pattern is exhibited by nearly all the air-fall tuffs less than 10 cm thick (Fig. 3). Tuff polygons were locally displaced by bioturbation while the underlying aeolian sediments were still unconsolidated. Thus the tuffs were cemented soon after deposition, and this early cementation is attributed to natrocarbonatite ash.

The finer-grained tuffs are finely crystalline calcite with subordinate amounts of biotite, melilitite and other minerals, which float in a matrix of calcite. About half these tuffs contain calcite globules 0.05–3.0 mm in diameter (Hay 1978, Figs. 5 and 6). The calcite has a pore-filling texture, and the globules were earlier interpreted as casts of natrocarbonatite droplets. It now seems more likely that they are the fillings of air bubbles formed near the surface (Hay & O'Neil 1983). Alternatively, they may be pseudomorphs after gregoryite. Neither crystals of primary carbonatitic calcite nor recognizable pseudomorphs after nyerereite were identified in the tuffs. The carbonatite was presumably alkali rich, like that of Ol Doinyo Lengai, in view of the apparent lack of primary carbonatitic calcite crystals.

The medium to coarse-grained tuffs consist of sand-size nephelinite and melilitite lava particles (Fig. 4) and crystals cemented by calcite and, less commonly, the zeolite phillipsite. The best evidence for the former presence of natrocarbonatite tuff is in the Footprint Tuff, which was studied in much more detail than the others. In the Footprint Tuff, tens of thousands of footprints are exposed on bedding planes (Hay & Leakey 1982). The tuff is generally 12–15 cm thick and is

FIG. 3. Melilitite-carbonatite air-fall tuff of the Laetolil Beds showing characteristic fracture pattern. Upper surfaces of tuff polygons were rounded before burial by aeolian tuff.

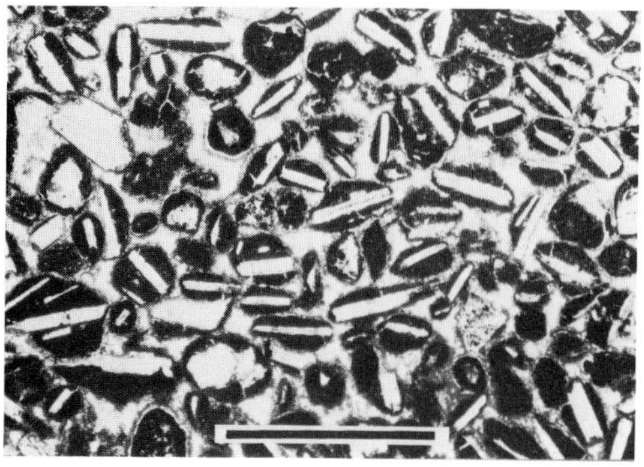

FIG. 4. Photomicrograph of air-fall tuff of the Laetolil Beds consisting of lava globules cemented by calcite. Tabular crystals in globules are melilitite altered to phyllosilicate. Relics of unaltered melilitite are found in some other samples. Bar scale is 1 mm long.

subdivided into 18 widespread laminae, each of which represents a fall of ash. The lower 14 layers were deposited in the latter part of the dry season, and the upper four after the rainy season had begun.

Evidence for natrocarbonatite ash is of several types. The earliest deposited calcite cement is brown and sparry, unlike the later calcite cement which is colourless or white, and micrite to spar in crystal size (Hay 1978). Crystals of brown calcite may be isolated and are commonly rounded, as if the early cement was originally patchy, and the crystals were partly dissolved before the later cement was deposited around them. Some of the isolated crystals were earlier misinterpreted as pseudomorphs after nyerereite (Hay & Leakey 1982). The brown calcite has a carbonatite chemical signature in the form of relatively high contents of Sr (0.57%) and Ba (0.14%) and a $\delta^{18}O$ value of 22‰, significantly less than that of later deposited Sr-poor calcite cement (26‰) (Hay & O'Neil 1983). Fenestral and cement-supported textures are both common and widespread in several layers of the tuff, indicating the former presence of a substantial amount of some constituent, most likely natrocarbonatite ash, that was dissolved and/or recrystallized. An early role of the natrocarbonatite may have been to provide a fine-grained matrix that made the sand-size lava globules cohesive enough to retain clear, steep-sided footprints, just as the addition of fine-grained material such as Portland cement or wood ash will make dry sand more cohesive. If the carbonatite had been fine grained initially, a suitable primary matrix would have been supplied at the time of deposition. Subsequent reaction of natrocarbonatite with water would have provided an even finer-grained secondary matrix of precipitated calcite. Meanwhile, unless the sodium carbonate was entirely leached by rainfall, it would crystallize into trona as the water evaporated, weakly cementing the surface layer of ash and stabilizing the footprints until they were buried by the next fall of ash. Later, altered natrocarbonatite ash of the Footprint Tuff and an overlying much thicker tuff were very likely the source of calcium carbonate required for calcification of dung and plant remains in the tuff, and for cementing both the tuff and *c.* 10 cm of the underlying aeolian tuffs.

Aeolian tuffs

The aeolian tuffs are massive to crudely bedded and have been extensively modified by bioturbation, especially by termites (e.g. Hay 1981, Fig. 4). They are dominantly of medium sand size, and particles are chiefly lava globules and crystals similar to those of the melilitite-carbonatite air-fall tuffs. The vertebrate fauna is both rich and varied, and includes many small, delicate bones. Well-preserved bird eggs occur widely and land snails are common. The assemblage is characteristic of an upland savannah, which is further supported by the lack of hippo and crocodile (Leakey *et al.* 1976).

Aeolian tuffs accumulated at an average rate of 20–25 cm ka^{-1}, on the basis of K-Ar dating of tuffs near the bottom and top of the fossiliferous deposits (Leakey *et al.* 1976). Many if not most of the fossiliferous deposits were weathered to a significant extent before burial. An alkaline soil environment, saturated with respect to calcium carbonate, was required for the preservation of delicate calcium carbonate fossils in weathered ash deposits. A supply of calcite was required to cement the tuffs, for only about a third of the calcite in the tuffs could have been derived from weathering of the silicate tephra (Hay 1978). Natrocarbonatite ash would have provided both an alkaline soil environment and a supply of calcium carbonate. Moreover, if evidence of natrocarbonatite ash is accepted for the air-fall tuffs, then natrocarbonatite ash must be considered a probable constituent of the aeolian tuffs, which differ from the air-fall tuffs principally in being reworked.

Conclusions

Many East African faunal sites are in subaerial deposits of volcanic ash, both primary and reworked. Examples include Napak, Uganda; Songhor, Koru and Rusinga, Kenya; and Laetoli, Tanzania. Several sites at Napak are in air-fall tuffs, as are some of the sites of Songhor and Koru. Footprints at Laetoli are in air-fall tuffs, and the faunal remains are in aeolian tuffs. In these examples, subaerial tuffs have preserved remains of faunas unlike those of riverine and lacustrine sediments. The Laetoli fauna is that of an upland savannah, relatively distant from a major river or lake.

Carbonatite eruptive rocks are widespread in East Africa, and carbonatite eruptions contributed ash to some fossiliferous deposits. Some, and perhaps most, of this eruptive carbonatite magma was alkali rich, or natrocarbonatite. In some examples, the former presence of natrocarbonatite can be determined on textural and chemical grounds. In less favourable examples, the likelihood of natrocarbonatite in tuffs can be inferred to some extent on field relations and the content of calcite and associated materials.

Natrocarbonatite ash can aid in fossilization when altered by contact with water. It can provide (1) an 'instant' cement of sodium carbonate (= trona); (2) an alkaline environment in which bone apatite and calcium carbonate fossils are relatively insoluble; (3) a fine-grained matrix for preserving footprints and moulds of soft-bodied organisms; and (4) a supply of calcium carbonate for calcification or coating of bones, replacement of plant material, and filling of fossil moulds.

ACKNOWLEDGEMENTS: The writer is indebted to W.W. Bishop for a visit to fossiliferous sites at Napak in 1964. Isotopic measurements on Kerimasi carbonatites and related materials were generously provided by J.R. O'Neil of the US Geological Survey.

References

BISHOP, W.W. 1958. Miocene Mammalia from the Napak volcanics, Karamoja, Kenya. *Nature* **182,** 1480–2.

—— 1962. Tertiary mammalian faunas and sediments in Karamoja and Kavirondo, East Africa. *Nature* **196,** 1283–7.

—— 1963. The later Tertiary and Pleistocene in eastern equatorial Africa. In: HOWELL, F.C. & BOURLIERE, F. (eds) *African Ecology and Human Evolution.* Aldine Publishing Co., Chicago, pp.246–75.

—— 1968. The evolution of fossil environments in East Africa. *Trans. Leics. Lit. Philos. Soc.* **62,** 22–44.

DAWSON, J.B. 1962. The geology of Oldoinyo Lengai. *Bull. Volc.* **24,** 349–87.

DEANS, T. & ROBERTS, B. 1984. Carbonatite tuffs and lava clasts of the Tinderet foothills, western Kenya; a study of calcified natrocarbonatites. *J. Geol. Soc.* **141,** 563–80.

DEINES, P. & GOLD, D.P. 1973. The isotopic composition of carbonatite and kimberlite carbonates and their bearing on the isotopic composition of deep-seated carbon. *Geochim. Cosmochim. Acta* **37,** 1709–33.

DIETRICH, W.D. 1942. Altesquartare Saugetiere aus der südlichen Serengeti, Deutsch-Ostafrica. *Palaeontographica* **94,** 43–133.

HALSTEAD, L.B. 1982. *The Search for the Past.* Doubleday & Co., Garden City, NY.

HAY, R.L. 1978. Melilitite-carbonatite tuffs in the Laetolil Beds of Tanzania. *Contrib. Mineral. Petrol.* **67,** 357–67.

—— 1981. Paleoenvironment of the Laetolil Beds, northern Tanzania. In: RAPP, G.R. & VONDRA, C.F. (eds) *Hominid Sites: Their Geologic Settings.* American Association for the Advancement of Science, Selected Symposium No. 63, pp.7–24.

—— 1983. Natrocarbonatite tephra of Kerimasi volcano, Tanzania. *Geology* **11,** 599–602.

—— & LEAKEY, M.D. 1982. The fossil footprints of Laetoli. *Sci. Am.* **246,** 50–7.

—— & O'NEIL, J.R. 1983. Carbonatite tuffs in the Laetolil Beds of Tanzania and the Kaiserstuhl in Germany. *Contrib. Mineral. Petrol.* **82,** 403–6.

KING, B.C. 1949. The Napak area of Southern Karamoja, Uganda, *Mem. Geol. Surv. Uganda* **5,** 59 pp.

LEAKEY, M.D., HAY, R.L., CURTIS, G.H., DRAKE, R.E., JACKES, M.K. & WHITE, T.D. 1976. Fossil hominids in the Laetolil Beds. *Nature* **262,** 460–6.

LE BAS, M.J. 1977. *Carbonatite-Nephelinite Volcanism.* J. Wiley & Sons Inc., London and New York.

O'NEIL, J.R. & HAY, R.L. 1973. O^{18}/O^{16} ratios in cherts associated with the saline lake deposits of East Africa. *Earth Planet. Sci. Lett.* **19,** 257–66.

PICKFORD, M. 1982. The tectonics, volcanics, and sediments of the Nyanza Rift Valley, Kenya. *Z. Geomorph. Suppl. Bd.* **42,** 1–33.

R.L. HAY, Dept of Geology, 245 Natural History Building, 1301 W. Green Street, University of Illinois, Urbana, Illinois 61801, USA.

Sedimentation and fossil preservation in the Nyanza Rift System, Kenya

M. Pickford

SUMMARY: Two major factors led to the accumulation of fossiliferous strata in the Nyanza Rift System—volcanism and tectonism. Several volcanoes have been active in, or near, the Nyanza Rift Valley during the past 20 Ma. Arranged in rings round these volcanoes were 3-D annular lithofacies dependent for their origin on various volcanic, geomorphic and pedogenic factors. Near-vent facies are characterized by predominantly coarse clastic deposits which accumulated on steep slopes (> 10°) in upland areas. They usually show few signs of pedogenesis. Further from the vent (4–15 km radially) the apron facies are represented by finer-grained volcaniclastic strata, often with significant amounts of pedogenic overprint but little evidence of deposition from, or in, water. These apron facies accumulated on gentler (3–10°) slopes, both by subaerial fallout and by radial reworking of the near-vent facies. Even further from the volcanic centre are the arena facies which are generally fine-grained and waterlain in virtually flat-lying countryside (0–3° slopes). Primary volcanic ashes are occasionally present in the arena facies. The bulk of these volcanogenic sediments accumulated on dry land. Their deposition was not especially related to existing sedimentary basins. The second major factor which was responsible for the genesis of at least part of the Nyanza sediments was the formation of basins by tectonic activity associated with rifting. Depressions were formed in which fluvial and lacustrine sediments accumulated. The bulk of these sediments represent the products of erosion of the scarp shoulders of the rift, the sloping half-graben blocks, and the upstanding volcanic forms within the rift and on its shoulders. The distribution of fossil vertebrate species in western Kenya is closely related to depositional facies. Terrestrial assemblages are particularly distinct from aquatic faunas and generally occur in strata which accumulated subaerially as part of the apron and arena facies. Aquatic faunas are found in the fluvial and lacustrine sediments. In addition, the geochemical composition of the source rocks is an important factor in fossil preservation. Sediments derived from rocks rich in lime and low in silica contain abundant vertebrate and invertebrate fossils, while those that are silica-rich and lime-poor yield remarkably few faunal remains.

Early interpretations of sediments in the Nyanza Rift Valley (Fig. 1) (Kent 1944; Shackleton 1951) were strongly influenced by the expectation that the strata had accumulated in depressions, and that these depressions had contained water. This preconception, by which all stratiform rocks in the area were called 'lake-beds', led Kent (1944) to postulate the existence of a Miocene lake in western Kenya. The Miocene lake hypothesis was widely accepted by palaeontologists with the results that most of the fossils from Nyanza were treated as though they were contemporary, and all the sites were considered to have had similar ecological conditions at the time of deposition. The 'unit fauna' and 'unit ecology' concepts greatly influenced interpretation of the fossil record. Because of this, the recorded distribution patterns were apparently confusing and synonymy problems proliferated. This, in turn, affected the biostratigraphic and phylogenetic reconstructions.

Fresh light was thrown on the problem by Bishop (1967, 1968), followed by Van Couvering (1972) and Pickford (1981, 1982, 1983). They have shown that most of the strata in western Kenya are not lacustrine in origin, and that they represent a period of 6–8 Ma during which much evolution and major environmental changes took place. It is now known that a great deal of the Miocene fossiliferous strata of western Kenya accumulated prior to the onset of local rifting, and was the result of accumulation of volcanic debris on and around volcanoes of carbonatitic-nephelinitic composition. Subsequently, in the Middle Miocene, rift depressions began forming, first as simple downwarps in which sediments accumulated (e.g. Maboko) and later as fractured half-graben in which lavas and sediments accumulated (e.g. plateau phonolites of the Kisumu area, the Kano Plains sediments, and the Nyando Ashes; Pickford 1982). With this clarification of the stratigraphic sequence, the geochronology, and the depositional environments, the fossil record from western Kenya was open to re-interpretation.

Pickford (1981) established a preliminary biostratigraphy for the area and recorded facies-controlled faunal distribution patterns. Subsequently, palaeoenvironmental interpretations were made and major environmental changes were documented (Pickford 1985). In particular, a major desiccation and vegetation change took

From FROSTICK, L.E. *et al.* (eds) 1986, *Sedimentation in the African Rifts*, Geological Society Special Publication No. 25, pp. 345–62.

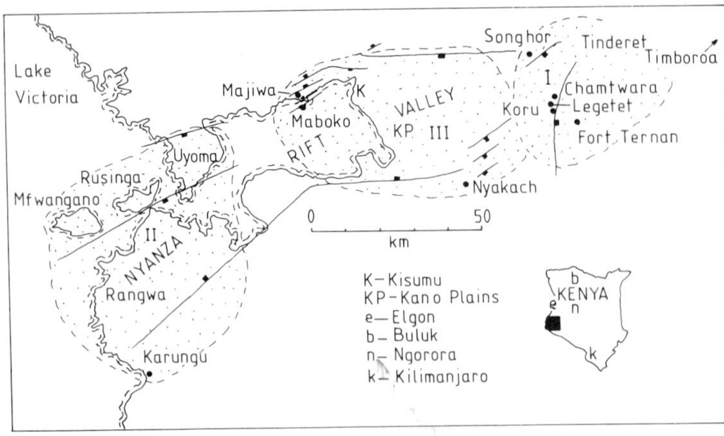

FIG. 1. Locality map of western Kenya showing principal localities mentioned in text. I. Timboroa volcanic province; II. Rangwa volcanic province; III. Maboko-Nyakach tectonic basin.

place in western Kenya, *c.* 16–17 Ma, accompanied by significant changes in both the vertebrate and invertebrate faunas.

This paper documents the depositional environments and faunal distribution patterns of western Kenya in greater detail. It also examines the underlying causes of these patterns, including the presence of the annular facies that surround volcanoes and paravolcanic depressions. Furthermore, the importance of source rock geochemistry in preservation of fossils is assessed.

Volcanogenic stratigraphy

Volcanoes generate sedimentary sequences in two ways. Firstly, they are a primary source of material that accumulates by subaerial fallout and by radial redistribution of clasts from high relief centres. Secondly, they are topographic irregularities on regional slopes that dam existing drainage both outside (e.g. Kilimanjaro, Fig. 2) and within the rift system (e.g. Silali, Suswa in the Gregory Rift), and sediments accumulate behind them. Another factor of lesser importance is the isostatic depression associated with large volcanoes such as Kilimanjaro and Elgon, which can produce sinks and encourage sediment accumulation. This generally happens much later than the active period of the volcano, since such crustal adjustments take a long time to happen.

Sedimentary patterns in and around carbonatite-nephelinite central volcanoes

In western Kenya, depositional environments are related to the geomorphology of central volcanoes and their surroundings (Fig. 3). In the Koru-Songhor area, the bulk of fossiliferous strata accumulated on the lower slopes of one such volcano (possibly Timboroa, not Tinderet as is so often implied in the literature since this is an Upper Miocene volcano sitting unconformably on Lower Miocene strata). The main topographic feature comprises the vent facies, which are coarse-grained agglomerates, and usually poorly fossiliferous except for abundant wood. These agglomerates seldom have pedogenic alteration, but have been extensively altered by zeolitization.

Surrounding the vent facies come the apron facies usually some 4–11 km from the vent and on slopes of 3–10°. At the time of formation they represented sedimentation on the lower slopes of the volcano, both by subaerial accumulation from pyroclastic fallout and by the reworking of a radial drainage, with sediments being deposited where transporting power declined as streams left the steep upper slopes of the volcano to flow over more gentle lower slopes. Sediment accumulation on the apron would have been enhanced by the presence of vegetation, which would not only slow erosion on these lower slopes, but also actively trap sedimentary particles (Pickford 1982). Under such circumstances pedogenesis took place, but was interrupted by each pyroclastic episode. Thus a superposed series of primary and reworked volcanogenic strata with variable amounts of pedogenic alteration accumulated.

Comparable volcanic apron deposits are known at Laetoli in Tanzania, and on Mfwangano and to a lesser extent at Rusinga. The latter two being associated with the central volcano Kisingiri (Rangwa). The bulk of the fossiliferous strata on Rusinga, as well as at Uyoma, Karungu

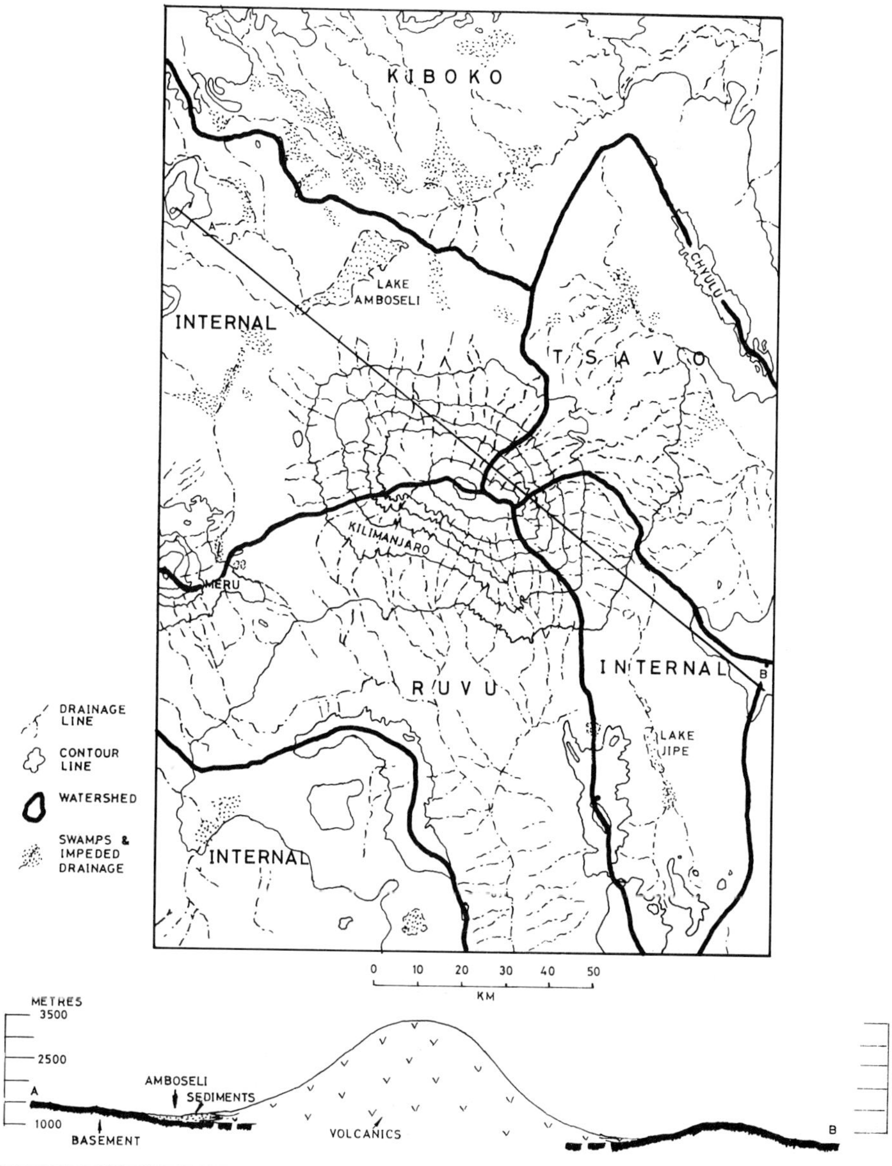

FIG. 2. Amboseli basin at the foot of Kilimanjaro central volcano. The huge volcano has interrupted the regional coastward slope forming a dam in which the Amboseli Beds have accumulated. Internal drainage basins are indicated. Lower Miocene sediments associated with the Rangwa volcano may have occupied an analogous basin. Note the isostatic downwarp. Location of Kilimanjaro is given in Fig. 1.

and at Kakesio in Tanzania accumulated, however, in situations radially further from the centres than the apron facies. I have called these the arena facies. These accumulated in even flatter countryside (slopes of $<3°$) well away from the volcanic centre, but still close enough to be subjected to occasional pyroclastic fallout, especially in downwind positions. The bulk of the sediment which accumulated as arena deposits was, however, epiclastic in origin, and the arena facies consequently exhibit abundant evidence of both running and still water deposition. Braided stream, floodplain, and lacustrine sediments all occur, as do occasional fine-grained pyroclastic intercalations.

As time passes, and as the edifice grows with

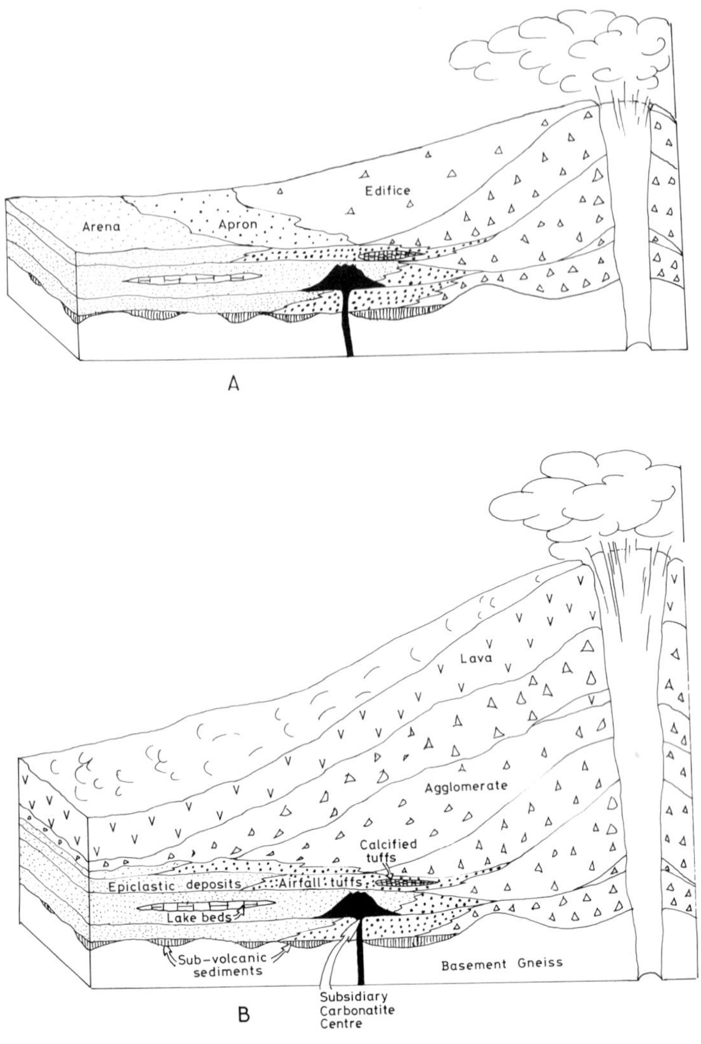

FIG. 3. Development of lithofacies in the vicinity of a carbonatite-nephelinite central volcano. A. Arranged round a central volcano are three major geomorphological/depositional zones: (1) the edifice, comprising coarse near-vent pyroclastics; (2) further out radially are the apron facies, comprising finer-grained pyroclastics and reworked edifice sediments; there are varied amounts of pedogenic alteration; and (3) even further from the centre are the arena facies, comprising predominantly waterlain strata with minor pyroclastic horizons. B. As the volcano grows, each facies expands radially, migrating over pre-existing deposits and preserving them. In the final stages of both the Rangwa and Timboroa volcanoes, lavas flowed from the vent and covered most of the pre-existing pyroclastic rocks.

each eruption, the three facies migrate outwards covering earlier deposits (Fig. 3). Thus the arena facies is often covered by an apron facies, which in turn is buried by the coarse vent facies. Such 'ideal' sequences are known at Koru, Mfwangano, Rusinga, Karungu, Uyoma and Elgon, and seem to be standard depositional successions. Localized differences occur, but they are related to local features such as parasitic vents and minor tectonism, the frequency of eruptions, and the duration of quiescent periods.

Paravolcanic depressions

Clark (1969), King (1970) and Van Couvering (1972) suggested that Miocene sediments at

Elgon, Rangwa and elsewhere accumulated in paravolcanic basins which formed early during the development of the centres. This is because the sediments are usually basal to the volcanic sequence. The tendency to label all these sediments as 'lake beds' demanded the existence of water-filled depressions. However, the existence of closed depressions has not been proven; neither is it necessary to postulate their existence in every case. In the first place, much of the sediment is not lacustrine, but accumulated on dry land as a result of perivolcanic processes, including pyroclastic activity and outward radial reworking of sediments. Secondly, the development of a large mountain on an irregular surface will almost invariably dam the pre-existing drainage somewhere round its flanks, as has happened on the N side of Mt Kilimanjaro.

To call these basins paravolcanic 'depressions' may be misleading in that they were not necessarily depressed as a result of paravolcanic tectonic activity, as implied by King *et al.* (1972). The consequence of natural damming on a regionally sloping substrate would be accumulation of an asymmetrical sedimentary wedge. This is a feature common to several of the Miocene central volcanoes of western Kenya and eastern Uganda. Furthermore, any sediments accumulating behind such volcanic dams would automatically appear to be at the base of the local sequence because of the topographic position of the basin at the foot of the 'dam wall'. It is possible that the Miocene sediments of Rangwa and Koru accumulated in settings analogous to those of Amboseli, N of Kilimanjaro.

Isostatic depression of the basement appears to have occurred at Kilimanjaro (Fig. 2) and also possibly at Elgon. Depression is a result of long term downwarping due to the superimposed volcanic load, and is, therefore, nothing to do with sediment accumulation during early phases of volcanic activity. The present-day condition of these downwarped areas has, however, led to the perception that the sediments filled paravolcanic basins already in existence, as for example in King (1949) and Bishop & Trendall (1967). In fact, the downwarping post-dates much of the sediment accumulation.

The geochemical role of volcanoes in bone preservation

Although some fossiliferous sites in East Africa occur far from any volcanoes, more than 98% of the fossils housed in the Kenya National Museums were collected from sediments which are genetically related to volcanic rocks. A tally of these fossils reveals that sediments associated with carbonatitic-alnoitic volcanic source rocks yield the most fossil bones, while sediments derived from rhyolitic sources yield fewest. The crucial factors are the percentage of silica and calcium oxide in the source rocks.

Fossil bones are rare or non-existent in sediments of rhyolitic and trachytic areas, where silica content generally exceeds 60% (Fig. 4). As the silica content drops to 50%, fossil bones become more common, while at compositions of 40% or less, they are generally very common. Conversely, when the amount of lime (CaO) in a source rock is as low as 5%, fossil bones are generally rare or absent. As the content of lime increases to 10% bones become common, while at concentrations above 30%, fossil bones are almost invariably present in abundance, as for example, in the carbonatitic-alnoitic deposits of the Koru area (Fig. 5).

Despite the fact that carbonatitic-alnoitic volcanic rocks are volumetrically the least important of the Kenyan igneous field (Williams 1972), sediments derived from them have yielded the greatest number of fossil bones and teeth. Sediments derived from basalts, which are volumetrically the most important volcanic rocks in Kenya, have yielded large quantities of bone, but in comparison with carbonatites and alnoites are appreciably less fossiliferous.

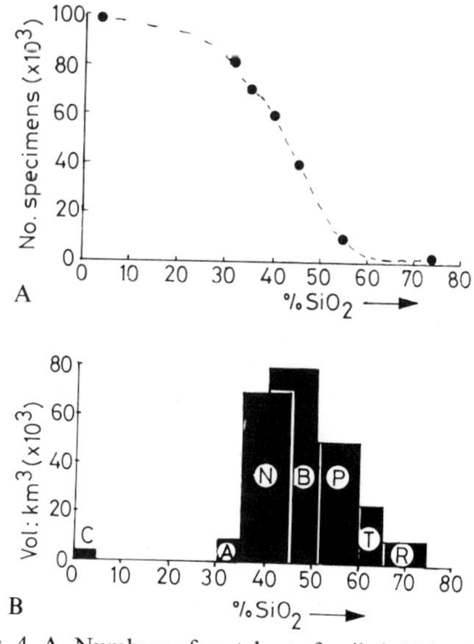

FIG. 4. A. Numbers of vertebrate fossils held in the National Museums of Kenya, classified according to composition of silica content of source rocks.
B. Volume of major volcanic rock types in Kenya. C = carbonatite; A = alnoite; N = nephelinite; B = basalt; P = phonolite; T = trachyte; R = rhyolite.

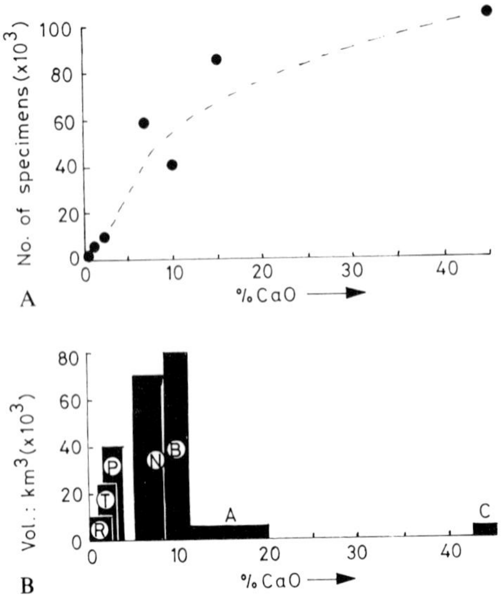

FIG. 5. A. Numbers of vertebrate fossils held in the National Museums of Kenya classified according to the lime content of volcanic source rocks. B. Volume of major volcanic rock types in Kenya.

For terrestrial gastropods, the chemical composition of the sediments is even more crucial to preservation than it is for bones and teeth. Apron and arena facies associated with carbonatitic-alnoitic centres yield abundant gastropods. Sites with basaltic, phonolitic and more siliceous sediments have yielded virtually no terrestrial gastropods.

In contrast, fossil plant material is found in association with virtually all sediments. The richest deposits, regardless of source rock composition, are normally those in which plants were buried as a result of volcanic activity. Near-vent agglomerates ranging in composition from low-silica nephelinites to high-silica rhyolites frequently contain abundant fossil tree trunks. Such occurrences provide compelling evidence for the former existence of large trees, probably in woodland or forest tracts on the upper slopes of Miocene volcanoes. In fact, fossil wood associated with acid volcanics is usually much better preserved (silicified) than wood found with basic agglomerates where it is usually calcified. Fossil leaves are generally found in abundance in sand to silt grade air-fall tuffs regardless of the chemical composition of the volcanic rock. For example, leaves are particularly abundant in the air-fall alnoitic tuffs of Rusinga and Mfwangano, the phonolitic tuffs at Ngorora and the trachytic tuffs of the Nakuru areas. However, contrasting with the wide distribution of fossil wood and leaves, fruits and seeds seems to be closely dependent upon the chemical composition of the sediments. It is true that a few fruits and seeds are known from areas of basaltic composition (Buluk, East Turkana), but the only rich fossiliferous fruit and seed deposits known are those associated with the alnoitic-carbonatitic sediments of Rusinga and Mfwangano (Chesters 1957).

Tectonogenic stratigraphy

Tectonic activity produces irregularities in the Earth's surface, uplifting some parts which then undergo erosion and downwarping others which act as sediment traps. In western Kenya, tectonic activity started well after the region had become blanketed by the volcanic ejecta from Timboroa and Rangwa. These are two central volcanoes which formed considerable topographic features (Van Couvering 1972) in a planar landscape (Shackleton 1951). Perivolcanic sedimentation had already occurred, giving rise to the richly fossiliferous and predominantly subaerial facies at Koru, Songhor, Rusinga and Mfwangano.

Tectonic downwarping which began *c.* 16 Ma resulted in the formation of a shallow basin along what was eventually to become the axis of the Nyanza Rift Valley (Fig. 6). In this shallow basin, which was *c.* 60 km long by *c.* 30 km wide, accumulated the Middle Miocene Maboko and Nyakach Formations. Sediments of this age outcrop not only within the rift, but also extensively along its shoulders, indicating that the mid-Miocene downwarp was wider than the rift valley of today. In fact, the rifting led to better definition of the depression as a pair of mirror-image half-graben some 30% narrower than the original downwarp. The western, or Maboko, half-graben dips from N to S, and is bounded on the S by the E–W Kendu fault (Pickford 1982). The eastern, or Nyakach, half-graben dips from S to N, and is bounded on the N by the E–W Nyando fault. The boundary between these two half-graben seems to be on strike with the Nandi fault, a major sinistral strike-slip fault which trends N–S in the vicinity of the Nyanza Rift Valley. Similar half-graben seem to occur in parts of the Gregory Rift Valley, for example in the Samburu Hills, W of Baragoi in northern Kenya.

Sediments accumulated in the initial downwarp as well as later in the topographic lows of the half-graben. Indeed, sediments are still accumulating in the Nyakach and Maboko half-graben. Accumulation rates depend on local erosion, which seems in the short term to be related to slope rather than relief. This means that source areas are confined to the immediate vicinity of the rift shoulders and to the tilt blocks

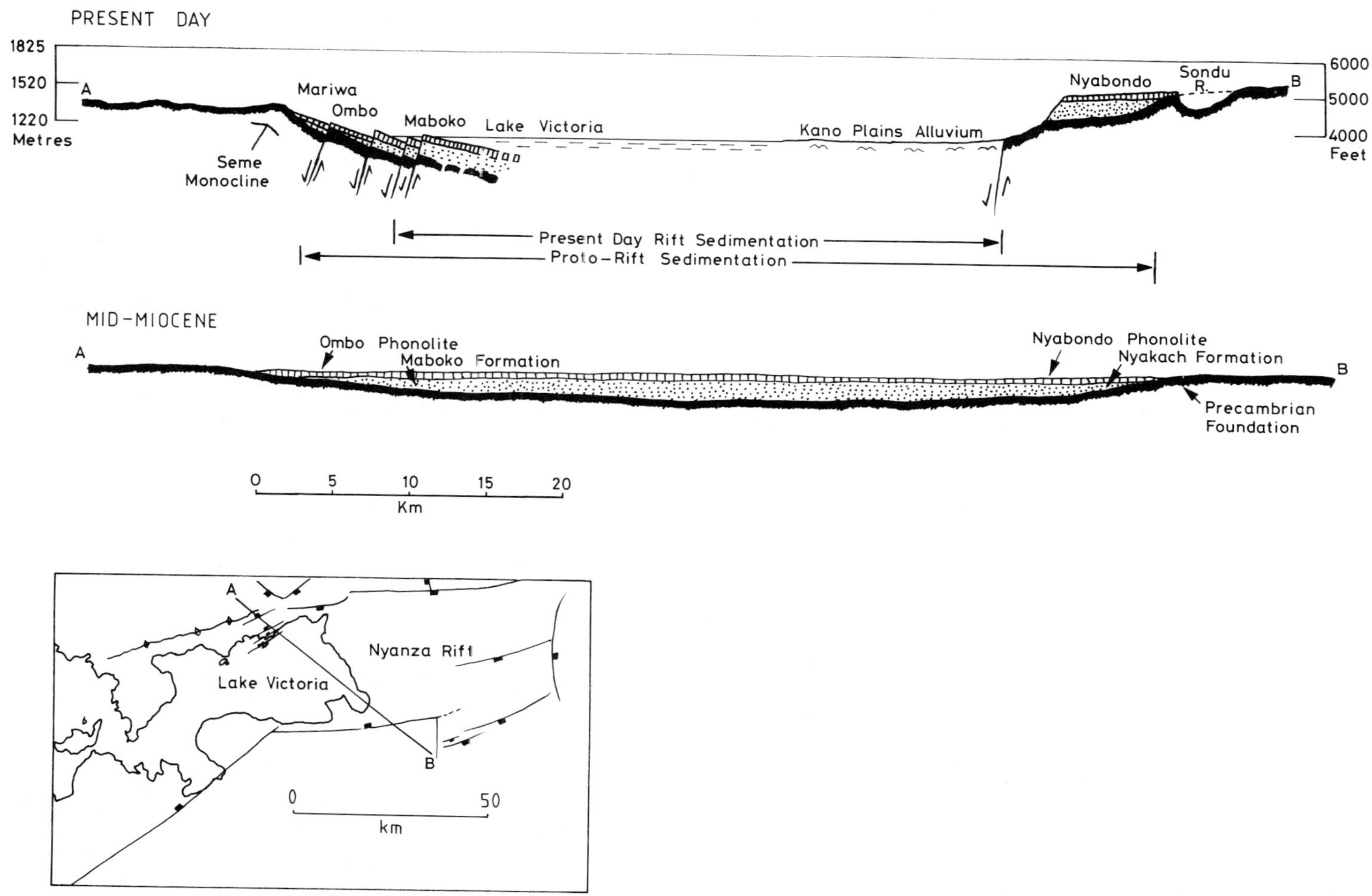

FIG. 6. Early evolution of the Nyanza Rift Valley. About 16–17 Ma the Maboko-Nyakach area underwent regional downwarping in an area 60 km E–W by 30 km N–S. Marginal slopes were of the order of 1°. Subsequent faulting defined the present rift system, which consists of two half-graben side by side but facing in opposite directions.

TABLE 1. *The determination and characteristics of depositional environments of Miocene sediments in western Kenya*

Facies No.	Depositional environment	Sediment type and colour	Diagenesis	Faunal/floral content	Examples of localities
1	Near-vent edifice of central volcano	coarse agglomerates with minor tuff lenses, massive, poorly bedded	zeolitization; little or no pedogenesis	large pieces fossil wood, very few gastropods and mammals	Kabewa, Kapurtay and many areas on Elgon, Rangwa, Timboroa
2	Apron of central volcano, intermittent deposition with pedogenesis	calcified tuffs, weathered clay tuffs; bedding clear to weak, cm to m scale, pinks, reds and greys predominate; kunkar nodules common, little evidence of aquatic deposition	calcification and pedogenesis common	exclusively terrestrial fauna; infauna common, includes cocoons, termite foraging tunnels, root casts	Koru, Songhor (part), Mteitei valley, Legetet, Chamtwara (Fort Ternan?), Walangani Beds (Mfwangano), Kunya
3	Apron of central volcano	grey to green tuffs; sand to silt grade bedded on cm to m scale; accretionary lapilli occur	feldspathization; zeolitized; little pedogenesis	fossil leaves common in selected sections, some cocoons as infauna; otherwise poor in fossil remains	R106 (part), R5 (part), Songhor (part), Makira, Kaniamwia, Nyenga
4	Channel in apron of central volcano	cross-bedded silts, sand and conglomerate, yellow predominant	clay formation; little pedogenesis	many gastropods and mammals including articulated skeletons	Meswa Bridge
5	Braided stream system (ephemeral) near upper limits of volcanic arena	silts, clays and conglomerates interbedded at 0.5 m scale; cross-bedded, cut and fill structures, abrupt lithological changes, predominantly pinks, greys and reds	calcification of silts; minor pedogenesis	terrestrial fauna with very few aquatic elements; little if any infauna, trees *in situ* and transported	R1, R114

6	Dry part of floodplain in volcanic arena	clays, silts, conglomerates, cross-bedding, cut and fill structures, channels, greys predominate	calcrete pedogenesis, and clay formation common	mainly terrestrial fauna, but turtles and crocodiles rare; termite foraging tunnels, root casts occur; cocoons as infauna	Kaloma, Majiwa, Nyakach, Maboko (part), Makira (Mfwangano), R106 (part), R5 (part), R113 (part), R107 (part), R117
7	Wet part of floodplain in volcanic arena	clays and silts with conglomerate lenses, cross-bedding, cut and fill structures, channels, fining upwards sequences	limited calcrete pedogenesis; clay formation	mixed aquatic and terrestrial fauna; turtles and crocodiles common; aquatic gastropods and crabs occur	Maboko (part), R5 (part), Nyakach (part), Karungu (part), Simenya, Chianda (part), Nyagwena, Kadianga
8	Large river system in volcanic arena (permanent)	interfingered conglomerate silts and clays, well-bedded, cross-bedded, cut-and-fill, channels in channels; *Etheria* banks common, generally bright red	little or no pedogenesis, clay formation, slight calcification	aquatic and amphibious fauna predominate; fossils often rolled and polished; bivalves and fish common	Gumba, Gunda, R113, R73, Wayondo, Chianda (part), Karungu (part), Ombo, Kiune
9	Lake margin in volcanic arena	sands, silts and inter-fingered shales, browns and greens predominate	zeolitization; feldspathization; little or no pedogenesis	predominantly aquatic and amphibious fauna	Kulu (part), Magare beach
10	Lake in volcanic arena	shales often white or pale green, pink or yellow	feldspathization; no pedogenesis	complete fish skeletons, crocodile and fish coprolites, little else	Kulu (part), Wakondu, Sena, Rangoye, Nyakongo, Nyamuga, Kosala

within it (Pickford 1982). Transport distances seem in general to be short so that sorting of sediments can be uneven or poor. The result is that rapid lateral and vertical changes in lithology can occur.

The sedimentary record of western Kenya is thus in two phases. Early sedimentation resulted from perivolcanic deposition, both by volcanic activity and by surface processes. Later sedimentation followed subsequent tectonic activity.

Bioturbation and pedogenesis of Miocene strata

A great deal of the sediment in western Kenya accumulated subaerially. Beside air-fall tuffs and agglomerates, a considerable proportion of water-transported sediment also came to rest on floodplains or accumulated in ephemeral lakes, or remained on depositional hillslopes (Pickford 1982). A significant proportion of strata accumulated in positive relief settings such as on the flanks of volcanoes, e.g. Rangwa and Timboroa.

Bioturbation

In all these environments, bioturbation took place. Plants established themselves in subaerial conditions, even on fresh ash deposits, with the result that root traces and disturbed bedding are common features of the sedimentary record throughout western Kenya. In many areas, such as Songhor, Rusinga, Koru and Uyoma, the traces of various animal burrows, together with insect brood cells and cocoons, and rich concentrations of gastropods which died while aestivating, show that post-mortem transport of these fossils was insignificant. Furthermore, the infaunal remains, such as termitaries, solitary bee cells, and hatched moth cocoons and their escape burrows, suggest that the substrate was friable and relatively dry. It is important to bear in mind that many of the sediments of western Kenya obtained their fossil content through colonization after deposition, generally during pedogenic episodes. Because of this, it is possible to infer that post-mortem transport of the vertebrate fossils from one facies to another was insignificant, except perhaps in braided stream and large river deposits, such as those of the Gumba facies. At some sites, such as Koru and Songhor, various burrowing mammals (*Bathyergoides*, *Megapedetes* and *Orycteropus*) occupied the area, articulated remains occasionally being found in ancient burrows. This may be the reason why few examples of mixed faunas are known (Pickford 1983). It explains why fresh ash deposits are usually devoid of faunal remains, while palaeosols are richly fossiliferous.

Pedogenesis

Pedogenesis depends on the rates of sedimentation and the frequency of volcanic activity. In western Kenya palaeosols are extremely common, ranging in maturity from 'incipient' to the stage IV calcretes of Leeder (1975). They are frequently fossiliferous. Conservative estimates suggest that at least 75% of the fossils from western Kenya, including primates and other mammals, come from palaeosols of one kind or another. Important sites are Koru, Legetet, Chamtwara, Songhor, Fort Ternan, Rusinga, Mfwangano and Uyoma.

Calcretes are rare in the rift valley, because few areas were stable for long enough to allow pedogenesis to reach this stage. Outside the rift valley they are relatively common. For example, they occur widely in the semi-arid areas N and E of Mt Kenya. However, in areas fed by carbonatitic ash, pseudo-calcretes can form in areas which might not normally encourage their genesis. Examples of such pseudo-calcretes are known at Songhor, Koru, Majiwa, Maboko and Mfwangano in the Nyanza Rift, where they are frequently rich in fossils.

In other parts, interruptions in volcanic activity were long enough to allow relatively mature palaeosols of other kinds to form. Examples are the red soils at Koru, Songhor and Chamtwara, and the grey soils at Fort Ternan (Bishop & Whyte 1962). Where volcanic activity was more frequent, pedogenesis was arrested by each volcanic deposit, and various immature palaeosols have been preserved. These range from barely altered ash to relatively mature soil profiles.

Depositional environments in western Kenya

Until the late 1960s there was little assessment of the depositional environments of western Kenya, mainly because Kent's (1944) Miocene lake hypothesis had obscured the possibility of other sorts of environment. Bishop (1967, 1968) was the first to recognize that fossiliferous deposits in East Africa were not all lacustrine. He recognized two broad sedimentary categories: (1) non-volcanic fluviatile sediments; and (2) volcanic-primary and derived pyroclastic rocks. He listed a number of sites belonging to each type.

Subsequent examination of over 300 sites in western Kenya (Pickford 1984) has resulted in the recognition of ten categories of Miocene depositional environments. Each of these is characterized by a suite of lithological features, a faunal content, and a set of recognizable post-depositional changes, etc. (Table 1). The categories are shown in their relative geomorphic position in Fig. 7, and are as follows: (1) near-vent agglomerates; (2) subaerial slopes with palaeosols (no matter how immature); (3) subaerial ashes (airfall tuffs with no pedogenic alteration); (4) channel deposits in predominantly subaerial facies; (5) braided stream deposits; (6) dry floodplains; (7) wet floodplains; (8) large river systems; (9) lake margins (and swamps); and (10) lakes.

Distribution of fossils in sedimentary facies of western Kenya

There are several readily recognizable types of faunal assemblage in the Miocene sequence of western Kenya (Figs. 8–13) (Pickford 1981). Each type of assemblage occurs at a number of sites, but nearly always with characteristic sedimentary deposits. The vent or edifice facies (facies 1 in Figs. 7–13) usually contains abundant fossil wood but seldom yields vertebrates. Conversely, wood tends to be rare in facies 2–10. In apron palaeosols (facies 2 in Figs. 7–13), terrestrial vertebrates and invertebrates are usually abun-

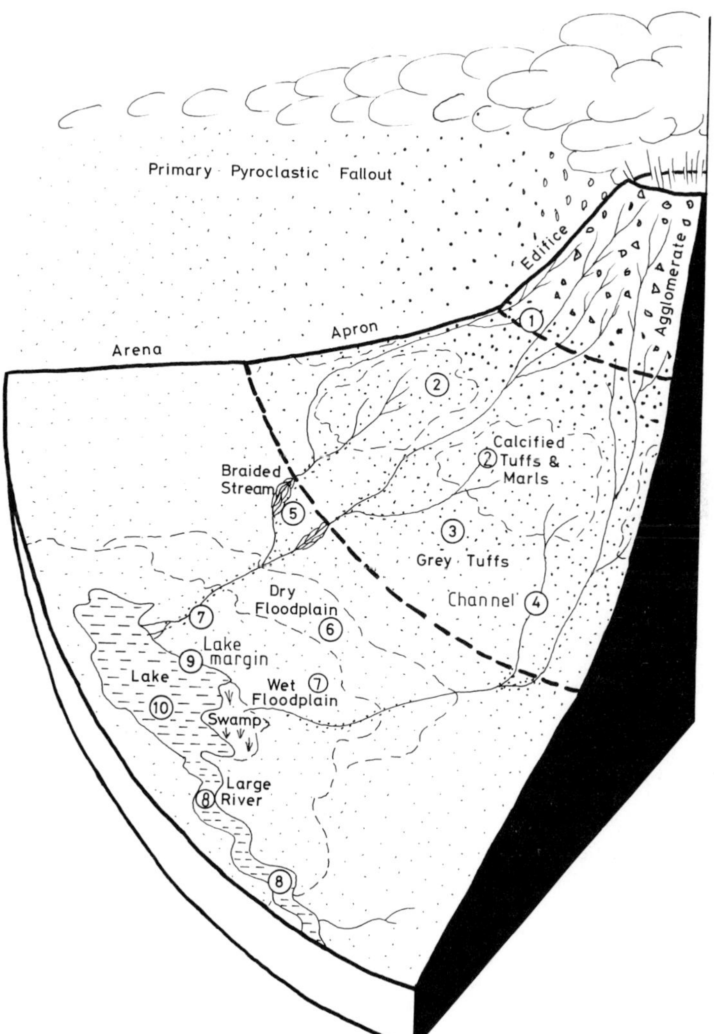

FIG. 7. Depositional environments in the vicinity of carbonatite-nephelinite central volcanoes. For explanation of facies 1–10 see Table 1.

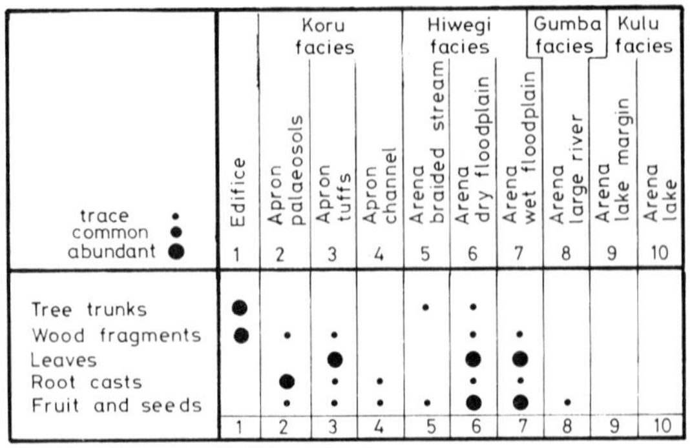

FIG. 8. The distribution of fossil plant remains in various depositional environments of western Kenya. 1–10 are the lithofacies defined in the text and in Table 1.

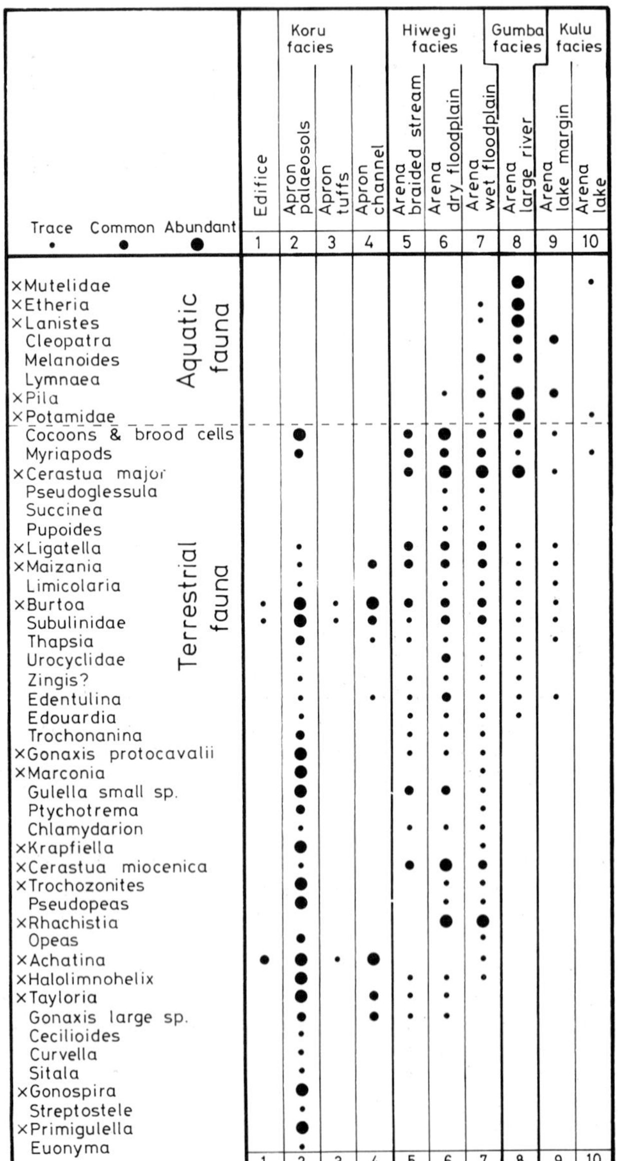

FIG. 9. The distribution of invertebrate fossils in the various depositional facies of western Kenya. 1–10 are lithofacies defined in the text; X = diagnostic taxa.

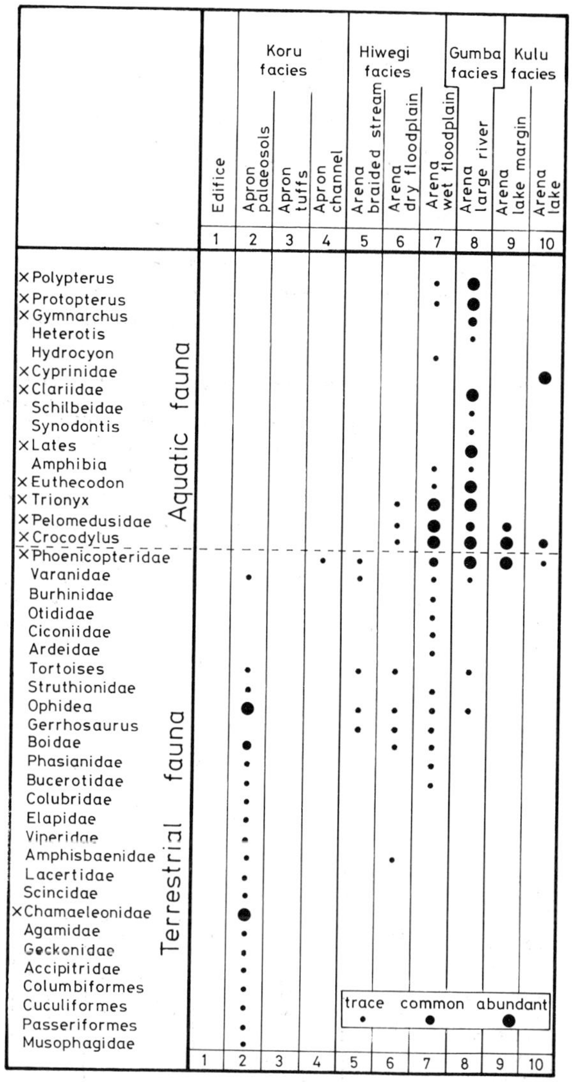

FIG. 10. The distribution of lower (non-mammal) vertebrate fossils in the various depositional environments represented in western Kenya. 1–10 are facies defined in the text; X = diagnostic taxa. Taxa are arranged so as to reflect their biofacies rather than their taxonomic positions.

dant, while aquatic or amphibious taxa are conspicuously absent. Fossil chameleons are common in facies 2 but rare elsewhere, while, among the mammals, the rodents *Bathyergoides* and *Afrocricetodon* are particularly characteristic. Large mammals are exceptionally rare in the apron palaeosols, but primates are common, comprising up to 50% of the non-micromammal fauna. Among the primate species confined to apron palaeosol deposits are *Micropithecus clarki*, *Proconsul major* and *Dendropithecus songhorensis*. It is probable that these taxa lived on the slopes of ancient volcanoes but seldom ventured into the surrounding arena where *Proconsul nyanzae* and *Dendropithecus macinnesi* were common.

In contrast, apron tuffs which have *not* undergone soil forming processes are remarkably free of fossil vertebrates, but often contain fossil leaves and twigs. This suggests, firstly, that the vertebrate fossil content accumulated largely during the periods of pedogenesis; and, secondly, that the absence of fossil leaves in the palaeosols may be due to destruction during pedogenic bioturbation.

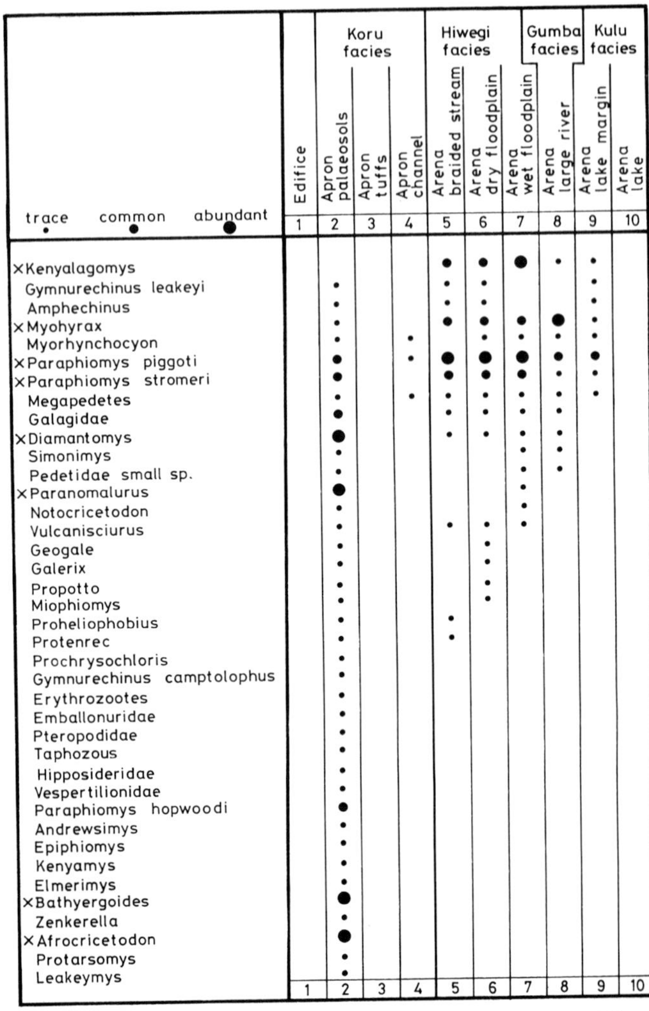

FIG. 11. The distribution of micromammals relative to depositional environments in Miocene deposits of western Kenya. 1–10 are depositional environments; X = diagnostic taxa. Arrangement of taxa is according to depositional facies rather than taxonomic order.

Apron channel deposits (facies 4 in Figs. 7–13) are rare, but one particular example, at Meswa Bridge was a remarkable assemblage of bones and gastropods that resembled a log jam. Many of the bones are partly articulated and they range in size from those of rodents to those of proboscideans (Pickford & Tassy 1980). However, generally speaking, such deposits seem to be devoid of vertebrates. Facies 1–4 have been combined into a single geomorphological category which I have called the Koru facies, because it is best developed at Koru. Basically it comprises all the volcanic apron deposits (Fig. 7). Its fossil content is characterized by an exceptional abundance of small mammals, by a rarity of large mammals and by the complete absence of aquatic or amphibious taxa.

In contrast the Hiwegi group of facies (5–7 in Figs. 7–13) yields abundant large fossil vertebrates, fewer micromammals and a variety of aquatic and amphibious forms. In braided stream deposits (facies 5) and in both dry and wet floodplain strata, a variety of characteristic taxa occur including hyracoids, suids, rhinocerotids and proboscideans. In addition to these, aquatic vertebrates are sometimes present, while turtles and crocodiles are usually present in the wet floodplain deposits (facies 7). Primates are relatively common in the Hiwegi facies, but form a less conspicuous portion of the non-micromam-

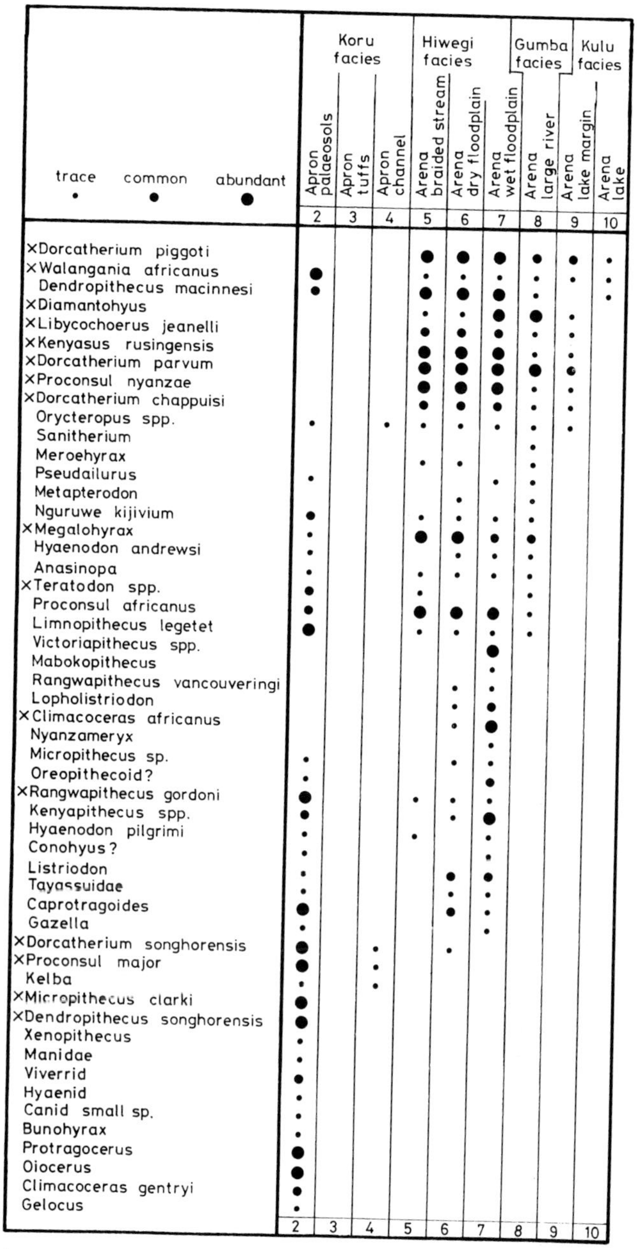

FIG. 12. The distribution of medium-sized mammals in Miocene strata of western Kenya. 2–10 are the various depositional facies; X = diagnostic taxa. Arrangement of species is by environment of deposition rather than zoological placement.

mal remains. The commonest primates in facies 5–7 are *Proconsul nyanzae* and *Dendropithecus macinnesi*.

In contrast, primates are very rare in the Gumba facies (facies 8 in Figs. 8–13) and the Kulu facies (facies 9 and 10) while aquatic and amphibious taxa are abundant. In deposits of large rivers, facies 8, the freshwater oyster *Etheria* is usually present, sometimes forming conspicuous reefs. Fossil bones are often broken, or rolled and polished indicating significant post-mortem transport. In the same way, fossil fish

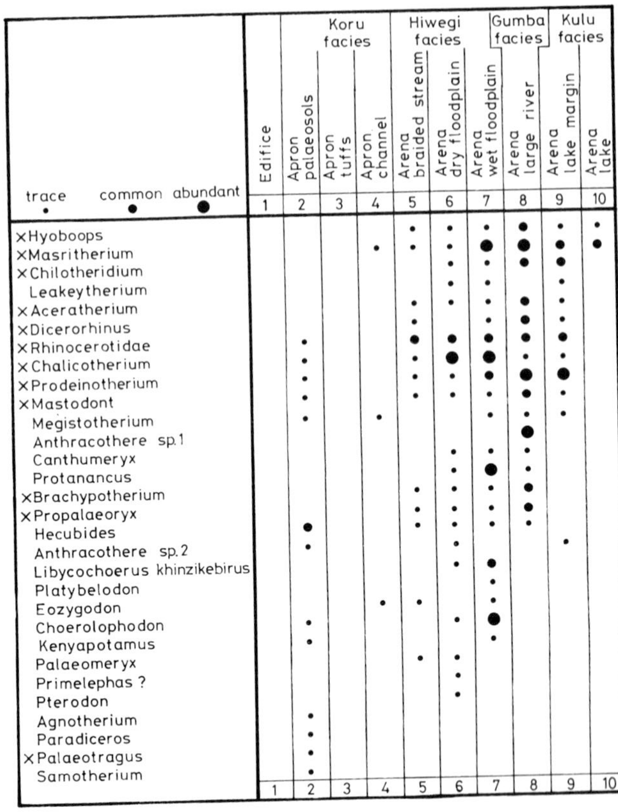

FIG. 13. The distribution of large mammals arranged by depositional environment in western Kenya. 1–10 are depositional facies; X = diagnostic taxa. Arrangement of taxa is by biofacies rather than taxonomic position.

remains, which are generally common in the Gumba facies, are invariably disarticulated and broken.

In lake margin (facies 9) and lacustrine (facies 10) deposits, such as the Kulu facies, are abundant aquatic faunal remains, e.g. crocodiles (and their coprolites), turtles and fish. Mammals are less common. However, the proboscidean *Prodeinotherium* is found more often in the Gumba and Kulu facies than elsewhere, suggesting that it may have inhabited the margins of these large rivers and lakes. Similarly, the genus *Diamantohyus* and the anthracotheres, *Brachyodus* and *Hyoboops*, occur predominantly in these Gumba and Kulu facies.

In summary, the distribution of vertebrates and invertebrates in the W Kenyan Miocene deposits reveals definite patterns. In the subaerial facies near volcanic centres, fossils are understandably almost exclusively of terrestrial species. In contrast, waterlain deposits found further from the volcanic centres yield abundant aquatic faunal elements, while the intermediate Hiwegi facies (braided stream and floodplain deposits often with palaeosol development) yields a somewhat intermediate assemblage, but which is made up predominantly of terrestrial forms.

Conclusions

The sedimentary record in western Kenya is the product of two controlling factors, volcanic and tectonic, and has been variably modified by post-depositional bioturbation and pedogenesis. The earliest parts of the succession, comprising the Lower Miocene sediments, were dominated by processes associated with the building of central volcanoes. There was little influence from tectonism. The Middle Miocene sediments, in contrast, owe their origin predominantly to tectonic activity, firstly with respect to initial downwarping in the Maboko-Nyakach area, and later with the better defined structural basins of the two mirror-image half-graben that comprise the present-day Nyanza Rift Valley.

Freed from the old model, in which all Kenyan sediments were considered to be lacustrine, research has followed increasingly interesting lines of enquiry. For example, the study of the distribution patterns of animals and plants relative to depositional facies has revealed that certain species of mammal appear to have lived preferentially in selected areas. The hominoid Primates *Proconsul major* and *Micropithecus clarki* are found exclusively in the apron facies of carbonatite-nephelinite central volcanoes, while *Proconsul nyanzae* and *Kenyapithecus africanus* are found predominantly in arena facies where relief was lower and vegetation probably more open. Comparable distribution patterns occur with many other vertebrate and invertebrate species.

Within these broad relationships between palaeontology and ancient landforms it is evident that the chemical composition of the sediment source rocks was of major significance in the preservation of bones, teeth and invertebrate remains. In general, it has been observed that fossil content varies inversely with silica content, and directly with lime content.

This geochemical control on fossilization seems to hold true irrespective of depositional environment, of which ten major categories have been recognized in western Kenya. These ten depositional environments cluster into three groups and are related to geomorphic position on and around an active central volcano. Thus in areas near the centre, subaerial facies predominate because sediments were accumulating in positive relief settings. Further from the centre in the volcanic apron, sediments accumulated by a combination of volcanic fallout and perivolcanic surface processes. Pedogenic activity was of major significance, not only in altering the sediments but in the generation of the vertebrate and invertebrate fossil record, all of which is non-aquatic in affinities. Even further removed from the volcanic centres, and subject to volcanic activity only at infrequent intervals, is the volcanic arena, a low-lying, low relief region in which aquatic influences in sedimentation are much more evident.

ACKNOWLEDGEMENTS: I am particularly grateful to the late W.W. Bishop who got me interested in western Kenya, its numerous problems and its great potential. Mr Norman Brooks, the managing director of the Homa Lime Co. at Koru, was the epitome of the helpful, interested and enlightened landowner. Funds to carry out the fieldwork came from a variety of sources, including the L.S.B. Leakey Foundation, the Boise Fund, the L.S.B. Leakey Trust, and the Foundation for Research into the Origins of Man. Finally I must acknowledge the help of the Director of the National Museums of Kenya and the National Council for Science and Technology of the Republic of Kenya.

References

BISHOP, W.W. 1967. The later Tertiary in East Africa—volcanics, sediments and faunal inventory. In: BISHOP, W.W & CLARK, J.D. (eds) *Background to Evolution in Africa*. University of Chicago Press, Chicago, pp.31–56.

—— 1968. The evolution of fossil environments in East Africa. *Trans. Leics. Lit. Philos. Soc.* **62,** 22–44.

—— & TRENDALL, A.F. 1967. Erosion surfaces, tectonics, and volcanic activity in Uganda. *J. Geol. Soc.* **122,** 385–420.

—— & WHYTE, F. 1962. Tertiary mammalian faunas and sediments in Karamoja and Kavirondo, East Africa. *Nature* **196,** 1283–7.

CHESTERS, K.I.M. 1957. The Miocene flora of Rusinga Island, Lake Victoria, Kenya. *Palaeontographica* **101B,** 30–67.

CLARK, L. 1969. The North Elgon depression, Eastern Uganda. *J. Geol. Soc.* **125,** 39–46.

KENT, P.E. 1944. The Miocene beds of Kavirondo, Kenya. *J. Geol. Soc.* **100,** 85–118.

KING, B.C. 1949. The Napak area of southern Karamoja, Uganda. *Mem. Geol. Surv. Uganda* **5,** 1–57.

—— 1970. Volcanicity and rift tectonics in East Africa. In: *African Magmatism and Tectonics*. Oliver & Boyd, Edinburgh, pp.263–83.

——, LE BAS, M.J. & SUTHERLAND, D.S. 1972. The history of the alkaline volcanoes and intrusive complexes of eastern Uganda and western Kenya. *J. Geol. Soc.* **128,** 173–205.

LEEDER, M.R. 1975. Pedogenic carbonates and flood sediment accretion rates: a quantitative model for arid-zone lithofacies. *Geol. Mag.* **112,** 257–70.

PICKFORD, M. 1981. Preliminary Miocene mammalian biostratigraphy for western Kenya. *J. Human Evol.* **10,** 73–97.

—— 1982. The tectonics, volcanics and sediments of the Nyanza Rift Valley, Kenya. *Z. Geomorph. N.F.* **42,** 1–33.

—— 1983. Sequence and environments of the Lower and Middle Miocene hominoids of western Kenya. In: CIOCHON, R.L. & CORRUCCINI, R.S. (eds) *New Interpretations of Ape and Human Ancestry*. Plenum Publishing Co., London, pp.421–39.

—— 1984. *Kenya Palaeontology Gazetteer. Western Kenya*. National Museums of Kenya Special Publication, vol. 1, pp.1–208.

—— 1985. A new look at *Kenyapithecus* based on recent discoveries in Western Kenya. *J. Human Evol.* **14,** 113–43.

—— & TASSY, P. 1980. A new species of *Zygolophodon*

(Mammalia, Proboscidea) from the Miocene hominoid localities of Meswa Bridge and Moroto (East Africa) *N. Jb. Geol. Palaeont. Mh.* **4,** 235–51.

SHACKLETON, R.M. 1951. A contribution to the geology of the Kavirondo Rift Valley *J. Geol. Soc.* **106,** 345–92.

VAN COUVERING, J.A. 1972. *Geology of Rusinga Island and correlation of the Kenya mid-Tertiary faunas.* Unpublished PhD thesis, Cambridge University.

WILLIAMS, L.A.J. 1972. The Kenya Rift volcanics: a note on volumes and chemical composition. *Tectonophysics* **15,** 83–96.

M. PICKFORD, Institut de Paléontologie, 8 rue Buffon, 75005 Paris, France.

Tectonic and environmental control on rodent diversity in the Plio-Pleistocene sediments of the African Rift System

C. Denys, J. Chorowicz & J.J. Tiercelin

SUMMARY: Tectonic, sedimentological, climatic and zoo-geographic data are combined in order to explain faunal development in the East African Rift Valley during Plio-Pleistocene times. The main factors controlling the evolution of rodents are the tectonic events which influence local sedimentation and climate, and not only the global climatic glacial events which appear to have reduced effects in the rift valley.

Many types of geological studies have been carried out in recent years in the African Rift. These include, for example: tectonic analysis (Mohr 1974, 1983; Chorowicz *et al.* 1983), isotopic dating of volcanic activity (Baker *et al.* 1971, 1972; Barberi *et al.* 1972; Logatchev *et al.* 1972; Chessex *et al.* 1975; Baker & Mitchell 1976), sedimentological analysis (Tiercelin 1981), studies of mammalian faunas, especially rodents (Jaeger 1976; Denys 1981, 1983, 1986; Sabatier 1982; Wesselman 1984), and palynological studies (Bonnefille 1976, 1979, 1983; Bonnefille & Riollet, 1980; Vincens 1982). Some of the relationships between these different topics have been considered: (1) the evolution of climates as inferred from pollen and rodent studies (Bonnefille 1976, 1979, 1983; Jaeger 1976, 1979); and (2) the correlations between the tectonic development of the rift and the evolution of rodent and other mammalian faunas (Denys 1984). But as yet, there has been no attempt to integrate all of these data in order to determine the causal relationships for all the possible correlations between different groups.

In this paper, the tectonic, geomorphological, volcanic and climatic factors that control modern sedimentation and the distribution of faunas are described and an attempt is made to explain their interrelationships. These factors are then discussed for the Plio-Pleistocene record, and an overall model for the evolution of East African faunas is formulated.

Modern tectonics, volcanism, climate, sedimentation and faunas

The rift system is the main physical feature of East Africa. It is composed of two distinct regions, the eastern and the western branches, which are linked by the Asswa lineament that appears to act as a transform fault zone (Chorowicz & Mukonki 1979; Kazmin 1980) (Fig. 1). The environments are quite different in the two regions.

The eastern branch

This comprises, from N to S, the Afar depression, the Ethiopian Rift, the Gregory Rift and the N Tanzanian Divergence. It is generally a narrow rift valley, 30–65 km wide, bordered by escarpments 1000–1500 m high in some places. The troughs are filled with sediments and volcanic material, or contain lakes of various size, Lake Turkana being the largest and the deepest.

The eastern branch has an abnormally low annual precipitation rate despite its position within an equatorial zone. It has been called the 'arid corridor' by Balinksy (1962). There are many reasons for the aridity:

1 The distribution of rainfall is seasonal. In most places the evaporation exceeds precipitation. Monsoon winds are disturbed in East Africa by the rift structure itself and also by the Lake Victoria low pressure area (Vincens 1982).

2 The rainfall distribution is also unequal between the rift shoulders, which receive higher precipitation, and the rift floor. Thus in the northern Gregory Rift, the top of escarpments in the Baringo-Bogoria region receive 1000–1300 mm a^{-1} while the adjacent rift floor receives only 800 mm a^{-1} (Tiercelin 1981). The mean annual rainfall is <500 mm a^{-1} for the low altitude margins of Lake Turkana (<800 m), whereas on the Ethiopian and Kenyan Plateaus (>1500 m high) it reaches 1000 m a^{-1} (Vincens 1982).

3 Precipitation falling on the Eastern Rift escarpments does not feed the drainage network of the rift floor to any great extent. This is due to rivers flowing from the crest of the rift shoulders towards the Indian Ocean in the E and to Lake Victoria or to the Nile basin in the W (King 1978; Tiercelin 1981).

These climatic conditions explain the covering of the rift floor by *Acacia* savanna (Western 1980) and the presence of rodents adapted to arid conditions (Fig. 2). The rift margins show a denser vegetation and quite different faunas, but not a heavy forest cover (Fig. 3).

From Frostick, L.E. *et al.* (eds) 1986, *Sedimentation in the African Rifts*, Geological Society Special Publication No. 25, pp. 363–72.

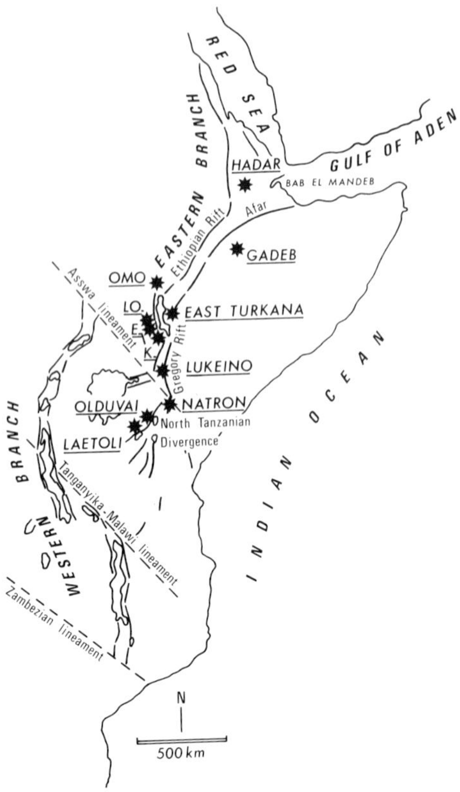

FIG. 1. Main geological features of the rift. Stars = positions of the Plio-Pleistocene rodent faunas of East Africa; Lo = Lothagham; E = Ekora; K = Kanapoi. (From Baker *et al.* 1972; Chorowicz & Mukonki 1979.)

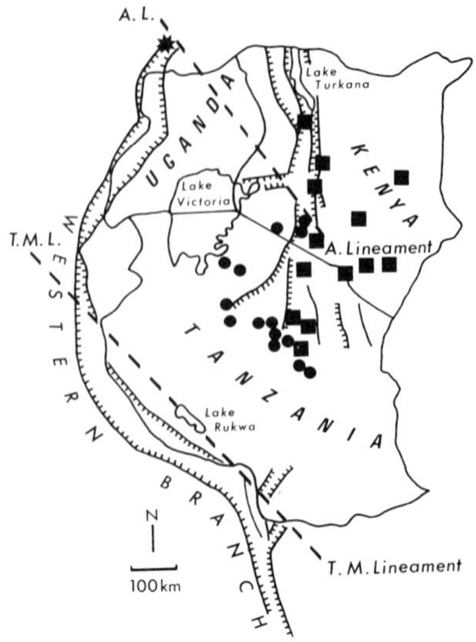

FIG. 2. Distribution of the different species of *Gerbillus* in East Africa. *G. harwoodi* (circle) is restricted to the S of the Asswa lineament (AL), while *G. pusillus* (square) does not cross the rift valley toward the western regions of Kenya. *G. gerbillus* (star) occurs only in the N of the western rift valley. This partition is due to the fact that *Gerbillus* sp. are well adapted to aridity and prefer the dry bottom of the rift valley than its moister margins (from Baker *et al.* 1972; Kingdon 1973).

The western branch

From N to S, the Western Rift is occupied by a series of large lakes, the largest and deepest being Lakes Tanganyika and Malawi. It displays deep troughs (500–1500 m), bordered by high and steep escarpments like the Congo-Nile crest line which is 3000–4000 m high. Successive troughs are separated by swells, for example, the mountainous area which is the result of tectonic compression between Lakes Kivu and Tanganyika (Chorowicz *et al.* 1979).

The western branch is more humid than the eastern. With higher precipitation, the hydrologicai balance is positive and allows large lakes to exist. For instance, the annual rainfall is between 850 mm and 1500 mm in Malawi. The distribution of rodent faunas is poorly known, but the diversity appears greater than in the Eastern Rift. Development of endemism is clearly shown by the fish of Lakes Tanganyika and Malawi (Fryer *et al.* 1983).

The Asswa lineament

Initially known in its western part (Almond 1969; Mohr 1974), the Asswa lineament has now been traced to the SE using satellite imagery. It is interpreted as a late Precambrian lineament, possibly partly reactivated as a transform zone of the rift system during the Cenozoic (Chorowicz *et al.* 1983). This zone of transcurrent faults tectonically links the two branches of the rift system, but from the geomorphological point of view it is an area characterized by upland ranges, plateaus and large volcanoes (e.g. Mt Elgon) that act as obstacles to the exchange of arid-adapted faunas. It seems to coincide with the 'Sclater line' (Sclater 1896) separating the northern and southern Sudano-Zambezian savannas and faunas (Fig. 4).

Despite the lack of detailed information concerning this area, it seems that tectonics, geomorphology, volcanism, climate and sedimentation are linked together, and have influenced the

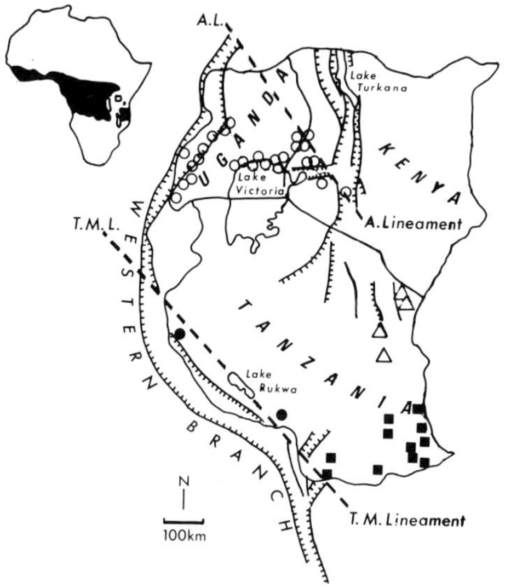

FIG. 3. Distribution area map of different forest living *Anomalurus derbianus* in East Africa. (From Baker *et al.* 1972; Kingdon 1973.) The different species of this flying squirrel avoid the bottom of the rift valley, and some are restricted S of the Asswa lineament. Open circle = *Anomalurus derbianus jacksoni* (Uganda, western Kenya), observed in the Kavirondo Rift escarpments, and also seen in the Asswa lineament highland zones; square = *A. derbianus cinereus* (southern Tanzania); open triangle = *A. derbianus orientalis* (Usambara, Ulguru mountains and southern highlands). Closed circle = *A. derbianus* unknown race, observed in the escarpments at the margin of the Western Rift Valley, near the Tanganyika-Malawi lineament.

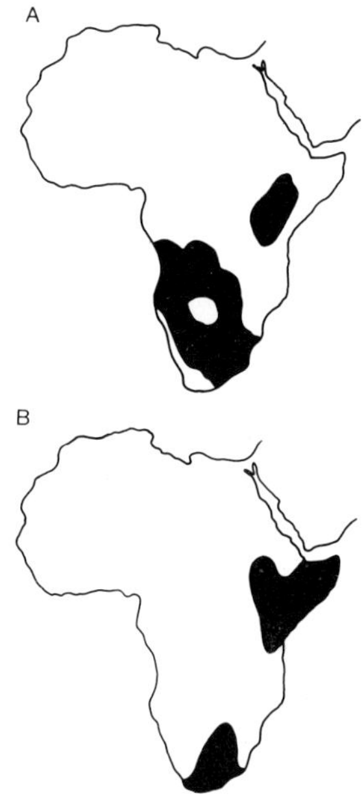

FIG. 4. Distribution of two rodent species: A. *Pedetes capensis* (Spring Hare); B. *Acomys subspinosus* (Spiny Mouse). These maps show the distinction between northern and southern savanna domain and the separation of the two species areas by the Eastern and Western Rift Valley barriers (from Kingdon 1973).

distribution of the modern faunas, especially rodents.

Plio-Pleistocene morphotectonics, volcanism, climate, sedimentation and evolution of the rodent faunas

Lower Pliocene (5–4 Ma)

A major phase of tectonic activity is recorded between 5 and 4 Ma in the eastern branch of the rift (Baker *et al.* 1972; Logatchev *et al.* 1972; Mohr 1983). This led to the development of small basin and swell structures which progressively developed from Oligocene times onwards with the opening rift.

This period is poorly known from the faunal viewpoint. The 4 Ma Kanapoi fauna (Behrensmeyer 1976) (Fig. 1) shows two Afro-Asiatic ubiquitous rodents. The first is the genus *Hystrix* known from the Upper Miocene of the Pakistan Siwaliks (Black 1972; Flynn & Jacobs 1982), North Africa (Arambourg 1959), Lukeino, East Africa (dated at 6.5 Ma, Pickford 1978) and Europe (Bachmayer & Wilson 1970, Black *et al.* 1980). The second is the genus *Tatera*, a representative of a family known from the Miocene of North Africa (Jaeger 1977), Anatolia and Greece (Sen 1982), and the Siwaliks (Flynn & Jacobs 1982), and living today in Pakistan, Arabia and Africa. At the same time, however, the Langebaanweg fauna from South West Africa (4.5 Ma) (Hendey 1981) contains a different assemblage of rodents that already characterize the southern African province. This faunal difference may indicate the initiation of the isolation of the southern rodent faunas, following tectonic movements in different parts of the rift, but especially along the Asswa lineament.

These fault movements, by creating barriers associated with environmental changes, prevented faunal exchanges between northeastern and southern Africa.

Middle Pliocene (4–3 Ma)

Four well-studied rodent faunas occur in several regions of the rift valley.

N of eastern branch: Hadar

Situated in the Afar depression, the Hadar site (3.3–2.8 Ma, Brown 1982; 3.9–3.6 Ma, Walter & Aronson 1982) has yielded the genera *Millardia* and *Golunda* (Sabatier 1982) (Table 1) that also live today in Pakistan (Flynn & Jacobs 1982). A species very similar to the living *Golunda ellioti* from Pakistan has been described at 1.8 Ma from the Siwaliks (Jacobs 1977; Flynn & Jacobs 1982). One species of this genus, which is very similar to the Hadar form, has been described from the Omo Member B fauna (Ethiopia) at 2.94 Ma (Jaeger & Wesselman 1976; Wesselman 1984). Moreover, the Hadar fauna is also characterized by *Saidomys*, which is known from the Pliocene of Egypt and Afghanistan (Sen *et al.* 1979; Slaughter & James 1979; Brandy *et al.* 1980; Sen 1983). *Hystrix* and *Tatera* have also been recorded from this site, and the other rodents are similar to the modern Ethiopian Highlands assemblage.

The 11 species of rodents described by Sabatier (1982) indicate a mountainous mixed forest environment very similar to that suggested by the palynological studies of Bonnefille *et al.* (1984). From 4 to 2.9 Ma, the Hadar region was a depression 50 km wide, occupied by a lake and bounded in the W by the Ethiopian escarpment and in the E by a volcanic barrier and the escarpment of the Danakil block (Taieb & Tiercelin 1980; Tiercelin, this volume). It was separated from the Omo valley by the Ethiopian Highlands. After 2.9 Ma, the basin became infilled, possibly following tectonic reactivation of the Ethiopian escarpment (Tiercelin, this volume).

The isolation of this region, which offered various biotopes for rodents, may explain why Hadar fauna is both qualitatively and quantitatively distinctive when compared with other faunas from the rift (Denys 1983, 1984). Moreover, the similarities of faunal composition with those recorded in the Asiatic region, suggests close relationships and faunal exchanges with western Pakistan. Two gaps in the sedimentation record of the Red Sea are known at 5–4 Ma and at 3–2 Ma (DSDP 1974, Leg 23, cores 225, 227 and 228). This suggests the possibility of a continental passage through the Bab el Mandeb Strait and the Danakil Horst.

Mid-eastern branch: Omo and East Turkana

The Omo Member B fauna (Turkana depression, 3 Ma, Fig. 5) has a different assemblage from that at Laetoli and Hadar except that *Golunda* has been recorded at Hadar. The 14 species described by Jaeger & Wesselman (1976) and Wesselman (1984), are known for their rather wide distribution (*Mastomys*, *Lemniscomys* and *Paraxerus* association). *Thallomys* is more ecologically restricted but is known by two species or subspecies in both northern and southern savannas (Denys 1983).

TABLE 1. *Faunal lists of the rodents from Laetoli (Denys 1983, 1986), Shungura Omo Member B (Wesselman 1984), Hadar (Sabatier 1982)*

Laetolil Beds (Laetoli)	Omo Member B	Hadar
Xerus cf. *janenschi*	*Xerus erythropus*	*?Xerus*
Xerus sp.		
Paraxerus sp.	*Paraxerus ochraceus*	
Gerbillinae gen. indet.		
Tatera cf. *inclusa*	*Tatera* sp.	*Tatera* sp.
Thallomys sp. nov. A	*Thallomys jaegeri*	
Mastomys sp. nov.	*Mastomys minor*	
Dendromus sp.	*Arvicanthis* sp. indet	*Saidomys afarensis*
Steatomys sp.	*Mus* aff. *minutoides*	*Mus* sp.
Saccostomus sp. nov.	*Murinae* sp. indet.	*Millardia taiebi*
	Golunda gurai	*Golunda gurai*
	Lemniscomys aff. *striatus*	*Praomys* sp.
Heterocephalus quenstedti		
	Thryonomys gregorianus	*Tachyoyctes pliocaenicus*
Hystrix sp. nov.	*Hystrix cristata*	*Hystrix* aff. *cristata*
Hystrix cf. *makapanensi*		
Xenohystrix crassidens	*Xenohystrix crassidens*	*Xenohystrix crassidens*

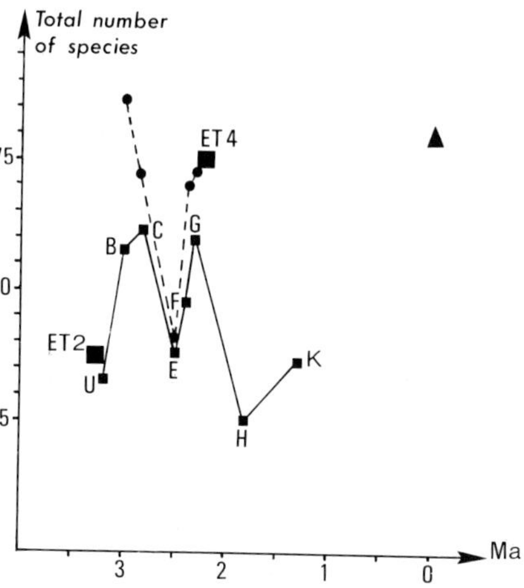

FIG. 5. Evolution of the species diversity at Shungura Omo Members B–K at East Turkana (from Coppens *et al.* 1976; Leakey & Leakey 1978; Wesselman 1984). There is a strong diversity decrease in the Member H fauna (*c.* 1.85 Ma) following the very important values of Members F and G. The peaks of Members F and G correspond to the immigration of *Equus* and Jerboa. The high value for Member B could be due to a mixing of faunas of Member B (3.18 Ma) and Member B-10 (2.94 Ma). At East Turkana, the fossiliferous localities are not well placed in the absolute time scale. Triangle = modern Kenya faunas (all ecologic niches together, without rodents, bats and insectivoras); large square = East Turkana faunal units 2 and 4 (from Leakey & Leakey 1978) (micromammals not included); small square = Omo large mammal faunas (Coppens *et al.* 1976); circle = Omo with rodents, bats, insectivora. New dates proposed by Brown *et al.* (1985) for Omo F and G have been applied.

The Omo Shungura site lies in a downwarped and downfaulted tectonic depression in which fluvial, deltaic and lacustrine sediments have been recorded (De Heinzelin *et al.* 1976). According to Brown (1981), the Lower Omo valley has been occupied by a large perennial stream flowing southwards into a lake since 4.5 Ma. Three high lacustrine phases have been described (4–3 Ma, then 1.85 and 1.6–1.4 Ma). The valley subsided during Plio-Pleistocene times, implying that a thick sequence of sediments is present.

Bonnefille (1976) described a more or less wooded savanna environment for the Member B and Member C deposits, with some riverine woodland and mountainous forest in the highlands. The mosaic of habitats existing along the river floodplain is reflected in the high species diversity recorded from this site.

No rodents are known below the KBS Tuff, either in the East Turkana sediments or in the Usno Beds. The large mammal faunas from these sites, compared with Hadar, Laetoli and Omo Shungura Member B, are also very different (Denys 1984).

S of eastern branch: Laetoli

The Laetoli fauna, situated S of the Asswa lineament in the northern Tanzanian divergence (Fig. 1), has yielded 14 rodent species. The fossiliferous levels are dated at between 3.77 and 3.59 Ma (Leakey *et al.* 1976). This fauna shares only two genera with Hadar (*Tatera*, *Hystrix*) (Fig. 5) and no 'Asiatic' genera. Most of the species described at Laetoli (Denys 1983, 1986) share more characteristics with the modern southern African rodent genera than with those of eastern Africa (the closest similarity is with the Kalahari rodent fauna). However, these species differ from the southern ones in displaying specialized characteristics that become exaggerated in the upper levels of the Laetoli site (upper Ndolanya Beds, 3–2.5 Ma) and in the Olduvai Bed I (1.75–1.65 Ma), situated 50 km to the N (Hay 1976; Jaeger 1976).

The Laetoli site is surrounded by large volcanoes whose products have provided exceptional environments for fossilization (Hay, this volume). It is separated from the northern part of the rift by fractures of the Asswa lineament and other fault escarpments (Sonjo-Eyasi). According Hay & Leakey (1982) and Hay (1986), uplift in the Laetoli region began *c.* 3 Ma, and the Sonjo-Eyasi fault which affects the Olduvai Bed II levels (1.71–1.41 Ma) began to move at 1.2 Ma.

The Middle Pliocene period (4–3 Ma) is characterized by a high level of endemism amongst the rodent faunas. Their morphotectonic, climatic, sedimentological and volcanic environment is well known in each site. All the data suggest that the Eastern Rift once contained a series of small basins, isolated from each other by tectonic or volcanic swells. Recent tectonic studies (Ebinger *et al.* 1984; Chorowicz *et al.* 1983) show that the large Tanganyika and Malawi basins resulted from the coalescing of many small isolated elementary basins.

Upper Pliocene (3–2 Ma)

The different faunas of this period continued speciation and increased in faunal diversity. Between 3 and 2.5 Ma, in the upper part of the Laetoli succession (Upper Ndolanya Beds), a change in the composition of the rodent faunas and other animals indicates a more humid environment. The presence of a species of *Thryonomys* that lives near water today implies the proximity of a river. This environmental change could be due to the diversion of local drainage, following the onset of tectonic uplift in the Laetoli region (Denys 1983; Hay 1986).

However, at the same time the Gaded site, which is in the Ethiopian Highlands on the margins of the rift and is dated at 2.5 Ma (Bonnefille 1983), provides evidence of a cooler, drier climate. It lacks the mountainous forest vegetation that exists at the site today. This change can be correlated with a major interval of cooling between 2.6 and 2.4 Ma in the southern ocean and northern hemisphere (Zagwin 1960; Shackleton & Kennett 1975; Frakes 1979; Suc 1984).

This increase in aridity has also been observed at the Omo site between tuff D and tuff E, and initially dated at between 2.5 and 2 Ma (Bonnefille 1976). Brown *et al.* (1985) revised the date to between 2.8 and 2.5 Ma. This event could therefore be correlated with global cooling, but might also be explained in terms of local uplift of the Gadeb region, implying modifications of the drainage and patterns of precipitation, as well as a regression of the lake independent of global climatic change.

The Omo Member F rodent fauna is composed of 16 species (Wesselman 1984). Initially dated at 2.08 Ma (Coppens *et al.* 1976) the tuff F is now dated at 2.4 Ma (Brown *et al.* 1985). Some of the rodents like the Gerboa (*Jaculus*, *Gerbillus*, *Heterocephalus* and *Acomys*) appear to be very well adapted to aridity and are first recorded at this site.

The Omo Member G fauna, now dated at 2.33 Ma (Brown *et al.* 1985), is characteristic of extremely arid conditions and very open environments (Wesselman 1984). Some modern species like *Equus*, *Phacochoerus aethiopicus*, *Kobus ellipsiprymnus* and *Crocuta crocuta* begin appearing at this time (Coppens *et al.* 1976; Eisenmann 1976, 1981). This could be the result of the opening up of a continuous rift corridor between North Africa and the northern part of the rift and even East Asia. This is a result of the suppression of swells, favouring migrations towards southern parts of the rift.

Plio-Pleistocene boundary (2–1.6 Ma)

An important tectonic episode is known at about 2 Ma (Baker *et al.* 1972) in the Afar region and in the Ethiopian and Gregory Rifts. This episode, approximately at the Plio-Pleistocene boundary, heralds a more advanced stage of rifting, characterized by successive vertical faulting, collapse of the swells separating the basins and intense volcanism.

Mid-eastern branch

At East Turkana, Vincens (1982) describes, from palynological studies, a very arid period at 1.89 Ma, which is correlated with the presence of rodents adapted to aridity like *Heterocephalus*, *Jaculus* and many Gerbillinae.

S of eastern branch

In this area the main events are changes of environment. In the Olduvai Bed I (western margin of the Eastern Rift and northern Tanzanian Divergence), Jaeger (1976) and Bonnefille (1976) showed qualitative and quantitative changes in the composition of the rodent faunas and the floras, reflecting an environmental change between the middle and the top of Olduvai Bed I (from humid to drier environments). From the lower part of Bed I, Hay (1976) described a lake (1.75 Ma) that was *c.* 7–25 km wide and bordered by grasslands. It was not far from a mountainous highland forest growing on the surrounding volcanoes or rift escarpments. Towards the top (1.65 Ma) a strong regression is recorded and then, in Bed II, a new lake flooded the basin. It was *c.* 9–15 km wide but extending from 16 to 24 km in times of flood. This lake disappeared at the top of Bed II, following the downfaulting of the basin and the uplift of its margins. This rapid subsidence implies that the lake progressively filled and the swamps dried up (Hay 1976).

This second change is also recorded in rodent and pollen records. Bonnefille (1976) and Jaeger (1979) both found evidence for a return to more humid conditions at the base of the Bed II (toward 1.6 Ma). This was followed by the development of a fluviatile environment within a savanna (Hay 1976), reflected by the presence of *Thryonomys* at the top of Bed II (1.1 Ma).

Recent studies of Pleistocene deposits in Lake Natron have shown the existence of a rodent fauna with the same species as found at Olduvai upper Bed I or lower Bed II. These studies indicate an environment very similar to that of the modern environment, i.e. dry acacia savanna (Denys, in press).

The arid phases recorded for the Pleistocene in the various basins do not appear to be totally synchronous. Biostratigraphic correlations are very difficult to establish because of the development of endemism amongst the faunas (see the KBS and Hadar controversies, Hay 1980; Brown 1982; Denys 1984). In this particular case, the events cannot be correlated well with a global climatic event. For example, Haq *et al.* (1977) found a major cooling phase near the top of the Olduvai palaeomagnetic event at *c.* 1.6 Ma, but

Hays *et al.* (1969) and McIntyre *et al.* (1967) record warming at the same time in the Pacific Ocean and near the equator.

Although a general climatic change cannot be denied, the apparent shift of this arid phase in time and space can be explained better by tectonic factors. This phase coincides with important tectonic events in some segments of the rift. It seems to follow the progressive opening of the rift corridor due to the disappearance of the swells. This implies decreasing rainfall with the disappearance of forested uplands, decreasing lake surface areas, changes in drainage, and the consequent infilling of the basins by renewed sedimentation. Such climatic modifications are responsible for environmental changes and development of dry savanna on the floor of the rift. Evidence of progressive change is further supported by the progressive advance towards the S of the rift of taxa adapted to arid conditions, for example, *Jaculus* which arrived at Olduvai at *c.* 0.4 Ma. This represents the most southerly point of its distribution (Jaeger 1979). This event seems to occur at about the same time as the last tectonic movements in the northern Tanzanian Divergence.

Model of evolution of the East African faunas

The integration of the present and the Plio-Pleistocene tectonic, sedimentological and faunal data leads to the development of mechanisms that explain the speciation of rodent faunas. These faunas tend to occur as a result of local changes of environment especially during the early stages of rift evolution when tectonism plays a most important role upon which other factors are dependent. In this paper the faunal evolution in the plateaus on each side of the rifts is ignored, mainly because the rate of biotope change is low compared with that within the rift. For the eastern branch two major stages can be distinguished.

Initial formation of a rift

The young rift develops immediately after the pre-rift volcanic outpourings of continental tholeiites (Kanika *et al.* 1981). During the late Oligocene, after an initial rifting stage, small basins opened in a plateau area covered with forest. This then developed into a recognizable rift (Chorowicz *et al.* 1983) bordered by high, steep margins. Rapid subsidence led to the separation of many deep basins, isolated from each other by mountainous swells, that eventually more or less filled with sediments and volcanic materials.

The doming of the rift shoulders then led to an increase in the amount of rainfall along the escarpments but the rift floor remained dry because of high evaporation, high temperatures and lower precipitation, as well as the diversion of streams away from the crests of the escarpments. Savannas developed on the rift floor plains at an early stage. Nevertheless, lakes were able to appear depending on the hydrological balance between rainfall, runoff and evaporation.

The existence of lake and river margins increased the number of available ecological niches and gave rise to greater faunal diversity than that found in a homogeneous savanna biotope. This is particularly true of the Pliocene in which there were a wide range of environments, implying a great diversity of rodent ecological niches (Misonne & Verschuren 1966; Delany & Happold 1979). Diversity is favoured by the isolation of small populations in basins separated by geographic barriers (allopatric speciation) or by ecological barriers (sympatric speciation). Geographic barriers could be, for example, rift escarpments, mountainous swells or volcanoes; while true rain forest (often developed at high altitudes on rift margins) represents an ecological barrier to rodents. This might explain the occurrence of endemism both for rodents (Denys 1983) and for some large mammals like hippopotames (Denys 1984) during the Pliocene. A modern model of parapatric speciation is described by Patton (1969) for *Perognathus* in Mexico. Patton described six chromosomic races (two species) which have different habitats, separated by rivers, each race being adapted to a precise part of the riverine environmental mosaic. Brown (1973) shows that the rodent diversity is directly related to habitat diversity and the availability of food resources.

Late formation of a rift

In the eastern branch of the rift a late stage of rift evolution can be recognized: troughs are filled and barriers between the initial small basins have disappeared. The rift shoulders are lower than during the previous stages. Decreasing rainfall, following erosional lowering of the uplands together with modifications of the drainage, disturb the hydrological balance and lead to climatic change by inducing drier conditions. As a consequence, sedimentation also changes, escarpments become less steep, the soil and lake water composition changes, and the lakes are infilled.

For faunas, these changing conditions induce a decrease in the number of ecological niches available, by diminution of forest and mosaic environments and by the introduction of a uniform dry savanna biotope. This implies a reduction of available food resources and leads to more competition for limited niches which, in turn, result in decreased diversity. Many extinctions and replacements with better adapted species occur, while at the same time there is a tendency for faunal homogenization due to the opening up of new migration paths in a rift from which major obstacles have been removed.

Conclusions

In the absence of palaeontological data outside of the rift valley sediments, this model assumes that the rift itself was the main structure affecting the evolution of faunas during the Plio-Pleistocene in Africa, especially rodents that are particularly dependent on the environment. This is not unreasonable since N–S migrations along the rift are recognized today but there appear to be very few E–W movements across the rift. It is known, for example, that modern bovids cannot cross the White Nile (Grubb, pers. comm.) and that the rift valley is a barrier, especially for forest taxa coming from the W.

This model correlates faunal evolution with the tectonic phases of rift valley formation and its geomorphic, climatic and volcanic consequences. It differs from the European Plio-Pleistocene speciation models because it is not closely related to global climatic oscillation. This can be explained by the small cross-sectional size of the rift valley compared to other settings. Particularly structure and very active tectonic evolution could have modified more general climatic oscillations, attenuating the effects of wet periods and exaggerating the effects of dry periods.

References

ALMOND, D.C. 1969. Structure and metamorphism of the basement complex of North-East Uganda. *Overseas Geol. Miner. Resour.* **10,** 146–63.

ARAMBOURG, C. 1959. Vertébrés continentaux du Miocène supérieur de l'Afrique du Nord. *Publ. Serv. Carte Géol. Algérie* **3,** 130–9.

BACHMAYER, F. & WILSON, R.W. 1970. Die fauna der altpleistozänen Höhlen und Spaltenfüllung bei Kohdifisch (Österreich). *Ann. Naturhist. Mus. Wien* **74,** 533–87.

BAKER, B.H. & MITCHELL, J.G. 1976. Volcanic stratigraphy and geochronology of the Kedong-Olorgesaille area and the evolution of the South Kenya Rift Valley. *J. Geol. Soc.* **132,** 467–84.

——, MOHR, P.A. & WILLIAMS, L.A.J. 1972. Geology of the eastern rift system of Africa. *Spec. Pap. Geol. Soc. Am.* **136,** 67pp.

——, WILLIAMS, L.A.J., MILLER, J.A. & FITCH, F.J. 1971. Sequence and geochronology of the Kenya rift volcanics. *Tectonophysics* **11,** 191–215.

BALINSKY, B.I. 1962. Patterns of animal distribution on the African continent. *Ann. Cape Prov. Mus.* **2,** 299–310.

BARBERI, F., BORSI, S., FERRARA, G., MARINELLI, G., SANTACROCE, R., TAZIEFF, H. & VARET, J. 1972. Evolution of the Danakil depression (Afar, Ethiopia) in light of radiometric age determinations. *J. Geol.* **80,** 720–9.

BEHRENSMEYER, A.K. 1976. Lothagham Hill, Kanapoi, Ekora: a general summary of stratigraphy and faunas. In: COPPENS, Y. *et al.* (eds) *Earliest Man and Environments in the Lake Rudolf Basin.* University of Chicago Press, Chicago, pp.163–70.

BLACK, C.G. 1972. Review of fossil rodents from the Neogene Siwaliks Beds of India and Pakistan. *Paleontology* **15,** 238–66.

——, KRISHTALKA, L. & SOLOUNIAS, N. 1980. Mammalian fossils of Samos. 1. The Turolian rodents and insectivores of Samos. *Ann. Carnegie Mus.* **49,** 359–78.

BONNEFILLE, R. 1976. Palynological evidence for an important change in the vegetation of the Omo basins between 2.5 and 2 Myr ago. In: COPPENS, Y. *et al.* (eds) *Earliest Man and Environments in the Lake Rudolf Basin.* University of Chicago Press, Chicago, pp.421–31.

—— 1979. Méthode palynologique et reconstitutions paléoclimatiques au Cénozoïque dans le rift est-africain. *Bull. Soc. Géol. Fr.* **21,** 331–42.

—— 1983. Evidence for a cooler and drier climate in the Ethiopian Uplands towards 2.5 Myr. *Nature* **303,** 487–91.

—— & RIOLLET, G. 1980. Palynologie, végétation et climats de Bed I et Bed II à Olduvai, Tanzanie. *Actes 8th Congr. Panafr. Préhist.* Et. *Quat. Nairobi.* TILLMIAP. Nairobi, pp.123–7.

——, VINCENS, A. & BUCHET, G. 1984. Vegetation history of the Hadar Hominid site (Abstract). *Eighth International Conference of Ethiopian Studies*, Addis Ababa, Nov. 1984.

BRANDY, L.D., SABATIER, M. & JAEGER, J.J. 1980. Implications phylogénétiques et biogéographiques des dernières découvertes de Muridae en Afghanistan, au Pakistan et en Ethiopie. *Geobios* **13,** 639–43.

BROWN, F.H. 1981. Environments in the lower Omo basin from one to four million years ago. In: RAPP, G. & VONDRA, C.F. (eds) *Hominid Sites: Their Geologic Settings.* Westview Press, Boulder, Colorado, pp.149–63.

—— 1982. Tulu Bor Tuff at Koobi Fora correlated with the Sidi Hakoma Tuff at Hadar. *Nature* **300,** 631–3.

——, MCDOUGALL, I., DAVIES, T. & MAIER, R. 1985. An integrated Plio-Pleistocene chronology for the Turkana basin. In: DELSON, E (ed) *Ancestors.* A.R. Liss Inc., New York, 82–90.

BROWN, J.H. 1973. Species diversity of seed-eating rodents in sand dune habitats. *Ecology* **54,** 775–87.

CHESSEX, R., DELALOYE, M., MULLER, J. & WEIDMANN, M. 1975. Evolution of the volcanic region of Ali Sabieh (TFAI), in the light of K/Ar age determination. In: PILGER, A. & ROSLER, A. (eds) *Afar Depression of Ethiopia*. E. Schweizerbart'sche Verlagsbuchhandlung, Stuttgart, pp.221–7.

CHOROWICZ, J., LE FOURNIER, J., LE MUT, C., MUKONKI, M. na B., ROSENDAHL, B., SPY-ANDERSON, F.L. & TIERCELIN, J.J. 1983. Tectonique et sédimentation dans différents bassins et modélisation du rift est-africain. *Colloque ATP géodynamique* **1**, CNRS (eds), III.1.

—— & MUKONKI, M. na B. 1979. Linéaments anciens, zones transformantes récentes et géotectonique des fossés de l'Est africain d'après la télédétection et la microtectonique. *Mus. Roy. Afr. Centrale Rapp. ann. 1979*, 143–67.

——, —— & POTTIER, Y. 1979. Mise en évidence d'une compression horizontale liée à l'ouverture des fossés est-africains (branche occidentale), dans le seuil entre les lacs Kivu et Tanganyika. *C. R. Somm. Soc. Géol. Fr.* **5–6**, 231–4.

COPPENS, Y., HOWELL, F.C., ISAAC, G.L. & LEAKEY, R.E.F. (eds) 1976. *Earliest Man and Environments in the Lake Rudolf Basin*. University of Chicago Press, Chicago.

DE HEINZELIN, J.P., HAESAERT, J. & HOWELL, F.C. 1976. Plio-Pleistocene formations of the lower Omo basin, with particular reference to the Shungura formations. In: COPPENS, Y. *et al.* (eds) *Earliest Man and Environments in the Lake Rudolf Basin*. University of Chicago Press, Chicago, pp.24–9.

DELANY, M.J. & HAPPOLD, D.C.D. 1979. *Ecology of African Mammals*. Longman, London and New York, 434pp.

DENYS, C. 1981. Premiers résultats paléoécologiques fournis par les rongeurs pliocènes de Laetoli (Tanzanie). *Quaternaria* **23**, 1–17.

—— 1983. *Les rongeurs du Pliocène de Laetoli (Tanzanie). Evolution, paléoécologie et paléobiogéographie. Approche qualitative et quantitative*. Unpublished PhD thesis, Université P. et M. Curie, Paris, 595pp.

—— 1984. Biostratigraphic revision of the faunal dates at Hadar (Ethiopia). *Newslet. Stratig.* **13**, 175–83.

—— 1986. Rodentia and Lagomorpha 6.1. Fossil rodents (other than Pedetidae). In: LEAKEY, M.D. & HARRIS, J.M. (eds) *The Pliocene Site of Laetoli, Northern Tanzania*. Oxford University Press, Oxford, 118–70.

—— (in press). Micromammals from the West Natron Pleistocene deposits (Tanzania). Biostratigraphy and palaeoecology. *Sci. Géol. Strasbourg*.

EBINGER, C.J., CROW, M.J., ROSENDAHL, B.R., LIVINGSTONE, D.A., & LE FOURNIER, J. 1984. Structural evolution of Lake Malawi, Africa. *Nature* **308**, 627–9.

EISENMANN, V. 1976. Equidae from the Shungura Formation. In: COPPENS, Y. *et al.* (eds) *Earliest Man and Environments in the Lake Rudolf Basin*. University of Chicago Press, Chicago, pp.225–33.

—— 1981. The arrival of *Equus*. *Nature* **292**, 865.

FLYNN, L.J. & JACOBS, L.L. 1982. Effects of changing environments on Siwaliks rodent faunas of northern Pakistan. *Palaeogeogr. Palaeoclimatol. Palaeoecol.* **38**, 129–38.

FRAKES, L.A. 1979. *Climate Throughout Geologic Time*. Elsevier, Amsterdam.

FRYER, G., GREENWOOD, P.H. & PEAKE, J.F. 1983. Punctuated equilibria, morphological stasis and the paleontological documentation on speciation: a biological appraisal of a case history in an African lake. *J. Linnean. Soc.* **20**, 195–205.

HAQ, B.U., BERGGREN, W.A. & VAN COUVERING, J.A. 1977. Corrected age for the Pliocene/Pleistocene boundary. *Nature* **269**, 483–8.

HAY, R.L. 1976. *Geology of the Olduvai Gorge*. University of California Press, Berkeley, 203pp.

—— 1980. The KBS Tuff controversy may be ended. *Nature* **284**, 401.

—— 1986. Geology of the Laetoli area. In: LEAKEY, M.D. & HARRIS, J.M. (eds) *The Pliocene Site of Laetoli, Northern Tanzania*. Oxford University Press, Oxford (in press).

—— & LEAKEY, M.D. 1982. Les empreintes de pas fossiles de Laetoli. *Pour la Science* **54 (Avril)**, 27–38.

HAYS, J.D., SAITO, T., OPDYKE, N.D. & BURCKLE, L.H. 1969. Pliocene-Pleistocene sediments of the Equatorial Pacific. Their paleomagnetic, biostratigraphic and climatic record. *Bull. Geol. Soc. Amer.* **80**, 1481–514.

HENDEY, Q.B. 1981. Paleoecology of the late tertiary fossil occurrences in 'E' Quarry, Langebaanweg, South Africa and a re-interpretation of their geological context. *Ann. South Afr. Mus.* **84**, 1–104.

JACOBS, L.L. 1977. Fossil rodents (Rhizomyidae and Muridae) from Neogene Siwaliks deposits, Pakistan. *Bull. Mus. Ariz.* **52**, 1–103.

JAEGER, J.J. 1976. Les rongeurs (Mammalia, rodentia) du Pleistocene inférieur d'Olduvai Bed I (Tanzanie). 1: Les Muridés. In: SAVAGE, R.J.G. & CORYNDON, S.C. (eds) *Fossil Vertebrates of Africa*. Academic Press, London, pp.58–120.

—— 1977. Les rongeurs du Miocène moyen et supérieur du Maghreb. *Paleovertebrata* **8**, 166pp.

—— 1979. Les faunes de rongeurs et de Lagomorphes du Pliocène d'Afrique orientale. *Bull. Soc. Géol. Fr.* **21**, 301–8.

—— & WESSELMAN, H.B. 1976. Fossil remains of micromammals from the Omo group deposits. In: COPPENS, Y. *et al.* (eds) *Earliest Man and Environments in the Lake Rudolf Basin*. University of Chicago Press, Chicago, pp.351–60.

KANIKA, M., KAMPUNZU, A.B., CARON, J.P.H. & VELLUTINI, P.J. 1981. Données nouvelles sur le volcanisme de la haute Ruzizi (Sud Kivu, Zaire). *C. R. Acad. Sci. Paris* **292**, 1277–82.

KAZMIN, V. 1980. Transform faults in the East African Rift system. In: *Geodynamic Evolution of the Afro-Arabic Rift System*. Accad. Nazionale dei Lincei, Rome, Atti dei Convegni Lincei, **47**, 65–73.

KING, B.C. 1978. Structural and volcanological evolution of the Gregory Rift Valley. In: BISHOP, W.W. (ed) *Geological Background to Fossil Man*. Scottish Academic Press, Edinburgh, pp.29–54.

KINGDON, J. 1973. *East African Mammals. An Atlas of evolution in Africa*. Academic Press, London.

LEAKEY, MARY D., HAY, R.L., CURTIS, G.H., DRAKE, R.E., JACKES, M.K. & WHITE, T.D. 1976. Fossil hominids from the Laetolil Beds. *Nature* **262**, 460–6.

LEAKEY, M. & LEAKEY, R.E.F. 1978. *The Fossil Hominids and an Introduction to Their Context, 1968–*

1974. Koobi Fora Research Project, Vol. 1. Oxford University Press, Oxford.

LOGATCHEV, N.A., BELOUSSOV, V.V. & MILANOVSKY, E.E. 1972. East African rift development. *Tectonophysics* **15,** 71–81.

MCINTYRE, A., BE, A.W.H. & PRIEKSTAS, R. 1967. Coccoliths and the Pliocene-Pleistocene boundary. In: SEARS, M. (ed) *Progress in Oceanography*. Pergamon Press, Oxford, pp.3–25.

MISONNE, X. & VERSCHUREN, J. 1966. Les rongeurs et lagomorphes de la région du parc national du Serengeti (Tanzanie). *Mammalia* **30,** 517–37.

MOHR, P.A. 1974. Mapping of the major structure of the African Rift System. *Spec. Rep. Smithson. Astrophys. Obs.* **361,** 85pp.

—— 1983. Volcanotectonic aspects of Ethiopian rift evolution. *Bull. Centres Rech. Explor. Prod. Elf-Aquitaine* **7,** 175–89.

PATTON, J.L. 1969. Chromosome evolution in the pocket mouse: *Perognathus goldmani* Osgood. *Evolution* **23,** 645–62.

PICKFORD, M.H.L. 1978. Stratigraphy and mammalian paleontology of the late Miocene Lukeino Formation, Kenya. In: BISHOP, W.W. (ed) *Geological Background to Fossil Man.* Scottish Academic Press, Edinburgh, pp.263–78.

SABATIER, M. 1982. Les rongeurs du site Pliocène à Hominidés de Hadar (Ethiopie). *Paleovertebrata* **12,** 1–56.

SCLATER, W.L. 1896. The geography of mammals. IV. The Ethiopian region. *Geogr. J.* **7,** 282–96.

SEN, S., 1982. Biogéographie et biostratigraphie du Néogène continental de la région Egéenne. Apports des rongeurs. *Mém. Spec. Geobios* **6,** 465–72.

—— 1983. Rongeurs et lagomorphes du gisement Pliocene de Pul-e-Charkhi, bassin de Kabul, Afghanistan. *Bull. Mus. Nat. Hist. Paris* **1C,** 33–74.

——, BRUNET, M. & HEINTZ, E. 1979. Découverte de rongeurs 'africains' dans le Pliocène d'Afghanistan (bassin de Sarobi), implications paléobiogéographiques et stratigraphiques. *Bull. Mus. Nat. Hist. Paris* **1C,** 65–75.

SHACKLETON, N.J. & KENNETT, J.P. 1975. Late Cenozoic oxygene and carbon isotopic changes at DSDP site 284: implications for glacial history of the northern hemisphere and Antarctica. *Init. Rep. DSDP* **29,** 801–7.

SLAUGHTER, B.H. & JAMES, G.T. 1979. *Saidomys natrunensis* an arvicanthine rodent from the Pliocene of Egypt. *J. Mammal. Ann. Arbor* **60,** 421–5.

SUC, J.P. 1984. Origin and evolution of the Mediterranean vegetation and climate in Europe. *Nature* **307,** 429–32.

TAIEB, M. & TIERCELIN, J.J. 1980. La formation d'Hadar (Afar, Ethiopie): stratigraphie et paléoenvironnements sédimentaires. *Proc. 8th Congr. Panafr. Préhist. Quat. Stud. Et. Nairobi*, Sept. 1977, TILLMIAP, Nairobi, pp.109–14.

TIERCELIN, J.J. 1981. *Rifts continentaux. Tectonique, climats, sédiments. Exemples: la sédimentation dans le nord du rift Gregory (Kenya) et dans le rift de l'Afar (Ethiopie) depuis le Miocène.* Unpublished PhD thesis, Université Aix-Marseille II, 260pp.

VINCENS, A. 1982. *Palynologie, environnements actuels et Plio-Pleistocènes à l'Est du Lac Turkana (Kenya).* Unpublished PhD thesis, Université Aix-Marseille II, 225pp.

WALTER, R.C. & ARONSON, J.L. 1982. Revisions of K/Ar ages for the Hadar Hominid site, Ethiopia. *Nature* **296,** 122–7.

WESSELMAN, H.B. 1984. Omo micromammals. In: HECHT, M.K. & SZALAY, F.S. (eds) *Contributions to Vertebrate Evolution*, vol. 7. S. Karger, New York, 219pp.

WESTERN, D. 1980. Linking the ecology of past and present mammal communities. In: BEHRENSMEYER, A.K. & HILL, A., (eds) *Fossils in The Making.* University of Chicago Press, Chicago, pp.41–54.

ZAGWIN, W.H. 1960. Aspects of the Pliocene and early Pleistocene vegetation in the Netherlands. *Mededel. Geol. Sticht.* **5C,** 78pp.

C. DENYS, Laboratoire de Paléontologie des Vertébrés et de Paléontologie Humaine, Université de P. et M. Curie, 4, place Jussieu, 75252, Paris, Cedex 05, France.

J. CHOROWICZ, Dept de Géotectonique, Université de P. et M. Curie, 4 place Jussieu, 75252, Paris, Cedex 05, France.

J.J. Tiercelin, GIS Océanologie et Géodynamique, Université de Bretagne Occidentale, 29287 Brest, Cedex, France.

Index

Page numbers in italics refer to figures, those in bold to tables